AA007701

1st International Ninevah Conference on Engineering and Technology (INCET 2021)

Held online due to COVID-19

IOP Conference Series: Materials Science and Engineering
Volume 1152

Mosul, Iraq
5 - 6 April 2021

ISBN: 978-1-7138-3422-9
ISSN: 1757-8981

Printed from e-media with permission by:

Curran Associates, Inc.
57 Morehouse Lane
Red Hook, NY 12571

Some format issues inherent in the e-media version may also appear in this print version.

This work is licensed under a Creative Commons Attribution 3.0 International Licence.
Licence details: http://creativecommons.org/licenses/by/3.0/.

No changes have been made to the content of these proceedings. There may be changes to pagination and minor adjustments for aesthetics.

Printed with permission by Curran Associates, Inc. (2021)

For permission requests, please contact the Institute of Physics at the address below.

Institute of Physics
Dirac House, Temple Back
Bristol BS1 6BE UK

Phone: 44 1 17 929 7481
Fax: 44 1 17 920 0979

techtracking@iop.org

Additional copies of this publication are available from:

Curran Associates, Inc.
57 Morehouse Lane
Red Hook, NY 12571 USA
Phone: 845-758-0400
Fax: 845-758-2633
Email: curran@proceedings.com
Web: www.proceedings.com

TABLE OF CONTENTS

COMMUNICATION

CONTROL SYSTEM

COMPUTER SCIENCE AND ENGINEERING

SIGNAL PROCESSING AND IMAGING

IOP Conf. Series: Materials Science and Engineering 1152 (2021) 011001 doi:10.1088/1757-899X/1152/1/011001

INTRODUCTION

2021 1st International Ninevah Conference on Engineering and Technology (INCET2021) was successfully held at Ninevah University, Mosul, Iraq during April 05-06, 2021. Due to the pandemic of COVID-19, the organizing committee and scientific committee decided to follow the guidelines of World Health Organization (WHO), and local government agencies on the current situation. Thus, the INCET2021 was held virtually this year. All sessions that were held on day 1 and day 2 of the conference schedule were conducted successfully through online platform (Zoom Webinar).

The main target of INCET2021 was to provide an opportunity for researchers and professionals from electronics and electrical engineering (EEE) fields and with cross-disciplinary interests to circulate information on the latest developments of the electronics and electrical engineering. Keynote speakers from United State of America, United Kingdom and Australia were invited to deliver their latest information in electronics and electrical engineering fields. The keynote speakers presented various topics and discussed various challenges in electronics and electrical engineering fields.

In this proceeding, papers selected from various submitted papers to the conference and from many universities and industries. All submitted papers were subjected to double-bind peer-review by international reviewers and members of conference committees. In this proceeding, papers were selected based on their originality, novelty, quality and relevancy to the conference. This proceeding presents to the readers the recent various fields of electronics and electrical engineering such as Electronics Engineering, Communication Engineering, Wireless and Mobile Communication, Information Technology, Computer Science and Engineering,

Content from this work may be used under the terms of the Creative Commons Attribution 3.0 licence. Any further distribution of this work must maintain attribution to the author(s) and the title of the work, journal citation and DOI.
Published under licence by IOP Publishing Ltd

Embedded Systems, Control Engineering, Power Electronics, Solid state Engineering and Robotics and Automation.

The organizing committee would like to thank reviewers as they helped to improve the quality of this proceeding. The organizing committee also would like to thank all authors for their participations in this proceeding. The organizing committee would like to thank keynote speakers for their interesting topics and deliver their knowledge to audience from different background fields. The organizing committee also would like to encourage you to participant to our next event INCET2022.

Prof. Dr. Jafar Ramadhan Mohammed
Conference Co-Chairman
Vice-chancellor for Scientific Affiars of Ninevah University, Mosul, Iraq

Conference Chairman and Co-Chairman

Prof. Dr. Osama Ismaeel Almushhadany
Conference Chairman
President of Ninevah University

Prof. Dr. Jafar Ramadhan Mohammed
Conference Co-Chairman
Vice-chancellor for Scientific Affiars of Ninevah University

IOP Conf. Series: Materials Science and Engineering 1152 (2021) 011001 doi:10.1088/1757-899X/1152/1/011001

CONFERENCE COMMITTEES

Prof. Dr. Osama Ismaeel Almushhadany
Conference Chairman
President of Ninevah University

Prof. Dr. Jafar Ramadhan Mohammed
Conference Co-Chairman
Vice-chancellor for Scientific Affairs of Ninevah University

Organizing and E-Management Committee

- Prof. Dr. Jaafar Ramadan Mohamed / Chairman
- Prof. Dr. Khaled Khalil Mohammed / Member
- Dr. Aws Zuheer Yonis Al-Ashqar / Member
- Dr. Mohammed Abdul-Muttaleb Abdullah / Member
- Dr. Ahmed Sabaawi / Member
- Dr. Safwan Hafeedh Younus/ Member
- Lecturer Karam Mudhafar Younus / Member
- Lecturer Muhammad Abdul Jalil Ibrahim / Member
- Assist. Lecturer Mohannad Nihad / Member
- Assist. Lecturer Akram Talal Mohammed / Member
- Assist. Lecturer. Mohammed Ameer Abd Alahmed / Member

Scientific Committee

- Prof. Dr. Khalil H. Sayidmarie / Chairman
- Assist. Prof. Dr. Ahmed Thannon Younes / Member
- Assist. Prof. Dr. Muhammad Hazem Al Jammas/ Member
- Assist. Prof. Dr. Mujahid Fahmi/ Member
- Assist. Prof. Dr. Qais Thannon Najim/ Member
- Assist. Prof. Dr. Younes Mahmoud Aboush/ Member
- Assist. Prof. Dr. Dia Muhammad Ali/ Member
- Assist Prof. Maan Al-Adwany/ Member
- Dr. Ibrahim Khalaf Mohamed/ Member
- Dr. Sedki Younis/ Member

International Advisory Committee

- Prof. Jonathan Chambers, University of Leicester, United Kingdom
- Prof. Amin Abbosh, University of Queensland, Australia
- Prof. Raed Abd-Alhameed, University of Bradford, United Kingdom
- Prof. Hussain Al-Rizzo, University of Arkansas at Little Rock, United States of America
- Prof. Claudio Paoloni, Lancaster University, United Kingdom
- Prof. Dr. Wai Lok Woo, Northumbria University, United Kingdom
- Prof. Dr. MOHAMMAD FAIZ LIEW ABDULLAH, Universiti Tun Hussein Onn Malaysia, Malaysia
- Prof. Zahriladha Zakaria, Universiti Teknikal Malaysia Melaka, Malaysia
- Dr. Mustafa Bakr, University of Oxford, United Kingdom
- Dr. Yasir Noori, University of Southampton, United Kingdom
- Dr. Zaid Abdullah, University of Leicester, United Kingdom
- Prof. Ghassan Hameed, Ministry of Higher Education and Scientific Research, Iraq
- Prof. Alyaa Abbas Al-Attar, President of the Northern Technical University, Iraq
- Prof.Hanaa Abdul Hussein Altai, Ministry of higher education and scientific research, Iraq
- Prof. Dr. Basil Mohammed Saied, University of Mosul, Iraq
- Asst. Prof. Dr.Hassan S. Al-Abbas, Dean of AL-Hadba University College, Iraq
- Dr. Ali Ahmed Alzobaidi, President international Union of Creators in Iraq, Iraq

IOP Conf. Series: Materials Science and Engineering 1152 (2021) 011002 doi:10.1088/1757-899X/1152/1/011002

Peer review declaration

All papers published in this volume of *IOP Conference Series: Materials Science and Engineering* have been peer reviewed through processes administered by the Editors. Reviews were conducted by expert referees to the professional and scientific standards expected of a proceedings journal published by IOP Publishing.

- **Type of peer review: Double-blind**

The reviewers (at least two for each paper) received the manuscript and a review report form, prepared for the conference, with certain questions and a field for comments. The authors, on the other side, received anonymous reviews with the following possible recommendations: Accepted without modifications/ Accepted with minor corrections/major revisions/reject. The paper was revised by the authors accordingly and then final decision was taken by the editors. All Papers were checked for plagiarism using Turnitin.

- **Conference submission management system**:

The papers were submitted to conference website, handled by the editors. The handling editor managed the correspondence between reviewers and authors.

• Number of submissions received: 45

•Number of submissions sent for review: 45

•Number of submissions accepted: 34

•Acceptance Rate (Number of Submissions Accepted / Number of Submissions

Received X 100): 75%

•Average number of reviews per paper: 2

•Total number of reviewers involved: 90

- **Any additional info on review process**:

The papers were evaluated based on the technical criteria, quality criteria and presentation aspects. The referees argued the paper findings in details throughout their reviews and provided an overall evaluation based on a scoring criteria from 1 to 5. There was also a pre-screening process of the papers by the editors, prior to the submission of the final papers.

- **Contact person for queries:**

Prof. Dr. Jafar Ramadhan Mohammed
Vice-chancellor for Scientific Affairs of Ninevah University
jafar.mohammed@uoninevah.edu.iq

Content from this work may be used under the terms of the Creative Commons Attribution 3.0 licence. Any further distribution of this work must maintain attribution to the author(s) and the title of the work, journal citation and DOI.
Published under licence by IOP Publishing Ltd

Interference Mitigation in the Wireless Communication Systems Using Adaptive Filters

Jafar Ramadhan Mohammed[1*]

[1]College of Electronic Engineering, Ninevah University, Mosul-41002, Iraq

*E-mail: jafar.mohammed@uoninevah.edu.iq

Abstract. In the current fifth generation (5G) wireless communication systems, the interference mitigation issue becomes a major and most important task for improving the performance of such systems due to corded spectrum and increased demands for new applications. Antenna arrays controlled by adaptive algorithms are proved to be a good solution for such problem. However, there are still some practical limitations in implementing such fully adaptive arrays especially when consisting a large number of the controllable elements. Other disadvantages of such arrays include long convergence time of the algorithm and the complexity of the beamforming network. In this paper, an adaptive algorithm is connected to only limited number of the array elements instead of all elements. We found that the elements that located on the center of the array could be chosen as adaptive, while the other elements which have less effect on the array pattern are left without adaptation process. Thus, the proposed array has many advantages with compared to those of the fully adaptive arrays while maintaining the same performance in suppressing the interference signals through null steering. Simulation results show that the effectiveness of the proposed array in generating deep nulls with very short convergence time.

Index Terms — Rectangular Arrays, Partially Optimized Elements, Constraint Optimization.

1. Introduction

The performance of the wireless communication systems can be significantly reduced under highly interference environment due to the crowded spectrum. This situation is expected to become worse in the near future when emerging the fifth generation (5G) wireless communication systems [1-2].

One of the simplest and oldest method for noise suppression is by using median filters [3] for image restoration. Other advanced methods use adaptive filters to control the complex weights of the array elements whose reshape the corresponding array patterns according to the required radiation characteristics such as steered nulls toward undesired direction and maintaining the main beam undistorted [4].

In some simple scenarios, the interfering signals may originate from fixed directions, thus, the array weights won't need to be adaptive or controllable. However, the uncontrollable array elements become useless under the conditions of rapidly changing interfering directions. In this case, it is necessary to use an adaptive algorithm to keep updating and recalculating of the optimum array weights and accordingly the corresponding array pattern must allow for the continuous adaptation to any changing in the electromagnetic environment (i.e., pointing the main beam of the array pattern toward desired signal and the nulls in the direction of undesired interfering signals) [5-9]. The adaptive variables (controllable elements) are different from fixed weights (uncontrollable elements) where it allows for continuously updated [10-12].

In this paper, the pros and cons of the adaptation process of the adaptive weights under various optimization algorithms such as conjugate gradient (CG), least mean squares (LMS), sample matrix inversion (SMI), recursive least squares (RLS) and the constant modulus algorithm (CMA) are investigated and compared. Then, these adaptive algorithms are applied only to the limited number of

Content from this work may be used under the terms of the Creative Commons Attribution 3.0 licence. Any further distribution of this work must maintain attribution to the author(s) and the title of the work, journal citation and DOI.
Published under licence by IOP Publishing Ltd

array elements instead of all elements. Thus, not all of the array elements are adaptive. Indeed, there are many elements are nonadaptive or fixed. Therefore, the beamforming network is not too complex and the algorithm may converge very fast. It is worthy to mention that the use of only few elements for null steering have been previously presented in [13]. However, the concept of the controllable elements by means of adaptive algorithms is new and worthy to be investigated due to many advantages and applications.

2. Optimization Algorithms for Adaptive Arrays

Figure 1 shows the block diagram of the popular adaptive array system where it consists two major parts, namely, adaptive controller, and the controllable complex-weights, i.e., amplitude and phase excitation of each array elements. These complex weights are controlled by the adaptive controller via an optimization algorithm. The criteria of the optimization algorithm could be minimizing the means square error (MSE), maximizing the signal to interference ratio, shaping the array pattern by steering the main beam toward desired signal and nulls toward interfering directions, and minimizing the variance. In this section, the nulling capability for interference suppression of different adaptive algorithms is examined under full control of all array elements.

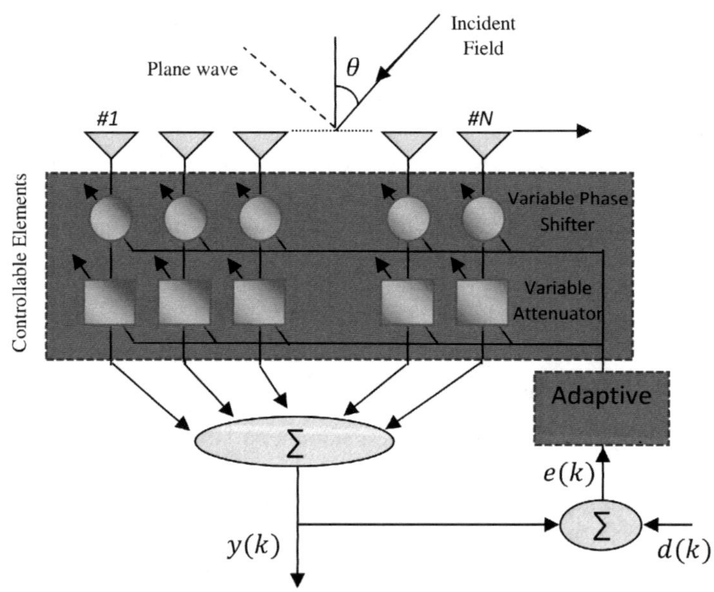

Figure 1. Block diagram of the adaptive array.

2.1 Least Mean Squares (LMS)

The adaptive control shown in Figure 1 updates the array weights (amplitude and phase) to optimally nulling the array pattern. For the LMS algorithm, the updating process of the controllable weight vector is [14-16]:

$$\bar{w}(k+1) = \bar{w}(k) + \mu\, e^*(k)\, \bar{x}(k) \quad (1)$$

Where $\bar{w} = [w_1\ w_2\ \ldots .\ w_N]^T$ are the complex weights, μ is the step size of the algorithm, and $\bar{x}(k)$ is the input signals including the desired signal, $\bar{s}(k)$, interfering signals, $i_1(k), i_2(k), \ldots, i_I(k)$, I is the total number of the interfering signals, and the Gaussian noise signal, $\bar{n}(k)$.

IOP Conf. Series: Materials Science and Engineering 1152 (2021) 012001 doi:10.1088/1757-899X/1152/1/012001

$$\bar{x}(k) = \bar{a}_0\bar{s}(k) + [\bar{a}_1\ \bar{a}_2\ \ldots.\bar{a}_I]\,.\begin{bmatrix} i_1(k) \\ i_2(k) \\ \vdots \\ i_I(k) \end{bmatrix} + \bar{n}(k) \quad (2)$$

In the vector form, it can be written as

$$\bar{x}(k) = \bar{x}_s(k) + \bar{x}_i(k) + \bar{n}(k) \quad (3)$$

Where $\overline{a_0}$ is the desired steering vector, and $\bar{a}_1\ \bar{a}_2\ \ldots.\bar{a}_I$ are interfering steering vectors. From Figure 1, the error signal can be written as

$$e(k) = d(k) - \bar{w}^H(k)\,\bar{x}(k) \quad (4)$$

As mentioned, the main concern in this paper is to adaptively shape the array pattern to place the required nulls. The array factor (AF) can be found by

$$AF(\theta) = \sum_{n=1}^{N} w_n^H e^{j(n-1)\frac{2\pi}{\lambda}d\sin(\theta)} \quad (5)$$

The above-mentioned equations are simulated and the converged LMS array pattern is shown in Figure 2. For comparison, the array pattern of the uniformly excited (i.e., all the weights' value are unity) linear array is also shown in Figure 2. In this example, one desired signal impinging on the array from direction-40°, and two interfering signals from directions -5° and -26°, respectively is considered. The number of array elements is N=12 (Note that all of these array elements are considered here to be controllable). The step size value is chosen to be $\mu = 0.006$ which is within the acceptable range according to [14] and ensures that the LMS algorithm perform well. It can be seen that the LMS array pattern accurately placed the required nulls. However, the number of the controllable elements, N=12, is much larger than the number of the interfering signals, I=2. This may add unnecessary complexity to the feeding network.

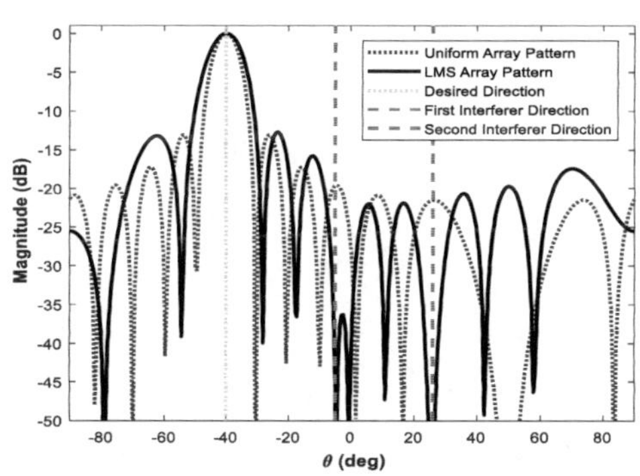

Figure 2. LMS array pattern for fully controllable array elements.

2.2 Conjugate Gradient (CG)

The LMS algorithm uses the steepest descent method to converge to the optimum weight values. However, this process needs to pass through many iterations before get converged. Thus, it is generally classified as a slow algorithm which is undesirable in the applications that require fast tracking capability [8].

The convergence speed can be improved by the use of conjugate gradient method instead of steepest descent method and thus the algorithm is referred to as conjugate gradient (CG) [16]. In CG, the array weights are updated iteratively such that the new paths are perpendicular to the old paths. The updating process of the controllable weight vector is [16, 17]:

$$\overline{w}(k+1) = \overline{w}(k) - \mu(k)\,\overline{D}(k) \quad (6)$$

Where $\overline{D}(k)$is the direction vector which gives the new perpendicular paths during the updating process, and $\mu(k)$ is the variable step size which can be found according to [9]. The array pattern based on CG algorithm is shown in Figure 3. The desired and interfering directions are same as in the previous example. Also, the number of array elements is chosen to be N=12 where all of them are controllable. Their optimum values after CG convergence are w_1 =0.7448 - j0.6673, w_2 = 0.4549 + j2.8912, w_3 =-2.8671 -j2.2878, w_4 =3.3819 - j0.0546, w_5 =-2.5254 +j0.7198, w_6 =-2.0850 -j1.5462, w_7 =4.2223 -j0.5139, w_8 =-0.5865 + j2.4628, w_9 =-3.0749 - j1.2352, w_{10} = 3.2749 - j2.8870, w_{11} =1.1888 +j2.6427, w_{12} =-3.8234 - j0.1385. As can be seen, the desired nulls in the CG array pattern are accurately placed. However, this result cannot be released in practice due to unavailability of RF components to release such accurate weight values.

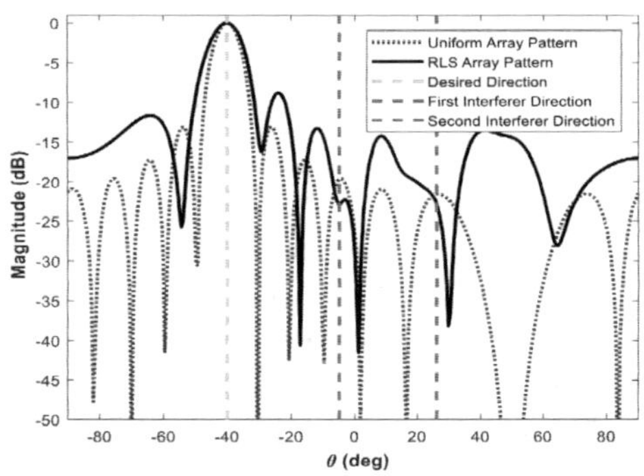

Figure 3. CG array pattern for fully controllable array elements.

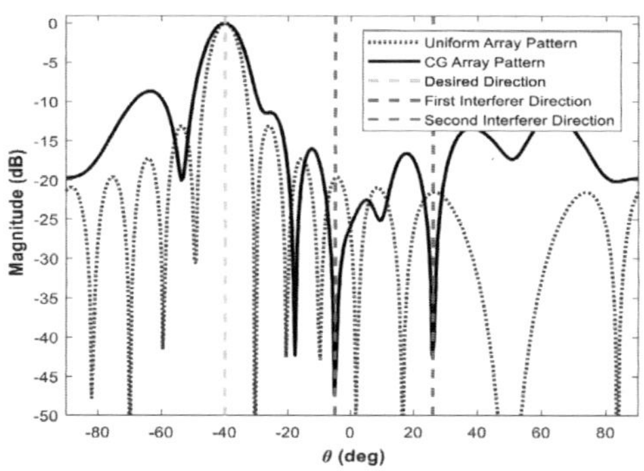

Figure 4. RLS array pattern for fully controllable array elements.

2.3 Recursive Least Squares (RLS)

Another way to accelerate the convergence speed of the adaptive controller that shown in Figure 1 is by applying recursive least squares (RLS) algorithm instead of LMS. For exponentially weighted RLS, the update error and weight equation are [15, 17]

$$e(k+1) = d(k+1) - \overline{X}^T(k+1)\overline{w}(k) \quad (7)$$

$$\overline{w}(k+1) = \overline{w}(k) + \phi^{-1}(k+1)\overline{x}(K+1)e(k+1) \quad (8)$$

where $\phi^{-1}(k)$is the inverse correlation matrix $\phi(k) = X^T(k)X(k)$. Note that instead of doing an update in the direction of the input vector X(k)as in the LMS algorithm, the input signal is first whitened by multiplied it by the inverse correlation matrix. More details about RLS algorithm and its derivation could be found in [14] and [16]. The array pattern based on RLS algorithm is shown in Figure 4. Even though the algorithm is converging fast, its nulling capability for interference suppression is unsatisfactory.

2.4 Constant Modulus (CM)

In this algorithm, the desired signal should ideally have a constant magnitude. However, some difficulties in maintaining the signal's amplitude constant arise in the multipath fading channels. Also, the frequency selective channels destroy the constant magnitude of the signal. Generally, to keep the constant modulus property an amplitude equalizer should be used to restore the amplitude of the original signal.

The error signal in CM algorithm can be written by [18]

IOP Conf. Series: Materials Science and Engineering 1152 (2021) 012001 doi:10.1088/1757-899X/1152/1/012001

$$e(k) = y(k)|y(k)|^{P-2}(R_P - |y(k)|^P) \quad (9)$$

Where P is chosen to be 1. This error signal can replace the previous error signal that was shown in Figure 1 and equation (4). The updating process of the controllable weight vector of this algorithm remains same as that of the LMS algorithm. The only difference between CM and LMS is the error signal.

The array pattern based on the CM algorithm is shown in Figure 5. Like other examples, a desired signal from direction -40°, and two interfering signals from directions -5° and -26°, are assumed to impinge on an array with N=12 controllable elements. From this figure, it can be seen that the required nulls are not accurate enough and thus the interfering signals cannot be suppressed.

2.5 Sample Matrix Inversion (SMI)

In this method, a K-length block of data is used. Thus, it is usually called a block-adaptive approach where the weights are adapted block-by-block basis.

For kth block of length K, the weights can be calculated as [19]

$$\bar{w}(k) = \left[\bar{X}_K(k)\bar{X}_K^H(k)\right]^{-1}\bar{d}(k)\bar{X}_K(k) \quad (10)$$

This method has some advantages in the application where the characteristics of the desired signal may change rapidly. The SMI algorithm has the ability to track such changes due to its high rate of convergence. Unlike the LMS algorithm, the SMI estimates the covariance matrix using K-time samples. The nulling capability of this algorithm is shown in Figure 6 for block length equal to K=100 and N=12 controllable elements. It can be seen that its performance for interference suppression is satisfactory.

Although all or most of the above-mentioned algorithms are able to reshape the array pattern to reduce the effect of interfering signals, but they still suffer from low convergence speed due to many unavoidable iterations and high complexity in the feeding network due to the fact that each controllable element in the array needs at least one set of variable attenuator and variable phase shifter. These RF components need to be very accurate to achieve the required nulls. Any approximation in the weight's value will be at the cost of moving the nulls from their desired directions. Thus, none of the above-mentioned methods are desirable in practice. The need for a simple and effective solution to this important problem is potentially required. One solution is addressed in the following section.

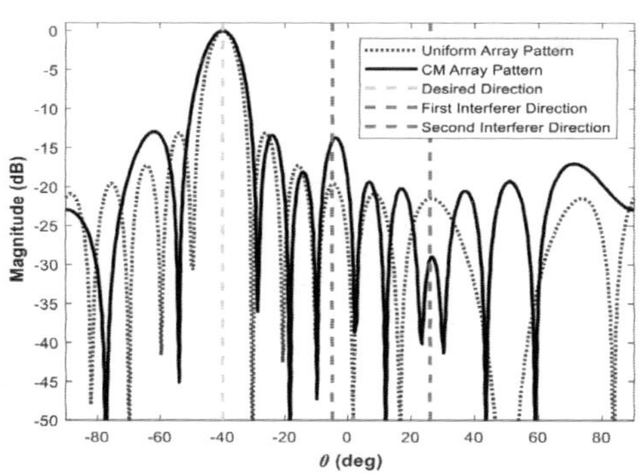

Figure 5. CM array pattern for fully controllable array elements.

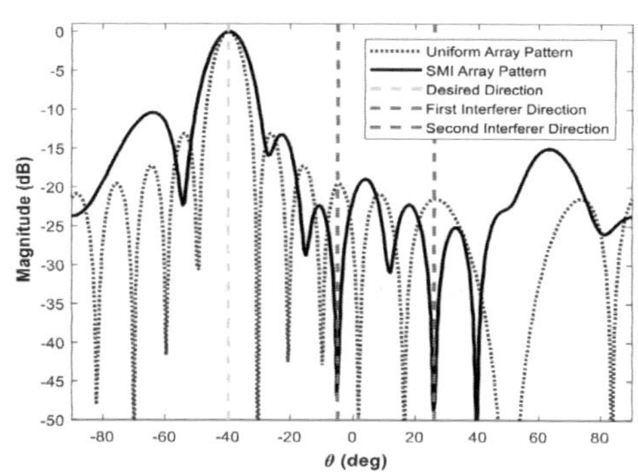

Figure 6. SMI array pattern for fully controllable array elements.

3. The Proposed Adaptive Array

IOP Conf. Series: Materials Science and Engineering 1152 (2021) 012001 doi:10.1088/1757-899X/1152/1/012001

Figure 7 shows the block diagram of the proposed adaptive array. It implies an adaptive controller based LMS algorithm, and only a small number of the controllable elements, say M elements, with another number of uncontrollable elements, N-M elements. For the purpose of reducing the complexity of the feeding network and maximally exploiting the number of degrees of freedom for null steering, the number of the uncontrollable elements is selected to be much larger than the number of controllable elements, i.e., $(N - M) > M$. This selection is mainly relay on the interference environment specifically on the number of the interfering signals where it should be noted that for efficient suppression of the interfering signals the number of the controllable elements should be at least equal to the number of interfering signals, i.e., $M \geq I$.
In some applications that use large arrays, the number of the interfering signals may be much smaller than the number of the array elements. In this case, only a small number of the degrees of freedom instead of all of them are sufficient to achieve the required nulling. Thus, selecting a small and sufficient number of the elements to be controllable is one of the effective solutions to significantly reduce the complexity of the feeding network and accelerate the convergence speed of the adaptive algorithm while maintaining excellent suppression for interfering signals.
Although the concept of adaptive algorithms is not new, the idea of controlling only a small number of the array elements via LMS algorithm is new and it has not been previously investigated to the best of author's knowledge. Many other advantages of the proposed adaptive array are explained on the following section.

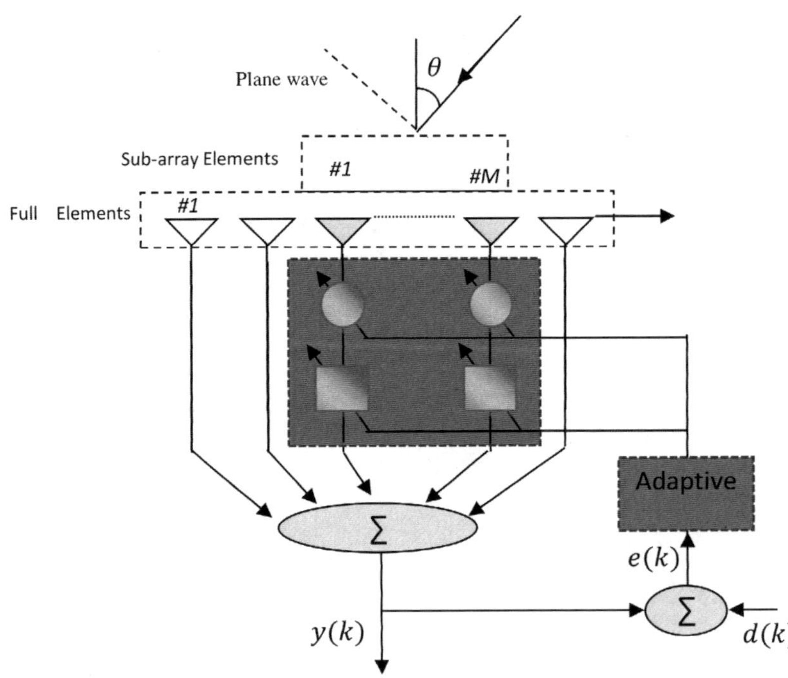

Figure 7. Block diagram of the proposed adaptive array.

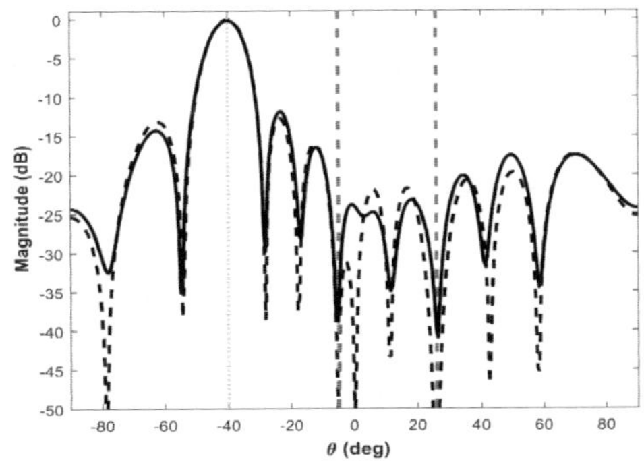

Figure 8. Proposed array pattern for only 4 controllable elements and 8 uncontrollable elements (solid) and Popular LMS

4. Simulation Results

Extensive numerical examples are provided in this section to illustrate the performance, in terms of pattern nulling, accurate null steering, null depths, convergence speed of the adaptive algorithm, and the complexity of the feeding network of the proposed array. The considered antenna array consists of M controllable elements and N-M uncontrollable elements as shown in Figure 7. The uncontrollable elements are excited with uniform excitation. All of the array elements are assumed to lie along the x-axis with an element distance of $d = \lambda/2$.

In the first example, we used the proposed array that presented in section III to steer the required nulls with only M=4 controllable elements. Figure 8 shows the radiation pattern of the proposed array. Compare Figure 8 with that of Figure 2 for fully controllable array elements, it can be seen that the proposed array pattern with only 4 controllable elements is comparable to that of the popular LMS based fully controllable array elements for placing the required nulls. More important, the main beam shape of the proposed array is very close to that of the desired uniform array pattern (the half power beam widths (HPBW) of the proposed array and the uniform array patterns are 11.66^{o} and 11.6^{o} respectively).

In the second example, the updating process of the controllable elements in both arrays is shown in Figure 9. It can be seen that the four controllable weights are much easier to update and control over the method that consists of twelve controllable elements.

In the last example, the convergence speed of the adaptive algorithm in both arrays are compared and shown in Figure 10. It can be seen that the MSE of the proposed array is much smaller than that of the popular LMS array. Moreover, the proposed array requires only 15 iterations to converge while the popular LMS array needs at least 60 iterations to converge. These results fully confirm the effectiveness of the proposed array.

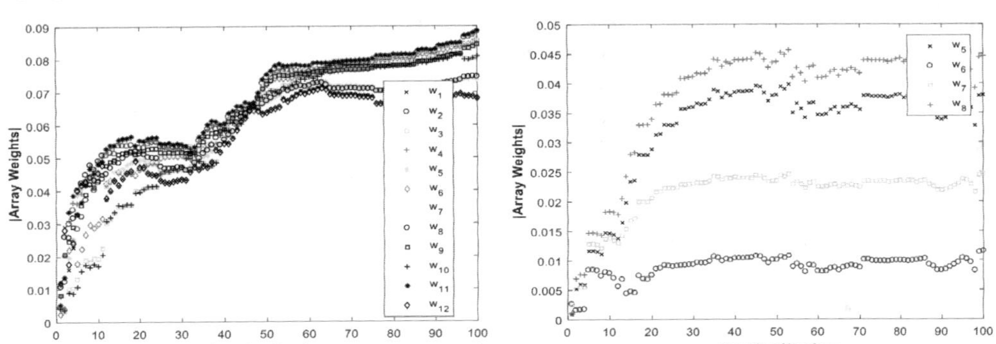

Figure 9. Updating of the controllable weights in the popular LMS array (left), and the proposed array (right).

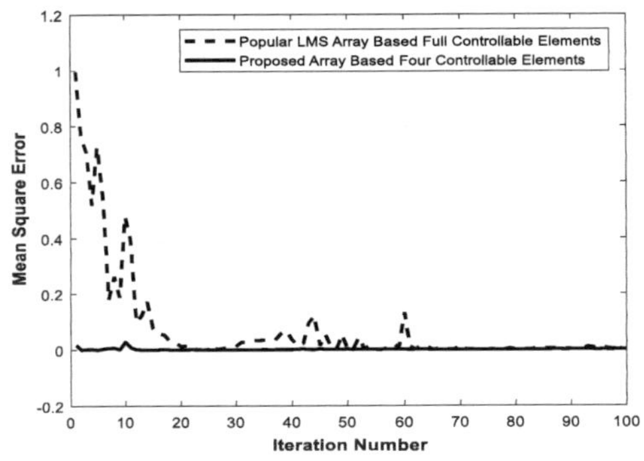

Figure 10. Convergence speed of the tested arrays.

5. Conclusions

Adaptive arrays are most powerful methods for interference suppression in the rapidly changing environments. However, there are some difficulties in implementing them in practice. These difficulties include complex feeding network and slow convergence speed of the adaptive algorithm especially when arrays imply large number of elements. In this paper, these difficulties are overcome by considering only a small subset of the array elements to be controllable. Simulation results have shown that the performance of the proposed array is comparable to that of fully controllable array with less complexity in the feeding network and faster convergence speed. Moreover, the subset of the controllable elements is capable to steer the required nulls.

6. References

[1] Steven Hong, Joel Brand, and Jung Choi 2014 Applications of self-interference cancellation in 5G and beyond IEEE Communications Magazine, Vol.52, Issue 2, pp. 114 – 121.

[2] Mohammed, J. R. 2008 An Improved Median Filter Based on Efficient Noise Detection for High Quality Image Restoration 2nd Asia International Conference on Modeling and Simulation (AMS), Kuala lampur, Malaysia, 13-15 May, 2008. PP. 327-331.

[3] W. Li, Y.-P. Li, and W.-H. Yu 2012 On Adaptive Beamforming for Coherent Interference Suppression via Virtual Antenna Array Progress in Electromagnetics Research, Vol. 125, 165-184.

[4] Mohammed, J. R. 2014 Low Complexity Adaptive Noise canceller for Mobile Phones Based Remote Health Monitoring International Journal of Electrical and Computer Engineering (IJECE), Vol. 4, No. 3, pp. 422~432. ISSN: 2088-8708

[5] Mohammed, J.R. 2018 Thinning a Subset of Selected Elements for Null Steering Using Binary Genetic Algorithm Progress in Electromagnetics Research M, vol. 67, pp. 147-157.

[6] Mohammed, J.R and Sayidmarie, K. H. 2017 Performance Evaluation of the Adaptive Sidelobe Canceller with various Auxiliary Configurations," AEÜ International Journal of Electronics and Communications, vol. 80, pp. 179–185.

[7] Mohammed, J. R. 2013 Design of Printed Yagi Antenna with Additional Driven Element for WLAN Applications", Progress in Electromagnetics Research C, Vol. 37, 67-81.

[8] Mohammed, J.R. 2019 Obtaining Wide Steered Nulls in Linear Array Patterns by Controlling the Locations of Two Edge Elements AEÜ International Journal of Electronics and Communications, vol. 101, pp. 145–151.

[9] Mohammed, J.R.2018 Element Selection For Optimized Multi-Wide Nulls in Almost Uniformly Excited Arrays IEEE Antennas and Wireless Propagation Letters, vol. 17, issue 4, PP. 629-632.

[10] R. L. Haupt and M. Leifer 2019 Adaptive Antenna Arrays 2019 IEEE Radar Conference (RadarConf), Boston, MA, USA, pp. 1-138. doi: 10.1109/RADAR.2019.8835686

[11] Mohammed, J.R, and Sayidmarie, K. H. 2014 Sidelobe Cancellation for Uniformly Excited Planar Array Antennas by Controlling the Side Elements IEEE Antennas and Wireless Propagation Letters, vol.13, pp. 987 - 990.

[12] Liu, M.; Ma, L.; Wang, N.; Zhang, Y.; Yang, Y.; Wang, H. 2019 Passive multiple target indoor localization based on joint interference cancellation in an RFID system. Electronics, 8, 426.

[13] Mohammed, J.R and Sayidmarie, K. H. 2017 Synthesizing Asymmetric SideLobe Pattern with Steered Nulling in Non-uniformly Excited Linear Arrays by Controlling Edge Elements International Journal of Antennas and Propagation, vol. 2017, Article ID 9293031, 8 pages.

[14] R. Monzingo, R. Haupt, T. Miller 2011 Introduction to Adaptive Arrays, Scitech Pub.

[15] Haykin, S. 2002 Adaptive Filter Theory Fourth Edition, Person Education, India.

[16] Mohammed, J. R.2007 A New Simple Adaptive Noise Cancellation Scheme Based On ALE and NLMS Filter" 5thAnnual International Conference on Communication Network and Service Research (CNSR), Fredericton, New Brunswick, Canada, PP. 245-254.

[17] Gross, F. B. 2015 Smart Antenna Using Matlab Second Edition, McGraw-Hill Education.

[18] Haykin, S.; Sayed, A.H.; Zeidler, J.R.; Yee, P.; Wei, P.C. 1997 Adaptive tracking of linear time variant systems by extended RLS algorithms IEEE Transactions on Signal Processing, Vol.45, no.5, pp. 1118 -1128.

[19] Godara, L 2004 Smart Antennas CRC Press, Boca Raton, FL.

IOP Conf. Series: Materials Science and Engineering 1152 (2021) 012002 doi:10.1088/1757-899X/1152/1/012002

A Beam Steering System Design based on Phased Array Antennas

Karam Mudhafar Younus[1*], Ahmed Abdulbari Jasim[1] and Roger W. Clarke[2]

[1] College of Electronics Engineering, Ninevah University, Mosul, 41002, Iraq.
[2] Faculty of Engineering & Informatics, University of Bradford.

*E-mail: karam.younus@uoninevah.edu.iq

Abstract- This paper presents a low-cost, compact, beam-steering phased array antenna (PAA) system based on a novel approach to integrating low-cost, high-performance phased shifters with their controlling boards within an antenna structure. Consequently, the system has been reduced to one structure for the antenna array and integrated phase shifters, rather than the sophisticated PAA with two separate boards which are connected by RF coaxial cables. The system has been achieved by the following stages; Firstly, the steering vector weights matrix was calculated by MATLAB. Then, four elements of uniform linear Microstrip patch antenna array with their feeding network power divider were designed and simulated in the CST simulation program. As a result, a practical model has been implemented. The designed RF-PCB for the 4-bit phase shifters MAPS_010143 were integrated within the antenna panel. The system works at a resonance frequency of (2.4 GHz), with ($\pm 45^0$) steering capability. The results are presented for the three models. The practical results match the simulation results very closely, fence proving the accuracy of the model and demonstrating a highly effective design.

1. Introduction

The use of steerable Phased Array Antenna (PAA) was started in military applications several decades ago. More recently, interest has been shown in PAAs to enhance the development of telecommunications for many commercial and industrial applications, such as military radar systems, weather surveillance radars, RFIDs, and late generation of cellular systems (5G), due to the significant performance advantage that they provide compared with traditional mechanical radars. PAAs usually required complex integration of high-cost solid-state and phase shifters. Thus, reduction in the complexity, cost, and size is the solution to facilitate their usage in wider areas [1-5].

One of the proposed approaches is to create several sub-arrays within the antenna. Each group is controlled by a phase-shifter. This will reduce the phase shifters in the structure as shown in figure 1. Each of these sub-arrays is controlled by a phase shifter. Each group is seen as a single element to another group. The total Array Factor of the system considers all the elements. However, the limitation of this idea is the increase in the side lobes and grating lobes, also the structure size issues [6-8].

Content from this work may be used under the terms of the Creative Commons Attribution 3.0 licence. Any further distribution of this work must maintain attribution to the author(s) and the title of the work, journal citation and DOI.
Published under licence by IOP Publishing Ltd

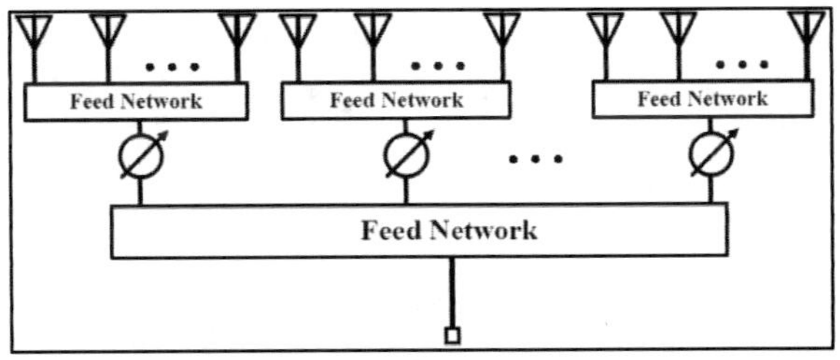

Figure 1. Grouping Antennas into a Sub-Array [8]

Another approach is by using frequency control to achieve the beamforming [9-12]. These types of systems are not suitable for many applications because most of the applications require a fixed frequency. Beam steering of Microstrip Patch Antenna (MPA) was investigated by Thongsopa and Krairiksh in [13], where beam steering of a dual-feed patch antenna depends on the difference between the input feed frequencies. This idea did not achieve the steering effectively, having many issues with side lobes. Sangjun and Won in [14] presented a wearable patch antenna, where the main beam can be switched between (0^0) and ($\pm30^0$). It was proposed that the antenna should steer three beams in the same direction at the same frequency. However, the results show that the antenna has one beam and no practical results were presented. Attia and Ramahi in [15] proposed a novel method to steer the main lobe of an MPA by partially covering the single MPA with a high refractive index substrate. They achieved ($\pm25^0$) overall steer. The steered beam had low-directivity due to the use of a single element. Indeed, the physical structure of the antenna is quite massive comparing with a single element antenna. Wenquan et al. in [16] have achieved a maximum beam steer of (48^0) by introducing complementary split-ring resonators in the ground plane of the patch. The issue here is that the switching of rings is not accurate due to the use of switching diodes. Hussaini et al. in [17], have proposed a novel beam steering for a uniform circular six elements antenna by using a reactive loading and time modulated switching technique. They proposed a Genetic algorithm (GA) to optimize the results. The steering was effective, however, there were some issues in the return-loss of the antenna. In [18], a novel idea had been suggested based on a cascaded reconfigurable defected microstrip structure. PIN diodes were inserted in the defected area. Each stage performs ($\pm7^0$) phase shifting. The proposed idea tries to steer the main beam of a four MPA around ($\pm20^0$). Scott and Fusco in [19] have proposed a new method to perform the phase-shifting by integrating two parts of electronically phase shifters in series with the feeding lines. Two dipole array antennas have been used. The total achieved steering was ($\pm12^0$) by changing the biasing voltages to the components. The total ($\pm12^0$) has limited steering application. However, the idea was to find an alternative cheap way to perform the steering. As has been presented, all attempts tended to find alternative ways to replace the expensive phase shifters. Accordingly, low-cost, small-size, and high-performance phase shifters have been chosen for this project. Additionally, they can be integrated within the antenna panel which leads to eliminating the need to create two separate boards for the antenna array and the phase shifter. In contrast, limited to one compact PAA which has a multilayer board consisting of an antenna array, feeding network, and phase shifters with their controllers.

This paper presents a novel, low-cost and compact size PAA beam steering system. The system has high beam steering performance and the novelty is to use low-cost digital phase shifters, which have been integrated into the antenna panel. The results show a compact PAA system, working at a resonance frequency of (2.4 GHz), with ($\pm60^0$) steering capability. Besides, there is a reduction in the cables that feed the PAA system, i.e., only one RF cable and the controlling lines, instead of the sophisticated traditional PAA systems in which the phase shifters board is separated from the antenna array panel, and RFs coaxial cables are used to feed each element in the antenna array. The system methodology is explained in section two. While the results and the discussion illustrated in section 3. Finally, the conclusion is presented.

2. System Design Methodology

The proposed system has been achieved by the following steps; Designing and implementing a uniform linear array (ULA) of four elements MPA starting with a single MPA. Then, designing the corporate feeding network based on Wilkinson Power Divider, to feed the antenna array elements. After that, designing a Radio Frequency Printed Circuit Board (RF-PCB) for the phase shifters and then integrate the RF-PCB with the antenna array structure. Finally, developing a MATLAB code to calculate the steering vector weights matrix and test the system performance.

2.1 Design and implement a ULA

The radiation pattern of a single element antenna is wide and has relatively low-directivity [20-23]. To meet the demands of a long communication link, it is recommended to design an antenna with high directivity. This can be achieved by modifying the electrical size of the antenna which leads to enhancing the directivity. An alternative way to meet high directivity is to design an antenna with an assembly of elements that can be designed in a suitable geometrical and electrical configuration. This type of antenna is known as an Array. Array's elements are usually identical [20]. The total array field is equal to the multiplication of the single element positioned at the center by a factor which is known as Array Factor (AF) and can be generalized as:

$$\mathrm{AF} = \sum_{n=1}^{N} e^{j(n-1)\psi} \tag{1}$$

Where $\psi = kd\cos\theta + \beta$

The antenna array main lobe can be steered by controlling the excitation of the phase or amplitude, or both, of each element in the array [2]. Assuming that scanning pattern within $\theta^0(0^0 \leq \theta^0 \leq 180^0)$, this can be performed by adjusting the (β) among the array elements as in equation (2).

$$\psi = kdcos\theta + \beta|_{\theta=\theta_0} = kdcos\theta_0 + \beta = 0 \rightarrow \beta = -kdcos\theta_0 \tag{2}$$

This is the basis of scanning PAA. Generally, the PAA performance increases when the number of elements increases in terms of directivity, HPBW, and scanning capability.

2.2 MPA Array Design

The project has been designed to work at (2.4 GHz) which falls in ISM unlicensed band. The single MPA (2.4 GHz) was designed and simulated in CST. FR-4 substrate with permittivity of (4.3) and a thickness of (1.6 mm) was chosen for the design. Antenna array design depends on the single MPA.

The initial space was chosen to be (0.5λ) which is equal (62.5 mm) for (2.4 GHz), then optimized. It is worth pointing out that the radiation pattern of the antenna array will differ when the feeding lines are constructed. The differences occur due to the mutual coupling among the feeding lines and the radiated elements. The antenna array structure of four MPA with spaces of (0.5λ) is shown in figure 2 and table1.

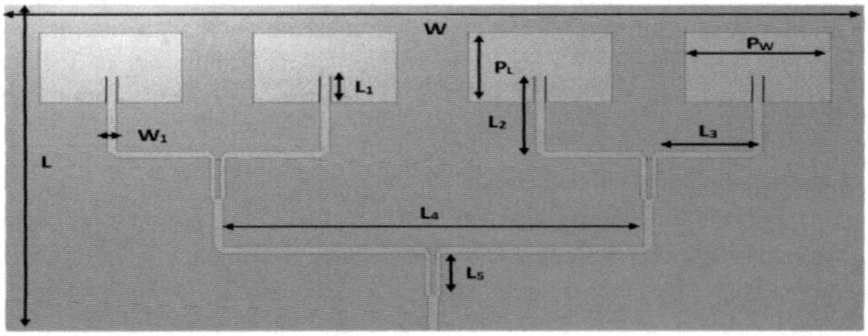

Figure 2. Antenna Array (A): Front View. (B) Back View, CST

Table 1. Antenna dimensions

Parameter	W	W_1	L	L_1	L_2	L_3	L_4	L_5	P_W	P_L
Value (mm)	250	2.88	139	10.85	30.6	27.4	122.11	16.78	41.8	28.8

IOP Conf. Series: Materials Science and Engineering 1152 (2021) 012002 doi:10.1088/1757-899X/1152/1/012002

2.3 Antenna Feeding Networks and Power Divider
The feeding network (power divider) should have high performance and be compact. This network should have a special shape to fit the integration of the phase shifters as will be seen later. The corporate feeding network has been designed based on the Wilkinson power divider (WPD). The straight structure of the WPD was expanded from (2:1) to (4:1) splits as shown in figure 2.

2.4 Phase Shifter (MAPS_010143)
The project aim is to perform the beam steering by changing the phase shifter weights which are integrated within the antenna panel. The challenge is to use accurate and low-cost phase shifters. MACOM Technology Solution Company has produced a phase shifter with interesting specifications. The MAPS-010143, figure 5, is a 4-bit phase shifter that has an integral CMOS driver build in a (4 mm) PQFN-SMD packet, 24-pins which perform high phase-shifting accuracy with minimum losses [23].

Figure 3. 4-bit Phase Shifter PQFN SMD[23]

The phase shifter has low-DC power consumption and minimal attenuation variation over the phase shift range. It provides a phase shift from (0^0) to (360^0) in (22.5^0) steps. It has two controlling modes: serial and parallel [23]. The serial mode is designed to connect the phase shifter circuit to a computer system to perform the clocking and change the phase. In the parallel mode, control can be achieved by connecting the four controlling bits to a normal 4-bit switch. The small size of these phase shifters can be exploited by integrating them with the antenna panel, which is one of the project's aims.

2.4.1 Phase Shifters' RF-PCB Design
Phase shifters' RF-PCB controlling circuit has been designed by using a modeling program (Proteus). PCB is fabricated from a substrate such as epoxy/glass-fiber plate and the conducting layer which is usually copper of a thickness of (35/70) μm. PCB may have single-side or double-sided layers and this depends on the design requirements, in some cases more than one layer can be used. This is called multilayer board such as computers mainboards. It is possible to connect the layers by using Vias. For this project, a single layer of RF-PCB board was used to mount all the components. The design started by modifying a 24-Pin SMD (4×4 mm) footprint and made it compatible with the IC configuration. For the practical model, it was decided to have separate parts of RF-PCB for each phase shifter, to allow more flexibility in integration with the antenna array panel. The design has the specification of an RF-PCB, in terms of substrate type and lines dimensions. The substrate was chosen to be FR-4. Each of the phase shifters has its feeding power and control switches on the board as shown in figure 4.

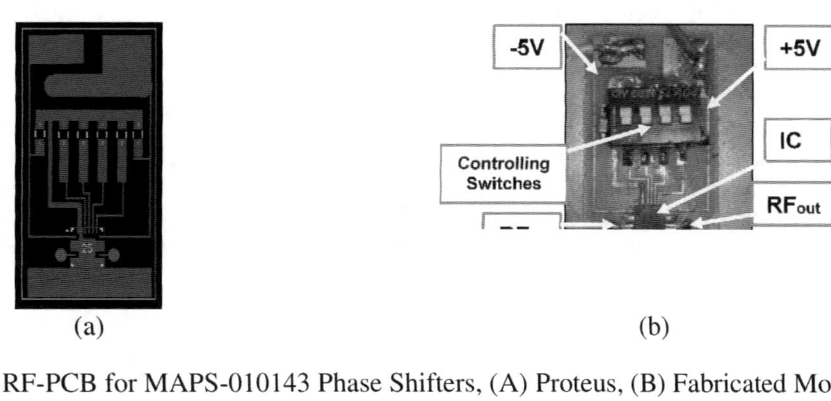

(a) (b)

Figure 4. RF-PCB for MAPS-010143 Phase Shifters, (A) Proteus, (B) Fabricated Model

2.5 Beam Steering Vector Weights Matrix Calculation

The PAAs can be controlled electrically to steer the main beam towards a specific direction. Beam steering is performed by changing the elements phase weights in the antenna array, however, there is no full control of the pattern shape, in terms of HPBW, nulls positions, and side lobes numbers and levels. The vector weights matrix calculation for beam steering in this project has been calculated by developing a MATLAB code. The main idea is based on AF derivation. The AF of ULA can be given as in equations (1 and 2). Where (β) represents the phase shift differences among the elements, and the main beam of the antenna array can be steered by changing the phase of the adjacent elements. The phase shifters that are used for the design give (16) phase shift of a step of (22.5^0). The MATLAB code asks for the required angle to be entered to steer the main beam towards it. Then, it will start calculating the exact phase weight for each element in the array. Theoretically, the maximum achievable beam steering is ($\pm 60^0$), i.e. (120^0). The system can perform the steering of (1^0) steps from (0^0) to (45^0) with very high accuracy. Above (45^0) to (60^0) this accuracy decreases in terms of the maximum main lobe peak pointing angle. The phase shifter was chosen to work in parallel mode; thus, the phases are changed digitally by switches. Each combination of digits gives a corresponding phase shift. Table 2 shows the combination of the switches to perform the (16) phases steps. (D_1) is the LSB and (D_4) is the MSB.

Table 2. The Combinations of Digits of the Phase Shifters Phases Steps

Phases	0	22.5	45	67.5	90	112.5	135	157.5	180	205.5	225	247.5	270	292.5	315	337.5
D1	0	1	0	1	0	1	0	1	0	1	0	1	0	1	0	1
D2	0	0	1	1	0	0	1	1	0	0	1	1	0	0	1	1
D3	0	0	0	0	1	1	1	1	0	0	0	0	1	1	1	1
D4	0	0	0	0	0	0	0	0	1	1	1	1	1	1	1	1

Five steering cases were chosen of a step of (10^0) from (15^0 to 45^0) and a (60^0), to illustrate the steering trends. All the cases show both, positive and negative steering directions.

Table 3 illustrates the phase weight of each element in the array and the corresponding combination of the digital switches for an example of a steering case towards $+15^0$.

Table 3. An example of Beam Steering Vector Weights Matrix

Steering For (+15) Degrees															
Element 1 Phase				Element 2 Phase				Element 3 Phase				Element 4 Phase			
315				270				225				180			
D1	D2	D3	D4	D1	D2	D3	D4	D1	D2	D3	D4	D1	D2	D3	D4
0	1	1	1	0	0	1	1	0	1	0	1	0	0	0	1

2.6 Integration of RF-PCB with Antenna Structure

Phase-shifters were integrated within the antenna array structure by fitting them on the antenna back panel. (RF_{in}) and (RF_{out}), for each RF-PCB were connected to the antenna elements' feeding lines. The connections were made by Vias of (0.7 mm) through the substrates of the RF-PCB and antenna array. All Vias have the same length to avoid any phase differences among the antenna array elements. The antenna array ground plane was connected to the RF-PCBs ground-plane, as in figure 5, which gave more stability for the phase shifter performance.

IOP Conf. Series: Materials Science and Engineering 1152 (2021) 012002 doi:10.1088/1757-899X/1152/1/012002

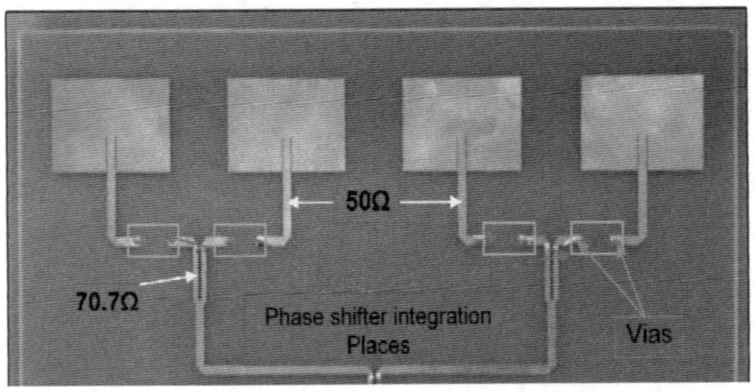

(a)

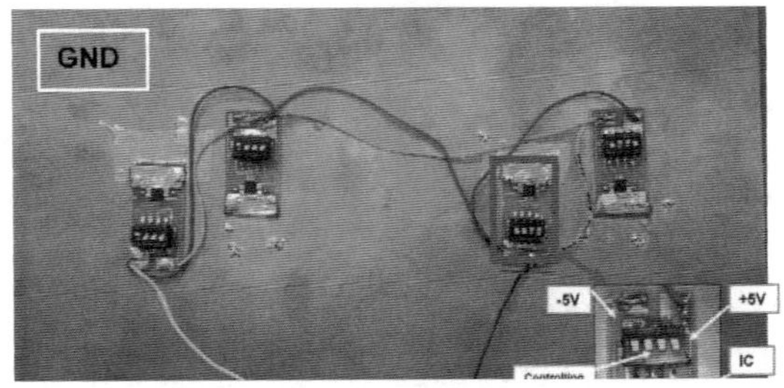

(b)

Figure 5. (a) PAA Design Front Panel. (b) Phase Shifter Integrated with the Antenna Array Back-Panel.

The Reflection coefficient (S_{11}) of the PAA after integrating the phase shifters is shown in figure 6. The figure compares the (S_{11}) for the three models, MATLAB, CST and practical model.

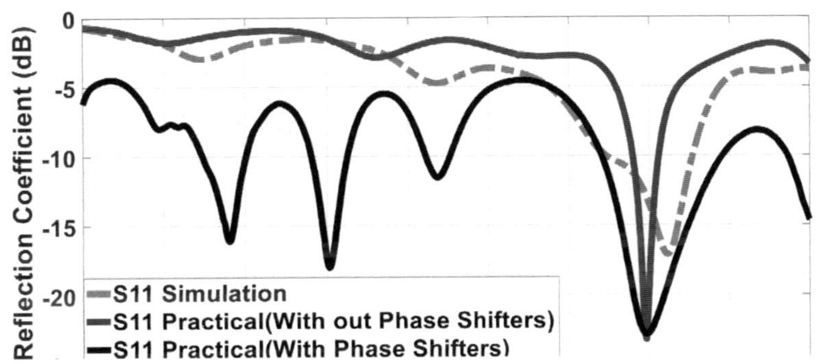

Figure 6. (S_{11}) for PAA Matlab, CST, and Practical

The effects of integrating phase shifters within the antenna array feeding lines are clear on the (S_{11}), however, the antenna still resonates at (2.4 GHz).

3. Beam Steering Results and discussion

The PAA has been implemented practically according to the CST design. The vector weights matrix was applied. The practical model was tested in the anechoic chamber as shown in figure 7.

Figure 7. PAA testing in the Anechoic Chamber

The results were as expected and verified the MATLAB and CST results. figure 8 to figure 13 show the selected steering cases.

IOP Conf. Series: Materials Science and Engineering 1152 (2021) 012002 doi:10.1088/1757-899X/1152/1/012002

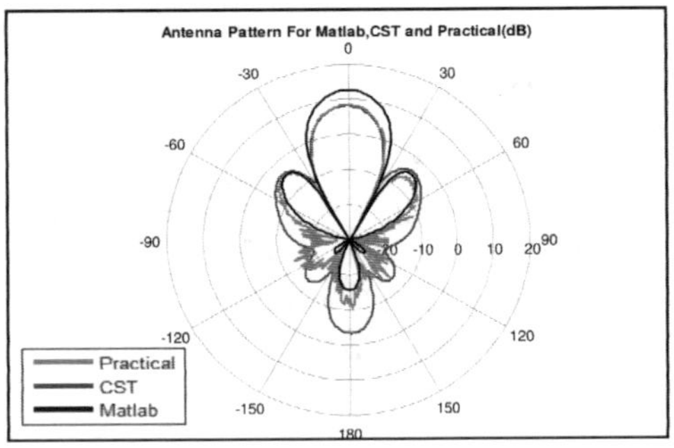

Figure 8. Antenna Pattern for Matlab, CST and Practical (no steering)

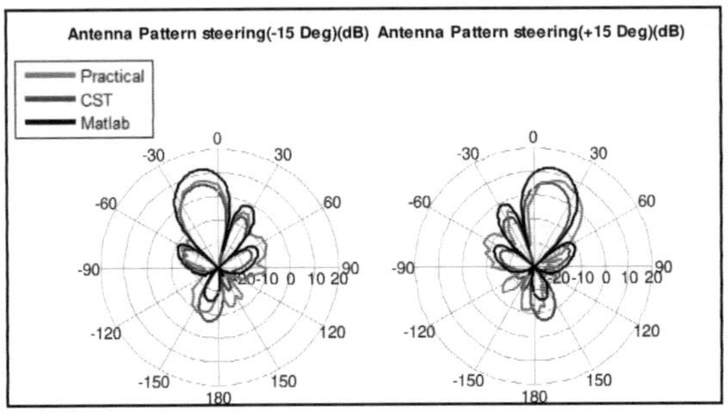

Figure 9. Scanning of (-15^0) Left Figure, (+15^0) Right Figure Matlab, CST and Practical

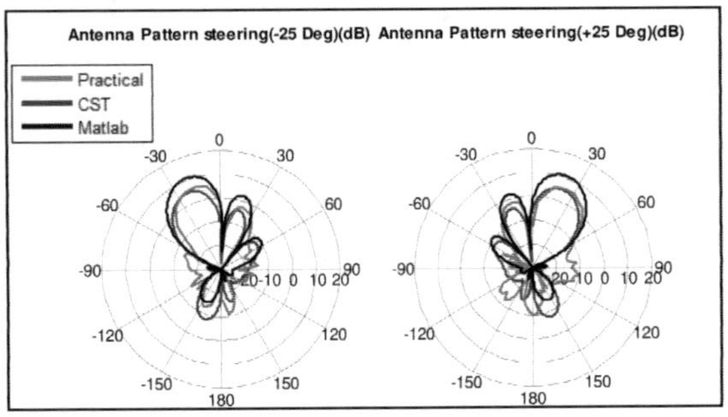

Figure 10. Scanning of (-25^0) Left Figure, (+25^0) Right Figure Matlab, CST and Practical

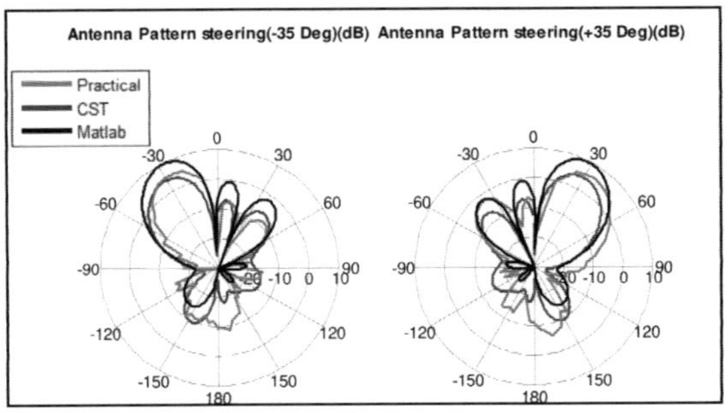

Figure 11. Scanning of (-35^0) Left Figure, (+35^0) Right Figure Matlab, CST and Practical

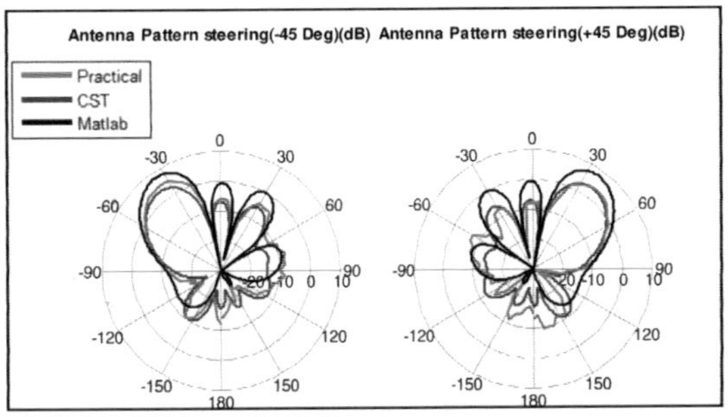

Figure 12. Scanning of (-45^0) Left Figure, (+45^0) Right Figure Matlab, CST, and Practical

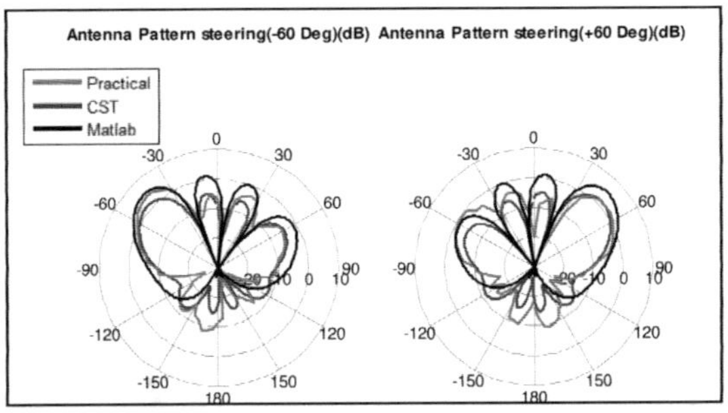

Figure 13. Scanning of (-60^0) Left Figure, (+60^0) Right Figure Matlab, CST, and Practical

Five steering cases were chosen in a step of (10^0) from (15^0 to 45^0) and a (60^0), to illustrate the steering trends. The practical model results were close to the MATLAB results and CST, which is the benchmark. figure 8 to figure 13show a comparison of the three models, MATLAB, CST, and practical model steering results. It is clear from the figures that there are minor differences between the MATLAB pattern and CST. These differences are to be expected because MATLAB results are based on mathematical equations, however, CST is closest to actual results and takes into consideration all the effects such as mutual coupling among elements, the substrate effects, and the ground plane effects. Thus, CST is considered the benchmark for practical model beam steering results. The mismatch losses of the phase shifters and the antenna array were expected. This problem is critical and is considered to be the main cause for any degradation in the system performance for two reasons; firstly, the phase shifter IC should be matched at the (RF_{in}) to ensure that it is receiving the suitable operation level value. Secondly, the (RF_{out}) matching, to ensure that it is matched with feeding lines without any reflections. The phase shifter has a characteristic impedance of (50Ω). Consequently, the antenna array feeding lines were designed to be as (50Ω) at the places where the

IOP Conf. Series: Materials Science and Engineering 1152 (2021) 012002 doi:10.1088/1757-899X/1152/1/012002

phase shifters were fitted. The Microstrip line width for this is (2.88 mm), according to the design specifications. However, when fitting the phase shifters on the back panel of the antenna array, the connection Vias (RF_{in}) and (RF_{out}) were (0.7 mm) through the two substrates, RF-PCB and antenna array respectively. Nevertheless, the results have the same shape and trends, which leads to concluding that the mismatches are as minimal as possible, the phase shifters are performing the steering correctly in all cases and the steering matrix is correct and it is suitable to use for the practical model.

3.1 Steering Accuracy

Theoretically, the maximum achievable beam steering is ($\pm 60^0$), i.e. (120^0). The system can perform the steering of (1^0) steps from (0^0) to (45^0) with high accuracy and with sidelobe levels around (-10 dB). Above (45^0) to (60^0) this accuracy decreases, in terms of steering the maximum of the main lobe peak pointing angle and sidelobe levels. There is more than one reason behind the decline in performance. For instance, the antenna array aperture size, the number of elements in the array, and the PAA steering concepts of changing only the phase or phase and the amplitude of each element in the array. These reasons will be discussed in the following sections.

3.2 Pattern Shaping and HPBW

HPBW is considered an important beam steering performance parameter. The narrower beam leads to better performance. However, as mentioned before, more than one factor can affect the antenna array HPBW. table 4 and figure 14 to figure 17 show a comparison of the HPBW of the main lobe for all steering cases in CST and practical model, also, the practical antenna gain taking into account the substantial sidelobes present at large steering angles.

Table 4. HPBW and gain for CST and Practical

Steering Angle (Deg)	0	-15	15	-25	25	-35	35	-45	45	-55	55	-60	60
HPBW Simulation (Deg)	24.8	25.7	25.7	26.9	26.9	28.9	28.9	30.3	30.3	31.4	31.4	31.6	31.6
HPBW Practical (Deg)	24.8	26.1	26.9	27.3	27	29.2	29.5	30.6	30.2	31.5	31.2	32	31.8
Antenna gain(dB)	8.39	5.6	5.73	5.07	5.07	4.2	4.33	3.6	3.53	2.5	2.26	2.1	1.58

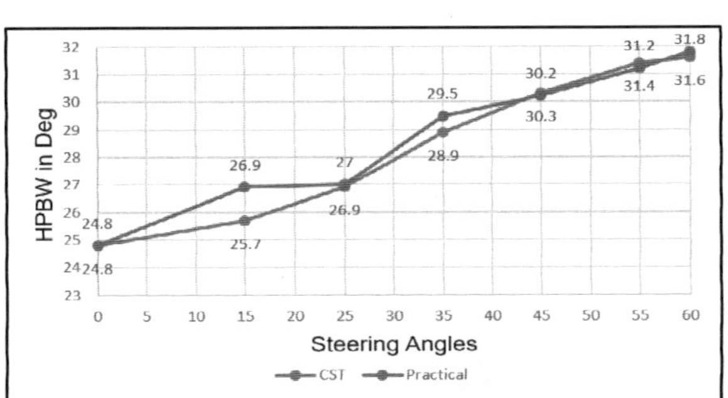

Figure 14. HPBW for CST and Practical (Positive Steering)

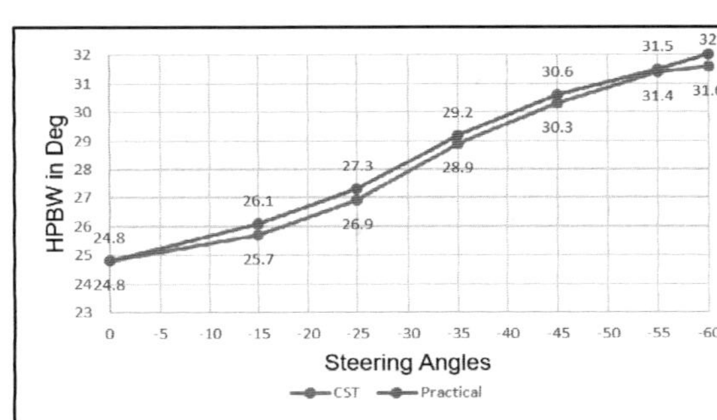

Figure 15. HPBW for CST and Practical (Negative Steering)

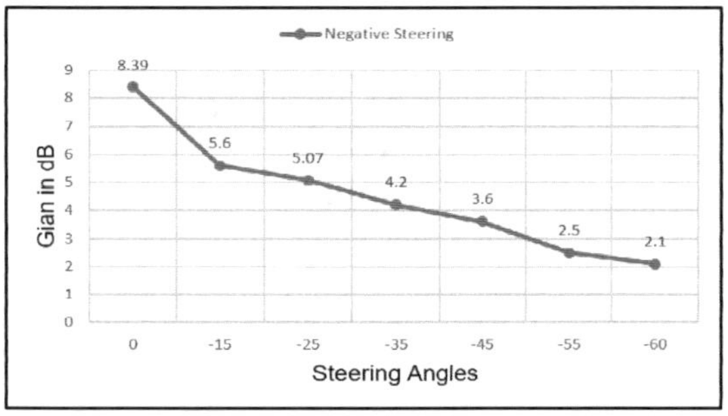

Figure 16. Main Lobe Gain, Negative Steering

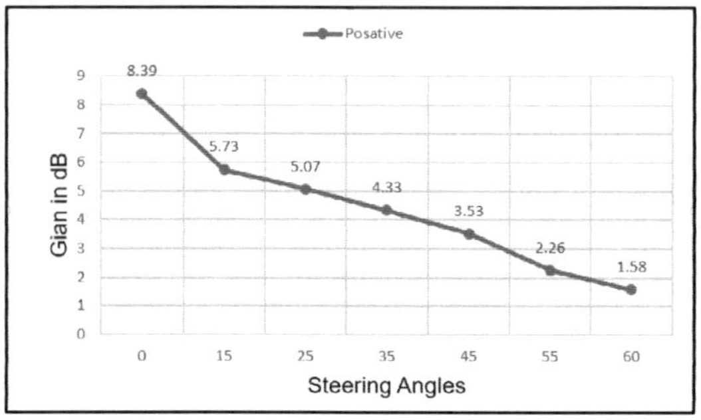

Figure 17. Main Lobe Gain, Positive Steering

It is clear that the values of steering for both the practical model and CST are close to each other. The noticeable change is that when increasing the scan angle, the HPBW angle increases which leads to broadening the antenna main lobe and at the same time increases the side-lobes numbers and levels decreases antenna gain and creates the grating lobes. The decrease in the main lobe level is expected because the power is lost in the increasing side lobes and grating lobes.

4. Conclusion

This paper presents a novel approach of beam steering system based on phased array antennas. The project aimed to build a novel, low-cost, and compact PAA. The system has been achieved by the following stages; Firstly, the steering vector weights matrix was calculated by Matlab. Then, four elements of uniform linear Microstrip patch antenna array with their feeding network power divider were designed and simulated in the CST simulation program. A prototype model has been implemented. Then, the specially designed RF-PCB for the 4-bit phase shifters MAPS_010143 were integrated within the antenna panel. The system has been tested and the results have been compared with the simulation results. The system has high beam steering. The result presents a compact PAA system, working at a resonance frequency of (2.4 GHz), with (120^0) steering capability, reducing the cables that feed the PAA system and limiting them to only one RF cable and the controlling lines, instead of the sophisticated traditional PAA systems in which the phase shifters board is separated from the antenna array panel and RFs coaxial cables are used to feed each element in the antenna array.

Acknowledgments

This work was supported by the Iraqi Ministry of Higher Education and Scientific Research (HCED) as based on a Master Dissertation Submitted in Partial Fulfilment of the Requirements for the Degree of Master by Advanced Study in Personal, Mobile and Satellite Communications for the (2014-2015) academic year and has been accomplished at the University of Bradford.

5. References

[1] Marconi G 1906 Directive Antenna*Proc. Royal Soc.***77A** 413

[2] Haupt R 2015 Timed and Phased Array Antennas*Wiley-IEEE Press eBook*

[3] Beverage H H, Rice C W and Kellog E W 1923 The wave antenna: A new type of highly directive antenna *American Institute of Electrical Engineers***42**pp372-381

[4] Kavitha L B S and Raglend I J 2013 A wide-scan phased array antenna for a small Active Electronically Scanned Array: A review *2013 International Conference on Circuits, Power and Computing Technologies (ICCPCT)* pp 1008-1016

[5] Friis H T and Feldman C B 1937A multiple unit steerable antenna for short-wave reception *Bell System Technical Journal***16** 3 pp337-419

[6] Xiaojing H,Guo Y J and Bunton J D 2010 A hybrid adaptive antenna array *Wireless Communications, IEEE Transactions on***9** 5 pp 1770-1779

[7] Xiaoyong L, Qi Z and Yuehua L 2011 Integrative design of power divider and phase shifter for phased array *Microwave Symposium Digest (MTT) 2011 IEEE MTT-S International* pp. 1-4

[8] Abbaspour-Tamijani A and Sarabandi KAn 2003 affordable millimeter-wave beam-steerable antenna using interleaved planar subarrays *Antennas and Propagation, IEEE Transactions* **51** 9 pp. 2193-2202

[9] Aamo K 1994 Frequency controlled antenna beam steering *Telesystems Conference, 1994. Conference Proceedings IEEE National* pp. 51-54

[10] Modelski J W and Yashchyshyn Y 2012 Microwave ferroelectric and reconfigurable antennas *Microwave Workshop Series on Millimeter Wave Wireless Technology and Applications (IMWS) IEEE MTT-S International* pp. 1-2

[11] Hong W, Kim Y G, Baek K and Lee Y Design and testing of a millimeter-wave beam-steering mesh-grid array for 5th generation (5G) mobile communication handset devices 2014 *USNC-URSI Radio Science Meeting (Joint with AP-S Symposium), Memphis, TN, USA* 2014 pp. 282-282

[12] Nishio T, Hao X, Yuanxun W and Itoh T 2004 A frequency-controlled active phased array *Microwave and Wireless Components Letters, IEEE* **14**3 pp. 115-117

[13] Thongsopa C, Krairiksh M, Srirattana N and Intarapanich A 2000 A single patch beam steering antenna *microwave ConferenceAsia-Pacific* pp. 1510-1513

[14] Sangjun H and Chang Won J 2011 Single patch beam-steering antenna with U-slot for wearable fabric applications*Antennas and Propagation (APSURSI), 2011 IEEE International Symposium* pp. 1560-1562

[15] Attia H, Siddiqui O and Ramahi O 2012 Theoretical and experimental demonstration of beam steering of patch antenna with superstrate *Antennas and Propagation Society International Symposium (APSURSI)*pp. 1-2

[16] Wenquan C, Yang X, Bangning Z, Aijun L, Tongbin Y and Daosheng G 2011 A Low-Cost Compact Patch Antenna With Beam Steering Based on CSRR-Loaded Ground*Antennas and Wireless Propagation Letters, IEEE* **10** pp. 1520-1523

[17] Hussaini S, Elfergani I, Abusitta M, Adebola A, Abd-Alhameed R Aand Rodriguez J 2013 Time modulated switching and reactive loading techniques applied to a circular array antenna using genetic algorithm optimization *Antennas and Propagation (EuCAP), 2013 7th European Conference* pp. 347-351

[18] Can D, Guo Y J, Pei-Yuan Q, Bird T S and Yintang Y 2013 A novel phase shifter based on reconfigurable defected microstrip structure (RDMS) for beam-steering antennas *Antennas & Propagation (ISAP), 2013 Proceedings of the International Symposium on*, 2013, **02** pp. 993-996

[19] Scott H and Fusco V F 2001 Antenna array beam-steering by the integration of a series phase shifter *High-Frequency Postgraduate Student Colloquium, 2001 6th IEEE* pp. 25-29

[20] Balanis C A 2005 *Antenna theory: analysis and design*, 3rd ed. Hoboken, N.J.: Wiley-Interscience

[21] Mailloux R J 2005 *Phased array antenna handbook*2nd ed.

[22] Visser H J 2005*Array and Phased Array Antenna Basics*: John Wiley & Sons, Ltd

[23] Fenn and Hurst P 2015 *Phased Array Antenna Theory*: MIT Press

[24] M. T. Solution, "MAPS-010143, Digital Phase Shifter4-Bit, 1.4 - 2.4 GHz," Rev. V4 ed

Unconventional and Irregular Clustered Arrays

Ahmed Jameel Abdulqader[1,2*], Jafar Ramadhan Mohammed[1] and Raad H. Thaher[2]

[1]College of Electronics Engineering, Ninevah University, Mosul-41002, Iraq
[2]College of Engineering, Mustansiriyah University, Baghdad-10052, Iraq

*E-mails: ah_8611@yahoo.com

Abstract. In this paper, two different structures based on fully and partially irregular clustered elements are presented to optimize the radiation patterns of the large arrays. In the Fully Irregular Clustered Elements (FICE) structure, all the elements of an ordinarily large linear array are divided into multiple irregular and unequal-size clusters, while in the Partially Irregular Clustered Elements (PICE) structure, only the elements those located at the edges of the ordinarily linear array are grouped into unequal-size clusters. In this structure, the central elements are left without clustering (i.e., excited individually). The PICE structure has several advantages over the FICE structure. Since the quantized amplitudes (i.e., discrete distribution) are used with the proposed clustered structures, the corresponding clustered array patterns are usually associated with the undesirable high periodic sidelobes. In order to overcome this problem, the elements in the clusters are distributed a periodically. Simulation results demonstrate the ability of the proposed two structures, FICE and PICE, to significantly reduce the high periodic sidelobe level to -35dB and -40dB respectively for an array with a number of elements N=100 and unequal-size clusters.

1. Introduction

Designing a large antenna arrays for modern communication systems such as 5G, radar and satellites is one of the most difficult challenges in the practice. Thus, techniques must be used to simplify the design process. Subarray techniques were widely used to simplify the practical implementation of the large arrays, in which a number of elements are connected to a common subarray to reduce the number of control points in the excitation network [1-9]. This reduction in the architecture's complexity allows a significant and reasonable simplification in array fabrication, and consequently lower cost. This simplification requires an unconventional design in a corporate feeding network, so, the calibration processes must be simplified theoretically, but it is still difficult to identify and correct the random errors that may affect the excitation of the elements during feeding [10]. The techniques that described in [11-13] aim to evaluate the effects of random errors in terms of amplitudes, phases and element locations on the array radiation pattern.

The concept of the subarray's technique involves dividing the radiating elements N in the large arrays into small groups called clusters, each cluster with M radiating elements, so that M≪N. Here, each cluster has only a common controller for specifying the amplitude or the phase (time delays) or both [14-15]. Although high directivity and half power beam width (HPBW) are still warranted, such a conventional architectural solution makes a minor contribution in terms of limited operating bandwidth and low reconfiguration potential due to the quantized distribution of amplitude or phase weights or both across the array aperture [15-21]. If the array aperture is divided into clusters with equal-sized, then the radiation patterns with undesired grating lobes are noticed (i.e., high periodic

Content from this work may be used under the terms of the Creative Commons Attribution 3.0 licence. Any further distribution of this work must maintain attribution to the author(s) and the title of the work, journal citation and DOI.
Published under licence by IOP Publishing Ltd

sidelobes) and they found to be more noticeable for increasing bandwidth. Thus, this structure with equal-sized clusters fails to provide a good solution. The authors in [5] suggested to place the clusters a periodically through the array aperture. It uses certain types of irregular cluster shapes (i.e., polyomino shaped) that are rotated in specific directions to fill the array aperture so that the periodicity of the arrangement of clusters is eliminated. However, this method needs a rather complex corporate feeding network.

Several optimization strategies [22-26] have been suggested in the literature to get an optimum architecture for the array clusters that meets the desired radiation pattern.

In this paper, the desired array pattern under the required constraints is obtained by using several irregular arrangements of the clustered elements. The first structure composes a number of clusters with each one having unequal number of array elements. This will be referred to as Fully Irregular Clustered Elements (FICE). In the second structure, the elements that located only on both ends of the linear array are divided into irregular clusters while the central elements are let without grouping into clusters. This will be referred to as Partially Irregular Clustered Elements (PICE). This structure provides more degrees of freedom than the first structure, FICE. Thus, the performance of the second structure may be better than that of the first structure. Also, more constraints may be flexibly applied with the second structure.

2. The Proposed Structure

In this section, the fully and partially irregular clustered structures are introduced. The array factors and the cost functions of these two structures are given. The genetic optimization is used to get the desired array patterns. Its specifications were: the number of populations is 50; rate of mutation is 0.15; and a single point crossover.

2.1 Fully Irregular Clustered Elements

This section includes two types of fully clusters, i.e., fully regular clustered elements (FRCE) and the fully irregular clustered elements (FICE). It should be mentioned that the FRCE was already presented in [14]. It will be shown that better results can be obtained by using FICE instead of the FRCE. Thus, this paper mainly focuses on the FICE. Suppose a linear array with an even number of elements N. The elements are arranged symmetrically around the centre, which means that only half of the elements are optimized. Thus, the number of the RF (attenuators and phase shifters) components is also halved. In the FICE structure, all the array elements, N, will be subdivided into multiple clusters, say Q, and each cluster, q, contains different M_q elements. Q is calculated by dividing N by M_q on the condition that the remainder of the division is zero, meaning that the c must be integer number. Figure 1 shows the structure of the FICE structure. From this figure, it can be seen that each cluster contains a number of elements equal to 2, 3, 4, and so on. By choosing different–size clusters, the problem of appearing quantized high sidelobes can be overcome.

The array factor of this structure can be written as

$$AF(u) = 2\sum_{q=1}^{Q/2} A_q \sum_{n=1}^{N/2} \delta_{cnq} w_n \cos\{\frac{(2n-1)}{2} kdu\} \tag{1}$$

where $\delta_{cnq} = \begin{cases} 1 & if\ the\ nth\ element\ belong\ to\ the\ mth\ cluster \\ 0 \end{cases}$

A_q represents the amplitude of each clustered output, w_n represents the complex weights of the elements in the original (without clustering) linear array and equals $a_n e^{jp_n}$, a_n is the amplitude excitation, p_n is the phase, k=2π/λ, where λ is the wavelength, u=sin(θ), where θ is the direction of main beam. d is the spacing distance between any two elements within array.

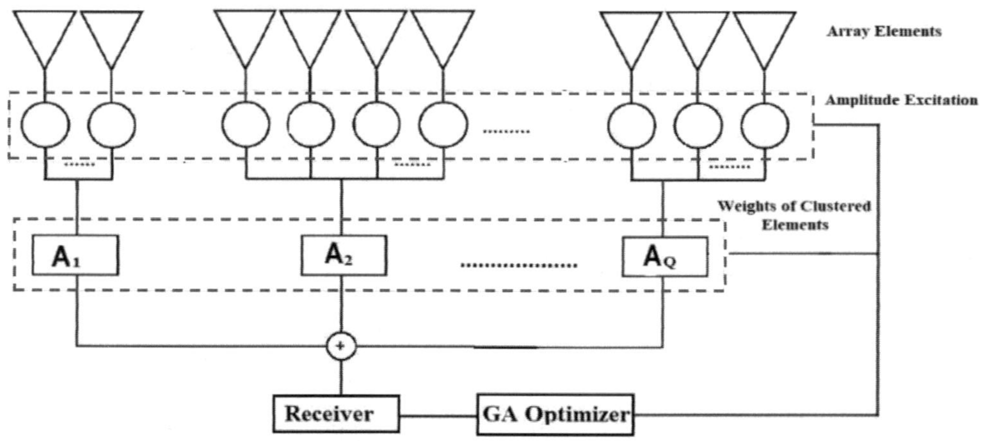

Figure 1. Structure of the FICE.

To further simplify the feeding network, amplitude only excitation is used and the phase is considered to be zero. Thus, equation (1) is simplified to:

$$AF(u) = 2\sum_{q=1}^{Q/2} A_q \sum_{n=1}^{N/2} \delta_{cnq} a_n \cos\{\frac{(2n-1)}{2} kdu\} \quad (2)$$

This equation is used in the optimizing process in order to obtain the optimum weights of the clusters. In this equation, A_q is calculated based on the values of a_n, where at the beginning the optimized values of a_n are calculated to obtain the desired pattern with required sidelobes, then the average values of the amplitudes of the optimized elements $a_1, a_2, a_m, \ldots\ldots a_M$ within each cluster, Q, are computed. As a result, the excitation of the amplitudes of the $a_1, a_2, a_m, \ldots\ldots a_M$ for each cluster will be quantized to the value of A_q. To control the level of the sidelobes and prevent them to be higher than the required sidelobe level, an additional condition is added to the cost function which draws a specified line to control the sidelobes. Any point above the line contributes a value to the cost function equal to the power difference between the clustered array pattern and the desired pattern of the uniform linear array. Thus, the cost function can be written as follows:

$$CostFunction = \sum |AF - Constraints\ (\text{Mask limit})|^2 \quad (3)$$

As the inputs for this function include the array factor with its A_q and a_n, the desired array pattern, as well as the specified mask constraints. The constraints of this function can be identified as follows:

$$Constraints_{\text{FICE}} = \begin{cases} uppermasklimit = -35\ dBExemptMainBeam \\ lowermasklimit = -60\ dB \end{cases} \quad (4)$$

2.2 Partially Irregular Clustered Elements

Another new structure presented in this paper is based on the partially irregular clustered elements. Here in this structure, the elements that located at the ends of the array are grouped into different-size clusters followed by the common weights Aq as shown in Figure 2. Whereas, the elements that located in the centre of the array are left without clustering (i.e., excited individually).

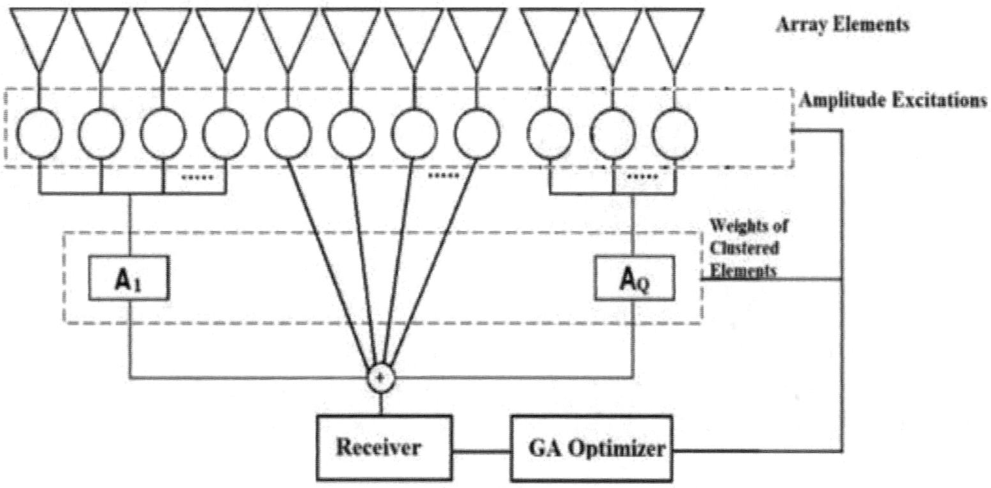

Figure 2. Structure of the PICE

The main benefit of this structure is its ability to provide more degrees of freedom than the previous one. Thus, the constraints have more flexibility on the peak sidelobe level. The array factor of this structure can be written as

$$AF(u) = \underbrace{2\sum_{n=1}^{L/2} b_n \cos\left(\frac{(2n-1)}{2} kdu\right)}_{\frac{L}{2}\text{Individually feeding elements}} + \underbrace{2\sum_{q=1}^{Q/2} A_q \sum_{n=\frac{(N-L)+1}{2}}^{N/2} a_n \cos\left(\frac{(2n-1)}{2} kdu\right)}_{\frac{Q}{2}\text{Subarrays on each side of the array}} \quad (5)$$

The mask constraints are

$$Constraints_{\text{PICE}} = \begin{cases} uppermasklimit = -40\ dBExemptMainBeam \\ lowermasklimit = -60\ dB \end{cases} \quad (6)$$

3. Simulation Results

In order to verify the effectiveness of the proposed structures (i.e., FICE and PICE), several examples are illustrated for each structure. In all examples, the element phases are made zero (i.e., element amplitudes are only used). The number of the elements in each side of the linear array is chosen to be 50 (the total number of the array elements is 100). The number of the elements in each cluster Mq may be chosen as 2, or 3, or any other value (when Mq=1 it represents a non-clustering state). Since a symmetrical linear array has been used in all examples, then, only a one side of the element amplitudes is displayed.

In the first example, the advantages of using the proposed FICE with compared to the previous structures such as fully regular clustered elements (FRCE) that were presented in [14] is addressed. Figure 3 shows the results of the FRCE and the proposed FICE. For the FRCE structure, the number of the clusters is Q=5, the number of the elements in each cluster is Mq =10. For the FICE structure, we used different sizes of Q, and different number of elements in each cluster Mq as can be seen in Figure 3. From the results of the FRCE structure, it can be seen that a periodic high sidelobes can be clearly noticed. This undesirable effect was found to be proportionally increased with the increased size of the clusters. As can be seen from the results of FICE (see Figure 3(c)), the problem of appearing a periodic high sidelobes is completely eliminated with the use of FICE structure.

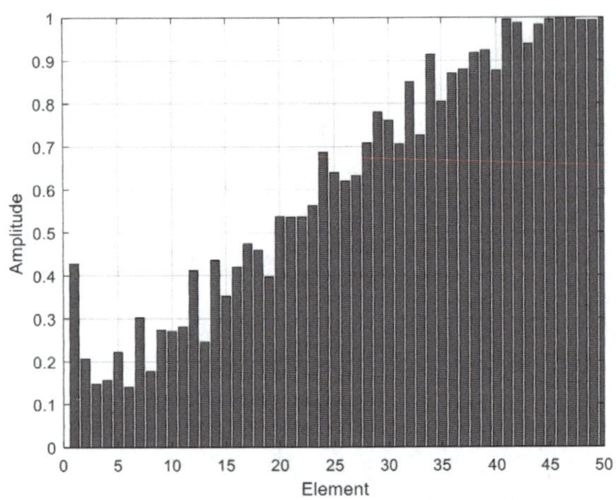

a. Fully optimized amplitude distribution

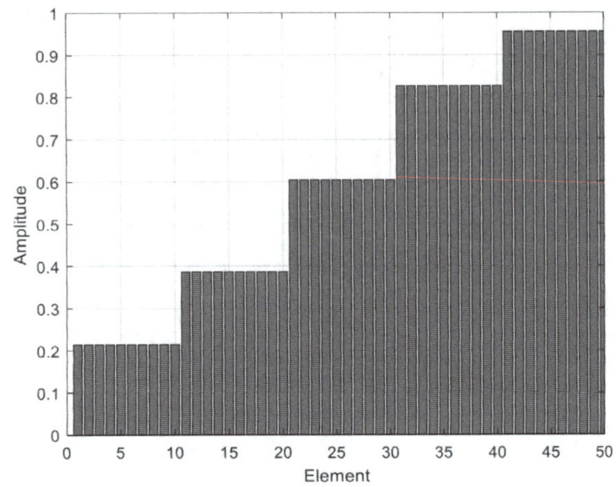

b. The FRCE amplitude distribution

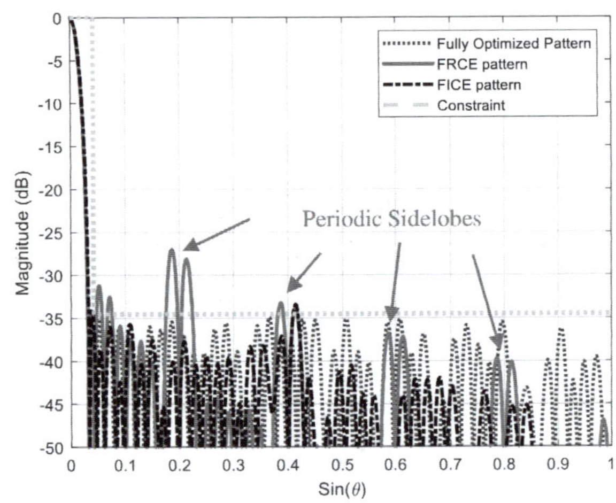

c. Radiation patterns

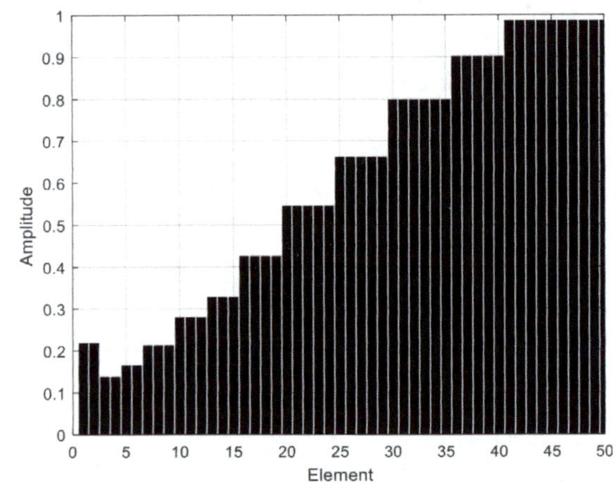

d. The FICE amplitude distribution

Figure 3. Results of FRCE and FICE Structures.

In the second example, the idea of PICE is illustrated. To highlight its performance, it is compared to the partially regular clustered elements (PRCE) that were presented in [14]. Among these four structures (i.e., FRCE, FICE, PRCE, and PICE), we found that the new two structures (i.e., FICE and PICE) give best results in terms of lower sidelobes.

Figure 4 shows the results for PRCE with 2N=100, M=6, the number of elements that are excited individually L=20, and for PICE with different sizes of clusters as shown in figure. From this figure, it is observed that the peak sidelobe level can be reduced to more than -34 dB in the case of PRCE, and to more than -40 dB for the structure of PICE.

IOP Conf. Series: Materials Science and Engineering 1152 (2021) 012003 doi:10.1088/1757-899X/1152/1/012003

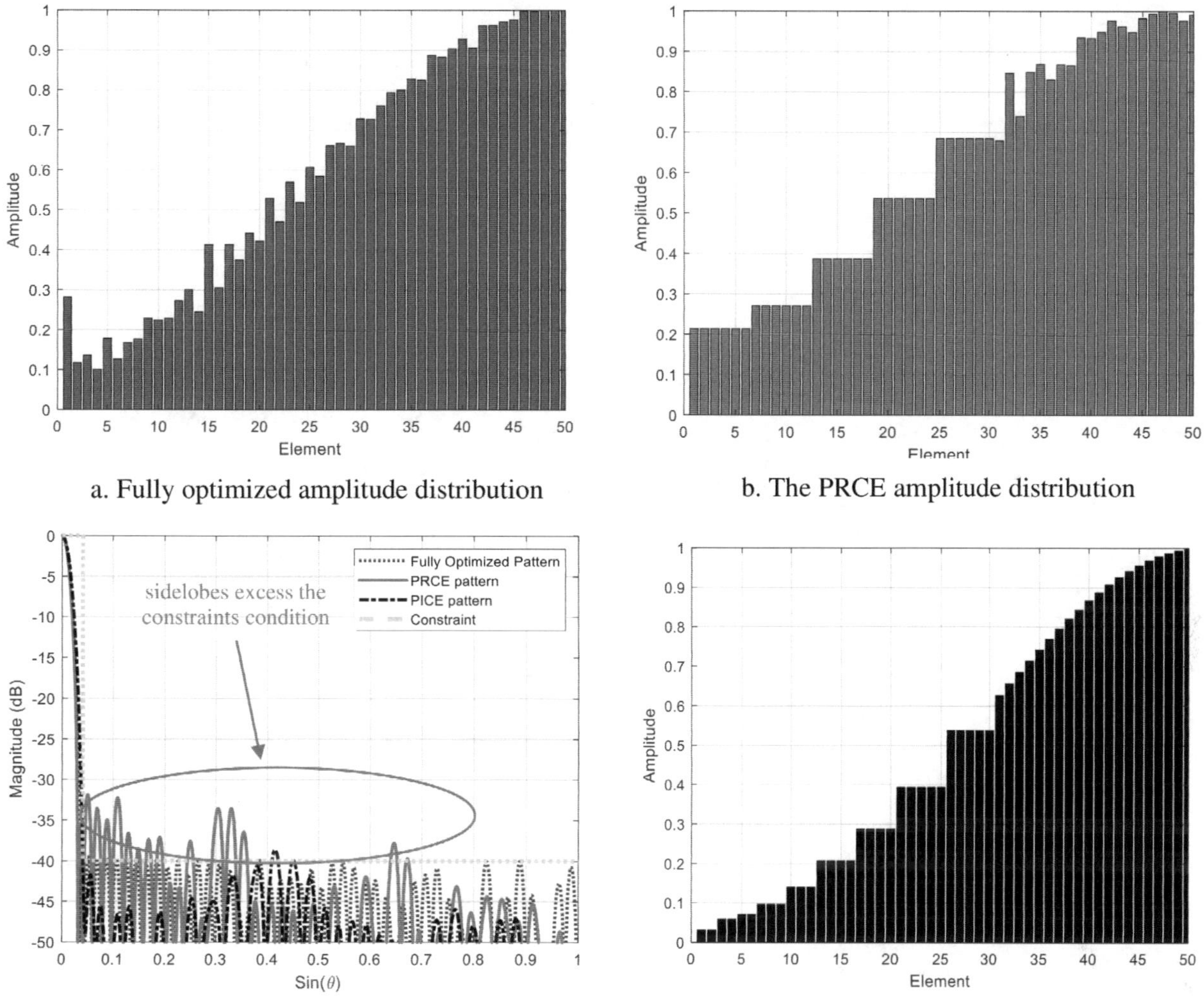

a. Fully optimized amplitude distribution

b. The PRCE amplitude distribution

c. Radiation patterns

d. The PICE amplitude distribution

Figure 4. Results of PRCE and PICE Structures.

Figures 5 and 6 show the variations of the cost functions according to (3), (4) and (5) versus the number of iterations during the optimization process for the FRCE, FICE, PRCE, and PICE structures. It is clear from these two figures that the two fully clustered structures need less number of iterations to converge with compared to that of the partially clustered structures. This is mainly due to the lesser number of degrees of freedom in the fully clustered structures.

To further highlight the effectiveness of the proposed structures with compared to some existing methods such as the I-CPE, M-CPE structures in [27]. Figure 7 show the results of all tested structures. The PICE structure provides lower sidelobes.

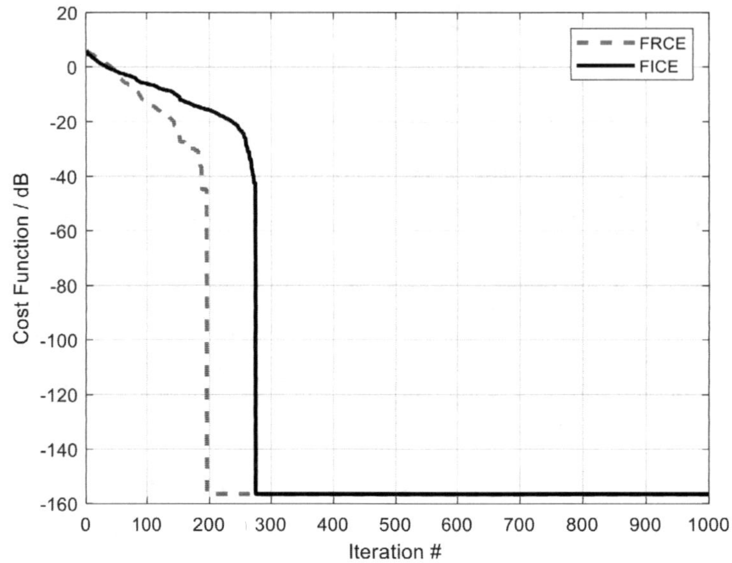

Figure 5. Comparisons between FRCE and FICE

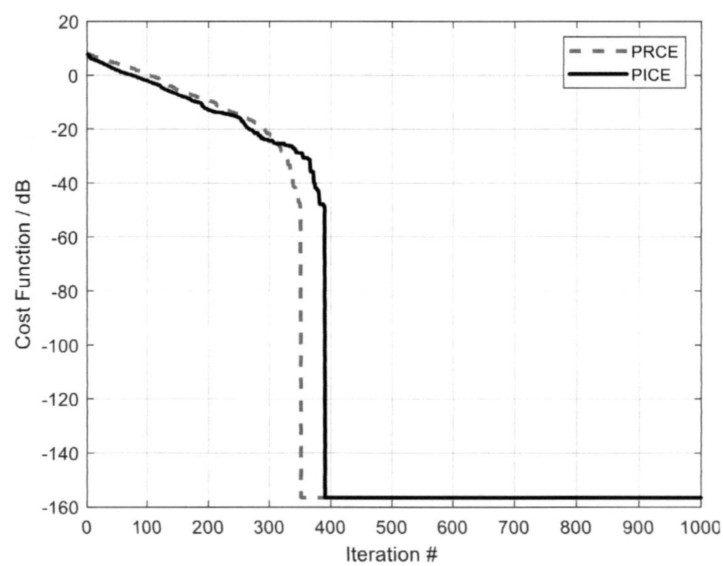

Figure 6. Comparisons between PRCE and PICE

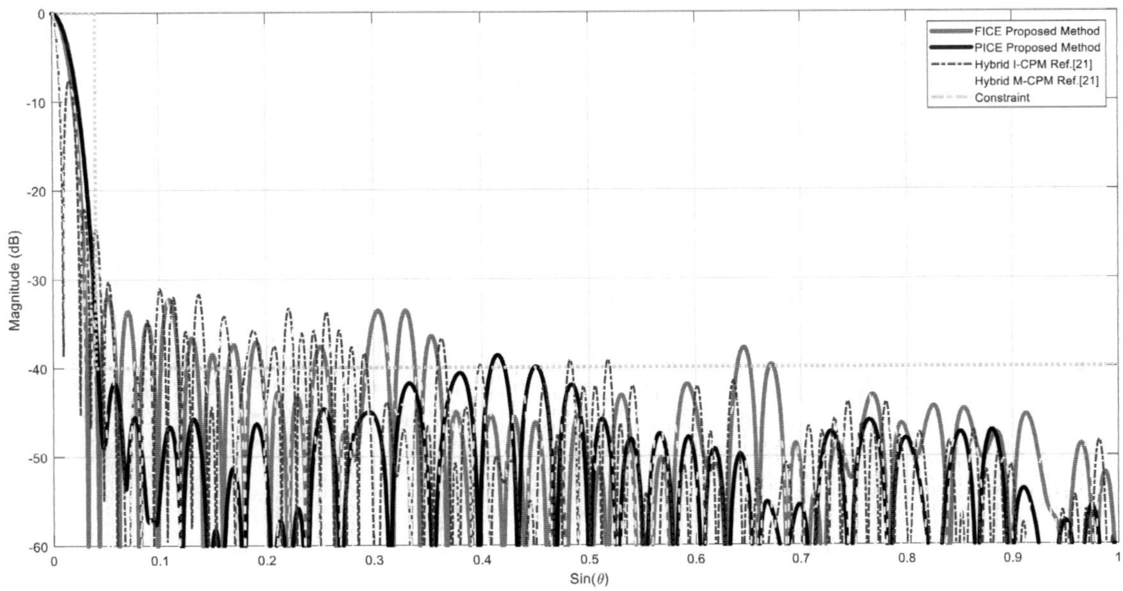

Figure 7. Comparisons between the proposed FICE, PICE structures and the existing structures in [27]

Furthermore, some numerical results are also shown with compared to the structures that were presented in [14] as shown in Table 1.

Table 1 Comparisons between FICE and PICE with proposed methods in Ref. [14]

Clustered Method	*Clustered Type*	*Size of Large Array, Clusters and Individually Elements*	*Level of Peak Sidelobe (dB)*
FRCE [14]	Regular	2N=100, Q=5(M=10) and L=0	≤ -30
FICE (this paper)	Irregular	2N=100, Q=2,3,4,5, and 10, and L=0	≤ -35
PRCE [14]	Regular	2N=100, Q=5(M=6) and L=20	≤ -32
PICE(this paper)	Irregular	2N=100, Q=2,3,4,5, and 10, and L=20	≤ -40

4. Conclusion

It is clear from the current investigation that the desired array patterns with required constraints can be obtained by using both the fully and partially irregular clustered element structures. The use of irregular clustered helps to eliminate the undesirable periodic high sidelobes that were unavoidable with the regular clustered. Other advantages include a significant reduction in the array complexity of the feeding network.

5. References

[1] Brockett T. J. and Y. Rahmat-Samii, "Subarray design diagnostics for the suppression of undesirable grating lobes," *IEEE Trans. Antennas Propag.*, **vol. 60**, no. 3, pp. 1373–1380, Mar. 2012.

[2] Mohammed, J. R., "Design of printed yagi antenna with additional driven element for WLAN applications", *Progress in Electromagnetics Research C*, **Vol. 37**, 67-81, 2013. ISSN:19378718

[3] Xiaowen Zhao ,Qingshan Yang , and Yunhua Zhang, "*Synthesis of Subarrayed Linear Array via l1 normMinimization Compressed Sensing Method*" IEEE Asia-Pacific Conference on Antennas and Propagation (APCAP 2018).

[4] Haupt R. L. "Optimized weighting of uniform subarrays of unequal sizes", *IEEE Trans. Antennas Propag.*, **vol. 55**, no. 4, pp. 1207-1210, 2007.

[5] Mailloux R. J., S. G. Santarelli, T. M. Roberts, and D. Luu, "Irregular polyomino-shaped subarrays for space-based active arrays," *Int. J. Antennas Propag.*, **vol. 2009**, pp. 1-9, 2009.

[6] Oliveri G., M. Salucci, and A. Massa, "Synthesis of modular contiguously clustered linear arrays through a sparseness-regularized solver," *IEEE Trans. Antennas Propag.*, **vol. 64**, no. 10, pp. 4277–4287, Oct. 2016.

[7] Sayidmarie, K. H. and Mohammed, J. R., "Performance of a wide angle and wideband nulling method for phased arrays", *Progress in Electromagnetics Research M*, **vol. 33**, pp. 239-249, October 2013.

[8] D'Urso M., T. Isernia, and F. Meliado, "An effective hybrid approach for the optimal synthesis of monopulse antennas," *IEEE Trans. Antennas Propag.,* **vol. 55**, no. 4, pp. 1059-1066, 2007.

[9] Oliveri G. and A. Massa, "Fully-interleaved linear arrays with predictable sidelobes based on almost difference sets," *IET Radar, Sonar & Navigation*, **vol. 4**, no. 5, pp. 649-661, 2010.

[10] Kristopher R. Buchanan and Carlos Flores-Molina," *Investigation of a Novel Subarray Nullsteering Technique for Distributed Random Arrays*" IEEE International Symposium on Antennas and Propagation (APSURSI), 2016.

[11] Mohammed J. R., A. J. Abdulqader, and R. H. Thaher, "Array pattern recovery under amplitude excitation errors using clustered elements," *Progress In Electromagnetics Research M*, **Vol. 98**, 183-192, 2020.

[12] Amir, Z. and A. R. Bijan, "GA-based approach to phase compensation of large phased array antennas," *Journal of Systems Engineering and Electronics*, **Vol. 29**, No. 4, 723–730, Aug. 2018.

[13] Mohammed, J. R. and K. H. Sayidmarie, "Sensitivity of the adaptive nulling to random errors in amplitude and phase excitations in array elements," *International Journal of Telecommunication, Electronics, and Computer Engineering*, **Vol. 10**, No. 1, 51-56, Jan.–Mar. 2018.

[14] Abdulqader A. J., J. R. Mohammed, and R. H. Thaher, "Antenna pattern optimization via clustered arrays," *Progress In Electromagnetics Research M*, **Vol. 95**, 177-187, 2020.

[15] Joerg Eisenbeis, Jonas Pfaff, Christian Karg, Jerzy Kowalewski, Yueheng Li, Mario Pauli, Thomas Zwick, "Beam pattern optimization method for subarray-based hybrid beamforming eystems", *Wireless Communications and Mobile Computing*, **vol. 2020**, 7 pages, 2020.

[16] Mohammed, J. R., "Phased array antenna with ultra-low sidelobes" *Electronics Letters*, **vol. 49**, issue 17, pp. 1055-1056, August 2013.

[17] Mohammed, J.R, and Sayidmarie, K. H., "Sidelobe cancellation for uniformly excited planar array antennas by controlling the side elements," *IEEE Antennas and Wireless Propagation Letters*, **vol.13**, pp. 987 - 990, 2014.

[18] Mohammed, J. R. and Sayidmarie, K. H., "A null steering method by controlling two elements", *IET Microwaves, Antennas & Propagation*, **vol. 8**, issue 15, pp.1348-1355, 2014.

[19] Mohammed, J.R.," Optimal null steering method in uniformly excited equally spaced linear array by optimizing two edge elements", *Electronics Letters*, **vol. 53**, issue 11, May 2017.

[20] Mohammed, J.R.," Element Selection For Optimized Multi-Wide Nulls in Almost Uniformly Excited Arrays ", *IEEE Antennas and Wireless Communication Letters*. **vol. 17**, issue 4, PP. 629-632, April 2018.

[21] Mohammed, J.R." Thinning a subset of selected elements for null steering using binary genetic algorithm", *Progress in Electromagnetics Research M*, **vol. 67**, pp. 147-157, March 2018.

[22] Toyama N., "*Radiation pattern optimization of aperiodic arrays consisting of sub-arrays*," IEEE Antennas and Propagation Society International Symposium, Albuquerque, NM, pp. 4229-4232, 2006.

[23] Abdulqader, A, J., J. R. Mohammed, and R. H. Thaher, "Phase-only nulling with limited number of controllable elements," *Progress In Electromagnetics Research C*, **Vol. 99**, 167–178, 2020.

[24] Alvarez-Folgueiras, M., J. A. Rodr´ıguez-Gonz´alez, and F. Ares-Pena, "High-performance uniformly excited linear and planar arrays based on linear semiarrays composed of subarrays with different uniform spacings," *IEEE Trans. Antennas Propag.*, **Vol. 57**, No. 12, 4002–4006, Dec. 2009.

[25] Yang, K., Y. Wang, and H. Tang, "A subarray design method for low sidelobe levels," *Progress In Electromagnetics Research Letters*, **Vol. 89**, 45–51, 2020.

[26] Zhao, X., Q. Yang, and Y. Zhang, "Synthesis of minimally subarrayed linear arrays via compressed sensing method," *IEEE Antennas Wireless Propag. Lett.*, **Vol. 18**, No. 3, 487–491, Mar. 2019.

[27] Manica, L., P. Rocca, and A. Massa, "Design of subarrayed linear and planar array antennas with SLL control based on an excitation matching approach," *IEEE Trans. Antennas Propag.*, **Vol. 57**, No. 6, 1684–1691, Jun. 2009.

Acknowledgments
The authors are grateful to the both Nineveh and Mustansiriyah universities for providing the research facilities which help to complete this research work.

IOP Conf. Series: Materials Science and Engineering 1152 (2021) 012004 doi:10.1088/1757-899X/1152/1/012004

Beam Tracking Using Least Mean Squares Algorithm for single-cell Massive MIMO Communication System

Ban A. Asi[1*], Farhad E. Mohmood[1] and Mohanad A.AL-Ibadi[1]

[1] Computer Engineering Department, College of Engineering, University of Mosul, Mosul, Iraq
[2] Electrical Engineering Departments, College of Engineering, University of Mosul, Mosul, Iraq
[3]AL-FURAT Al-Awsat Technical University

*E-mail: ban.enp1@student.uomosul.edu.iq

Abstract. Millimetres wave (mm-wave) is an attractive option for high data rate applications in the 5G wireless communication systems that require proper beamforming and channel tracking. In this paper, we study, analyse and compare the performance of two closely related stochastic gradient descent-based approaches, namely the least mean square (LMS) algorithm, and the normalized least mean square (NLMS) algorithm, for tracking the transmit array beam in addition to the channel status. These adaptive filters usually result in a trade-off between convergence and accuracy. We found that the quality of the tracking results, measured in mean squared error (MSE) sense, are heavily dependent on the present step- the size of the gradient descent.

1. Introduction

Recently, Massive multiple-inputs multiple-outputs (MIMO) frameworks are used in the fifth generation of mobile communication systems (5G) to enhance system efficiency and to reduce interference from multiple users as a large number of antennas are provided to the base station [1]. With the increasing demand for high spectral efficiency and high bandwidth, millimetre-wave (mm Wave) technology is adopted with massive MIMO technology in 5G. Millimetre wave's short-wavelength helps shorten the spacing of antennas. Beamforming is a major enabler of mm Wave communication [2][3] The beams must be synchronized between the base station (BS) and the associate user equipment (UEs) due to the high directivity when using beam modulation [4]. To get hold advantage of beamforming technology, in (mm- wave) range, accurate angle tracking is required. As the wireless user is moving in different directions, it is important to monitor the channel gain and the beam angle simultaneously. As the angle of the beam follows a non-linear function, non-linear tracking systems need to be used [5].

One way to track the beam is via the Ray tracing system. However, these methods involve a high computational complexity. Another way to track the beam is through some filters. These filters can be divided into two types: adaptive filter and fixed filter. The static filter is helpful when the signal and channel parameters are known. On the other hand, adaptive filters are useful when the dynamics of a signal or channel are unpredictable and change over time. An adaptive filter is a filter that works under the control of some algorithms such as the least mean square (LMS), and the recursive least square (RLS) [6].

Researchers Md. Faisal Rahman and A.H.M. Asadul Huq developed an adaptive noise canceling device by applying LMS and NLMS algorithms. Because LMS has a fixed step size, it is not suitable

Content from this work may be used under the terms of the Creative Commons Attribution 3.0 licence. Any further distribution of this work must maintain attribution to the author(s) and the title of the work, journal citation and DOI.
Published under licence by IOP Publishing Ltd

for working in an unstable environment. But for NLMS, is suitable for working in an unstable environment as well as a static environment [7], Dhaval N. Patel, and B.J. Makwana used algorithms least mean square (LMS), Sample matrix inversion (SMI), Recursive least square (RLS). It was observed that with an increase in the number of elements, an improvement in beamwidth and reduced interference between users could be obtained [8]. Researchers Vutha Va and Haris Vikalo proposed the Kalman beam tracking algorithm as a solution to reduce beam training overhead. An interesting effect was the size of the matrix. At the same SNR, the array must be chosen to be large enough for optimal performance [9]. Researchers Saurabh R. Prasad1 and Bhalchandra B. Godbole2 studied the effect of step size and the number of iterations on filter performance such as squared error, estimation accuracy, etc. The decrease in step size reduces steady-state error but at the same time increases convergence time.[6], in the mm Wave mobile communication case, Yavuz Yapici and Ismail Guvence derived the LMS algorithm and the Bidirectional least mean square BiILMS extension algorithm is derived from the steeper main algorithm. The numerical results show that LMS and BiLMS rapidly converge with the best performance for increased SNR. On the other hand, with an increasing SNR, the RLS algorithm exhibits a relatively slow convergence [10].
In this paper, the algorithms of (LMS, NLMS) to track the channel in communications (mm- wave) were studied and analysed. The results show that the algorithm of (LMS) is less complex compared to the algorithm of (NLMS) it reduces the difference between the desired signal and the actual signal of an unknown system where the MSE standard is used, and the time of implementation of the algorithm.
The remainder of this paper is organized as follows in Section II, the system model is presented with beam forming, in Section III, the algorithm (LMS, NLMS) is derived, in Section IV, we produced the numerical result, finally, some conclusions are present in Section V.

2. System models
In this section, we present the channel and the beamforming model that we used in this work.

2.1. Time-Varying (mm -wave) Channel Model
Here, the geometrical time-varying channel model can be used to model the mm Wave channel due to its high directional nature. In a given coherence time, represented by time index k, it can be represented as follows [10]:

$$H_k = \sum_{l=1}^{L} \alpha_{k,l}\, a_R(\theta_{k,l}) a_T(\Phi_{k,l})^T, \tag{1}$$

where L is the number of paths received by the antenna array at a time k, $\alpha_{k,l}$ is the complex path gain, and here we assume that it follows a Gaussian distribution with zero mean and covariance R_α, $\theta_{k,l}$ and $\Phi_{k,l}$ respectively, the angle of departure (AOD) and the angle of arrival (AOA) of the lth path at time index k.
Assuming a uniform linear array (ULA) along the Z-axis, the response vector of the receive arrays $a_R(.)$ and transmit array $a_T(.)$ can be expressed as follows [9]:

$$a_R(\theta) = \frac{1}{\sqrt{M}}[1 \;\; e^{-2j\pi\frac{d}{\lambda}\cos(\theta)} \; \ldots . \, e^{-j2\pi\frac{d}{\lambda}(M-1)\cos(\theta)} \;]^T, \tag{2}$$

$$a_T(\Phi) = \frac{1}{\sqrt{N}}[1 \;\; e^{-2j\pi\frac{d}{\lambda}\cos(\Phi)} \; \ldots . \, e^{-j2\pi\frac{d}{\lambda}(N-1)\cos(\Phi)} \;]^T, \tag{3}$$

Where $[.]^T$ is the complex conjugate of $[.]$. λ Is the signal wavelength, d is the inter-element spacing, N is the number of transmit antennas and M is the number of receive antennas.

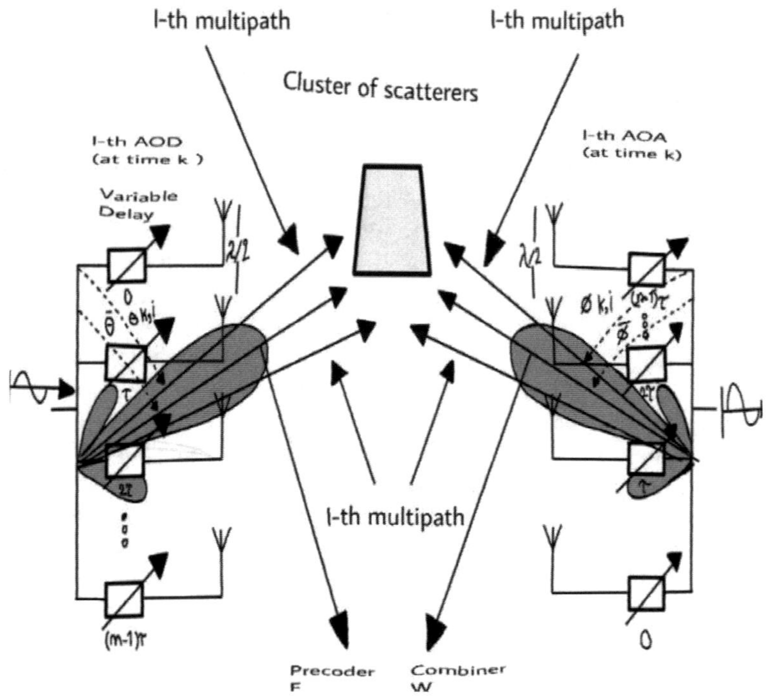

Figure 1. precede and combiner vector with mobile mm Wave communication setting and a single cluster of scattering in.

Using the channel model in (1), there are four-channel parameters that are possibly changing over time and can be grouped into a state vector, $x = [R_{(\alpha)}, J_{(\alpha)}, \theta, \Phi]^T$, where $R(.)$ and $J(.)$ are the real and imaginary operators. Assuming that x is a realization of a random-walk process, the dynamics model for tracking x can be stated as follows:

$$x_{k,1,l} = yx_{k,l} + u_k \quad (4)$$

Where y is a constant diagonal matrix, and given as:

$$y = diag\{\rho\ \rho\ 1\ 1\} \quad (5)$$

and ρ is constant that represents the amount of correlation between the path gains at two consecutive times [3][5]. u_k is the amount of additive noise at time index k, and is assumed to follow a zero-mean Gaussian distribution with covariance.

$$R_{uk} = diag\ \{\sigma^2{}_{\theta k}\ \sigma^2{}_{\Phi k}\ \sigma^2{}_{\alpha k}\ \sigma^2{}_{\alpha k}\}. \quad (6)$$

2.2. Formatting Beamforming Model

In massive MIMO communication systems, transmit and receive combining are normally used to steer the beam into a known user location, in addition to making the equivalent channel scalar. A beamforming vector f, and the combiner w, which are steering vectors at pre-defined pointing angles, are used for this purpose. Mathematically, for a ULA, these vectors are defined as follows:

$$f = \frac{1}{\sqrt{M}}\ [1\ \ e^{-2j\pi\frac{d}{\lambda}\cos\bar{\theta}}\ \ \dots\dots e^{-j2\pi\frac{d}{\lambda}(M-1)\cos\bar{\theta}}]^T, \quad (7)$$

$$w = \frac{1}{\sqrt{N}}\ [1\ \ e^{-2j\pi\frac{d}{\lambda}\cos\bar{\Phi}}\ \ \dots\dots e^{-j2\pi\frac{d}{\lambda}(N-1)\cos\bar{\Phi}}\ ]^T, \quad (8)$$

Where $\bar{\theta}$ and $\bar{\Phi}$ are the pointing angles of the receive and transmit pointing beams, respectively.

3. Tracking of The Adaptive Beam and Channel

In this section, the algorithms (LMS) and (NLMS) will be presented to estimate unknown mm wave channel parameters AOA, AOD, and path gain.

3.1. Least Mean Square Algorithm (LMS)

In signal processing applications, the least mean square (LMS) adaptive algorithm commonly utilizes the gradient vector estimation from the steepest descent. To determine the optimal weight vector, an iterative process is required to update the weight vector in direction of the negative gradient vector, the mean square error value is minimized by this algorithm, it is relatively easy and does not require any matrix inversion calculation or correlation function calculation [11].

Instead of estimating the weight, we will use the (LMS) algorithm to adaptively track the states x_k, which represents the measured states of the beam directions θ and Φ the components of the complex channel gain: the real α_R and the imaginary α_I as

$$x_k = [\alpha_{R,k}^T \;\; \alpha_{I,k}^T \;\; \theta_k^T \;\; \Phi_k^T \;]$$

In order to track the channel and adaptive beam for the scenario of the communication, we will first consider the (LMS) algorithm, in the state vector x_k since the measurement is not linear, we, therefore, want to derive (LMS) using the original algorithm of the steepest descent algorithm given as [10].

$$\hat{x}_{(k+1)} = \hat{x}_k - \mu \nabla_{\hat{x}_k} J_k, \tag{9}$$

where $\hat{x}_k$ is the estimation of x_k , We reflect the algorithm's adaptation step-size in equation (9) by the diagonal matrix $\mu = [\mu_\alpha 1_{2L} \; \mu_\theta 1_L \; \mu_\Phi 1_L]$, where μ_α , μ_θ and μ_Φ represent the particular channel path gain, AOD, and AOA respectively, the gradient operator is ∇, the mean square error at time k is J_k.

The estimation error is $e_{(k)} = y_{(k)} - h_{(\hat{x}_k)}$, has a real part $e_{R,k}$ and an imaginary part $e_{I,k}$, the MSE gave as $J_k = E\{\|e_k\|^2\}$ with $e_k = [e_{R,k} \; e_{I,k}]^T$

in our paper we will define the cost function, and we can express as:

$$\text{cost function} = \arg\min \, (J_K)$$

The gradient of the MSE is given as [10]:

$$\nabla_{\hat{x}_k} J_k = -2 e_k^T \frac{\partial h(\hat{x}_k)}{\partial \hat{x}_k} \tag{10}$$

$$Z_k = \frac{\partial h(\hat{x}_k)}{\partial \hat{x}_k} \tag{11}$$

$$\hat{x}_{(k+1)} = \hat{x}_k + 2\mu e_k^T Z_k \tag{12}$$

3.2. Numbering Normalized Least Mean Square Algorithm (NLMS)

The normalize least mean square (NLMS) algorithm can be considered to be a special implementation of the (LMS) algorithm that takes account of the change in signal level at the filter input, the relationship is the same for adaptively changing states in the (LMS) algorithm, in the least mean square it's difficult to select a learning rate that ensures algorithm stability. The only adjustment in (NLMS) is to normalize the input in order to solve the stated problem [12].

Normalizing the input vector provides the (NLMS) the ability to solve the problem by letting the step size adapt with the status of the vector:

$$\hat{x}_{(k+1)} = \hat{x}_k + \frac{\mu}{\|x_k\|^2} e_k^T Z_k, \tag{13}$$

Where μ is the step size of the adaptive filter, $\|x_k\|^2$ is the normalization norm vector of x_k.

4. Numerical results and discussion

Depending on the program (MATLB) to simulate the performance of the (LMS) and (NLMS) algorithms in the (mm-wave) beam tracking in this section. we suppose the multi-antenna system with N=16, M=16, and SNR= 30dB, we assume multipath L=1 for a narrow physical beam this is very likely, the time evolution of the channel underlying mm wave its regulated by $\rho = 0.995$ and σ_θ^2 and σ_Φ^2=(0.1) ^{2}, and the vector of the precoder and combiner the arbitrary direction of 45° is pointed out ($\bar{\theta} = \bar{\Phi} = 45^{\circ}$). For (LMS, NLMS, RLS) for the adaptive $\mu_\alpha = 0.1$ and $\mu_\theta = \mu_\Phi = 0.0001$ we use the better step-size value , and the wavelength $\lambda = \frac{c}{f} = c = 3 * 10^3, \frac{meters}{sec} = 0.9$ cm.

5. Results and discussion

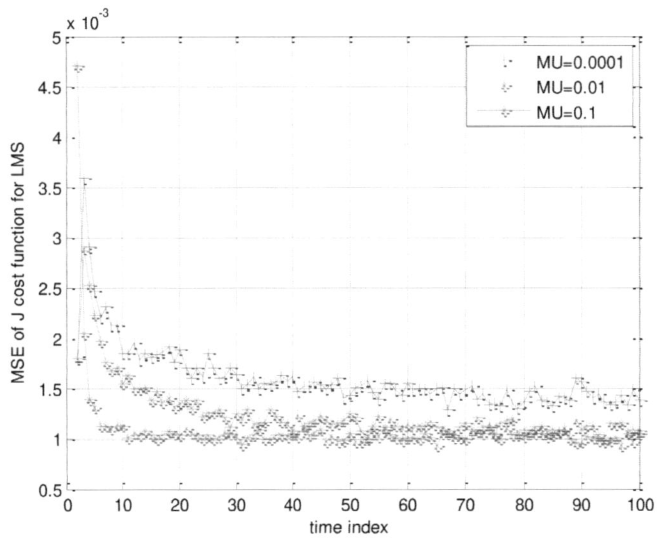

Figure 2. Comparison of the cost function of LMS algorithm for different values of μ.

Figure 2 shows the cost function for the algorithm (LMS) for different values of the step size μ, where μ is the step size parameter, and it controls the convergence characteristics of the LMS algorithm, and it is necessary to use an appropriate value for the performance of the LMS algorithm .from the figure 2, we can notice that the algorithm (LMS) at the lower value of μ, the algorithm converges slowly, then the cost function become high and gives high error rate. but for a large value of μ the algorithm converges faster and gives a lower cost function with a lower error rate.

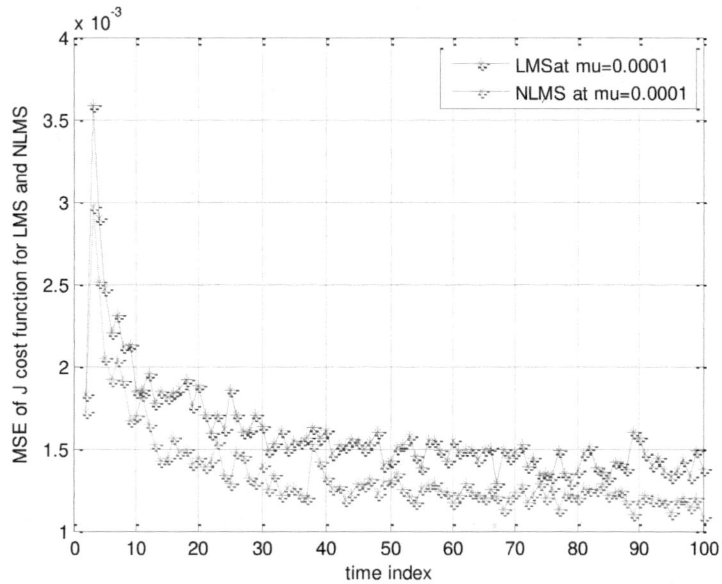

Figure 3. Compare performance between (LMS and NLMS) algorithm.

During complete execution, one of the main disadvantages of the (LMS) algorithm is having a fixed step size parameter. These calls for an understanding of the input signal statistics before the adaptive filtering process starts.

An extension of the LMS algorithm that bypasses this problem by choosing a different step size value μ for each iteration of the algorithm is the normalized least mean square (NLMS). The increase in step size leads to faster convergence, from the figure it can be seen that the algorithm (NLMS) converges faster than the algorithm (LMS) at the same step size parameter and the value of the MSE after converges will be low and higher stability can be obtained.

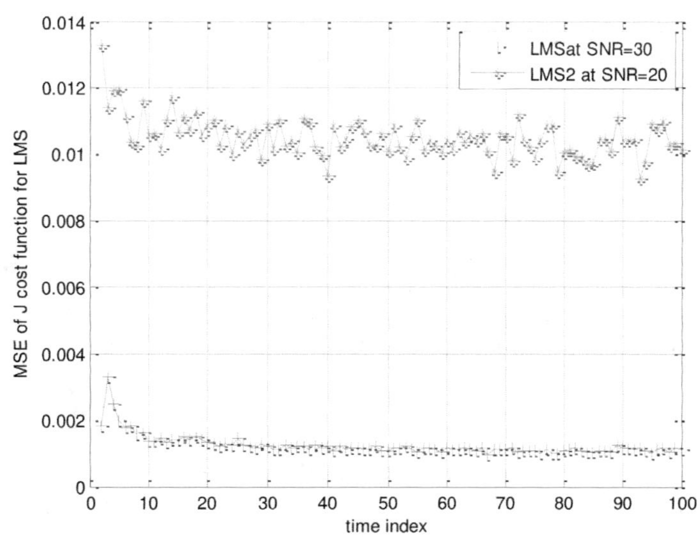

Figure 4. compares between cost function of the LMS algorithm at different values for SNR.

To tracking the best beam, we can use the algorithm that has the least error, In figure 4, we can note that the tracking algorithm LMS have a cost function at SNR=30 less than the cost function at

SNR=20, this means that the LMS algorithm at high SNR has a better performance with lower error and gives higher stability.

6. Conclusions

In this paper, channel tracking and adaptive beam formation were studied using adaptive algorithms (LMS, NLMS) algorithms (LMS, NLMS), the algorithms were derived under a non-linear control model. The results showed that the algorithm (LMS) is one of the easiest and simplest types of algorithms and needs large step size values to converge faster and thus obtain a stable system. The algorithm (NLMS) is considered a special implementation of the algorithm (LMS) when choosing a normal step size leads to a more stable and convergent adaptive algorithm.

Acknowledgments

Authors wishing to acknowledge the University of Mosul.

7. References

[1] Almohammedi, Ali, and Mohamed Deriche. "LMS for Truncate CSI Feedback in Massive MIMO." *2019 International Conference on Computing, Electronics & Communications Engineering (iCCECE)*. IEEE, 2019.

[2] Xia, Xiaoxuan, et al. "Low Complexity Angular-Domain Detection for the Uplink of Multi-User mmWave Massive MIMO Systems." *Electronics* 9.5 (2020): 795

[3] Sun, Shu, and Theodore S. Rappaport. "Millimeter wave MIMO channel estimation based on adaptive. compressed sensing." *2017 IEEE International Conference on Communications Workshops (ICC Workshops)*. IEEE, 2017.

[4] Chung, Hyeonjin, et al. "Adaptive beamwidth control for mmWave beam tracking." *IEEE Communications Letters* (2020).

[5] Lim, Jaechan, Hyung-Min Park, and Daehyoung Hong. "Beam Tracking Under Highly Nonlinear Mobile Millimeter-Wave Channel." *IEEE Communications Letters* 23.3 (2019): 450-453

[6] Prasad, Saurabh R., and Bhalchandra B. Godbole. "Optimization of LMS Algorithm for System Identification." *arXiv preprint arXiv:1706.00897* (2017).

[7] Rahman, Md Faisal, and AHM Asadul Huq. "Comparison between LMS & NLMS Algorithms in Adaptive Noise Cancellation for Speech Enhancement."2015.

[8] Patel, Dhaval N., B. J. Makwana, and P. B. Parmar. "Comparative analysis of adaptive beamforming algorithm LMS, SMI and RLS for ULA smart antenna." *2016 International Conference on Communication and Signal Processing (ICCSP)*. IEEE, 2016

[9] Va, Vutha, Haris Vikalo, and Robert W. Heath. "Beam tracking for mobile millimeter wave communication systems." *2016 IEEE Global Conference on Signal and Information Processing (GlobalSIP)*. IEEE, 2016

[10] Yapıcı, Yavuz, and Ismail Güvenç. "Low-complexity adaptive beam and channel tracking for mobile mmWave communications." *2018 52nd Asilomar Conference on Signals, Systems, and Computers*. IEEE, 2018.

[11] Jose, Anila. "Simulation of an Adaptive Digital Beamformer using Matlab." *Simulation* 2.12 (2016).

[12] Ghauri, Sajjad Ahmed, and Muhammad Farhan Sohail. "System identification using LMS, NLMS and RLS." *2013 IEEE Student Conference on Research and Developement*. IEEE, 2013.

IOP Conf. Series: Materials Science and Engineering 1152 (2021) 012005 doi:10.1088/1757-899X/1152/1/012005

Performance Investigations of the Uniformly and Non-Uniformly Excited Linear Arrays

Roua Moafaq Younis[1*] **and Jafar R. Mohammed**[1]

[1]Department of Communication Engineering, College of Electronics Engineering, Ninevah University, Mosul- Iraq.

[*]E-mail: roua.moafaq@gmail.com

Abstract. Uniform linear arrays include the arrangement of all array elements in one dimension with equal amplitude excitations. Such types of arrays usually have good directivity, narrow beam-width, and they suffer from high sidelobe levels that may cause interference and degrade the overall system performance. The problem of high side lobe levels may overcome by using non-uniformly excited arrays instead of its uniform counterpart. In this paper, the Dolph and Taylor excited arrays are adopted as a non-uniformly excited array. The performance in terms of half power beam width (HPBW), Peak sidelobe level (SLL), directivity (D), and first null-to-null beam width of the uniformly and the two non-uniformly excited arrays are investigated and compared. Simulation results show that the non-uniformly excited arrays can significantly reduce the peak SLL at the cost of lower directivity and wider HPBW. Thus, it is advised to use the non-uniformly excited arrays in the environments that borne high interference pollution.

1. Introduction

Many practical applications of the antenna arrays require high demand on the required radiation characteristics that cannot met by a single radiant antenna; thus, multiple elements can be used to fulfil such requirements. An antenna array is a collection of two or more space-separated antennas arranged in a specific structure. Generally, an antenna array can increase the overall gain, increase diversity, cancel interference from a set of directions and steer the main beam in a particular direction [1]. By appropriately exciting each individual element with certain amplitudes and phases, arrays can produce the desired radiation characteristics. They are referred to as linear antenna arrays when the antenna components are positioned in a straight line [2]. An array of identical elements of the same magnitude, equal spaced, and each with a constant progressive phase is known as a uniformly excited array, and if it's main beam is found to be perpendicular to the array axis, it's considered to be a broadside array. The authors in [3] studied the spacing and phase progression between the array elements. Arrays with equal spacing between their elements and unequal amplitudes are called non-uniformly excited arrays such as Dolph-Chebyshev and Taylor antenna arrays.

The authors in [4] introduced a backfire antenna pattern with non-linear excitation (like Taylor) and uniform space between array elements. The authors also show that the directivity and the peak SLL of the designed antenna where the directivity as a function of antenna length was plotted.

In [5], the authors studied a non-uniform antenna array for which the elements are equally spaced with unequal amplitude excitation using Chebyshev excitation method. They also investigated the effect of number of array elements, separation distance, and the amplitude excitations of the array elements on the radiation characteristics. The separation distance between array elements were allowed to vary from 0.1λ to 2λ in steps of 0.02λ for a number of elements up to 10 elements. Kailash et al in.[6], offered a comparative analysis of the effects of different spacing and the number of the array elements in terms of array factor, directivity and half-power beam-width (HPBW) for the Broadside case. They found that the directivity and the radiation characteristics can be improved proportionally with the increased number of the array elements. In [7-10], different methods were proposed to design Dolph

Content from this work may be used under the terms of the Creative Commons Attribution 3.0 licence. Any further distribution of this work must maintain attribution to the author(s) and the title of the work, journal citation and DOI.
Published under licence by IOP Publishing Ltd

and Taylor arrays, while in [11-18] other methods were used to synthesis an optimum radiation patterns with desired constraints.

2. Principles of the Uniform and Non-Uniform Arrays

2.1 Uniform Array

In this type of arrays, all the antenna elements are fed with equal amplitude and phase. Generally, there are two main types of such arrays namely: Broadside array and End-Fire array.

Broadside array designed to radiate its radiation pattern that is perpendicular to the direction of the axis of an array. The maximum radiation direction is always perpendicular to the line or plane of the array according to the elements on a line or a plane. The following equations of the Array factor (AF), half power beam width (HPBW), first null beam width (FNBW) and the side lobe level (SLL) for the broadside uniform linear array are:

$$AF_n = \frac{1}{N}\frac{\sin\frac{N\Psi}{2}}{\frac{N}{2}\Psi} \quad (1)$$

Where N is the number of array elements, $\Psi = kd\cos\theta + \beta$, $k=\frac{2\pi}{\lambda}$, d is the separation distance between array elements, β is the phase excitation difference (for broadside antenna, β=0).

$$FNBW = 2\left[\left[\frac{\pi}{2} - \cos^{-1}\left(\frac{\lambda}{Nd}\right)\right]\right] \quad (2)$$

$$HPBW = 2\left[\frac{\pi}{2} - \cos^{-1}\frac{1.391\lambda}{\pi nd}\right]$$
(3)

$$sll = 20\log_{10}\left(\frac{2}{3\pi}\right)$$
(4)

2.2 Non-Uniform Array

In non-uniform arrays the elements are fed with different amplitudes and/or phases. Some common and widely used Types of non-uniform linear amplitude arrays include Dolph-Chebyshev array and Taylor array [4].

A typical example of the non-uniformly excited arrays is the Dolph-Chebyshev array. The Dolph-Chebyshev method is a compromise between the uniform and binomial methods, as if no side lobes are produced by the Dolph-Chebyshev method; it decreases to the binomial model [5][6]. Dolph-Chebyshev produced a pattern for a uniformly spaced array that has a minimum possible beam width for a given maximum sidelobe level. The pattern of this array is obtained from the Chebyshev polynomial, and it has 2N nulls for an array of 2N + 1 elements. The level or height of all the sidelobes is equal. In some cases of the Dolph distributions, the element amplitude excitations of the last end element on each side of the array may gradually increase in their amplitudes with compared to the previous value to achieve low side lobe array pattern [7].

To design the Dolph-Chebyshev array, the following equations are used:

$$f = 1 + 0.636\left\{\frac{2}{R_o}\cosh((\cosh^{-1}R_o)^2 - \pi^2)^{1/2}\right\} \quad (7)$$

where f represents the broadening factor in the main beam of the Dolph excited array with compared to that of the uniformly excited array.

$$D_0 = \frac{2R_0^2}{1+(R_0^2-1)f\frac{\lambda}{(L+d)}} \quad (8)$$

Where $D_0$is the maximum directivity in the Dolph excited array.

$$R_{0\,(\text{voltage ratio})} = 10^{\frac{R_{0\,dB}}{20}} \quad (9)$$

Where R_0 is the major side lobe voltage ratio and L is the length of the array. The HPBW of the Dolph-Chebyshev array can be calculated by firstly compute the value of the HPBW for the uniformly excited array under the same number of the array elements and spacing between them. After that we may find the value of HPBW of the Dolph-Chebyshev array by multiplying the HPBW of the uniform array by the beam broadening factor (f) as follows: -

$$\mathrm{HPBW_{uniform}} = \cos^{-1}\left[\cos\theta_0 - 0.443\frac{\lambda}{L+d}\right] - \cos^{-1}\left[\cos\theta_0 + 0.443\frac{\lambda}{L+d}\right] \quad (10)$$

$$\mathrm{HPBW_{Dolph}} = \mathrm{HPBW_{uniform}} \times \mathrm{f} \quad (11)$$

Another type of the non-uniform arrays is the Taylor array, which is similar to the Dolph-Chebyshev array, but it is more applicable to the continuous distributions. An array pattern that is optimal compromise between the beam width and the side lobe level is the one that produced by the Taylor design [8]. The minor lobes are maintained at an equal and specific level in an ideal design. Since the minorlobes are of equal ripple and stretch to infinity, an infinite power is implied. The technique as applied by Taylor, however, realistically leads to a pattern whose first few minor ones are monotonically decaying. Practically, a small monotonic decay exhibits even the degree of the nearest minor lobes. This decay is a function of the u-space over which these minor lobes must be located. The rate of decay of the closest minor lobes increases as this space decreases. For a very large space of u (over which the closest minor lobes are required to have an equal ripple), the rate of decay is negligible [9].

To design the Taylor array, the following equation are usually used:

$$\mathrm{AF_{Taylor}} = \frac{\sin u}{u}\frac{\prod_{n=1}^{\bar{n}-1}\left[1-\left(\frac{u}{u_n}\right)^2\right]}{\prod_{n=1}^{\bar{n}-1}\left[1-\left(\frac{u}{n\pi}\right)^2\right]} \quad (12)$$

Where $u = \pi\frac{l}{\lambda}\cos\theta$, and $u_n = \pi\frac{l}{\lambda}\cos\theta_n$

The value of the HPBW of such type of array is computed by

$$\Theta_0 = 2\sin^{-1}\left\{\frac{\lambda\sigma}{\pi l}\left[(\cosh^{-1}R_0)^2 - \left(\cosh^{-1}\frac{R_0}{\sqrt{2}}\right)^2\right]^{1/2} \quad (13)$$

Where

$$\sigma = \frac{\bar{n}}{\left(A^2+\left(\bar{n}-\frac{1}{2}\right)^2\right)^{1/2}} \quad (14)$$

$$A = \frac{1}{\pi}\cosh^{-1}(R_0) \quad (15)$$

$$\mathrm{SLL} = -20\log(\cosh(\pi A)) \quad (16)$$

3. Results and Discussions

In this section, the radiation patterns and some other characteristics will be investigated to verify the principles of such arrays with highlight of main advantages and disadvantages. After completing the design of the aforementioned arrays in MATLAB software, The code implemented for the uniform arrays that have the following specifications; The number of the considered array elements, N is equal to 10, the separation distances between the elements is $d=0.5\lambda$, the progressive phase $\beta=0$, and the amplitude excitations of the uniformly excited arrays is made to be ones for all array elements and for non-uniformly excited arrays (Taylor and Dolph-Chebyshev) they made varied according to the Dolph

or Taylor distributions. The other specifications are assumed to be same as in the uniform arrays. For non-uniform arrays, the peak sidelobe level is made to be equal to -26 dB.
The following results are obtained as shown in figures from Figure 1 to Figure 6 as well as the numerical values listed in Table 1.
The results of the uniformly excited array are shown in Figures 1 and 2 where the radiation pattern and the corresponding amplitude and phase excitations are plotted. It can be seen that the sidelobe level of this array is relatively high about -13.23 dB which is undesirable and may cause many problems in practice. Other performance measures are listed in Table 1. The results of the Dolph excited array are shown in Figures 3 and 4. It can be seen that the sidelobe level was greatly reduced up to -26 dB at the cost of wider half power beam width and lower directivity. As it is clear, the Dolph pattern has equal SLL.
The results of applying the Taylor excited array are shown in Figures 5 and 6. It can be seen that almost the same performance as the Dolph was noticed. Also, it can be seen that the amplitude excitation of the Taylor array has a large difference between the center and the edges of array which makes the implementation issue of the feeding network more complex and expensive.
Finally, Table 2 compares the performances of various designed array that were presented in [14], [15], [17], and [18].

Table 1: comparison between uniform array and non-uniform array (Dolph-Chebyshev and Taylor).

N=10, β=0,d=0.5	Uniform array	Chebyshev array	Taylor array
FNBW (deg.)	23.2	32.1	32.1
HPBW (deg.)	10.19	10.9	10.9
Directivity (dB)	10	9.189	9.08
SLL (dB)	-13.46	-26	-26

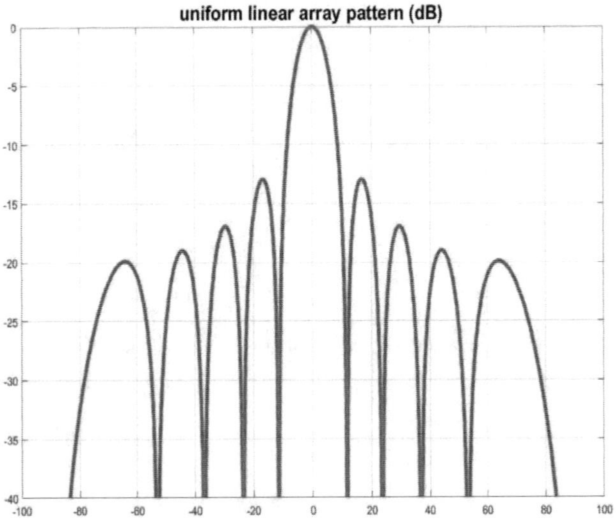

Figure 1. Radiation pattern of the uniformly excited array.

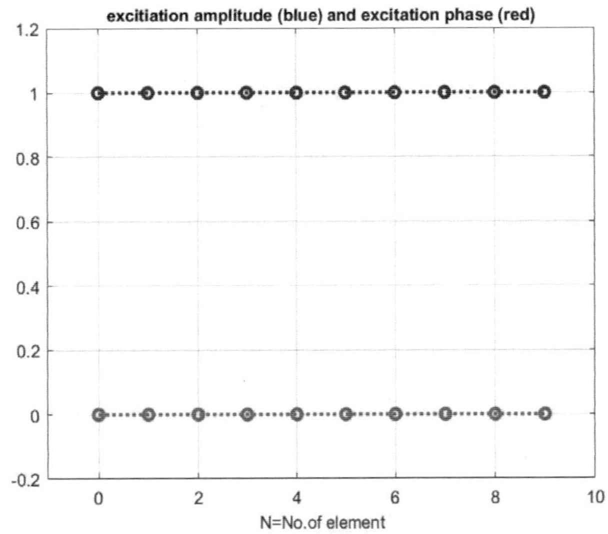

Figure 2. Amplitude excitation (blue) and the phase excitation (red) of the uniform array.

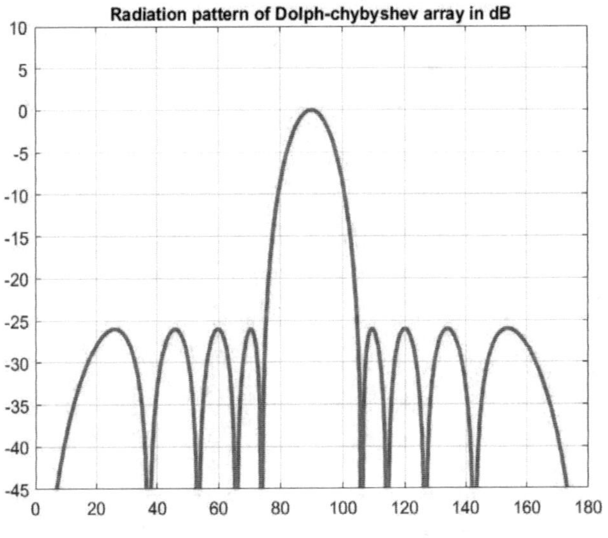

Figure 3. Radiation pattern of the Dolph excited array.

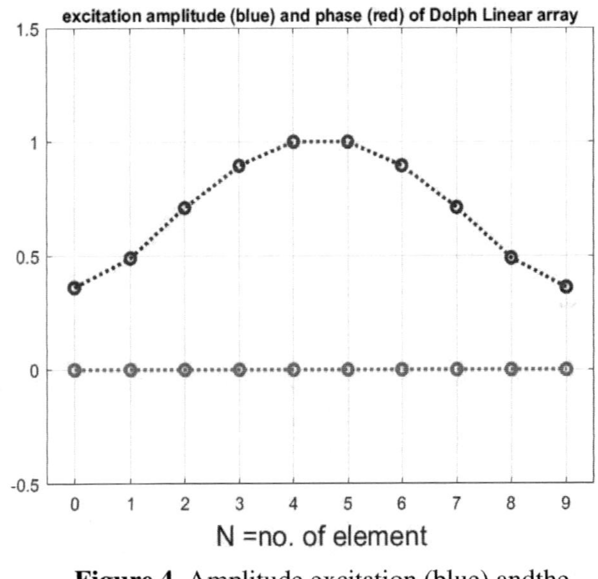

Figure 4. Amplitude excitation (blue) andthe phase excitation (red) of the Dolph array.

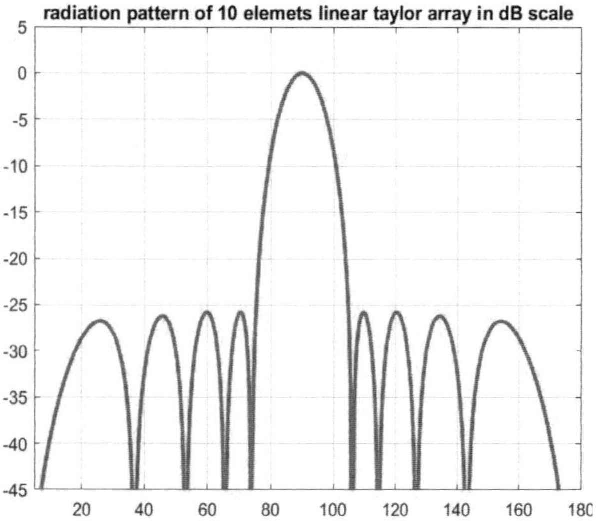

Figure 5. Radiation pattern of the Taylor excited array.

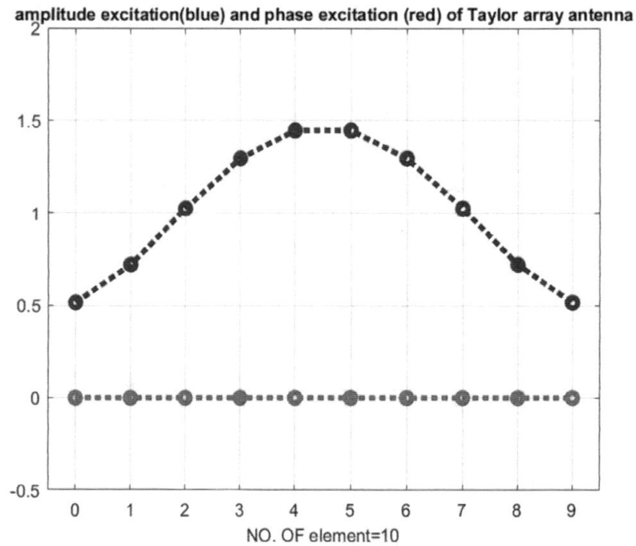

Figure 6. Amplitude excitation (blue) andthe phase excitation (red) of the Taylor array.

Table II: Comparison between various designed arrays (Uniform, partially element optimized [14] and [15], End element location [17], Fully element optimized [18]).

N=10, β=0,d=0.5	Uniform array	Designed array in [14] and [15]	Designed array in [17]	Designed array in [18]
FNBW (deg.)	23.2	33.4	32.21	35.0
HPBW (deg.)	10.19	10.4	10.25	11.2
Directivity, (dB)	10	9.6	9. 8	8.28
SLL (dB)	-13.46	-25	-18	-40

4. Conclusions

It has been shown from the presented results that the non-uniformly excited Dolph and Taylor arrays give lower sidelobe level with compared to that of the uniformly excited array. This reduction in the sidelobe level is come at the cost of wider beam width and lower directivity. Thus, the designer needs to carefully compromise between the required sidelobe level and the desired half power beam width and the directivity. More specifically, the results of the uniformly excited array show that the sidelobe level is at -13.24 dB, directivity was D =10 dB, and the HPBW was 10.19 degree, while for non-uniformly Dolph excited array the sidelobe level was -26 dB, D=9.08, and HPBW= 10.9 degree. For Taylor array the SLL=-26 dB, D=9.189, and HPBW=10.9 degree.

From these results, we may conclude that the uniform array, generally, gives higher directivity and sharper beam width, which is very desirable in the long-distance communication systems. For applications that require a robust performance against interfering signals we need to use, Dolph or

Taylor arrays or other arrays which they have lowest sidelobes to minimize the effect of interfering signals.
This study can be further extended by using an optimization (or adaptive) algorithm [19-20] to optimize the array performance. This left as a future work.

5. Reference

[1] E. B. Daher 2016 Analysis and design of non-uniform arrays for direction finding," Villanova University.
[2] M. A. Zaman and M. Abdul Matin 2012 Nonuniformly spaced linear antenna array design using firefly algorithm Int. J. Microw. Sci. Technol., doi: 10.1155/2012/256759.
[3] K. P. Dutta 2017 Study of Broadside Linear Array Antenna with Different Spacing and Number of Elements Int. J. Adv. Eng. Res. Sci., vol. 4, no. 5, pp. 190–194, doi: 10.22161/ijaers.4.5.30.
[4] A. T. Abed 2014 Study of Radiation Properties in Taylor Distribution Uniform Spaced Backfire Antenna Arrays Am. J. Electromagn. Appl., vol. 2, no. 3, p. 23, doi: 10.11648/j.ajea.20140203.11.
[5] A. Saeed, M. & Dlnya, A. Ibrahim, A. Saeed Mahmood, and & Dlnya 2016 Non-Uniform Chebyshev Excitation of a Linear Broadside Antenna Array Operating at 1.8GHz.
[6] S. SahajaM, V. Pathi RajuB, and P. ASrilakshmi 2017 Computation Of Uniform And Non-Uniform Antenna Arrays," Int. J. Eng. Trends Technol. (IJETT)-Special Issue, [Online]. Available: http://www.ijettjournal.org.
[7] A. El-Hajj, K. Y. Kabalan, and M. Al-Husseini 2003 Antenna array design using spreadsheets IEEE Trans. Educ., vol. 46, no. 3, pp. 319–324, doi: 10.1109/TE.2003.813518.
[8] A. Safaai-Jazi 1994 A New Formulation for the Design of Chebyshev Arrays IEEE Trans. Antennas Propag., vol. 42, no. 3, pp. 439–443, 1994, doi: 10.1109/8.280736.
[9] A. T. Villeneuve 1984 Taylor Patterns for Discrete Arrays IEEE Trans. Antennas Propag., vol. 32, no. 10, pp. 1089–1093, doi: 10.1109/TAP.1984.1143212.
[10] T. T. Taylor 1955 Design of Line-Source Antennas for Narrow Beamwidth and Low Side Lobes Trans. IRE Prof. Gr. Antennas Propag., vol. AP-3, no. 1, pp. 16–28, doi: 10.1109/TPGAP.1955.5720407.
[11] Mohammed, J. R. 2013 Design of Printed Yagi Antenna with Additional Driven Element for WLAN Applications Progress in Electromagnetics Research C, Vol. 37, 67-81, ISSN:1937-8718
[12] Sayidmarie, K. H. and Mohammed, J. R. 2013 Performance of a Wide Angle and Wideband Nulling Method for Phased Arrays Progress in Electromagnetics Research M, vol. 33, pp. 239-249.
[13] Mohammed, J.R, and Sayidmarie, K. H. 2014 Sidelobe Cancellation for Uniformly Excited Planar Array Antennas by Controlling the Side Elements IEEE Antennas and Wireless Propagation Letters, vol.13, pp. 987 - 990.
[14] Mohammed, J.R. 2017 Optimal Null Steering Method in Uniformly Excited Equally Spaced Linear Array by Optimizing Two Edge Elements", Electronics Letters, vol. 53, issue 11.
[15] Mohammed, J.R. 2018 Element Selection For Optimized Multi-Wide Nulls in Almost Uniformly Excited Arrays IEEE Antennas and Wireless Communication Letters. vol. 17, issue 4, PP. 629-632.
[16] Mohammed, J.R. 2018 Thinning a Subset of Selected Elements for Null Steering Using Binary Genetic Algorithm Progress in Electromagnetics Research M, vol. 67, pp. 147-157 ISSN: 1937-8726.
[17] Mohammed, J.R. 2019 Obtaining Wide Steered Nulls in Linear Array Patterns by Controlling the Locations of Two Edge Elements AEÜ International Journal of Electronics and Communications, vol. 101, pp. 145–151.
[18] Mohammed, J.R. 2019 An Optimum Side Lobe Reduction Method with Weight Perturbation Journal of Computational Electronics, US Springer, vol.18, no.2, pp.705-711.

[19] Mohammed, J. R. 2014 Low Complexity Adaptive Noise canceller for Mobile Phones Based Remote Health Monitoring International Journal of Electrical and Computer Engineering (IJECE), Vol. 4, No. 3, pp. 422~432. ISSN: 2088-8708

[20] Mohammed, J.R and Sayidmarie, K. H. 2017 Performance Evaluation of the Adaptive Sidelobe Canceller with various Auxiliary Configurations," AEÜ International Journal of Electronics and Communications, vol. 80, pp. 179–185.

IOP Conf. Series: Materials Science and Engineering 1152 (2021) 012006 doi:10.1088/1757-899X/1152/1/012006

Path Loss and Channel Modeling at 3.5GHz for 5G Cellular System

S Kh Al-Khero[1] and Y M Abbosh[1]

[1]Department of Communication Engineering, Electronics Engineering College, Ninevah University, Mosul, Iraq

*E-mail: semaa.nafi2019@stu.uoninevah.edu.iq

Abstract. Wireless channel characterization is the basis of the architecture of the wireless communication system. The detailed propagation models are indeed the requirement for fifth generation communication network design. This paper presents the characteristics of channel propagation for a fifth-generation system in Line-of - Sight (LOS) and Non-Line- of- Sight (NLOS) scenarios with co-polarization (V-V) and cross polarization (V-H) in a second floor/Communication Engineering Department/Electronics college building at 3.5GHz frequency. The comparison between two path loss models shows that the Close-In (CI) of free space reference distance path loss model is more suitable than Floating-Intercept (FI) path loss model. The results of both models show that the path loss exponent value (n) is ranged between 1.7 and 2.9 for the LOS scenario, while it was between 2.1and 3.1 for NLOS scenario for both co-polarization and cross polarization.

Keywords: 5G, Path loss, Wireless InSite, 3.5GHz, Indoor Environment.

1. Introduction

Wireless networking technology has progressed quickly to meet the demands of new techniques in a number of applications. However, the demand for high-speed data and connectivity is growing[1]. Compared to the fourth generation (4G), named Long-Term Evolution (LTE) and Long-Term Evolution Advanced (LTE-A), fifth generation (5G) introduces sever new devices scenarios that have diversified features such as ultra-high traffic volume density, ultra-high link density, and ultrahigh mobility[2]. Wireless channel characterization is the basis of the architecture of the wireless communication system. The detailed propagation models are indeed the requirement for 5G communication network design [3]. At present, 5G is the first cellular system designed to accommodate any 400MHz to 90GHz spectrum[4]. Mid-band is Sub-6 GHz bands in Fifth-Generation-New-Radio (5G-NR) included band numbers: number 77 (3.3-4.2 GHz) and number79 (4.4-5.0 GHz). However, just within the 3.4-3.7 GHz range could be the final spectrum bandwidth occupied exclusively for 5G systems[5]. The 3.5GHz frequency band has been suggested as a candidate frequency for the implementation of the 5G microwave band. As a result, channel characterizations at this carrier frequency are crucial in order to prepare for the launch of fifth generation[6].Thus, the aim of the research is to investigate path loss modelling by employing the shooting and bounding ray-tracing (SBRT) simulation in the long corridor indoor environment at 3.5GHz (in this case ;Line- of- Sight (LOS) and Non-Line-of-Sight(NLOS) conditions with co and cross polarization). The rest of the Paper is organized as follows: Section 2 presents the theory about path loss of channel characteristics. Related works are explored in section 3. Simulation setup has been discussed in section 4. Results have been analyzed in section 5, and finally, the conclusions of this paper have been synopsized in section 6.

Content from this work may be used under the terms of the Creative Commons Attribution 3.0 licence. Any further distribution of this work must maintain attribution to the author(s) and the title of the work, journal citation and DOI.
Published under licence by IOP Publishing Ltd

2. Path loss of Channel Characteristic (PL)

The path loss is decrease in an electromagnetic wave power density (attenuation) during cross-section propagation[7] .In the law of the path, several variables play a major role, the long distances between the transmitting and receiving antennas since the route can pass through multiple barriers, resulting in successive losses, as well as multipath, which cause reflections and collisions between the paths[8] .The Path Loss calculation is given as equation (1) [9].

$$L_{\text{path}}\,[\text{dB}] = P_T\,(\text{dBm}) - P_R\,(\text{dBm}) + G_T\,(\text{dBi}) + G_R\,(\text{dBi}) - L_S\,(\text{dB}) \quad (1)$$

Here G_T and G_R refer to transmit and receive antenna gains, respectively.P_T and P_R is referring to transmitted and received power, respectively. L_S refer to the other losses in the process. For the fifth generation system, path loss models are critical for network coverage and performance analysis[10] .For the single-frequency, both of the Close-In free space of reference distance and Floating-Intercept path loss models are generic all-frequency models that explain large-scale propagation path loss in a given scenario at all applicable frequencies[11]. Concerning computation complexity, the Close-In path loss model outperformed the Floating-Intercept model, The minimum mean-square error(MMSE) approach only needed to estimate one parameter for the CI path loss model[12], which was path loss exponent (n) utilizing a real physical anchor point that expresses the free space transmitted power from the TX antenna to the close-in distance d_0 [13]. Whereas, MMSE was used to estimate two parameters (α and β) in the FI path loss model[12], and does not take into account a physical anchor for the transmitted power[13]. Finally, the complexity of the FI model is two times greater than the complexity of the CI model [12].

3. Related Works

Over the past 25 years, there have been numerous studies of indoor wireless transmission and channel models for frequency below 6GHz. For example, authors in [14]employed half-wave dipole antenna for indoor analysis 433MHz. The findings demonstrated the comparison between the simulation and measurement results at a different location inside a laboratory building /Durban university. The measured and simulated results were slightly close to one another. Using a similar setting, another study demonstrated that the path loss increases with carrier frequency and load walls at a different frequency from 800MHz-6GHz[15]. Furthermore, an indoor path loss prediction for NLOS analysis was conducted by A. Bhuvaneshwari, R. Hemalatha et al. The study was carried out at wide corridor. The transmitter was stationary and the receiver moves at discrete distances along the corridor at 2.4GHz frequency band. The findings demonstrated that the loss increased as separation distances increased due to the propagation mechanisms of the signals in the propagation path[16]. The authors also indicated as a comparative analysis of the two bands, indoor measurement results obtained with channel sounder fitted using omnidirectional and horn antennas at 2.9 and 29 GHz frequency bands. The measurements were taken for two different floors within a Qualcomm building in Bridgewater, New Jersey, in the United States. The results showed that the path loss exponent (n) values at 29GHz is higher than the 2.9 GHz frequency bands due to the a number of partition walls between the links and the electrical properties of materials [17].

4. Simulation Setup

The case study used in this paper focuses on the second-floor corridor areas in the Electronics Engineering College/Department of Communication engineering building. The building was planned and simulated with software Wireless InSite (WI), the construction design layout is shown in Figure (1). In (WI), the materials are predefined where the walls are made up of brick with permittivity $\epsilon_r = 4.4$, and conductivity $\sigma = 0.001$ S/m, the ceiling and floor are made of concrete with $\epsilon_r = 15$, and $\sigma = 0.015$ S/m. The characteristics of the transmitter and receiver antennas used are described in table 1. Omnidirectional antennas with gain (2dBi) were vertically polarized for co-polarization, whereas the Tx antenna was vertically polarized and the Rx antenna was horizontally polarized for the cross-polarization cases at 3.5GHz frequency band. The Tx antenna was located at a stationary location along the corridor of the second floor of the building for the evaluation procedure, while evaluation started with the Rx moving along the corridor. The distance of the route ranges from (0.02-10) m. The total number of points obtained that have been analyzed in this work is 469 points with a separation distance of 0.02 m. Both LOS and NLOS scenarios were included in this research. For the whole inquiry, the serious effects of various construction materials were taken into consideration.

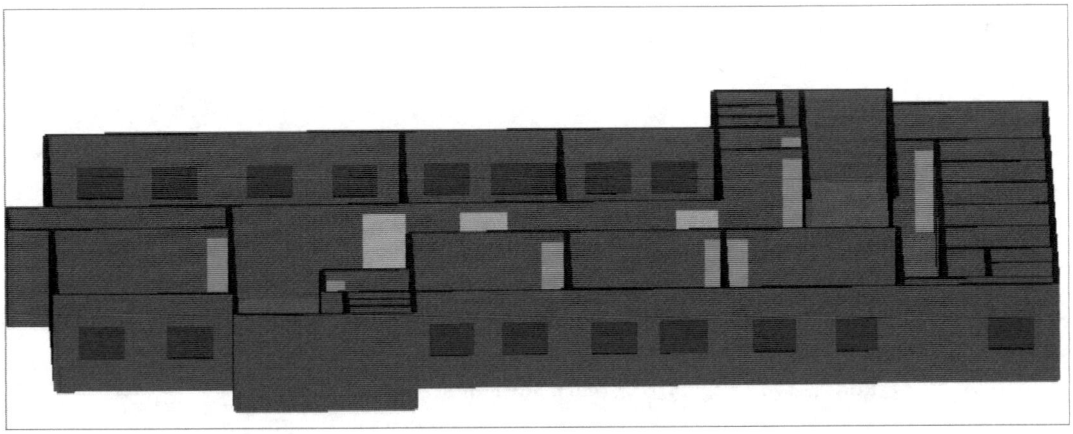

Figure 1. 3D view of the Second Floor Using floor Plan Editor.

Table 1. Antenna Specifications.

Item	Value
Frequency (GHz)	3.5
Transmit Antenna Power (dBm)	2.00
Tx-Antenna Gain (dBi)	2.15
Rx-Antenna Gain (dBi)	2.15
Tx-Antenna Height (m)	1.34
Rx-Antenna Height (m)	1.34

5. Results

The previously clarified case study was simulated using wireless InSite software. In this paper, the basic path loss models investigated are Close-In path loss model (CI) and Floating-Intercept (FI) path loss model.

5.1. Close-In Free Space Path loss Model (CI)

The CI model is described in equation (2) [18].

$$PL_{CI}(f, d)\,[dB] = PL\,(f, d_0) + 10n\log_{10}(d/d_0) + X\sigma_{CI} \quad (2)$$
$$\text{For } d \geq d_0, \text{ where } d_0 = 1.010 \text{ m}$$

$PL(f,d_0)$ refers to the free space path loss in dB at reference distance (d_0 = 1.010m), n is the path loss exponent, d is the T-R separation distance between the transmitter and receiver antennas and $X\sigma$ refers to a zero-mean Gaussian-distributed random variable with standard deviation σ dB.

Figures 2 and 3 include illustrations of the Close-In (CI) line plots at 3.5 GHz frequency band for the situation LOS and NLOS co-polarization and cross polarization with curve fitting. For co-polarization (Vertical-Vertical) antenna, the path loss exponent (PLE) values in the LOS and NLOS cases study were 1.7, 2.1, respectively. And for cross polarization (V-H) antenna in the LOS and NLOS cases, the path loss exponent (PLE) values were 2.9, 3.1, respectively. As seen for co-polarization(V-V) antenna LOS case PLE value is 1.7 which is less than the theoretical free space path loss exponent (n=2), whereas for cross polarization (V-H) antenna the PLE value is 2.1 larger than theoretical free space path loss also larger than co-polarization due to that cross polarization(V-H) refers to propagation in one state of polarization and reception in the state of orthogonal polarization unlike co-polarization transmitting and receiving in one state of polarization so there is no additional loss to the path loss. For NLOS environment, the PLE values for both polarizations are larger than LOS path exponent values due to the lack of a clear route in the NLOS case study. The CI path loss model parameters for LOS and NLOS cases are presented in table 2.

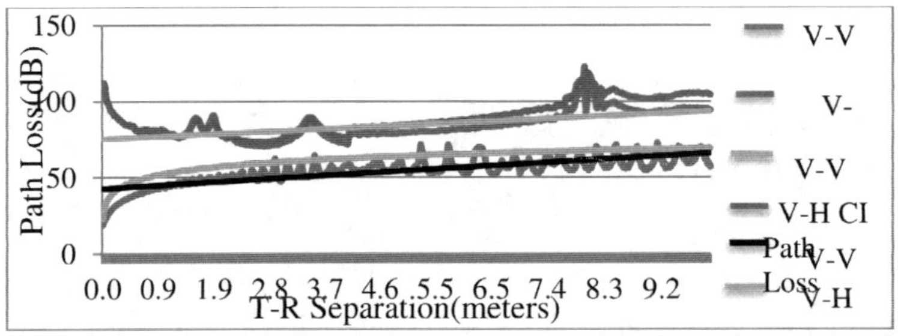

Figure 2. Close-In Path Loss Model, Simulation Path Loss with Curve Fitting for the LOS with Co-Polarization and Cross Polarization along the Corridor.

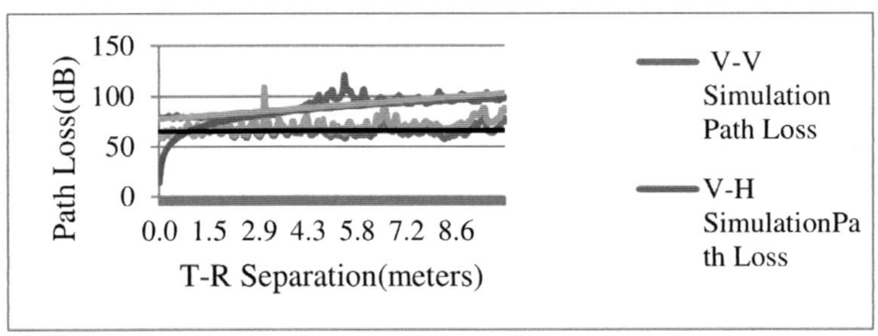

Figure 3. Close-In Path Loss Model, Simulation Path Loss with Curve Fitting for the NLOS with Co-Polarization and Cross Polarization along the Corridor.

Table 2. Close-In Path Loss Model Parameters for LOS and NLOS Cases.

	Co-Polarization (*LOS*)	Cross-Polarization (*NLOS*)	**Co-** Polarization (*NLOS*)	**Cross-** Polarization (*NLOS*)
PLE(n)	1.70	2.90	2.10	3.10
σ (dB)	0.97	3.20	3.29	6.02

5.2. Floating Intercept Path Loss Model (FI)

In the standard WINNER II and 3GPP, the floating-intercept (FI) path loss model is used. FI path loss model depend on the floating-intercept (α) value and line slope (β) as shown in equation (3) [13].

$$PL_{FI}\,(d)\,[dB] = \alpha + 10\beta\,\log_{10}(d) + X\sigma_{FI} \tag{3}$$

Here $X\sigma_{FI}$ is a zero mean Gaussian shadow fading random variable with standard deviation σ. The minimum mean-square error (MMSE) is used as a best-fit as in the Close-In model and needs to be solved for α and β to obtain the standard deviation. The results of the FI path loss model for the 3.5GHz frequency band in the line-of-sight (LOS) and non-line-of-Sight (NLOS) environments are shown in

Figures 4 and 5 for co-polarization (V-V) and cross polarization (V-H) respectively. In the co-polarization (V-V) LOS case compared to free-space path loss (FSPL), α values are different. It was 42.71dB compared to the theoretical free space path loss (FSPL) value was 43.39dB, while in cross polarization (V-H) LOS, α value was 75.28dB compared to the 43.39dB theoretical free space path loss (FSPL) at d_0= 1.010 m. In the case Vertical-Vertical (V-V) antenna polarization NLOS, α value was 65.09dB compared to the 43.39dB theoretical free space path loss (FSPL). In cross polarization antenna (V-H) NLOS case, α value was 77.50dB compared to the 43.39dB theoretical free space path loss (FSPL). Slope value (β) of the minimum-square mean fit line for the co-polarization antenna LOS case was 2.35 larger than free space (β=2) while close to the free space in the case co-polarization NLOS case. For the cross polarization (V-H) the β slope is 1.84 for the LOS and 2.57 for the NLOS. β only functions as a basic slope that provides the best fit for a data line plot and in any way has no physical basis or frequency dependence. This does not necessarily mean that the NLOS signals have more distance attenuation than the free-space signals. For the Floating-Intercept path loss model table 3, displays the parameters.

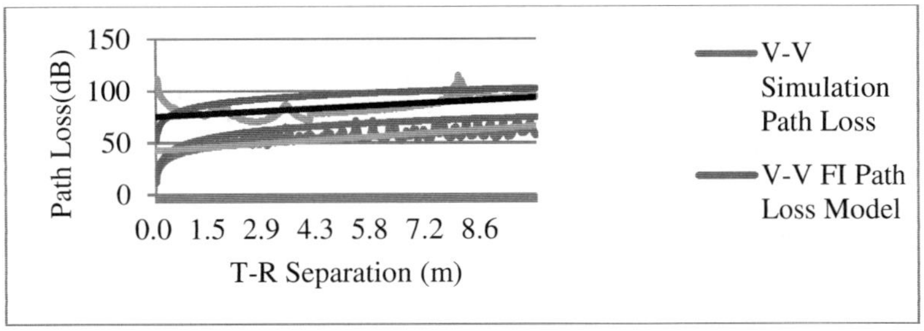

Figure 4. Floating-Intercept Path Loss Model, Simulation Path Loss with Curve Fitting for the LOS with Co-Polarization and Cross Polarization along the Corridor.

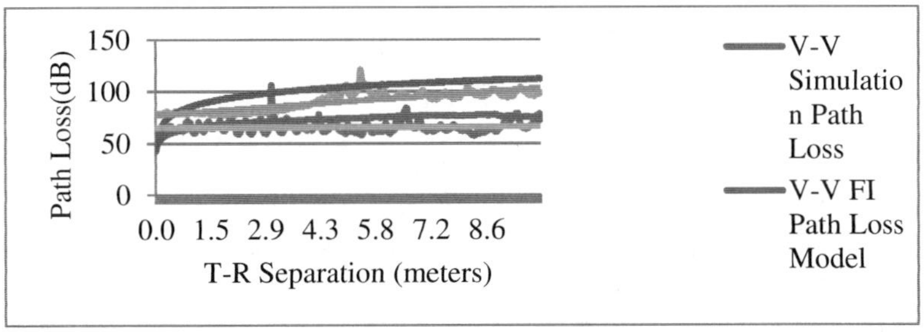

Figure 5. Floating-Intercept Path Loss Model, Simulation Path Loss with Curve Fitting for the NLOS with Co-Polarization and Cross Polarization along the Corridor.

Table 3. Floating-Intercept Path Loss Model Parameters for LOS and NLOS Cases.

	Co-Polarization (*LOS*)	Cross-Polarization (*LOS*)	Co-Polarization (*NLOS*)	Cross-Polarization (*NLOS*)
PLE(n)	1.70	2.90	2.10	3.10
$\boldsymbol{\beta}$	2.35	1.84	1.28	2.57
$\boldsymbol{\sigma}$ **(dB)**	8.76	8.72	5.10	9.01

6. Conclusions

This paper describes the path loss models for the indoor environment at 3.5GHz frequency band. The effects of path loss when transmitting 5G signals over the 3.5GHz frequency band were determined using simulation results. The comparison between two path loss models shows that by using the Close-In model-to-model path loss settings, the results were nearly similar to the simulated path loss, suggesting that the CI model is more reliable than the FI model. As a result, the findings indicate that the Floating-Intercept (FI) model is not physically adequate for the LOS and NLOS environment channels. Many techniques and investigations may be used for future research to enhance the emergence of the mid-band as a main frequency band for 5G connectivity and problems with uplink Coverage assistance.

Acknowledgments
Finally: The authors would like to thank the staff of communication engineering department specially Dr.Dia Ali and Mr. Mohammed Sameer for their help.

7. Reference

[1] Andrews J G, Buzzi S, Choi W, Hanly S V., Lozano A, Soong A C K and Zhang J C 2014 What Will 5G Be? *J. Sel. Areas Commun.* **32** 1065–1082
[2] Yang Y, Xu J, Shi G and Wang C-X 2018 *5G Wireless Systems;simulation and evalution techniques* ed Xuemin Sh Sh (Switzerland, S: Springer) 45-147
[3] Zhang K, Zhang R, Wu J, Jiang Y and Tang X 2019 Measurement and Modeling of Path Loss and Channel Capacity Analysis for 5G UMa Scenario *11th Int. Conf. on Wireless Communications and Signal Processing (WCSP)* 1–5
[4] Nokia 2017 5G deployment below 6 GHz *White paper* 1–12
[5] Liu J 2020 Channel Modeling and Tropospheric Effects on Millimeter Wave Communications for Aviation Applications *Doctoral Dissertation* (South Carolina, DC: University of South Carolina) 1-168
[6] Chrysikos T, Georgakopoulos P, Oikonomou I and Kotsopoulos S 2018 Measurement-based characterization of the 3.5 GHz channel for 5G-enabled IoT at complex industrial and office topologies *Wireless Telecommunications Symp.(WTS)* 1–9
[7] Maccartney G R, Samimi M K and Rappaport T S 2014 Omnidirectional path loss models in New York City at 28 GHz and 73 GHz *Int. Symp. Pers. Indoor Mob. Radio Commun. PIMRC* 227–231
[8] Muttair K S, Shareef O A and Mosleh M F 2020 Outdoor to Indoor Wireless Propagation Simulation Model for 5G Band Frequencies *IOP Conf. Ser. Mater. Sci. Eng.* **745** 1-14
[9] Liu J, Matolak D W, Mohsen M and Chen J 2019 Path Loss Modeling and Ray-Tracing Verification for 5/31/90 GHz Indoor Channels *USA 90th Vehicular Technology Conf.(VTC2019-Fall)* 1–6
[10] Zheng H, Li W, Tian L, Xu C, Huang F and Zhang J 2015 Path loss models for urban macro cell scenario at 3.35, 4.9 and 5.4 GHz *26th Annual Int. Symp. on Personal, Indoor, and Mobile Radio Communications (PIMRC)* 2229–3322
[11] Sun S, Rappaport T S, Rangan S, Thomas T A, Ghosh A, Kovacs I Z, Rodriguez I, Koymen O, Partyka A and Jarvelainen J 2016 Propagation Path Loss Models for 5G Urban Micro- and Macro-Cellular Scenarios *IEEE 83rd Vehicular Technology Conf. (VTC Spring)* 1–6
[12] Al-Samman A, Rahman T, Hindia M, Daho A and Hanafi E 2018 Path Loss Model for Outdoor Parking Environments at 28 GHz and 38 GHz for 5G Wireless Networks *Symmetry (Basel).* **10** 672
[13] Maccartney G R, Rappaport T S, Sun S and Deng S 2015 Indoor Office Wideband Millimeter-Wave Propagation Measurements and Channel Models at 28 and 73 GHz for Ultra-Dense 5G Wireless Networks *J. IEEE Access Electron* **3** 2388–2424
[14] Sobetwa M and Sokoya O 2019 Measurement Analysis of Indoor Parameters for an Indoor Wireless Propagation *Morocco Int. Conf. on Wireless Technologies, Embedded and Intelligent Systems (WITS)* 1–6\
[15] Kacou M, Guillet V, El Zein G and Zaharia G 2018 A Multi-wall and Multi-frequency Home Environment Path Loss Characterization and Modeling *London 12th European Conf. on Antennas and Propagation (EuCAP)* 1-5
[16] Bhuvaneshwari A, Hemalatha R and Satyasavithri T 2015 Path loss prediction analysis by ray tracing approach for NLOS indoor propagation *India Int. Conf.on Signal Processing and Communication Engineering Systems* 486–491

[17] Koymen O H, Partyka A, Subramanian S and Li J 2015 Indoor mm-Wave Channel Measurements: Comparative Study of 2.9 GHz and 29 GHz *USA IEEE Global Communications Conf. (GLOBECOM)* 1–6

[18] Majed M B, Rahman T A, Aziz O A, Hindia M N and Hanafi E 2018 Channel Characterization and Path Loss Modeling in Indoor Environment at 4.5, 28, and 38 GHz for 5G Cellular Networks *Int. J. Antennas Propagation* 1–14

On the Performance of Cooperative Relaying with Maximum-Ratio Combination (MRC) for mmWave Systems

Reyam H. Ali[1], **Mahmood A. Mahmood**[1*] **and Mohamad A. Ahmed**[1]

[1]College of Electronics Engineering, Ninevah University, Mosul, Iraq, P.O. 41002

[*]E-mail: mahmod.mahmod@uoninevah.edu.iq

Abstract. In this paper, cooperative communication using a two-way relaying network (TWRN) is considered for millimeter-wave (mmWave) communications. Two schemes of TWRN are employed for cooperation and performance evaluation, which are amplify-and-forward (AF) and decode-and-forward (DF) relays. Furthermore, maximum-ratio combination (MRC) is utilized in each node in the proposed systems. It is noteworthy that MRC is exploited in this paper as a channel equalization and also to maximize the strength of the received signals at each node. The performance is evaluated for the proposed systems over 28 and 73 GHz using one and two parallel AF and DF relays in the presence of the direct path. Moreover, compassion between the two frequencies is achieved, via obtaining the bit-error-rate (BER) against different values of signal-to-noise ratios (SNR)'s, by assuming the same distances between the source, relay/relays, and the destination. The results show the outperforming of 28 GHz in this performance metric, i.e. BER-SNR, by using AF relaying, due to the amplification factor added to the signal to compensate for the inherent attenuation in this frequency band. Moreover, the proposed DF relay schemes have limited compensation to the attenuated signals comparing to the AF relay.

1. Introduction

It is estimated that the number of devices connected to the network has grown by a factor of 100 to 10,000 by 2021 [1]. These devices range from tablets to computers, smartphones, and PDAs, that come under the concept of the Internet of Things (IoT) [1]. Besides, by 2021, requested mobile data is estimated to have risen by a factor of 1,000 per unit. This growth puts a lot of pressure on the 4G network. This causes increases latency, signal interference, and reduction in data rates [1]. New technologies are evolving to overcome these challenges to build the next generation 5G network [1]. Large-scale MIMO (massive MIMO), full-duplex, millimeter-wave (mmWave), and cloud radio access networks (CRAN), small-cell, beamforming are some of these technologies [2]. These technologies will deliver higher network capacity, serve more customers, enhance the energy efficiency, adaptability to the implementation of new networks and devices [1]. One of the key candidate technology for wireless mobile communications within the 5th and 6th generations [3]will be millimeter-wave (mmWave) communications. With frequencies in the mmWave band ranging from 30 GHz to 300 GHz, corresponding to wavelengths of 10 mm to 1 mm [3][4][5][6]. At these rates, the uncongested spectrum of mm-Wave can easily be 200 times greater than all today's cellular below 3 GHz [7] which will not have the ability to meet the next massive demand on services that use communication and the internet services. A mmWave communication system can support up to

Content from this work may be used under the terms of the Creative Commons Attribution 3.0 licence. Any further distribution of this work must maintain attribution to the author(s) and the title of the work, journal citation and DOI.
Published under licence by IOP Publishing Ltd

multiple gigahertz of bandwidth to enhance the capacity/data rate and can be used for mobile cellular access, indoor wireless communications, or outdoor communications such as wireless mesh networks [8]. It would also play an important role in cellular networks beyond 4G and 5G[7]. The mmWaves are slightly longer than X-rays or infrared waves, but shorter than radio-waves or microwaves[6]. Moreover, mmWave's very short wavelengths allow a huge number of antenna elements to be packed into small form factors[6][7]. Such multiple antenna systems can be used for electrically steerable shaping, very high-gain manufactured arrays at the base station, in the skin of a mobile phone, or even within a chip [7][9]. Most of the latest research emphasizes the characterization of the mmWaves sub-100 GHz domain with interest in the 28 GHz band, 38 GHz band, 60 GHz band, and E-band (71 to 76 GHz and 81 to 86 GHz). These bands are planned to be employed for mobile cellular 5G [6][10] to fulfill the data rate criteria at the Gbps. Various channel models have been developed by the research community, each meeting a particular contact scenario. For example, owing to its wide unlicensed spectral resources for indoor wireless networks, most of the early modeling efforts were carried out in the 60 GHz band [4][11][12]. However, until recently, the scarcity of measurements and knowing of other mmWave bands restricted the number of channel models available. A contiguous 7 GHz frequency band from 57-64 GHz has been assigned to short-range communications on an unlicensed basis by the Federal Communications Commission (FCC) since 2001 [12]. One remarkable feature of the 60 GHz band is its availability in most parts of the world [12]. However, mmWave and current lower-frequency communications have a variety of basic variations. This includes greater loss of propagation, loss of penetration, directivity, etc. Due to these challenges, the transmission range could be reduced to a few hundred meters or less [13] to reach the guaranteed ultra-high data rates of mmWave. A promising solution for the extremely high rates of data expected in future networks for wireless systems is relay-assisted communication. Intermediate (relay) nodes may be used by relaying the signal from the source to the destination instead of a long direct link from the source to the destination, to maximize the diversity and reduce the individual transmission time [13]. In such a case, when the distance between the nodes too long to satisfy the data rate demand, the source and destination will not communicate directly with each other, and/or certain barriers between them preclude direct contact. The relays break long connections to several short links but with very high-rate that can address the high propagation loss and blockage sensitivity of mmWave [13]. Decode-and-forward (DF) and amplify-and-forward (AF) [13][14] are the most frequent relay techniques. The relays use in this work are decode-and-forward (DF) and amplify-and-forward (AF) to compensate for attenuation and corruption of signals between the transmitter and receiver in mmWave backhaul networks. While the DF relay process the received signal, the steps of processing is decoded the received data re-encodes it, and then forwards it to the destination, the received signal is only amplified and forwarded by an AF relay[13]. The complexity of DF relays is considerably higher than AF relays and decoding and re-encoding often needs maximal processing power [13]. The channel considering in this work is the mmWave channel, the feature of this channel is that the signals pass over this channel fades beyond short distance according to the high attenuation of the waves due to path losses and the obstacle absolution, plus adaptive white Gaussian noise (AWGN) is also considered. Furthermore, in our channel model, The receiver inverts the operations found in the transmitter to decode the data followed by a decoder Maximal Ratio Combining (MRC) [15] in order to improve system performance by reducing the effects of fading on system performance[16]. Maximal Ratio Combining (MRC) also known as ratio-squared combining and redetections combining. In the MRC technique signals are combined from multiple diversity branches and get multiplied by a maximal ratio combining, it is used to add signals from each channel and the gain of that channel is proportional to the signal's RMS value and inversely proportional to the mean square noise level. Each channel uses various constants of proportionality[15]. The precise problem we study is how to obtain minimum complexity of detection with the best outage probability and error probability (reduce the bit-error-rate along with different signal-to-noise ratio). The remainder of this paper is arranged as follows. Section II presented the related works. Section III, we describe the system model, Numerical and simulation results are given in section IV, while the discussion is given in section V, and concluding remarks are presented in Section VI.

1.1 Related work

The mmWave channel modeling and cooperative communication form the basis for our study. The study of mmWave channel modeling is a very hot topic, and many universities and companies are working in this field. Some of the related literature on the topic is mentioned in this scope. The characterization of the mmWaves sub-100 GHz domain with a focus on the 28 GHz, 38 GHz, 60 GHz, and E-bands (71 to 76 GHz and 81 to 86 GHz)has been the topic of most recent studies [6][10]. Various channel models have been developed by the research community, each meeting a particular contact scenario. For example, owing to its wide unlicensed spectral resources for indoor wireless networks, almost all of the early modeling attempts in the 60 GHz band were carried out [4][11][12], However until recently, the scarcity of measurements and knowledge of other mmWave bands restricted the number of channel models available [4]. Akdeniz et al.[7] In New York, NY, USA, has been derived at 28 and 73 GHz spatial statistical channel models and channel parameters consist of angular dispersal, number of spatial clusters, outage, and loss of direction. It was discovered that powerful signals may be recognized 100-200m from possible sites for cells even in extremely NLOS environments, and it is possible to provide spatial multiplexing and variance at several Sites of multiple path clusters obtained. In New York City, an urban propagation campaign in the 28 GHz has been performed by Rappaport et al.[9], where the distance ranged from 75 m-125 m between the sender and the receiver. In Manhattan, New York, they have carried out an outage analysis [10].In Manhattan, New York, they have carried out an outage analysis [10]. The signal received by the RX was observed to be within 200 meters for all situations, and 57 percent of positions were outages due to obstruction with most of the outages above 200 meters from the TX [17]. With a reduction of the exponent of path loss and increase antenna gains, the overall distance coverage area was seen to climb. When the TX-RX antenna gain is 49 dBi combined[17], the overall coverage distance in a heavily obstructed area exceeds 200 m. In Canada, Korea, and the United States, the 57 to 64 GHz band is available, the band from 59 to 66 GHz is available in Japan, while the band available in Australia is from 59 to 63 GHz [11].

2. System model

We consider a mmWave wireless network consisting of one source (s), one destination (D), and intermediate relays. We assume that each node of them has a sender, receiver, and antennas. The antenna technique that is used in this work is a single input single output (SISO). The intermediate relays employed in the system model are either in the AF or DF modes to support mmWave wireless backhaul and to achieve higher data rates of mmWave links. Moreover, a direct connection between sender and recipient, i.e. the line-of-sight (LoS) path, is also considered. Different scenarios are taken into account to design the network. The first scenario is to use AF relays, and the second uses DF relays in the network scheme. Besides, at each node of the network, we derive the equations.

2.1 mmWave channel assisted with AF relays

We consider the channel of mmWave from source to destination, as long as a direct path between the source and destination terminals exists, the nodes communicate via that direct path with distance R_0. If the direct path is blocked by a barrier or because that the signal faced high attenuation due to high carrier frequency of mmWave between the source and destination, the node indirectly communicates with each other via intermediate AF relay node (R) (in figure 1 considering one relay) we assume the source transmits the signal X, and it passes through l_{th} mmWave channel.
The mmWave channel can be modeled as [18]

$$h_\ell = \frac{\lambda}{4\,\pi R_\ell}\, e^{\frac{-j2\pi R_\ell}{\lambda}} \tag{1}$$

Where h_ℓ denotes all channel paths (the amplitude of the ℓth mmWave channel gain) from source to destination also the path from source and destination via a relay, i.e. h_ℓ for (ℓ=0, 1,2,3,...., L).
Moreover, r_{SD} r_{SR} and r_{RD} are the distances from the source to the destination, the source and relay, and the relay and destination, respectively, where r_{SD} is considered to be 200m

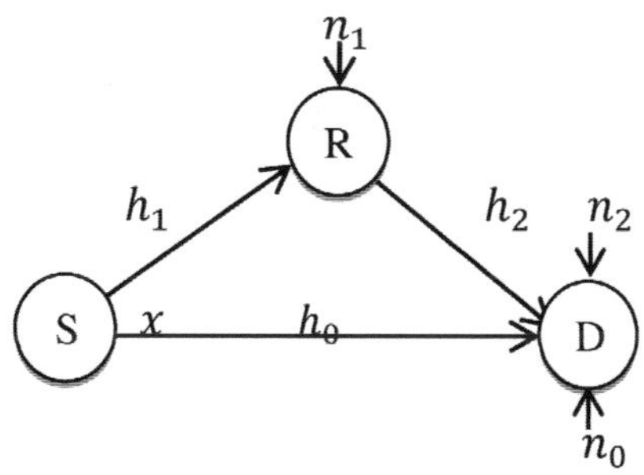

Figure1. mmWave channel assisted with one AF relay.

at frequency =28GHz as explained in [7], r_{SR} and r_{RD} are equal to $100m$, in which $\lambda = \frac{c}{f}$ is denoted for the wavelength with c and f to represent the speed of light and the operating frequency of the mmWave. The signal x passes through the mmWave channel (h_1), the complex additive white Gaussian noise (AWGN) added to the signal at the receiver antenna of the AF relay . The direct path from source to destination, the LoS path, received signals at relay and destination are

$$y_{SD} = h_0 x + n_0 \tag{2}$$
$$y_{sr} = h_1 x + n_1 \tag{3}$$
$$y_{rd} = h_2 B y_{sr} + n_2 \tag{4}$$

In the AF relaying scheme, the received signal in the relay is amplified by a gain factor B and then forwarded to the destination, where B is defined as the amplification factor scaling of the relay transmitted power. This indicates that the transmitted average power is constant at the relay (p_r) [13]. Therefore, B can be derived as

$$B \leq \sqrt{\frac{p_r}{h_1^2 E[|x|^2] + E[|n_1|^2]}}, \tag{5}$$

where pr is the signal power, and $E[.]$ represents the expectation of signal and noise. The received signals from each channel are added together at the received node, which can be written as equation (6), then decoded by MRC decoder to overcome the channel effects on the signal, by made the gain of each channel proportional to the root-mean-squared (r.m.s) value of the signal and made the gain in that channel inversely proportional to the mean square of the noise level [15]. Furthermore, the original shape of the signal can be restored by using this technique. The reconstructed signal at the receiver can be written as in equation (6), in which equalization by using the MRC technique is utilized to maximize the signal strength by combining all paths at the destination.

$$\hat{x} = h_0^* \, y_{sd} + h_1^* \, B^* h_2^* y_{rd} \tag{6}$$

The second scenario in this section is shown in figure 2, in which two AF relays in a parallel transmission model for the sake of improving the performance of the system by maximizing outage probability, minimizing the error probability, and enhancing the diversity gain. In this scenario the transmitted signal (x) is propagated through three paths to reaches the destination, the first is the direct path as illustrated in equation (2). The two other paths are indirect paths with intermediate AF relays between the source and destination to make them communicates with each other. The use of two relays in the system model instead of one relay is to achieve the lowest bit-error-rate (BER) at the destination. The received signals at the destination node from the relays R_1 and R_2 can be expressed as:

$$y_{sr1} = h_1 x + n_1 \quad (7)$$
$$y_{r1d} = h_2 B_1 y_{sr1} + n_2 \quad (8)$$
$$y_{sr2} = h_3 x + n_3 \quad (9)$$
$$y_{r2d} = h_4 B_2 y_{sr2} + n_4 \quad (10)$$

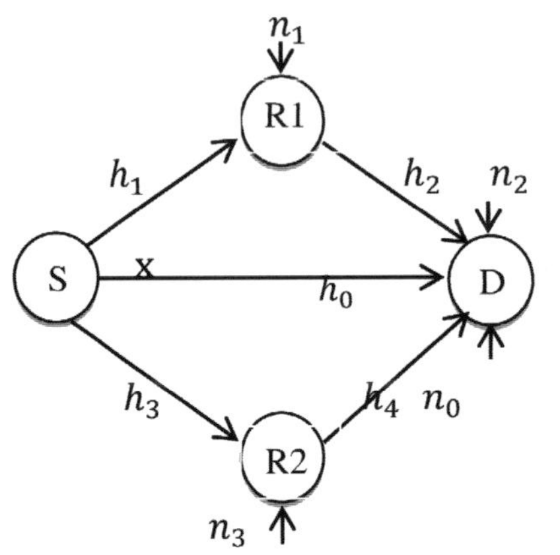

Figure 2. mmWave channel assisted with two AF relays.

The received signals in the first and second hop for AF relay are amplified by a gain factor B_1 and B_2, respectively, where the equation of B_1 is described in the first scenario in equation (5), While B_2 can be derived as

$$B_2 \leq \sqrt{\frac{p_r}{h_3^2 E[|x|^2] + E[|n_3|^2]}} \quad (11)$$

As we mentioned in the previous scenario that the signals that reach the destination are combined and decoded by the MRC decoder to reconstruct the original signal at the receiver as

$$\hat{x} = h_0^* \, y_{sd} + h_1^* \, h_2^* B_1^* y_{r1d} + h_3^* \, h_4^* B_2^* y_{r2d} \quad (12)$$

2.2 mmWave channel assisted with DF relays

In the network model that is shown in figure 3, we consider DF relay instead of AF relay that presented in the network model in figure 1. In DF relaying sketch, the received signal by the relay (y_{sr}) is processed and then send to the destination. The processing operation is achieved by decoding, re-encoding, and retransmitting it to the next node.

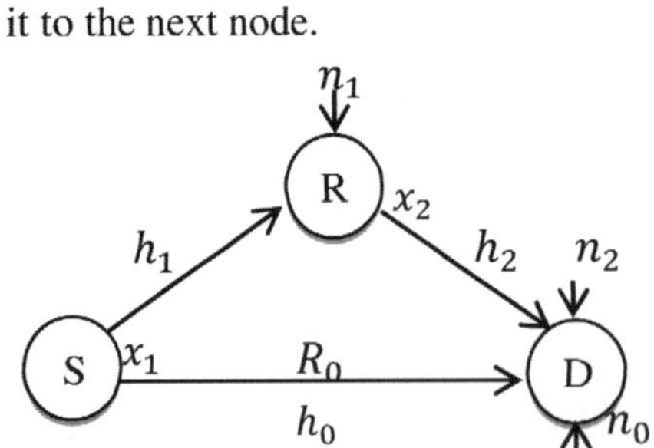

Figure 3. mmWave channel assisted with one DF relay.

In the direct path, the received signal at the destination, y_{sd}, can be written as:

$$y_{sd} = h_0 x_1 + n_0, \quad (13)$$

while the received signal at the relay, y_{sr}, is expressed as:

$$y_{sr} = h_1 x_1 + n_1, \tag{14}$$

where the DF relay performs equalization as in equation (15) to obtain the estimated symbol $\widehat{x_1}$, which is decoded, re-encoded, and transmitted to the destination.

$$\widehat{x_1} = h_1^* \, y_{sr} \tag{15}$$

The transmitted symbol, which is denoted as $x_2 = \widehat{x_1}$, is passed through a mmWave channel h_2 towards the destination. The DF relay utilizes the MRC equalizer to remove the channel effects on the signal, so the original signal $\widehat{x_1}$ is restored, decoded, and re-transmit again to reach the destination node as

$$y_{rd} = h_2 x_2 + n_2 \tag{16}$$

In this type of relay, AWGN is not accumulated due to the decoding process at the relay [19]. This feature is not found in AF relay so it helps to improve the achievement of the network. The reconstructed signal is given by

$$\widehat{x_1} = h_0^* \, y_{sd} + \, h_2^* \, y_{rd} \tag{17}$$

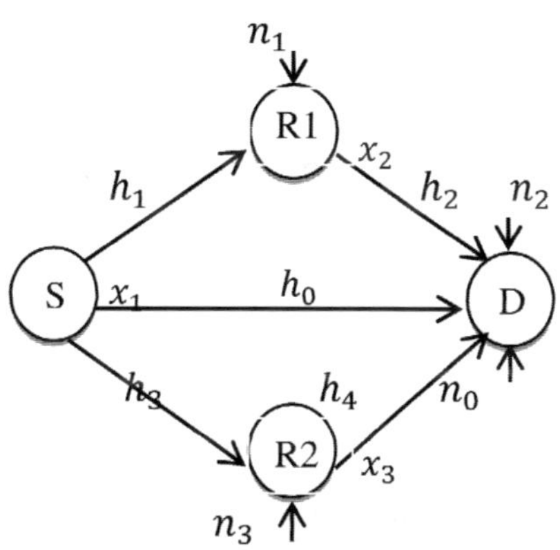

Figure 4. mmWave channel assisted with two DF relays.

The second scenario of DF relay is similar to that described in Figure 2, in which the signals received by two parallel relays R1, R2. The received signals at the two relays are expressed as

$$y_{sr1} = h_1 x_1 + n_1, \tag{18}$$
$$y_{sr2} = h_3 x_1 + n_3. \tag{19}$$

The signals that reach the destination node from the relays and the source are expressed as

$$y_{sd} = h_0 x_1 + n_{0,} \tag{20}$$
$$y_{r1d} = h_2 x_2 + n_2, \tag{21}$$
$$y_{r2d} = h_4 x_3 + n_{4,} \tag{22}$$

While the reconstructed signal is written as

$$\widehat{x_1} = h_0^* \, y_{sd} + \, h_2^* \, y_{r1d} + \, h_4^* \, y_{r2d} \tag{23}$$

IOP Conf. Series: Materials Science and Engineering 1152 (2021) 012007 doi:10.1088/1757-899X/1152/1/012007

3. Simulation Results

We simulate the proposed systems shown in figures (1-4) using Monte-Carlo simulations via Matlab programming. AF and DF relays are considered to assist the communication between a source and destination over 28 and 73 GHz mmWaves, in which binary phase-shift keying (BPSK) is employed as a modulation scheme in the proposed systems. Two scenarios are considered for the AF and DF relaying by utilizing one and two relays in parallel between the transmitter and receiver in the presence of the direct link between them. The distance between the source and destination is assumed 200m and the one or two relays is/are placed in the middle between them. Figure 5 shows the first scenario for the two frequencies mentioned above, in which one relay is used, as in Figure1 and Figure3 for the AF and DF relaying, respectively. This figure demonstrates two outcomes, the first that 28GHz has better BER-SNR performance than 73GHz, while the second shows the outperform of AF relay over DF relay for compensation of the inherent attenuation that existed in these frequency bands.

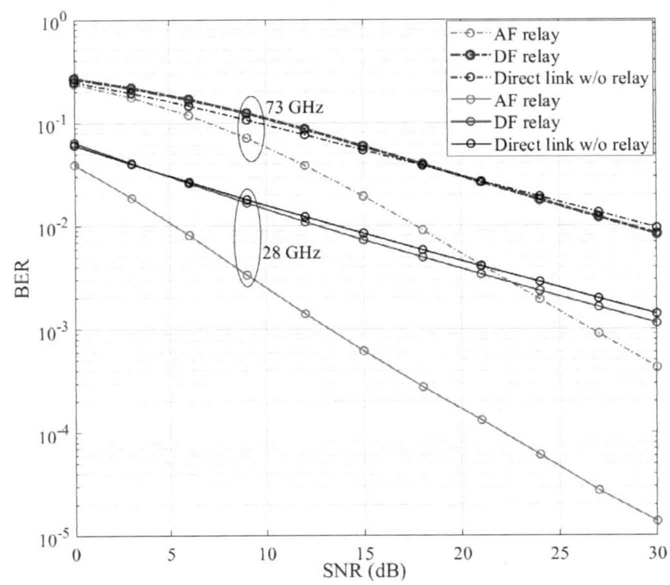

Figure 5 BER vs. SNR for cooperative relaying for mmWaves using one AF and DF relays over 28 and 73 GHz in the presence of the direct path.

Figure 6 shows the BER-SNR performance for the systems shown in figure 2 and figure 4, in which two parallel AF and DF relays are used in the presence of the LOS link, and for the same distances mentioned above. It is noteworthy, that all channels are created in this paper randomly, and no correlation between them.

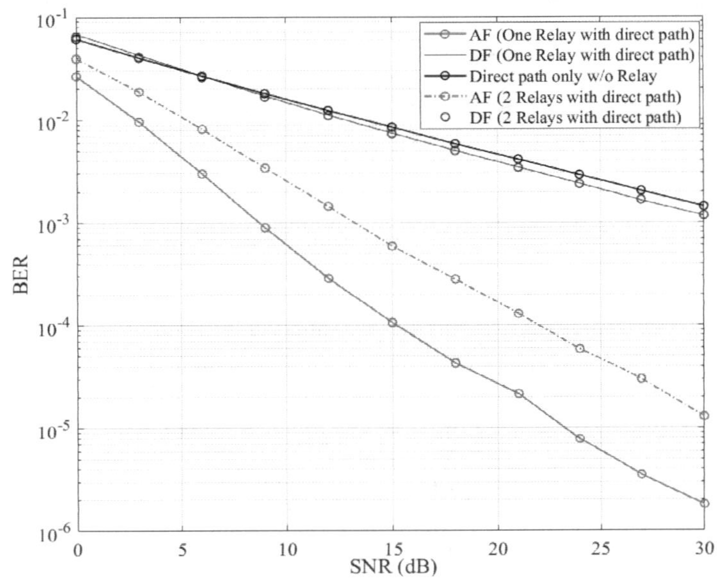

Figure 6 BER vs. SNR for cooperative relaying for mmWaves using one and two parallel AF and DF relays over 28 GHz in the presence of the direct path.

4. Conclusions:

Cooperative communications via relays have been considered in this paper to compensate for the inherent attenuation of the signal passed through mmWave channels. AF and DF relaying schemes have been employed, along with utilizing maximum-ratio combination (MRC) in each node in the proposed systems. The MRC technique has been exploited to equalize the mmWave channels and to maximize the strength of the received signals at each node. The performance has been evaluated, using one and two parallel AF and DF relays for the proposed systems, over 28 and 73 GHz in the presence of the LoS link. Moreover, compassion between the two frequencies has been achieved, via obtaining the BER-SNR performance, considering the same distances between the nodes for all scenarios. The results revealed 28 GHz has better BER-SNR by using AF relay, due to the amplification factor added to the signal to compensate for the inherent attenuation in this frequency band. Moreover, the two DF relay schemes did not add valuable compensation to the signals comparing to the AF relay.

5. References

[1] De Ree M, Mantas G, Radwan A, Mumtaz S, Rodriguez J and Otung I E 2019 Key Management for beyond 5G Mobile Small Cells: A Survey *IEEE Access* **7** 59200–36

[2] Koh J, Lim Y G, Chae C B and Kang J 2018 On the Feasibility of Full-Duplex Large-Scale MIMO Cellular Systems *IEEE Trans. Wirel. Commun.* **17** 6231–50

[3] Xiao Z, Dai L, Ding Z, Choi J, Xia P and Xia X G 2017 Millimeter-wave communication with non-orthogonal multiple access for 5G *arXiv* **7** 116123–32

[4] Hemadeh I A, Satyanarayana K, El-Hajjar M and Hanzo L 2018 Millimeter-Wave Communications: Physical Channel Models, Design Considerations, Antenna Constructions, and Link-Budget *IEEE Commun. Surv. Tutorials* **20** 870–913

[5] Seker C, Guneser M T and Ozturk T 2018 A Review of Millimeter Wave Communication for 5G *ISMSIT 2018 - 2nd Int. Symp. Multidiscip. Stud. Innov. Technol. Proc.* 1–5

[6] Uwaechia A N and Mahyuddin N M 2020 A comprehensive survey on millimeter wave communications for fifth-generation wireless networks: Feasibility and challenges *IEEE Access* **8** 62367–414

[7] Akdeniz M R, Liu Y, Samimi M K, Sun S, Rangan S, Rappaport T S and Erkip E 2014

Millimeter wave channel modeling and cellular capacity evaluation *IEEE J. Sel. Areas Commun.* **32** 1164–79
[8] Kwon S and Widmer J 2018 Relay selection for mmwave communications *IEEE Int. Symp. Pers. Indoor Mob. Radio Commun. PIMRC* **2017-Octob** 1–6
[9] Rappaport T S, Sun S, Mayzus R, Zhao H, Azar Y, Wang K, Wong G N, Schulz J K, Samimi M and Gutierrez F 2013 Millimeter wave mobile communications for 5G cellular: It will work! *IEEE Access* **1** 335–49
[10] Azar Y, Wong G N, Wang K, Mayzus R, Schulz J K, Zhao H, Gutierrez F, Hwang D and Rappaport T S 2013 28 GHz propagation measurements for outdoor cellular communications using steerable beam antennas in New York city *IEEE Int. Conf. Commun.* 5143–7
[11] Geng S, Kivinen J, Zhao X and Vainikainen P 2009 Millimeter-wave propagation channel characterization for short-range wireless communications *IEEE Trans. Veh. Technol.* **58** 3–13
[12] Va V, Shimizu T, Bansal G and Heath R W 2016 Millimeter wave vehicular communications: A survey *Found. Trends Netw.* **10** 1–116
[13] Yan Y, Hu Q and Blough D M 2018 Path Selection with Amplify and Forward Relays in mmWave Backhaul Networks *IEEE International Symposium on Personal, Indoor and Mobile Radio Communications, PIMRC* vol 2018-September (Institute of Electrical and Electronics Engineers Inc.)
[14] Sunil B and Nandana B T 2015 Performance analysis and comparison of AF and DF relaying systems in Rayleigh fading channel considering poisson interference field *ICIIECS 2015 - 2015 IEEE Int. Conf. Innov. Information, Embed. Commun. Syst.*
[15] Anitha M, Priya R S, Lakshmi M R and Vijayalakshmi S 2015 MIMO system performance using various modulations under different channels with STBC, ZF and MRC *Proc. 2015 IEEE Int. Conf. Electr. Comput. Commun. Technol. ICECCT 2015*
[16] Aalo V A 1995 Performance of Maximal-Ratio Diversity Systems in a Correlated Nakagami-Fading Environment *IEEE Trans. Commun.* **43** 2360–9
[17] Niu Y, Li Y, Jin D, Su L and Vasilakos A V. 2015 A survey of millimeter wave communications (mmWave) for 5G: opportunities and challenges *Wirel. Networks* **21** 2657–76
[18] Mesleh R and Alhassi A 2018 SMTs for Millimeter-Wave Communications *Sp. Modul. Tech.* 167–83
[19] Belbase K, Tellambura C and Jiang H 2019 Coverage, Capacity, and Error Rate Analysis of Multi-Hop Millimeter-Wave Decode and Forward Relaying *IEEE Access* **7** 69638–56

IOP Conf. Series: Materials Science and Engineering 1152 (2021) 012008 doi:10.1088/1757-899X/1152/1/012008

A Dual stopband two-sided frequency selective surface for mobile shielding applications

A T Mayouf[1*], K H Sayidmarie[2] and Y E Mohammed Ali[1]

[1]Department of Electrical Engineering, College of Engineering, University of Mosul, Mosul, Iraq
[2] Department of Communication Engineering, College of Electronic Engineering, Ninevah University, Mosul, Iraq

[*]E-mail: awsthamir@gmail.com

Abstract. A two-sided frequency selective surface (FSS) is proposed to shield two bands of the mobile frequencies (EGSM-900 and DCS-1800). The FSS consists of an array of square concentric-rings cells printed on both sides of the substrate to offer better shielding of the two mobile frequencies. The CST Microwave suit is used for the simulations. The equivalent circuit model for the two-sided FSS is analysed and the transmission frequency in terms of the two stop frequencies is estimated. The obtained shielding is better than 40 dB across the two bands. The proposed FSS can be used to reduce mobile radiation or to isolate halls from nearby mobile base stations. The bandwidth of the FSS cell can be increased by using ring pairs of unequal sizes on the two sides of the substrate.

1. Introduction

Many issues and problems of spectrum management have arisen due to the increasing use of devices and applications that operate at many frequencies. The radiation due to the mobile phone can cause many issues, such as the annoyance or interference of the device ringing tones. In some locations, like hospitals, examination halls, and places of worship, mobile equipment, tablet, and phones need to be inactivated. Thus, shieling or isolating a building or some parts of it from the mobile signals is of increasing concern. The use of jamming devices to overlay the interference of the mobile signals has become illegal in many countries, due to health issues. Another solution is by rejecting the unwanted mobile bands by employing Frequency Selective Surfaces (FSS). With the use of large glass windows in modern buildings, where waves can easily penetrate, adding an FSS can be a simple and cheap solution, while the use of window curtains [1-2] will offer a solution and isolate the mobile waves.

FSS have been extensively studied in the literature [3] and many applications were presented, such as shielding at certain frequencies [1-6]. The FSS were also used for transmitting waves of a certain band of frequencies through a conducting screen as in [7], or enhancing the transmission of mobile frequencies through the energy-saving glass windows [8], or to offer a bandpass property in a conducting screen [9]. The presented shape of the unit cell of the FSS had various geometries such as square [1-4], circular [5], hexagons [6], strips, U-shaped slots [7], crossed dipoles [8], and intertwined slots [9]. In [4], the GSM frequencies were shielded by a form of a wallpaper FSS, where square loops were used in the unit cell to provide a 20 dB attenuation. A double-ring FSS showed more than 20 dB of mobile frequencies shielding [5]. A curtain of FSS which has a double square loop unit cell on one side showed more than 35dB attenuation [1]. A fractal square loop FSS printed on FR-4 substrate

Content from this work may be used under the terms of the Creative Commons Attribution 3.0 licence. Any further distribution of this work must maintain attribution to the author(s) and the title of the work, journal citation and DOI.
Published under licence by IOP Publishing Ltd

provided a dual bandstop in the frequencies of GSM900, GSM1800, and IMT2000 [6]. The FSS unit cell of square, hexagonal, and circular loops were analysed for the purpose of a passband of the dual frequencies of GSM 900 & 1800MHz [10].

This paper investigates an FSS of double square rings printed on both sides of a dielectric sheet for GSM900 and GSM1800 bands shielding. The proposed two-sided design is aimed to increase the shielding of the single-sided FSS. The shielding of the e.m. wave is aimed for use as a curtain to provide the shielding at the wanted frequencies. The CST microwave suit is used in the simulations, where the filtering or blocking of certain frequencies is explained by the surface current distribution on the copper rings. Moreover, an equivalent circuit for the FSS unit cell having four rings is proposed, and formulas for the blocked and transmitted frequencies are derived. The effects of the shift in the alignment of the rings, and using rings of different sizes on the two sides of the FSS cell are investigated.

2. FSS Design

The concentric double-square rings that are known for their dual-band or wideband operation and having the same response for vertical and horizontal polarizations [11] were used on both sides of the FSS. The frequency bands of EGSM-900 (880-960 MHz) and the DCS1800 (1710-1880 MHz) are aimed at as they are used for mobile communications in Iraq.

The designed element structure of the FSS is shown 'in Figure 1'. The structure of the FSS is in the form of a fabric curtain with double copper rings printed on both sides of the sheet. The permittivity of the woven-fabric curtain (substrate) was chosen as 2.2 [12], and the thickness of 1 mm was assumed in the simulations. The ring resonance happens when the guided wavelength λ_g equals the ring average perimeter L_{av}, where

$$L_{av} = \lambda_g = \frac{\lambda_o}{\sqrt{\varepsilon_{eff}}} \quad (1)$$

Where curtain effective dielectric constant ε_{eff} is given by:

$$\varepsilon_{eff} = \frac{\varepsilon_r + 1}{2} \quad (2)$$

The calculated dimensions of the double ring on each side of the substrate are shown in Table 1, where the width of the rings is 1.4 mm. In this design, the centers of the two GSM mobile bands were set at 0.920 GHz and 1.795 GHz. The outer ring and inner ring dimensions were chosen according to the above equations to block the lower and upper mobile bands, respectively.

3. Simulation

The CST Microwave Studio was used to tune the double square rings dimensions and investigate the performance of the FSS that have been designed in the previous section. A linearly-polarized plane wave whose electric field is directed along the Y-axis is assumed normally-incident on the FSS. Figure 2 shows the simulated S-parameters of the double square rings, for the two cases of two-sided and single-sided designs. The response shows two stopbands at 0.935 GHz (transmission coefficient S_{21} of -40.6 dB, and reflection coefficient S_{11} approaching 0 dB) and at 1.9 GHz (S_{21}= -40.9 dB, and $S_{11} \approx 0$ dB) frequencies. These two resonance frequencies correspond to the two concentric rings. It can be seen that the two-sided FSS has a better blocking by about 3 dB. As seen in Figure 3, the surface current variation along the perimeter of the large rings forms one wavelength at the lower band, while that on the smaller ring forms one wavelength at the upper band. Such variations express that when a ring resonates the transmission through the FSS is blocked.

The other frequency at 1.382 GHz, which is seen in Figure 2, represents a transmission of the incident wave, ($S_{21} \approx$ 0dB and S_{11} = -33.3dB). At 1.382 GHz frequency, as seen in Figure 4, the opposite

IOP Conf. Series: Materials Science and Engineering 1152 (2021) 012008 doi:10.1088/1757-899X/1152/1/012008

directions of the currents on both rings result in almost elimination of the current effect and thus leads to transmission (no blocking) through the FSS at the 1.382 GHz frequency.

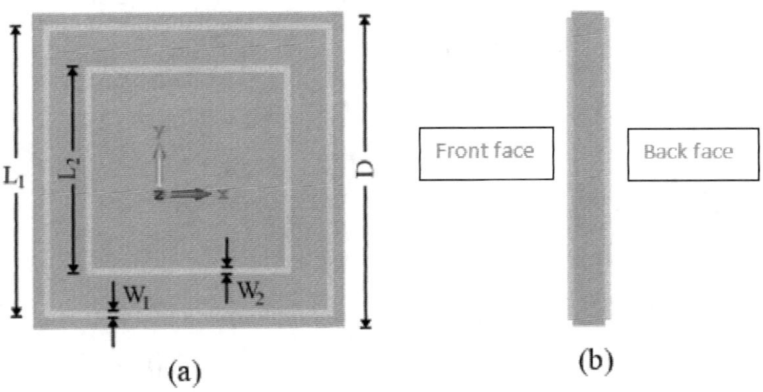

Figure 1. The two-sided FSS unit cell showing the double-square rings printed on both sides of the substrate (curtain). (a) front view, (b) side view.

Table 1. The calculated dimensions of the FSS cell compared to those obtained from the simulations. Dimensions are in mm.

Parameter	L_1	L_2	W_1	W_2	D
Calculated	65.65	1.4	46	1.4	70
Found from CST	65.5	1.4	46.5	1.4	70.5

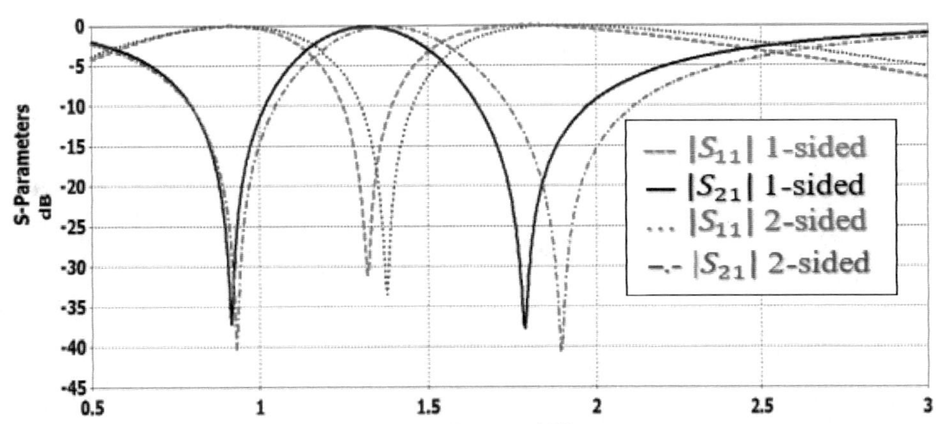

Figure 2. The S-parameters of the one-sided FSS compared to those of the two-sided FSS cells. The rings on the two sides are of similar sizes.

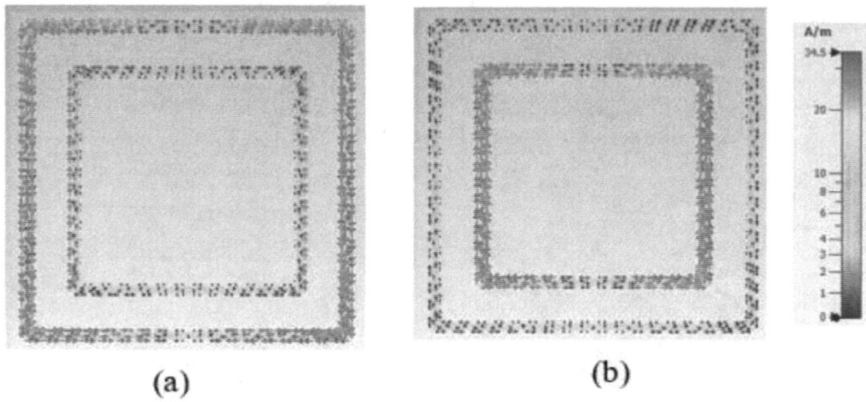

Figure 3. The current distribution on the surface of the double-square ring of two-sided FSS at; (a) 0.935 GHz, and (b) 1.9 GHz frequencies.

.

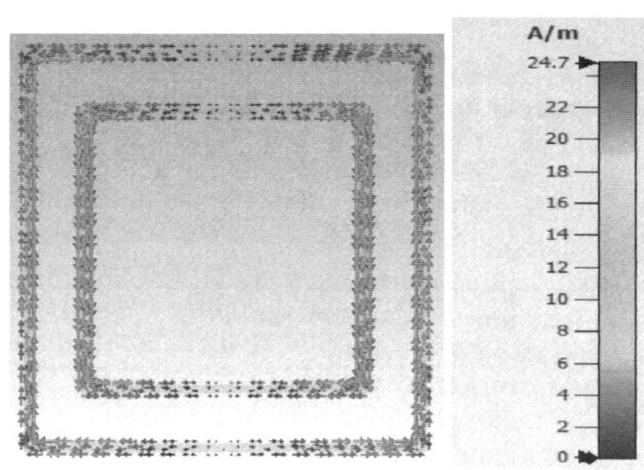

Figure 4. The current distribution on the surface of the double-square ring of two-sided FSS at 1.382 GHz frequency.

The two-sided FSS cell as shown in Figure 2 has a slight shift in the stopbands (0.86% and 6.44% at the first and second one respectively). The shift in the stop frequencies can be attributed to the mutual coupling between the double rings on one side with those on the other side. Table 2 lists the resonance frequencies that were obtained for the following designs of the unit cell:
a) double-square ring on one side only,
b) double-square ring on one side, and the inner ring on the other,
c) double-square ring on one side, and the outer ring on the other,
d) two-sided design having double-square ring on both sides.

Table 3 shows a comparison between the obtained S-parameters of the four FSS cells which are listed in Table 2. It can be seen that better shielding is obtained for the two-sided cell which has a lower S_{21} value by about 3dB as compared with the one-sided cell, and this is also evident in Figure 2.

Table 2. Values of the stop and pass frequencies for the shown four investigated FSS cells.

	1st stop frequency (GHz)	pass frequency (GHz)	2nd stop frequency (GHz)
double-square ring on one side	0.927	1.327	1.782
double-square ring on one side + outer ring on the other	0.935	1.357	1.777
double-square ring on one side + inner ring on the other	0.932	1.355	1.897
two-sided double square ring	0.935	1.382	1.9

Table 3. Reflection and transmission coefficients at the stop and pass frequencies for the shown four investigated FSS cells.

	@ 1st stop frequency		@ Pass frequency		@ 2nd stop frequency	
S-Parameter (dB)	S_{11}	S_{21}	S_{11}	S_{21}	S_{11}	S_{21}
double-square ring on one side	-0.12	-36.8	-30.7	-0.24	-0.12	-37.4
double-square ring on one side + outer ring on the other	-0.08	-40.2	-24.5	-0.26	-0.13	-36.3
double-square ring on one side + inner ring on the other	-0.12	-36.8	-41.7	-0.19	-0.07	-41.1
two-sided double square ring	-0.08	-40.6	-33.3	-0.18	-0.07	-40.9

For further insight into the operation of the 2-sided FSS cell, the FSS was modelled by an equivalent circuit comprising four parallel RLC circuits (each for one ring), as seen in Figure 5. The equivalent circuit is further developed by deriving formulas for the blocked and transmission frequencies. The substrate is represented here by a short section of transmission line having length h that is separating the two pair of rings that are printed on the two sides of the substrate. As the substrate is thin (h=1 mm), then his transmission line is very short (of length ≈ 0.01λe). Therefore, the impedance Z_{in} seen by the incident wave can be approximated by the parallel combination between Z_l and Z_{p2} (see Figure 5). The input impedance (Z_{in}) can thus be found as:

$$Z_{in} = \frac{Z_{p2}Z_1}{Z_{p2}+Z_1} \tag{3}$$

After some algebraic manipulations, the derived frequencies at which transmission is blocked through the FSS are:

$$\omega_1 = \frac{1}{\sqrt{L_1C_1}} = \quad \omega_3 = \frac{1}{\sqrt{L_3C_3}} \tag{4}$$

$$\omega_2 = \frac{1}{\sqrt{L_2 C_2}} = \quad \omega_4 = \frac{1}{\sqrt{L_4 C_4}} \tag{5}$$

Where the double rings (outer and inner rings) on the right side of the FSS has the stop frequencies ω_1, and ω_2, while the left side rings (outer and inner rings) of the FSS has stop frequencies of ω_3, and ω_4. The derivations were developed to find the two transmission frequencies for the right side and left side of the FSS cell which are ω_5, and ω_6 respectively. The FSS cell can be designed to have similar pair of rings on the two sides, so that:

$$\omega_5 = \omega_6 = \omega t$$

$$\omega_t = \sqrt{2}\,\frac{\omega_1\omega_2}{\sqrt{\omega_1{}^2 + \omega_2{}^2}} = \sqrt{2}\,\frac{\omega_3\omega_4}{\sqrt{\omega_3{}^2 + \omega_4{}^2}} \tag{6}$$

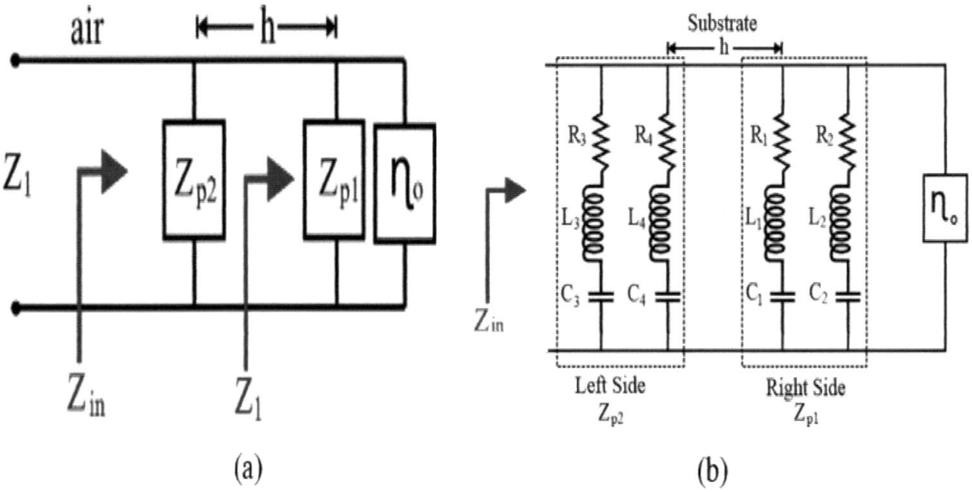

Figure 5. Equivalent circuit representing the two-sided double-square ring FSS cell (a), Equivalent lumped-element circuit of the FSS (b).

In the fabrication of the two-sided FSS cell, the rings on both sides may not be aligned perfectly, and there may be a shift, whose effect is investigated here. Figure 6 shows the scattering parameters of the FSS cell that were obtained from the simulations when there is a shift of 1mm and 2 mm in the diagonal direction to the rings, compared to the case where there is no shift. The first stop frequency is shifted slightly from 0.935 GHz to 0.917 GHz, and then to 0.892 GHz for the 1 mm and 2 mm shifts respectively. The -10 dB bandwidth has increased from 254 MHZ to 264 MHz and then to 271 MHz as the shift was increased from 0 mm, to 1mm, and then to 2mm. Thus, the spatial shift in the rings results in a small shift in the stopband frequency and a small increase in the response bandwidth.

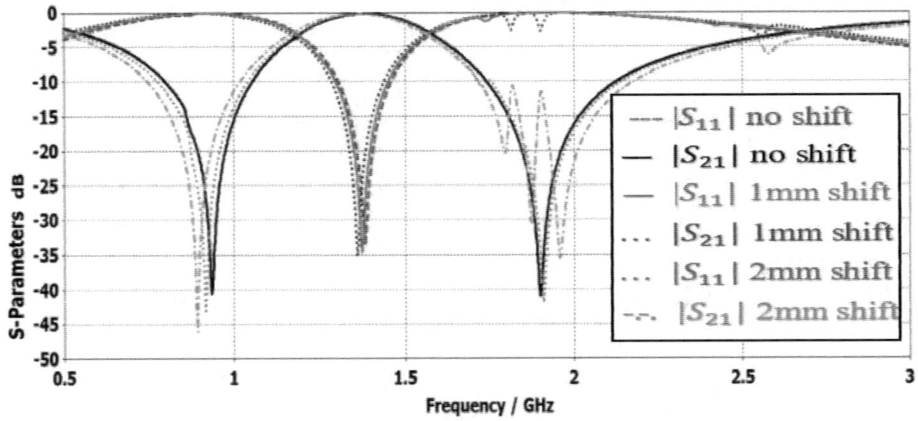

Figure 6. The reflection and transmission coefficients of the 2-sided double-square ring FSS cell, when the two pair of rings diagonally shifted by 0 mm, 1 mm, and 2 mm. The E-field of the incident wave is along Y-axis

Figure 7 shows the transmission coefficient (S_{21}) for (the double-square ring on each side) two-sided FSS cell for cases of;

a) the two sides have similar rings,
b) the rings on the other side are larger by 1%,
c) the rings on the other side are smaller by 1%,

The figure shows that the bandwidth has increased for both cases when the rings on the other side larger or smaller sizes. The bandwidth slightly changed (3.15 %), which can be attributed to the two resonances corresponding to the rings on both sides. Figure 7 also shows a slight change of about 2 % in the first stop frequency when the size of the rings on the other side was changed by 1 %. As there are two pairs of rings with different sizes, then each pair resonate at a different frequency. The overall response is a combination of the responses of the two pairs, and thus it will have an increase in the bandwidth and a shift in the centre frequency. The 2 % and 4 % differences in the sizes of the rings have shown the same trend, as can be seen from Figures 8 and 9, which show the S-parameters for the FSS cell for the two-sided FSS.

To explore the effect of farther differences in the size of the rings, the sizes of the rings on the backside were reduced by 15 % and 20 %, and the obtained S-parameters are shown in Figure 10 and Figure 11, respectively. The single passband in between the two stop frequencies has now split into two adjacent passbands. As the reduction factor was increased from 15% to 20%, the former two passbands have become three bands as shown in Figure 11. These results show that the sizes of the double rings on the other side can be chosen to obtain more than one passband while preserving the two reject bands.

4. Conclusions

A two-sided FSS formed of concentric double square conducting rings has been demonstrated to shield the GSM900/ 1800 MHz frequency bands. The FSS is built by printing copper rings on a fabric sheet that represents a curtain for offering a shield across windows. The CST simulations show an attenuation of better than 40 dB at the two GSM bands. The FSS was represented by an equivalent circuit comprising four parallel resonating-circuits, and the estimates of the blocked and transmit frequencies are in agreement with those found from the simulations. The bandwidth of the stopbands can be increased by employing rings of slightly different sizes on the two sides of the substrate. The larger difference in the size can lead to split the passband into two or three sub-bands.

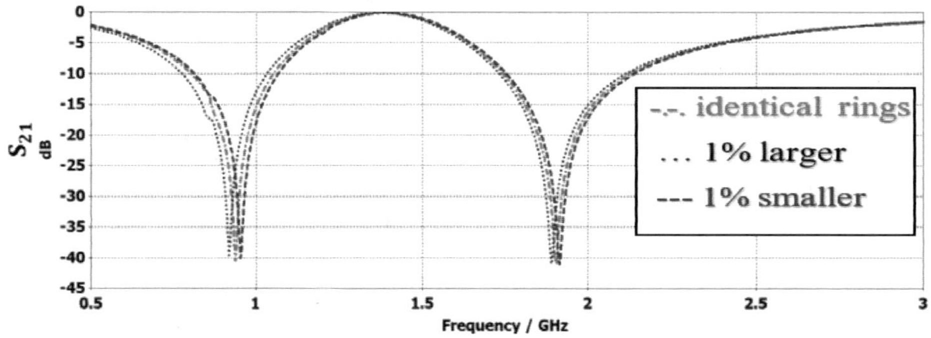

Figure 7. The obtained transmission factor of the 2-sided FSS cell for; the two pairs on both sides have similar sizes (-.-.-), the rings on the backside are 1% larger (…..), the rings on the backside are 1% smaller (-----).

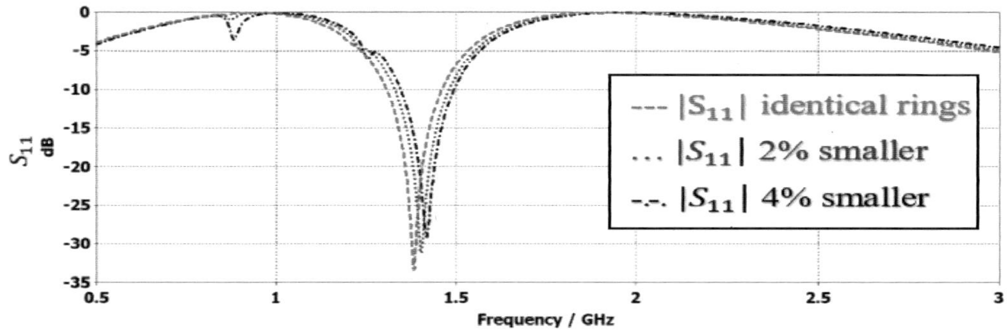

Figure 8. The obtained reflection coefficient of the 2-sided FSS cell for; the two pairs on both sides have similar sizes (----), the rings on the backside are 2 % smaller (…..), and 4 % smaller (-.-.-.).

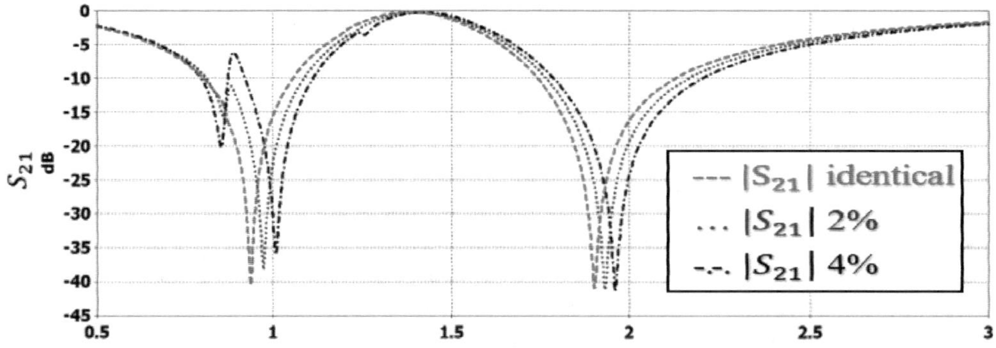

Figure 9. The obtained transmission factor for the 2-sided FSS cell for; the two pairs on both sides have similar sizes (----), the rings on the backside are 2 % smaller (…..), and 4 % smaller (-.-.-.).

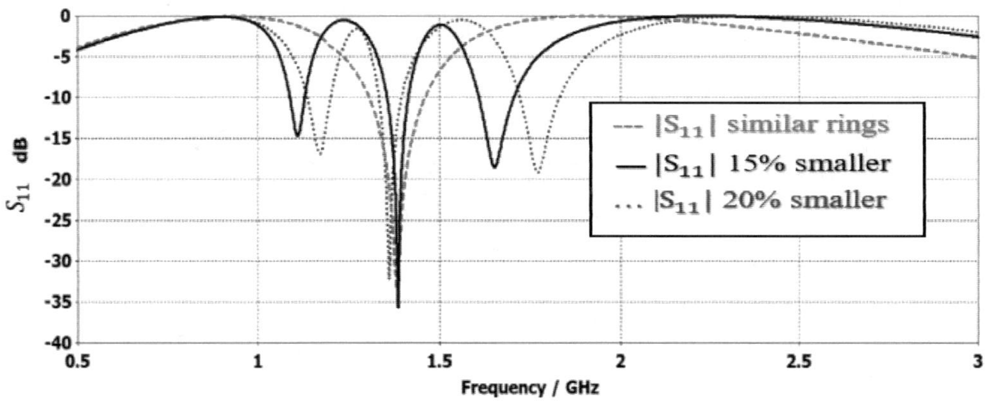

Figure 10. The obtained reflection coefficient of the 2-sided FSS cell for the cases; the two pairs on both sides have similar sizes (----), and the other side rings have 15 % smaller (___), and 20 % smaller (…..).

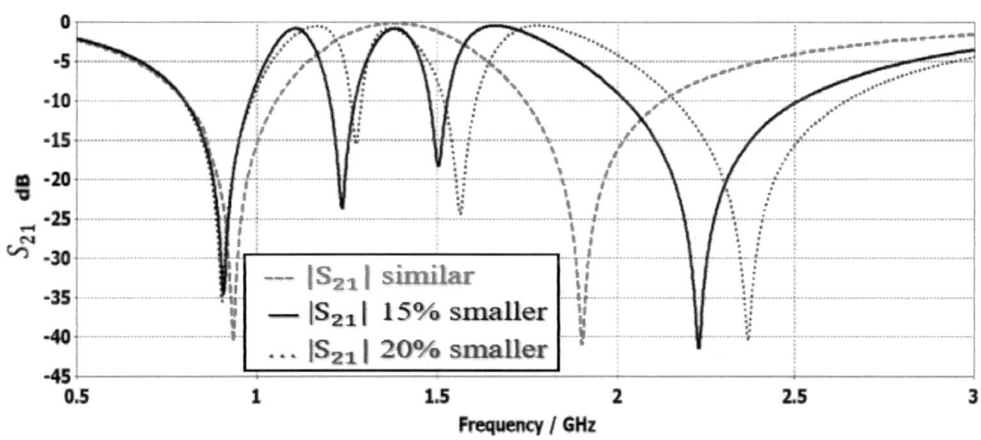

Figure 11. The obtained transmission factor of the 2-sided FSS cell for the cases; the two pairs on both sides have similar sizes (-----), the other side has rings has 15 % smaller (___), and 20 % smaller (…..).

5. References

[1] Mayouf A, Sayidmarie K H, and Ali Y 2020 A dual stopband frequency selective surface for mobile shielding applications 2020 *Proceedings of the 1st International Multi-Disciplinary Conference Theme: Sustainable Development and Smart Planning, Cyperspace*, pp.818-827, 28-30 June 2020.

[2] Mayouf A, 2020 Investigation of Frequency Selective Surfaces (FSS), *MSc Dissertation*, University of Mosul, Iraq.

[3] Munk B A 2000, *Frequency Selective Surfaces: Theory and Design* 2000, (Wiley and Sons).

[4] Sohail S I, 2016 Wi-Fi transmission and multi-band shielding using single-layer frequency selective surface, *IEEE Int. Symp. Antennas Propag.*, pp. 963–964.

[5] Manaa Y, and Aldhaheri R W, 2017 Dual-Band frequency selective surface for GSM shielding in modern buildings. *Aljouf University Science and Engineering Journal (AUSEJ),* Vol. 4, No. 2.

[6] Al-Atrakchii M A, Sayidmarie K H, Abd-Alhameed R A, 2020 Frequency selective surface using the metamaterial property of the U-Shaped strip 2020, *Proceedings of the 1st International Multi-Disciplinary Conference Theme: Sustainable Development and Smart Planning, IMDC-SDSP 2020, Cyperspace*, pp.818-827, 28-30 June 2020.

[7] Al-Atrakchii M A, Sayidmarie K H, Abd-Alhameed R A, 2021 A band pass frequency selective surface using U-shaped slots, 2021 *International journal of microwave and optical technology,* Vol. 16, No. 1, 2021, pp. 36-43.

[8] Kiani G I, Olsson L G, Karlsson A, Esselle K P, and Nilsson M 2011 Cross-dipole bandpass frequency selective surface for energy-saving glass used in buildings, *IEEE Transactions on Antennas and Propagation*, Vol. 59, No. 2, pp. 520-525, Feb. 2011.

[9] Vallecchi A, Langley A, and Schuchinsky A, 2017 Bi-state frequency selective surfaces made of intertwined slot arrays", *IEEE Transactions on Antennas and Propagation*, Vol. 65, no. 6, pp. 3093-3101.

[10] Singh J, Najim M, Agarwala V, Singh D, and Varma G D 2016 Critical analysis of frequency selective surfaces for dual band GSM-900 & 1800 MHz transmission. *Proceedings of Recent Advances in Electronics and Computer Engineering Conference RAECE 2015*, 13- 5 Feb. 2015, pp. 207–210.

[11] Bialkowski M E, and Sayidmarie K H, 2008 Investigations into phase characteristics of a single-layer reflectarray employing patch or ring elements of variable size, *IEEE Transactions on Antennas and Propagation*, Vol. 56, No. 11, PP. 3366-3372.

[12] Bal K and Kothari V K, 2010 Permittivity of woven fabrics: A comparison of dielectric formulas for air-fiber mixture, *IEEE Transaction on Dielectric Electrical Insulation*, vol. 17, no. 3, pp. 881–889.

MIMO to Enhance the Performance of VLC Systems

Ahmed Taher Ali [1], Safwan Hafeedh Younus[2] and Serkan Ozbay[3*]

[1]Department of Electrical and Electronics Engineering, Gaziantep University, Gaziantep, Turkey.
[2]College of Electronics Engineering, Ninevah University, Mosul, Iraq.
[3]Department of Electrical and Electronics Engineering, Gaziantep University, Gaziantep, Turkey.

[*]E-mail: sozbay@gantep.edu.tr

Abstract. In general, visible light communication (VLC) systems run at low data rates. By introducing multiple input and multiple output (MIMO) techniques, service at enhanced data rates can be viable for VLC systems. This work proposes MIMO technique to attain VLC system with high data rates. To attain MIMO method this system includes a number of photodiodes with a narrow field of view (FOV), which highlights the benefit of using MIMO technology in improving the data rate. The results show that using MIMO technique leads to enhance the VLC system when it is compared with the traditional system (without using MIMO). The results show that an improvement in the signal-to-noise ratio (SNR) level, which is the most attempt to determine communication quality, increasing the channel bandwidth and reducing the delay spread are achieved when MIMO technique is used.

1. Introduction

Visible light communication (VLC) utilizes the spectrum of light from 390 nm to 700 nm as a data communication medium. Light emitting diodes (LEDs) are used in VLC systems for communication along with illumination purposes [1]. Many applications, such as smart lighting, indoor communication, communication from vehicle to vehicle (V2V) and from vehicle to infrastructure (V2I), have been developed. Some studies often combine VLC with other strategies for communication. To increase the system bandwidth of conventional WiFi, WiFi-VLC hybrid indoor communication systems were proposed in [2]. VLC may be used as an alternative option because of the limitation of radio frequency (RF) in some locations, such as hospitals, ships, gas stations, etc. The demand for high-speed and ubiquitous wireless broadband connectivity has stimulated a massive increase in mobile data traffic. In combination with the worldwide adoption of social media and advanced multimedia applications, the growing number of mobile devices in various form factors and capabilities are the key contributors to this phenomenal growth. It proposed an optical wireless communication system that uses white lights for wireless indoor networks [3],[4]. A VLC system is the ideal choice for optical MIMO in respect of parallel transmission, as multiple LED lights are utilized for lighting and can transmit various data streams to optimize the throughput of each single LED [5],[6]. At the same time, there is a need for an array of photodiodes on the receiver side, and this configuration allows security, connection range and different distributors improvements, while the power needed is unchanged [7]. The existence of a large number of low bandwidths LEDs, makes MIMO techniques an appealing choice to achieve high data rates, and this paper explores the potential of this technique. MIMO is widely utilized in radio communications [8], where channels that are

Content from this work may be used under the terms of the Creative Commons Attribution 3.0 licence. Any further distribution of this work must maintain attribution to the author(s) and the title of the work, journal citation and DOI.
Published under licence by IOP Publishing Ltd

decorrelated around between are produced by dispersion and interference. This enables MIMO streams to provide greater capability than the counterparts of the single input single output (SISO), which provided a set total transmission power level. For short range systems the MIMO approach in indoor systems modeling is shown and the potential of the MIMO systems has been studied and a low-speed demonstration has been published [9], [10]. Working on space-time codes for MIMO is detailed in [11]. In this work the focus has been on the parallel data transfer by applying the MIMO technology approach. In the advanced generations of communication networks, MIMO technology has become attractive and more common in an increase a data rate. In this paper, the MIMO technique is suggested to increase the overall performance of the system. A significant improvement is being accomplished in SNR using the MIMO technique at high data rates. This significant enhancement in the SNR rate is attributed to the potential of the receiver to gather the VLC signal with reduced ISI using several photodiodes (PDs) due to its narrow FOV. However, due to the use of multiple receiving components, the MIMO technique has several drawbacks, including high cost and complexity compared with a traditional VLC system. As the increase in the elements or components of any system leads to its complexity. Thus, in this paper to keep the complexity at an acceptable level we used 4 x 4 MIMO VLC system, which leads to enhance the performance of VLC system. In this paper, the VLC non-return zero on-off keying (NRZ OOK) method is modeled and simulated by using white LEDs as transmitters. The VLC system simulation is performed using Matlab software. The section is organized as follows into parts. Section 2 gives the system set-up. Section 3 shows the simulation results. The conclusions are given in section 4.

2. INDOOR VLC GEOMETRICAL MODEL

In terms of some numerical analyses, we will address the potential implementation of the proposed system. In this work, an unfurnished environment was assumed to investigate our system, which was utilized in [4]. The room's size is 5m x 5m x 3m as shown in Figure 1. The room's lighting was provided by four lighting units which were used to attain an illumination level of 300lx to confirm that ISO and European standards [4]. The LEDs were placed at a distance of 2.5 m from the floor. The height of the receiver plane where the transceivers related with the user device were located at 0.85 m above the floor. In each light unit, there were (60 x 60) 3600 LEDs. The distance between each two adjacent LEDs was 1 cm. The half angle at half power of each LED chip is 70°, the center light intensity of an LED chips 0.73 cd. The optical power transmitted by an LED chip is 20 mW [4].

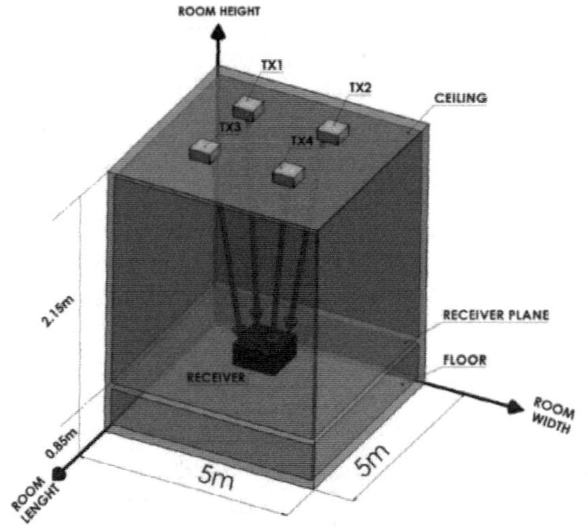

Figure. 1. Room configuration.

Reflections back to the second order was well-thought-out of this work. To model the reflection similar to that used in [12] and [13], we used ray tracing. The room is divided into many small areas for ray tracing and each area represents a second transmitter [14]. Each small area is 3.9 cm^2 in size, with a coefficient of reflection of 0.8 [4].

To depict the benefit of using MIMO, we compared our system with the traditional VLC system (without MIMO). Thus, two types of optical receiver have been used, the first being a single photodiode (PD) receiver with a wide FOV and a photo sensitive area of 1 mm^2, which was utilized with the traditional system. We assume small area of the photodetector, so the optical receiver can work at high data rates. The relationship between the area of the photodetector and the bandwidth can be found in [6]. For the MIMO system, we utilized a matrix of PDs, which consists of four PDs with FOV of 50° in order to highlight the benefit of using MIMO technology in improving the system performance and thus improving the data rate, which is the main objective of this research. Parameters were utilized in simulation are shown in Table I. Gfeller and Bapst [14] investigated the reflection coefficients of a variety of popular indoor materials. The reflection coefficients were found to range from 0.4 to 0.9. They also discovered that an ideal Lambertian pattern accurately approximated the power reflected by elements on the walls or ceiling. Thus, the reflection elements on the walls are viewed as a small transmitter in this paper, transmitting an attenuated version of the signals received from its center in a Lambertian pattern. In [4], authors investigated that the rate of reflected light is sufficiently low as compared to directed light. As a result, in VLC, the effect of directed light is significant, and it has a significant impact on system performance. For the sake of system analysis, we consider until the first reflection in this paper and neglecting higher reflection ranks from first reflection such as second reflection and beyond due to their very low impact on the system.

Table I: Simulation Parameters

Parameters	Values
Room Size	5.0 x 5.0 x 3.0 m^3
TX Location	TX1(1.25m,1.25m,3m) TX2(3.75m,1.25m,3m) TX3(1.25m,3.75m,3m) TX4(3.75m,3.75m,3m)
Transmitted optical power	20 mW
Semi angle at half power	70°
The LED number	3600 (60 × 60)
LED light size	0.01 m
Physical detector area of a PD	1.0 mm^2
Gain of an optical filter	1.0
Lens refractive index at a PD	1.5
Background light current	5100 μA
Noise bandwidth factors	0.562
PD responsivity	0.54 A/W
Noise current density	2.7 pA/$\sqrt{HZ}$

3. Results of Simulation

In the presence of propagation scattering and mobility, we assessed the results of the designed systems (MIMO-VLC system and traditional system) using a wide FOV receiver and a matrix of PDs receivers in an empty room. In eight separate places, the proposed structures were tested as the receivers traveled along the y-axis. On the basis of impulse response, delay spread, 3 dB channel BW, path loss and SNR rate, the results are described. The results for x = 3 and x = 4 are identical, depending on the similarity of the room, to the findings for x = 1 and x = 2 respectively, so only the results for x = 1 m and x = 2 m can be seen along the y-axis.

3.1. Analysis of Impulse Response

Figure 2 illustrates the impulse responses of the two systems (traditional VLC and MIMO-VLC systems). While the receivers were placed at the center, the impulse responses were obtained. It can be noticed that when the receiver is located in the center, the impulse response of each transmitter in the MIMO-VLC system is the same. This is due to the distances between the four LED lights (transmitters) and the receiver are equal. As it is seen in Figure 2(b), the impulse response of the MIMO system outperforms the traditional system in terms of signal distribution. The traditional system's impulse response (Figure 2(a)) involves several peaks relating to various process LOS components that arrive from various LEDs. The traditional system's impulse response also demonstrates that the component of LOS and first order reflection have a significant effect on the signal, as these components allow the signal to scatter over a wide temporal range due to the wide FOV of this receiver.

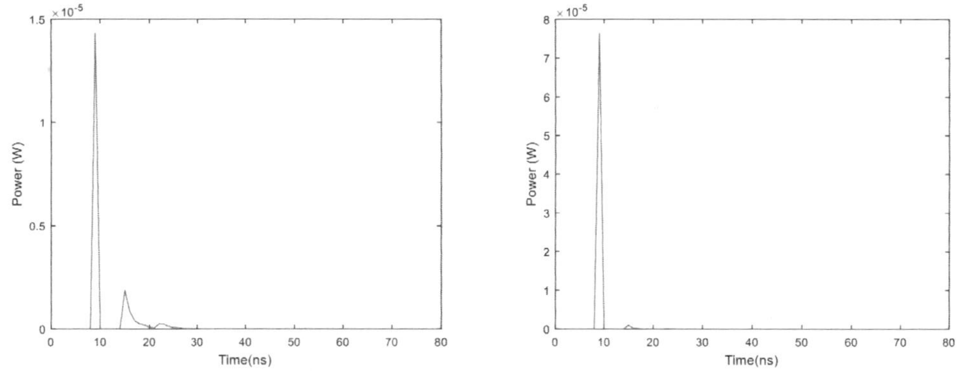

(a) (b)

Figure 2. Impulse response of the two systems at room center (x=2.5m, y=2.5m, z =0.85m). (a) traditional VLC system and (b) MIMO system.

3.2. Analysis of Delay Spread

The delay spread (D) offers a reasonable estimate of the susceptibility of the channel inter-symbol interference. The delay spread is given as [1]:

$$D_{rms} = (\mu^2 + (\mu)^2)^{1/2} \tag{1}$$

The mean excess delay (μ) is defined as:

$$\mu = \frac{(\sum_{i=1}^{M} P_{d,i}t_{d,i}+\sum_{j=1}^{N} P_{ref,j}t_{ref,j})}{P_{rT}} \tag{2}$$

and

$$\mu^2 = \frac{(\sum_{i=1}^{M} P_{d,i}t_{d,i}^2+\sum_{j=1}^{N} P_{ref,j}t_{ref,j}^2)}{P_{rT}} \tag{3}$$

M and N are defined here by the number of direct paths from LEDs to a given receiver and reflection paths towards the same receiver, $t_{d,i}$represents the time taken for a signal when travelling in a direct path from a transmitter to a receiver, $t_{ref,j}$ represents the time taken for a signal when travelling in a reflection paths from a transmitter to a receiver, $P_{d,i}$is the optical power received from the *ith*transmitter's direct path, and $P_{ref,j}$ is the optical power obtained from the *jth* surface element's reflected path. Due to LOS and NLOS components, P_{rT} is the total optical power obtained and is given as [1]:

$$P_{rT} = \sum_{i=1}^{M} P_{dir,i} + \sum_{j=1}^{N} P_{ref,j} \tag{4}$$

The delay spread linked to the two systems (traditional VLC system and MIMO system) is presented in Figure 3 while the optical receiver was walked on x = 1 m and x = 2 m. To receive and track each incoming signal from each transmitter separately, either directly or reflections from walls, a simulation was performed. For each transmitter, the difference in results is due to the fact that the FOV of the receiver permitted the receiver to receive a limited range of signals, which reduce the delay spread. Thus, in using the MIMO-VLC system instead of the traditional VLC, it offered great flexibility. In general, the results indicate that the delay spread of the MIMO system is lower compared to that of the wide FOV system. Therefore, it leads to an improvement in the performance of the system. This improvement is due to the limited range of signals received by the MIMO receiver that includes number of photodiodes have a lower FOV compare with the wide FOV. In traditional system, the highest delay spread is observed when the receiver is configured under one of the four transmitters (at position x= 1 and y= 1, 4), which are the worst points for delay spread, impacting the system's performance. This is a consequence of the signals arriving from other transmitters that reach the receiver for a longer time and a greater distance after traveling from the three other LED light units. While the lowest level of delay spread can be observed when the receiver is away from a certain distance from the lighting units, thus the distances traveled by the signal are reduced. For example, this consequence can be observed when the receiver is positioned in places (x= 1) and (y = 2, 3) where the delay spread level is observed and normalized at a certain level.

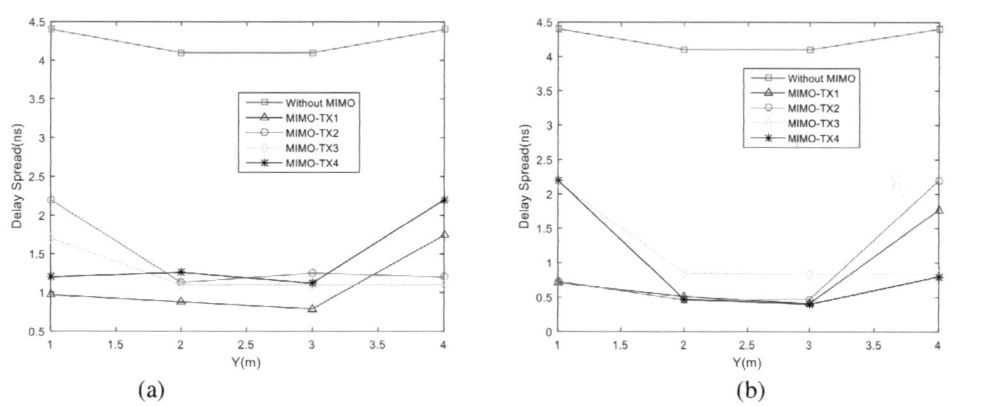

Figure 3. Delay spread of the two systems when the optical receiver was moved (a) along x = 1 m and (b) along x = 2 m.

3.3. Analysis of 3 dB Channel Bandwidth

The 3-dB channel BW estimates the amount of data rate which the indoor channel of the VLC systems can support. The channel bandwidth can be expressed as [1]:

$$B = 1/5D_{rms} \tag{5}$$

D_{rms} represents a delay spread defined in Equation (1). The 3 dB channel BW achieved by the two different systems is demonstrated in Figure 4. The results indicate that the MIMO-VLC system has a higher bandwidth relative to the traditional VLC system. The minimum communication channel bandwidth of the traditional system was 46 MHz at x = 1 m and y = 1 m. The results show that the usage of the MIMO system at each position contributes to an improvement in the 3-dB channel bandwidth, which leads to a high data rate enabled by the channel.

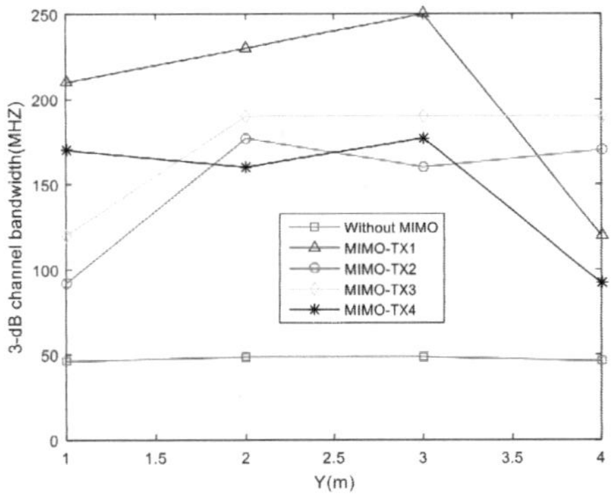

(a)

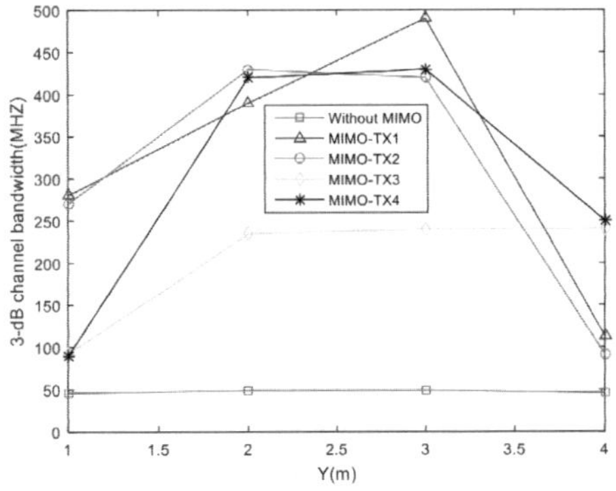

(b)

Figure 4. 3 dB channel BW of the wide FOV and MIMO-VLC systems when the receiver was moved: (a) along the y-axis and x=1, (b) along the y-axis and x=2.

3.4. Analysis of Path Loss

The path loss (*PL*) is identified as [7]:

$$PL = -10log_{10}(P_{rT}) \tag{6}$$

(P_{rT}) indicates the total power received by the receiver. The communication system path loss associated with the two systems is presented in Figure 5. With regard to the traditional VLC system, the results showed the stability of its level at close values at all locations. As in the traditional VLC system, the wide FOV was used and this enables it to receive a large range of signals, including unwanted signals resulting from reflections and ambient lights, as well as the effect of ISI, that does have a significant effect on the performance of the system. As far as the MIMO-VLC system is concerned, the results showed clear variations in the level of path loss at each location where the receiver is located. This variation is also due to the use of multiple photodiodes with a lower FOV than the wide FOV system, enabling a limited range of signals to be obtained.

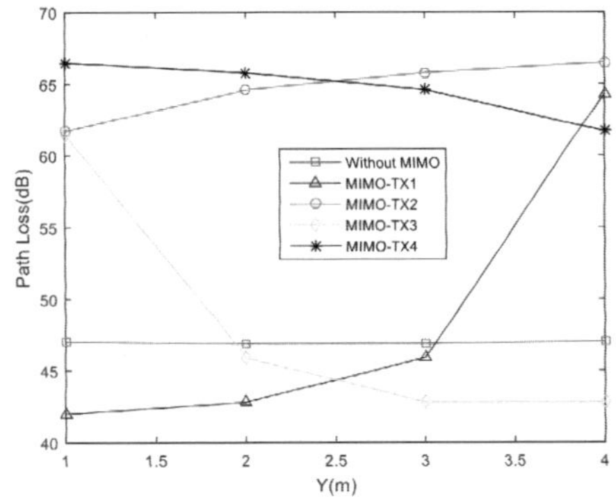

(a)

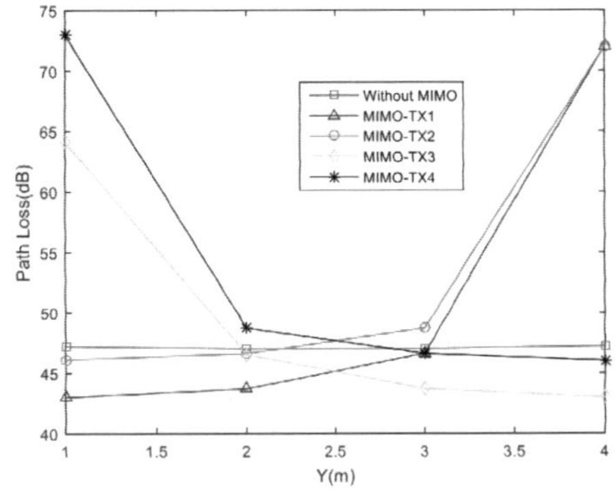

(b)

Figure 5. Path Loss of the two systems when the optical receiver was moved (a) along the y-axis and x=1, (b) along the y-axis and x=2

3.5. Analysis of SNR

The SNR correlated with the obtained signal can be determined by considering P_{s1} and P_{s0} in On-Off Keying (OOK), where P_{s1} represents power of logic1 and P_{s0} represents logic 0. The SNR is given as [15]:

$$SNR = \left(\frac{R(P_{s1}-P_{s0})}{\sigma t}\right)^2 \tag{7}$$

here σt represents standard deviation of the total noise. The total noise contains the shot noise, thermal noise and signal-dependent noise. R is the responsivity of the PD. The σt can be evaluated as [16]:

$$\sigma t = (\sigma_{shot}^2 + \sigma_{preamplifier}^2 + \sigma_{signal}^2)^{1/2} \tag{8}$$

here σ_{shot}^2 represents the component of the background shot noise, the component of the preamplifier noise is $\sigma_{preamplifier}^2$ and the shot noise associated with the received signal is σ_{signal}^2 .When working at 500 Mb/s, the SNR of the two systems (traditional and MIMO-VLC systems) can be seen in Figure 6. Using a MIMO-VLC system, a significant enhancement in the SNR rate was accomplished at 500 Mb/s. Moreover, it can be shown accurately that the MIMO VLC systems outperform the traditional VLC system. This is due to the reduction of the reflection element's impact through the use of a MIMO-VLC receiver that includes a number of narrow FOV (FOV=50°) photodiodes. To receive the appropriate signals and block the unwanted signals, the MIMO technology was used, which led to enhancing the SNR and performance of the system. A higher SNR gives our proposed system enough flexibility to significantly improve the data rate compared with the traditional system. It should be indicated that the MIMO-VLC system achieves around 12 dB SNR improvement across the traditional VLC system at the positions (1m, 1m, 0.85m) and (1m, 4m, 0.85m), which is the worst communication path for the MIMO-VLC system at x=1m.

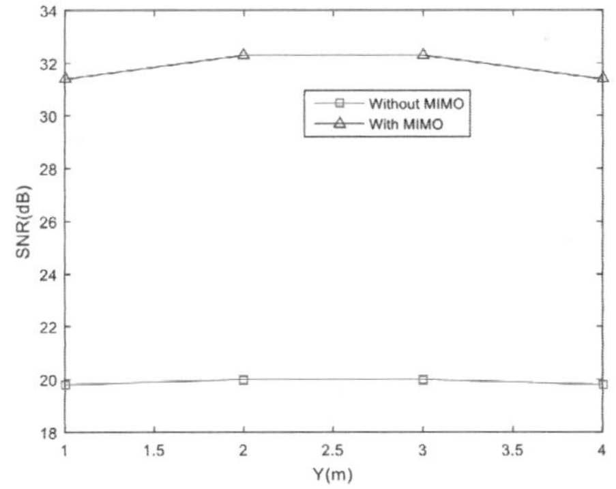

(a)

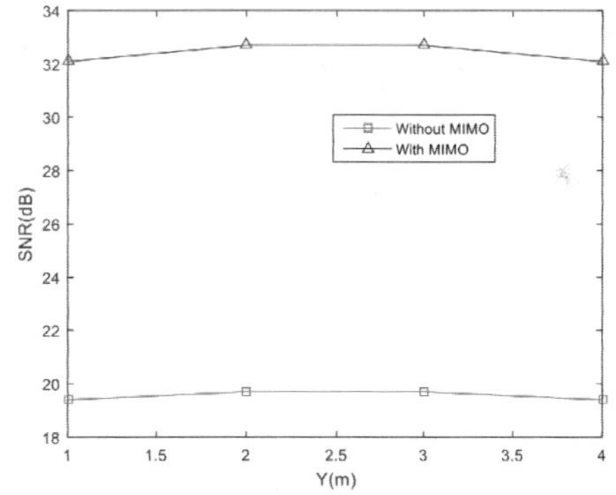

(b)

Figure 6. SNR of the two systems when the optical receiver was moved along (a) the y-axis and x=1 and (b) the y-axis and x=2

4. Conclusion

We introduced and developed high data rate single-user MIMO-VLC system in this paper and compare it with traditional VLC system. Analysis of delay spread, 3 dB channel BW and path loss demonstrated the effect of each transmitter on the signal received by the receiver when using the MIMO-VLC system. There are several different paths for these signals. The production of ISI was induced by multiple signal propagation, which clearly affects the system's performance. In the traditional system, the effect of ISI appears strongly, so it has been suggested to use the MIMO-VLC system to mitigate its effect. Compared to the traditional system, our proposed system helped raise and increase the SNR by about 12 dB in the worst communication path of the MIMO-VLC system. Improved SNR enhances the performance of the system and allows the system more flexibility to significantly improve the data rate.In general, all the results obtained showed an improvement in the system's performance compared to the traditional VLC system. In addition, the parallel transmission of data added to the system flexibility in increasing the data rate. The SNR levels show the perfect comparison between the two systems because it represents the most important factor in determining the quality of communications.

5. References

[1] Ghassemlooy Z, Popoola W and Rajbhandari S 2019 *Optical Wireless Communication*
[2] Wang Y and Hass H 2015 Dynamic Load Balancing with Handover in Hybrid Li-Fi andWi-Fi Networks *J. Light. Technol.* **33** 4671–82
[3] Harald H, Liang Y, Yunlu W and Cheng C 2016 What is LiFi? *J. Light. Technol.* **34** 1533–44
[4] Toshihiko K and Masao N 2004 Fundamental analysis for visible-light communication system using LED lights *IEEE Trans. Consum. Electron.* **50** 100–6
[5] Safwan H Y and Jaafar M H E 2017 WDM for high-speed indoor visible light
[6] communication system *Int. Conf. Transparent Opt. Networks* 1–6
[7] Safwan H Y, Aubida A A, Ahmed T H, Mohammed T A and Jaafar M H E 2019 Parallel Data Transmission in Indoor Visible Light Communication Systems *IEEE Access***7** 1126–38
[8] Safwan H Y, Ahmed T H, Mohammed T A and Jaafar M H E 2017 CGH for Indoor Visible Light Communication System *IEEE Access***5** 24988–5004
[9] David G, Mansoor S, Da S S, Peter J S and Ayman N 2003 From theory to practice: An overview of MIMO space-time coded wireless systems *IEEE J. SEL. Areas Commun.***21**208–302
[10] Svetla J, B A H and Mohsen K 2004 Power-efficient multispot-diffuse multiple- input-multiple-output approach to broad-band optical wireless communications *IEEE Trans. Veh. Technol.* **53** 882–9
[11] Steve H and Frank R. K 2006 A pixelated MIMO wireless optical communication system *IEEE J. Sel. Top. Quantum Electron.***12** 859–74
[12] Matt G, Chao L, Timothy P K and Kapil R D 2006 MIMO space-time coding for
[13] diffuse optical communication *Microw. Opt. Technol. Lett.***48** 1108–10
[14] Safwan H Y, Aubida A A and Jaafar M H E 2020 Multi-user High Data Rate Indoor
[15] VLC Systems *IETE J. Res.*
[16] Safwan H Y, Aubida A A, Ahmed T H, Mohammed T A and Jaafar M H E 2018 WDM for multi-user indoor VLC systems with SCM *arXiv*
[17] FRITZ R G and URS B 1980 Wireless in-house data communication via diffuse infrared radiation, RZ 941 (32513) *Inf. Syst.* **5** 248
[18] Safwan H Y, Aubida A A, Mohammed A and Ahmed T H 2020 Massive MIMO for indoorVLC systems *Int. Conf. Transparent Opt.* Networks **2020-July** 3–8
[19] John R B1994 Wireless Infrared Communications *Wirel. Infrared Commun.* 9219

IOP Conf. Series: Materials Science and Engineering 1152 (2021) 012010 doi:10.1088/1757-899X/1152/1/012010

A Method for Designing Unequal Power Divider for Implementing Antenna Array Excitations

Yaser Ahmed Ali[1], Ahmed Jameel Abdulqader[2,3] and Jafar Ramadhan Mohammed[3]

[1]Baiji Oil Training Institute, Ministry of Oil
[2]College of Engineering, Mustansiriyah University, Baghdad-10052, Iraq
[3]College of Electronics Engineering, Ninevah University, Mosul-41002, Iraq

[*]E-mail: ah_8611@yahoo.com

Abstract. In this paper, a method for designing a feeding network to obtain a low peak sidelobe pattern of an 8*1 linear antenna array is proposed. The optimized element weights were first obtained using genetic algorithm. Then two different networks based on equal and unequal power dividers are designed for the linear antenna arrays. In an equal power divider structure, the power is divided equally into all branches of the network to reach the final ports, while in the unequal structure; the division is only made equal in some branches and unequal in other branches to build an unequal power divider. In the two methods, the design process is based on two stages, the first stage uses a T-junction power divider, and the second stage a Wilkinson power divider is used. In an equal power divider, T-junction and Wilkinson divide the power equally, while in the case of an unequal power divider the T-junction divider is only unequal. In the two designs, the feeding network is designed according to the following specifications: the substrate (FR-4 with ϵ_r =4.3) and ground dimensions are 440.4mm x 103.8mm x1.6 mm and 440.4mm x 103.8mm x 0.035mm respectively, and at a resonant frequency of 2.36 GHz. The simulation results showed the effectiveness of the designed power dividers that to be used as a feeding network in the linear antenna arrays.

Keywords: Peak sidelobe pattern, equal power divider, unequal power divider, T-junction divider, Wilkinson power divider.

1. Introduction

Many modern communication applications, such as 5G wireless communications, satellites and radar require an antenna array with unique radiation characteristics such as high directivity and low sidelobe level. Reducing the high sidelobes is essential to suppress the interfering signals and maintaining the overall system performance at acceptable level of the output signal to noise plus interference ratio especially in the severe noisy environments that polluted with many signals which resulted in a very crowded spectrum. For instance, the tracking applications, require a low sidelobe to avoid the interfering from the other tracked targets [1]. The highest permissible limit for the sidelobe level is -13.2 dB in the uniformly excited feeding array (i.e., amplitude and phase excitations are ones and zeros respectively) [2]. However, this level is relatively high and it is not sufficient to avoid the interfering signals in the modern communication systems. Thus, this level should be reduced as much as possible. This can be achieved through unequal amplitude element excitations where it has maximum value at the centre of the array and gradually decaying at the edges of the array. The implementation of such taper in practice is most difficult task especially for large array sizes. This

Content from this work may be used under the terms of the Creative Commons Attribution 3.0 licence. Any further distribution of this work must maintain attribution to the author(s) and the title of the work, journal citation and DOI.
Published under licence by IOP Publishing Ltd

taper can be effectively configured by combining the unequal Wilkinson power divider and the T-junction power divider so that the output powers of each port is satisfying the required value of the amplitude element excitation.

Usually, in the equal power divider, the power is entered from a specific port, then it is divided into two or more ports, each of them carrying the same value of amplitude and all the ports are matched using a $\lambda/4$ transformer. The use of the equal power divider achieves uniform amplitude distribution on the array elements. Its array pattern has a peak sidelobe at -13.2 dB as mentioned before. In order to design a specific taper and then implement it practically, an optimization algorithm is first used to optimize the array amplitudes under some desired constraints such as the required side lobe level and undistorted beam width [3-8]. Then, the amplitudes of the optimized elements are implemented by using unequal T-junction power divider and Wilkinson power divider. The difference between equal and unequal power divider is through the power distribution of the port. In case of equal, the distribution is even, while in the case of unequal it is uneven [9].

Generally, there are several methods to reduce the sidelobes, and they can be divided into two parts, the first is based on optimizing the excitation of all elements using deterministic polynomials such as Dolph and Taylor [10-11]. The second method is based on the modification of the array excitation of some elements that located at the edges of the linear array [12-13]. In [14] the effect of amplitude errors, which is most important factor in practice, was investigated. The difference between the first and the second approaches is in the used strategy of designing the power divider. The first strategy requires a completely different design from the conventional equal power divider because the output of each port is different from the other, thus requires different calculations of impedances for each path of the output port. As for the second strategy, it is possible to rely on some port impedance paths whose output is of equal power, therefore this design is hybrid and requires fewer attenuators than the design in the first strategy.

Recently, many of the works in the literature have been concerned with reducing the level of sidelobes in the printed antenna array by relying on power divider methods such as parallel (or corporate) and series [15-17]. An innovative aperture printed array was introduced in [18], the array contained 100 microstrips distributed in two rows, each row of 50 elements with a distance of 0.15 between any two elements. In this case, the sidelobe level was found to be -20.9 dB. However, the size of the designed whole array was very large while the achieved reduction in the level of the sidelobe was not very low. In [19], Saputra and others introduced a novel feeding network to excite the elements at X-band with Dolph's distribution. The design was operating at a frequency of 9.3GHz and gave off -19.4dB. In [20], a lower sidelobe was obtained using an evolutionary algorithm to synthesis an array of 10 elements, after which a series feeding network was designed to excite the elements.

In this paper, a method for designing and implementing a specific taper (i.e., non-uniform amplitude excitations) that results in low side lobe level in the array pattern is presented. Two design strategies are introduced for the feeding networks of the linear array elements. The first strategy includes the use of the equal T-junction and Wilkinson power divider which is totally different from the conventional corporate equal power divider, whereas the second strategy includes the use of unequal power divider that realize the optimized amplitude excitations. The two designs are being built on a Microstrips board using CST 2019 version at the following specifications: 2.38 GHz frequency, FR4 substrate.

2. Architecture of the Designed Array

2.1 The 8:1 Equal Power divider

In a conventional equal power divider, the power is evenly divided among the ports that division out from the input port. In the case of a 2:1 power divider, the division is simple, especially in terms of matching, but by increasing the division branches leads to the further complexity of the feeding network and the problem of difficulty achieving the required matching appears.

In this section, the design of an equal power splitter is presented. The design is based on a T-Junction splitter with an equal Wilkinson splitter. The main advantage of this method is to deliver the impedance of 50 ohms to the point where Wilkinson is used, so that each port is integrated into an optimized Wilkinson dual port network. Thus, we can achieve a division of a power equally and appropriately. Finally, we can achieve the equal and matched power division. In the next two sub sections, the method of designing an equal power divider, which consists of two stages (T-junction and Wilkinson), is explained.

Equal T-junction Power divider

Since the weights are equal (i.e., amplitude and phase excitations are ones and zeros respectively) in the uniform antenna array, the T-junction design has an output amplitude proportional to the uniform weights. Figure 1 shows an equal T-junction power divider.

It is possible to find a relationship between the two power outputs P_A&P_B and the input power P_{input}, depending on the following equations:

$$P_A = \alpha P_{input} \tag{1}$$

$$P_B = (1 - \alpha)\, P_{input} \tag{2}$$

Where $0 < \alpha > 1$ The impedances are also calculated at each output port of the divider as follows:

$$Z_A = Z_o / \alpha \tag{3}$$

$$Z_B = Z_o / (1 - \alpha) \tag{4}$$

The α value decides the ratio of divided power at each output port. In the case of antenna arrays of uniform weights, the value of $\alpha = 0.5$ is chosen considering that the power division is equal. In order to simplify the feeding network for a future splitting (i.e., increasing the port branches), it is necessary to transfer the impedance in each output port to the input impedance. Therefore, a $\lambda/4$ transformer is usually used by which the resonant frequency of the feeding is determined in order to transfer the impedances (Z_A and Z_B) to the input impedance while maintaining the output power of each port.

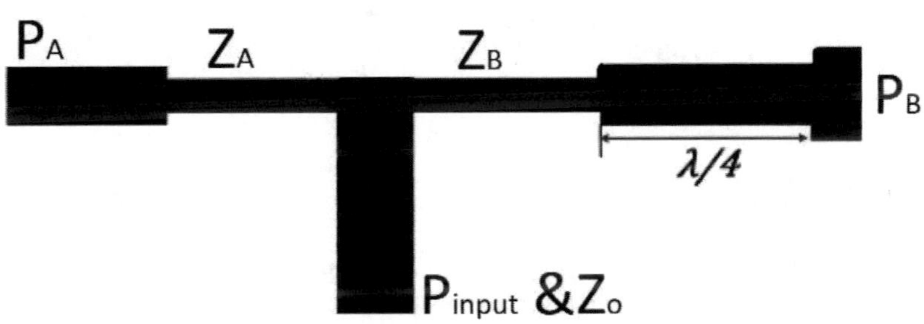

Figure 1. Equal T-junction power divider structure

Equal Wilkinson power divider

Equal Wilkinson power divider is the second stage in building an optimized integrated feeding network. In the case of the conventional equal Wilkinson power divider shown in [16], the calculation of impedances is easy, but by increasing the output ports, the complexity increases. A Wilkinson splitter can be constructed of an equal power divider in the form of a microstrip line as shown in Figure 2.

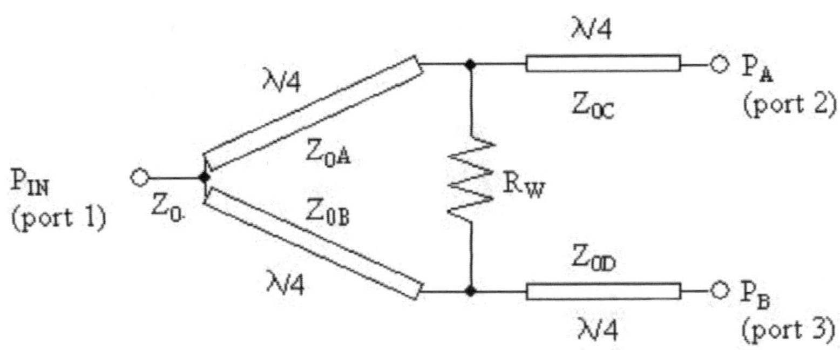

Figure 2. Equal Wilkinson power divider structure

The method of dividing the power between any two of the eight ports depends on setting a certain ratio of the power flow in any port, therefore, the equations shown in [17] can be used in the installation of the power divider as follows:

$$K^2 = \frac{P_A}{P_B} \quad (5)$$

$$Z_{0B} = Z_0\sqrt{\frac{1+K^2}{K^3}} \quad (6)$$

$$Z_{0A} = K^2 Z_{0B} = Z_0\sqrt{K(1+K^2)} \quad (7)$$

$$R_W = Z_0(K + \frac{1}{K}) \quad (8)$$

Here, a Wilkinson splitter is constructed by setting the value of k = 1, meaning that the ports have the same values of power.

2.2 The 8:1 Side Unequal Power divider

In this section, the methods used in the equal case are used to construct the unequal combination. The power outputs are changed to the ports located at the edges of the feeding network to achieve the desired peak sidelobe pattern level.

Unequal T-junction power divider

Since the weights of the elements in the array are predetermined using the theoretical methods. Designing a feeding network with an amplitude output proportional to the weights obtained from the optimization is achieved. In order to do this, an unequal T-junction divider is used as an initial step. Figure 3 represents an unequal T-junction power divider. Equations (1 to 4) are used in the calculations of the power values and impedances in the output of each port, with taking into account the control of the α value through which the unequal division is made.

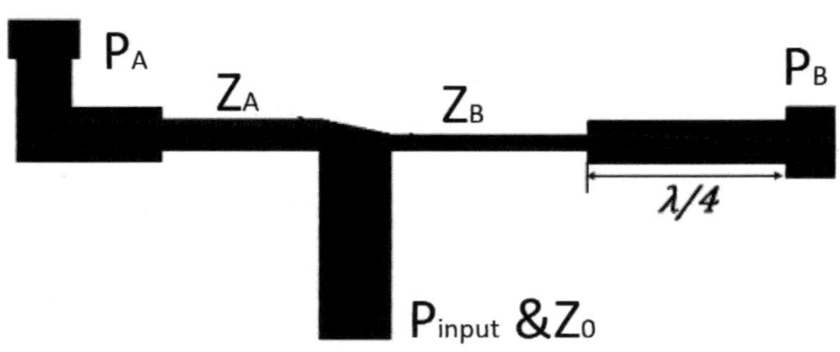

Figure 3. Unequal T-junction power divider structure

A side unequal power divider design includes several steps. As the design process of an unequal divider is implemented with three stages which are equal and unequal power dividers (the first and third stages are equal and the second is unequal). In the first and third stages, each port passes a power of -3dB, while the second stage, each port passes a power according to the designer at the value required to be designed. According to calculations and measurements, there are two important factors for developing an unequal power divider, the width of the strip line and the isolation resistance between the lines.

3. Simulation Results and Discussions

To validate the performance of the proposed power dividers, equal and unequal power feeding networks were designed and selected to be used in an 8*1 linear antenna array using CST Microwave

Studio with the following specifications: frequency of 2.36 GHz, substrate of FR-4 ($\varepsilon_r = 4.3$) with dimensions of (440.4 x 103.8 x1.6) mm^3 and ground of (440.4 x 103.8 x 0.035) mm^3. Simulation results for the two designs are exported and compared with the theoretical proposed results. As for the equal power divider, the excitations of the elements in the array need equal weights, thus, figure 4 shows the proposed uniform design. Table 1 shows the transmission line width dimensions of the parameters used in this design. The transmission line width equation shown in [18] has been adopted in calculating the transmission line width for our designs.

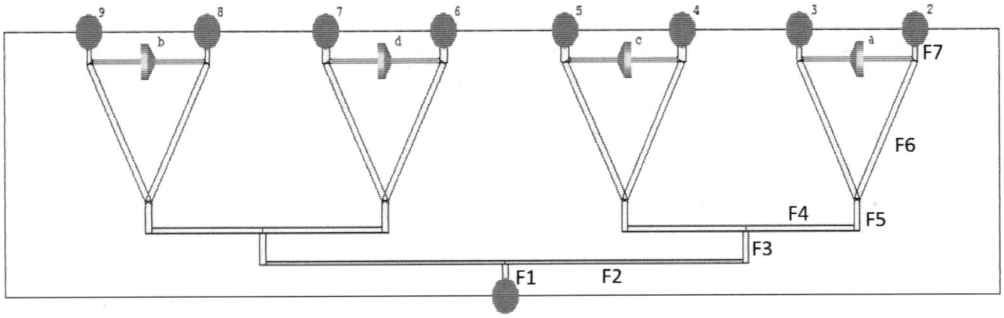

Figure 4. Complete structure of equal power divider

Table 1 Transmission line width of equal power divider structure

Transmission line width	F1	F2	F3	F4	F5	F6	F7
Value (mm)	3.1	1.6	3.1	2.4	3.1	3.1	3.1

As for unequal feeding network, figure 5 shows the design of an unequal power divider for an 8*1 linear array. Table 2 shows the dimensions of the parameters. Table 3 shows the values of the optimized weights of the elements in the array, on the basis of which an unequal feeding network was designed.

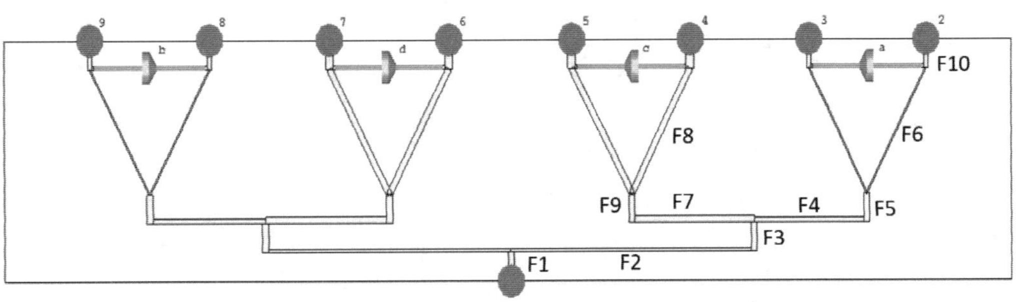

Figure 5. Complete structure of unequal power divider

Table 2 Transmission line width of unequal power divider structure

Transmission line width	F1	F2	F3	F4	F5	F6	F7	F8	F9	F10
Value (mm)	3.1	1.6	3.1	2.4	3.1	0.67	3.2	3	3.1	2.3

Table 3 Amplitude weights for 8×1 linear array with the inter-element spacing = 0.5λ (SLL = -20 dB)

Element No. (n)	**1**	**2**	**3**	**4**	**5**	**6**	**7**	**8**
Normalized amplitude (u_n)	0.5	0.5	1	1	1	1	0.5	0.5

It is observed through this structure that the elements in the centre of the array will be fed with equal powers as in the case of the first design. As for the elements on the sides, there is a need to design an unequal power divider in order to achieve the required perturbed weights depending on the unequal T-junction divider as well as the Wilkinson divider. In order to achieve an unequal power division, the value of the power ratio in the F4 branch is chosen with P4=0.25 P3 and the power in the F7 branch with P7=0.75 P3 (i.e., 3/4 of the power in the F7 branch compared to the F4 branch). Thus, the new powers are divided according to the next divisions, down to the end of the ports. In addition, the ports located at the edges of the feeding network that feeds the side elements will have less powers than the elements located in the centre.

Figure 6 shows the results of the S11 parameter for the proposed unequal power divider at 2.36 GHz, while Figure 7 shows the results of the S21 and S41. It is evident from Figure 5 that the branches fed with the same value of power will have the same values of S-parameters, meaning that the power parameter of the branches leading to the side elements carries the same amount of power as well as the centre branches. Therefore, S21=S31=S81=S91(the side ports) and S41=S51=S61=S71 (the centre ports). Table 4 shows the comparison of the theoretical element weights with the S-parameters results for each output port. It can be seen from this table the convergence of simulation results with the results of the assumed weights.

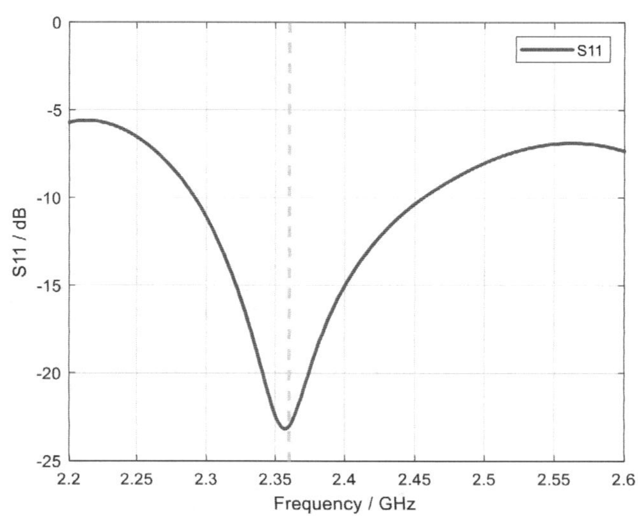

Figure 6. Simulated S_{11} for unequal power divider

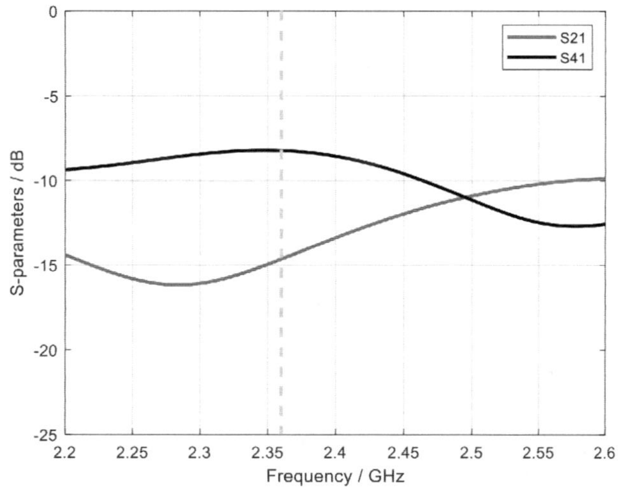

Figure 7. Simulated S-parameters for unequal power divider

Table 4 Comparison between assumed optimized weights and simulated S-parameters of the proposed unequal power splitter

S-parameter	**S_{21}**	**S_{31}**	**S_{41}**	**S_{51}**	**S_{61}**	**S_{71}**	**S_{81}**	**S_{91}**
Value (dB)	-13.82	-13.82	-7.75	-7.75	-7.75	-7.75	-13.82	-13.82
Value	0.204	0.204	0.41	0.41	0.41	0.41	0.204	0.204
Normalized S-parameter value	0.497	0.497	1	1	1	1	0.497	0.497
Normalized amplitude (u_n)	0.5	0.5	1	1	1	1	0.5	0.5

Finally, figure 8 shows the results of a comparison between the patterns in the case of equal and unequal divider output powers with theoretical pattern and how the unequal power divider was achieved in reducing the peak sidelobe pattern. It is clear that the simulated and assumed lines are completely homogeneous, and any difference between the patterns is a result of transmission line losses, which is not considered theoretically.

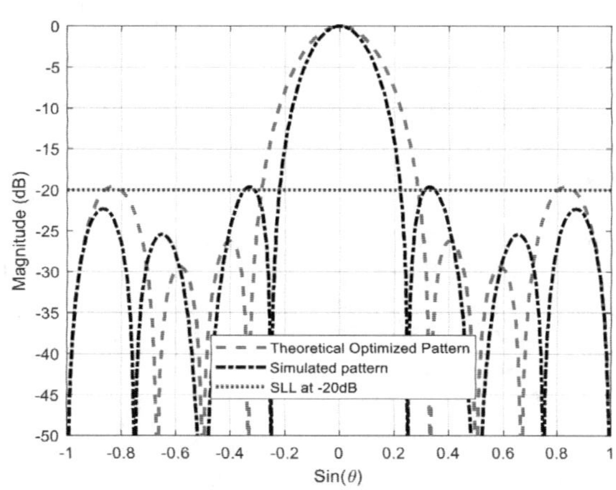

Figure 8 Comparison between the normalized theoretical and

4. Conclusion

In this paper, a method for designing the feeding network according to a specific weight distribution was implemented. The design considers a linear array that consists of 8 antenna elements in form of rectangular microstrips. The designed array was found to reduce the peak sidelobe pattern to -20dB with compared to the reference value -13.23 dB. Two methods have been proposed to fulfil the requirement of uniform and non-uniform linear antenna array weights by designing an equal and unequal power divider based on a T-junction power divider (equal and unequal) with an equal Wilkinson power divider. Simulation results showed that the output power in each port is proportional to the proposed weights resulting from the theoretical calculations, as shown in Table 4. Through these results, it has been proven that this array is a good candidate for applications such as point-to-point communication and a tracking system that requires low sidelobes.

5. References

[1] F. Y. Zulkifli, T. Hidayat, Basari and E. T. Rahardjo 2013 Sidelobe level suppression using unequal four-way power divider for proximity coupled microstrip antenna," 2013 Asia-Pacific Microwave Conference Proceedings (APMC), Seoul, pp. 1166-1168.
[2] C. A. Balanis 2016 Antenna Synthesis and Continuous Sources," in Antenna Theory Analysis and Design, N. Jersey, Ed., John Wiley & Sons, Inc, pp. 385- 421.
[3] Abdulqader, A, J., J. R. Mohammed, and R. H. Thaher 2020 Phase-only nulling with limited number of controllable elements," Progress In Electromagnetics Research C, Vol. 99, 167–178.
[4] Abdulqader A. J., J. R. Mohammed, and R. H. Thaher 2020 Antenna pattern optimization via clustered arrays Progress In Electromagnetics Research M, Vol. 95, 177-187.
[5] J. R. Mohammed, A. J. Abdulqader, and R. H. Thaher 2020 Array pattern recovery under amplitude excitation errors using clustered elements Progress In Electromagnetics Research M, Vol. 98, 183-192.
[6] Mohammed, J.R., "Simplified Rectangular Planar Array with Circular Boundary for Side lobe Suppression," Progress In Electromagnetics Research M, Vol. 97, page 57-68, October 2020.
[7] Mohammed, J.R., "Phased Sub-Arrays Pattern Synthesis Method with Deep Sidelobe Reduction and Narrow Beamwidth" IETE Journal of Research, Dec. 2020, DOI: 10.1080/03772063.2020.1853616.
[8] Mohammed, J.R., "Rectangular Grid Antennas with Various Boundary Square-Rings Array," Progress In Electromagnetics Research Letters (PIERL), vol.96, page 27-36, Jan. 2021.
[9] Lee D, Do H, Jang Y, Kim H, Han S-M, Lim J, Choi K and Ahn D 2016 A new design method of the unequal Wilkinsopower divider using an arbitrarily resistor value Microwave and Optical Technology Letters58(10) 2450–2452.
[10] Safaai-Jazi 1994 A new formulation for the design of Chebyshev arrays IEEE Trans. Antennas Propagat., vol. 42, pp. 439–443.
[11] Chatterjee, S., Chatterjee, S. 2014 Pattern synthesis of center fed linear array using Taylor one parameter distribution and restricted search Particle Swarm Optimization. J. Commun. Technol. Electron. 59, 1112–1127.
[12] Mohammed, J.R. 2018 Thinning a Subset of Selected Elements for Null Steering Using Binary Genetic Algorithm Progress in Electromagnetics Research M, vol. 67, pp. 147-157.
[13] Mohammed, J.R and Sayidmarie, K. H. 2017 Performance Evaluation of the Adaptive Sidelobe Canceller with various Auxiliary Configurations AEÜ International Journal of Electronics and Communications, vol. 80, pp. 179–185.
[14] Mohammed, J.R and and Sayidmarie, K. H. 2018 Sensitivity of the Adaptive Nulling to Random Errors in Amplitude and Phase Excitations in Array Elements International Journal of Telecommunication, Electronics, and Computer Engineering, vol. 10, no. 1, pp. 51-56.
[15] R. Bayderkhani and H. R. Hassani 2010 Wideband and low sidelobe slot antenna fed by series-fed printed array IEEE Trans. on Antennas and Propagation, vol. 58, no. 12, pp. 3898-3904.

[16] J. Yin, Q. Wu, C. Yu, H. Wang, and W. Hong 2017 Low sidelobe level series-fed microstrip antenna array of unequal inter-element spacing IEEE Antennas and Wireless Propagation Letters, vol. 16, pp. 1695-1698.
[17] R. Bayderkhani and H. R. Hassani 2009 Wideband and low sidelobe linear serises fed Yagi – like antenna array," Progress In Electromagnetics Research B, vol. 17, pp. 153-157.
[18] J. Lin, W. Shen, and K. Yang 2017 A low sidelobe and wideband series fed linear dielectric resonator antenna array IEEE Antennas and Wireless Propagation Letters, vol. 16, pp. 513-516.
[19] C. Niu, J. She, and Z. Feng 2007 Design and simulation of linear series-fed low sidelobe microstrip antenna array in Proceedings of Asia-Pacific Microwave Conference.
[20] Y. P. Saputra, F. Oktafiani, Y. Wahyu, and A. Munir 2016 Side lobe suppression for X-band array antenna using Dolph-Chebyshev power distribution in The 22nd Asia-Pacific Conference on Communications (APCC2016).
[21] M. R. Nikkhah, J. Rashed-Mohassel, and A. A. Kishk 2013 A low sidelobe and wideband series-fed dielectric resonator antenna array," presented at the in 21st Iranian Conference on Electrical Engineering (ICEE).
[22] Steer Michael 2019 Microwave and RF Design: Networks Volume 3. (Third Edition), NC State University.
[23] D R Khoirini, T Hariyadi and A B Pantjawati 2020 Design and realization of 4 way Chebyshev Wilkinson power divider at 3 GHz frequency for 3 dimensional radar application IOP Conf. Series: Materials Science and Engineering 850.

IOP Conf. Series: Materials Science and Engineering 1152 (2021) 012011 doi:10.1088/1757-899X/1152/1/012011

A Rectangular Monopole Filtenna with U-shaped Slot for 5G Applications

T H Al-Arajee[1] and K H Sayidmarie[1*]

[1] College of Electronics Engineering, Ninevah University, Mosul, Iraq

*E-mail: kh.sayidmarie@gmail.com

Abstract: A filtering antenna for 5G applications is proposed in this study, where the resonant element has a U-shaped slot embedded into the microstrip feed line so that no extra size is required. The proposedfiltenna has a very compact size of only 35 * 47 mm^2. Using CST Microwave Studio Suite, a comprehensive parametric analysis was done to evaluate and optimize the characteristics of the proposed filtenna. The filtenna is aimed for the 5G 3.4-3.8 GHz, as it covers a bandwidth of 412MHz (3.395 -3.807GHz) witha good reflection coefficient of -22.96dBat the center frequency. The U-shaped slot enhances the lower shoulder of the antenna frequency response and has no effect on the radiation at the desired band.

1. Introduction

In late 2016, the European Commission and the Radio Spectrum Policy Group (RSPG) identified the 3.4–3.8GHz band as the "primary band suitable for the introduction of 5G-based services in Europe even before 2020"[1], thus allocating this band for use in a rapidly growing system. This band offers adequate bandwidth for the 5G systems, and recent developments such as massive MIMO have further increased the importance of this band.

Filters and antennas are the most common and integral parts of all wireless communication systems, where transmitters and receivers are fitted with antennas. Some frequencies in the received signal need to be removed such that to reduce interference and/or noise. These signal processing functions are performed by filters that block particular frequencies while allowing others to pass.

The integration of filters and antennas, where the antenna acts as the filter's last resonator, is the filtering antenna. Filtering antennas can be used to reject a certain frequency band or to enhance bandwidth and selectivity [2].The design, fabrication, and testing of a compact filtering antenna with a second-order quasi-elliptic response was proposed in [3]. A U-shaped radiating patch is excited by an inset-coupled T-shaped resonator. The antenna achieved a fractional bandwidth of 2% at center frequency 5 GHz. In another work [4], two planar, efficient, wideband, compact, monopole filtennas were presented. The first one was developed from the common planar capacitive loaded loop (CLL)-based filter, while the second filtenna consisted of a driven element augmented with a CLL structure and with a slotted ground plane. The fractional bandwidth was expanded from 6.28% up to 7.9% for the first one at center frequency 2.34GHz.

In [5], a combined filtering planar dipole antenna having an edge-coupled feed was presented to show a second-order filtering response. The first stage of the filter is a U-shaped half-wavelength resonator that is followed by a simple planar dipole, whose bandwidth is controlled by the edge-coupling between the two stages. The antenna offered a21% bandwidth extending from 4.2 GHz to 5.2 GHz with a sharper roll-off than that of the conventional planar dipole. A low-profile planar dipole filtering antenna having an omnidirectional radiation pattern was presented in [6]. The antenna is

Content from this work may be used under the terms of the Creative Commons Attribution 3.0 licence. Any further distribution of this work must maintain attribution to the author(s) and the title of the work, journal citation and DOI.
Published under licence by IOP Publishing Ltd

formed of a planar dipole radiator, which is fed by a microstrip-to-slot transition structure. By adding non-radiative components, including a coupled U-shaped microstrip line and two I-shaped slots, to the feed network, the filtering characteristics are accomplished. Moreover, a novel dual-polarized filtering patch antenna without an extra circuit was suggested as filtenna[7]. With split-ring resonators (SRR) located within the antenna aperture, this proposed antenna achieved four radiation nulls. A wideband planar filtering dipole antenna is presented in [8]for 5G communication. Two rectangular radiating patches, that are fed by an inverted L-shaped feedline, compensate the dipole antenna. On the radiating rectangular patches, two complementary step impedance resonators (CSIRs) branches and two complementary split-ring resonators (CSRRs) are etched such that the antenna out of the operating band produces under and upper resonant zeros. The antenna has an operating frequency range of 3.08 to 5.15 GHz and gain > 2 dBi.

Low reciprocal coupling between antenna elements is required for the successful implementation of MIMO systems to reduce the envelope correlation coefficient (ECC). An open stub meandered (OSM) band-stop filter (BSF) architecture was used in [9]for MIMO applications, which implemented a mutual coupling suppression technique. A two-element patch array was optimized and applied on the ground-plane side of the OSM-BSF.

In this study, the analysis and design of a filtering antenna for the 5G (3.4-3.8GHz band) is presented. The antenna consists of a rectangular monopole fed by a microstrip line that offers more than the desired bandwidth. The filtering action is thus needed to limit the bandwidth to the desired 3.4-3.8 GHz range. The proposed filter is in the form of a U-shaped slot etched in the feed line, thus no extra space is needed, as seen in many conventional designs. The various stages of the design were investigated by the CST microwave Suite.

2. Monopole antenna

This section presents the monopole antenna aiming to cover 5G (3.4 to 3.8 GHz) band. The initial structure of the proposed antenna is shown in Figure1, where the FR4 (glass epoxy) substrate having dielectric constant ε_r = 4.3, loss tangent of tan δ = 0.025, and a thickness of 1.6 mm is used. The copper coating for the ground plane and monopole antenna is t=0.035 mm. The overall dimensions of the proposed antenna are L_s * W_s * h, where a rectangular radiating patch of L_p * W_p is denoted as MA1 and is connected to a 50-Ohm microstrip feed-line.

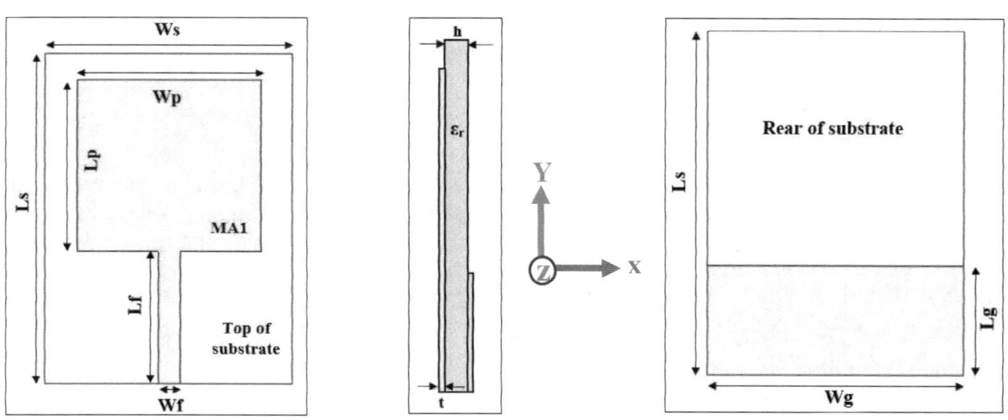

(a) (b) (c)

Figure 1. The geometry of the proposed monopole antenna; (a) front view, (b) side view, and (c) rearview.

The antenna parameters (patch width *W* and patch length *L)* can be found from the following equations (1- 5)[10]:

$$\varepsilon_{eff} = \frac{(1+\varepsilon_r)}{2} \quad (1)$$

$$W = \frac{c}{2*f_r} \times \left(\frac{2}{\varepsilon_r+1}\right)^{0.5} \quad (2)$$

$$\Delta L = 0.412 \times h \times \frac{(\varepsilon_{eff}+0.3)(W/h+0.264)}{(\varepsilon_{eff}+0.258)(W/h+0.8)} \quad (3)$$

$$L_{eff} = c/2 \times f_r \times \left(\varepsilon_{eff}\right)^{0.5} \quad (4)$$

$$L = L_{eff} - 2\Delta L \quad (5)$$

For the desired band of 3.4 – 3.8 GHz, the center frequency f_r= 3.6 GHz, thus, the patch width W_p= 25.6 mm, effective dielectric constant ε_{eff}= 2.65, and patch length L= 24 mm. The width of the feed line can be calculated from equation (6) to obtain line impedance of Z_o=50 Ohm[10]:

$$W/h = \begin{cases} 8e^A/e^{2A} - 2 & for\ W/h < 2 \\ \frac{2}{\pi}\left[B - 1 - \ln(2B-1) + \frac{\varepsilon_r-1}{2\varepsilon_r} \times \left(\ln(B-1) + 0.39 - \frac{0.61}{\varepsilon_r}\right)\right] & for\ W/h > 2 \end{cases} \quad (6)$$

Where

$$A = \frac{Z_o}{60}\sqrt{\frac{\varepsilon_r+1}{2}} + \frac{\varepsilon_r-1}{\varepsilon_r+1}\left(0.23 + \frac{0.11}{\varepsilon_r}\right) \quad (6.1)$$

$$B = \frac{377\pi}{2Z_o\sqrt{\varepsilon_r}} \quad (6.2)$$

Then equation (7) can be used to determine the characteristic impedance of the microstrip line[10] :

$$Z_o = \begin{cases} \frac{60}{\sqrt{\varepsilon_{eff}}}\left(\frac{8h}{W} + \frac{W}{4h}\right) & for\ W/h \le 1 \\ \frac{120\pi}{\sqrt{\varepsilon_{eff}}}\left[W/h + 1.393 + 0.667\ln(W/h + 1.444)\right] & for\ W/h > 1 \end{cases} \quad (7)$$

The calculated dimensions of the antenna are listed in Table 1. Figure 2 illustrates the effect of the ground plane length L_g or the effect of the gap length (L_f – L_g) between the radiating patch and the ground plane. This part of the feed-line works as a matching section between the radiating monopole and the 50-Ohm line. The matching is achieved by varying the gap length as has been used in [11,12, 13]. It can be seen that a shorter gap leads to shifting the frequency of best matching towards lower frequencies. The influence of the gap length on the frequency for best matching is larger than that on the bandwidth. The best value of L_g that realizes better matching at the required center frequency was found to be 8.8mm, where the minimum reflection coefficient S_{11} is -38.7 dB. However, the achieved bandwidth is 2GHz that is much larger than the desired value of 400 MHz.

The above design was developed by choosing other substrate parameters of (L_s=46mm, W_s=W_g=35mm, and L_f=18mm) to be compatible with the filtenna design which will be discussed in section 4. Simulations were then performed to find the best value for the gap length (L_f – L_g). The obtained reflection coefficient S_{11} is shown in Figure (3), the best value of L_g that realizes goodmatching at the required center frequency is 14.5 mm, where the reflection coefficient S_{11} is -32 dB and the bandwidth is approximately 2.28 GHz.

Table 1. The dimensions of the proposed antenna.

Parameter	Value in mm	Name
W_g	35.2	Ground plane width
W_f	3.1	Feed linewidth
W_s	35.2	Substrate width
W_p	25.6	Patch width
L_s	39	Length of the substrate
L_f	11.5	Length of the feed line
L_g	Variable	Length of the ground plane
L_p	24	Length of the patch
t	0.035	Thickness of copper

The simulated peak gain and the radiation efficiency of the designed antenna-1 are plotted in figure 4. The peak gain of 1.85 to 1.9 dBi in the operating band can be seen, while the efficiency is better than 85% across the band of operation, although a low-cost substrate of FR4 is in use. Higher values of efficiency can be obtained if another substrate of lower loss tangent has been used. The far-field radiation patterns in the E-Plane and H-Plane for the antenna are shown in Figure5 for the two frequencies of 3.5 GHz and 3.7 GHz. The antenna has the usual monopole pattern exhibiting the figure 8 shape in one plane and omnidirectional pattern in the other plane. The patterns are stable over the band of operation.

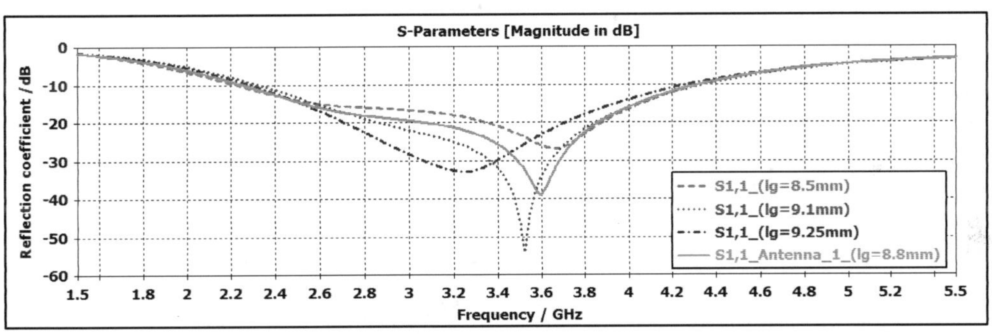

Figure 2. Variation of the reflection coefficient S_{11} with frequency for various values of L_g, (L_f= 11.5mm).

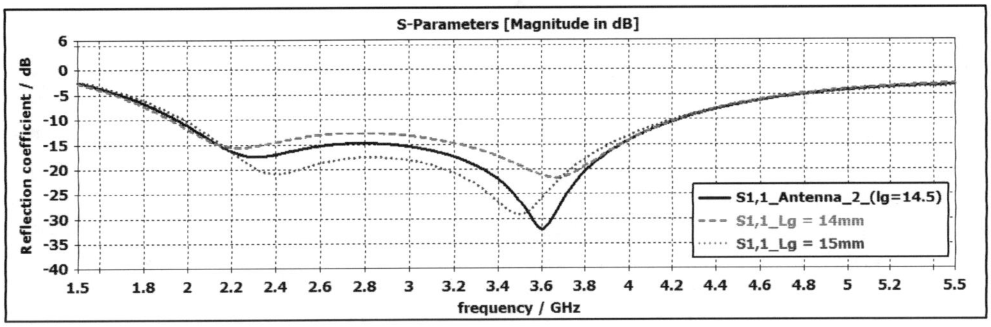

Figure 3. Variation of the reflection coefficient S_{11} with frequency for various values of L_g, (L_f= 18mm).

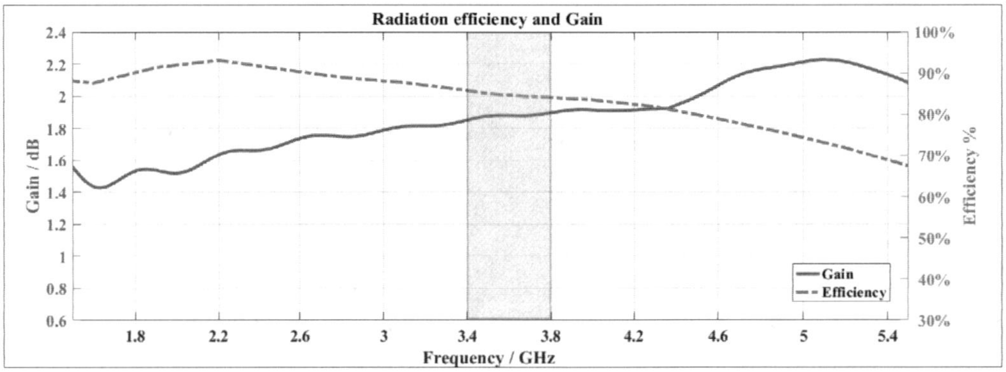

Figure 4. Variation of antenna gain and radiation efficiency with frequency for the monopole antenna-1, (Wp=25.6mm, Lp=24mm, and L_g = 14.5mm).

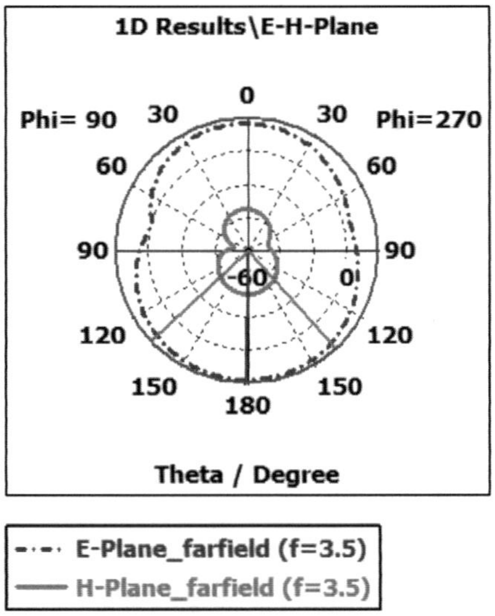

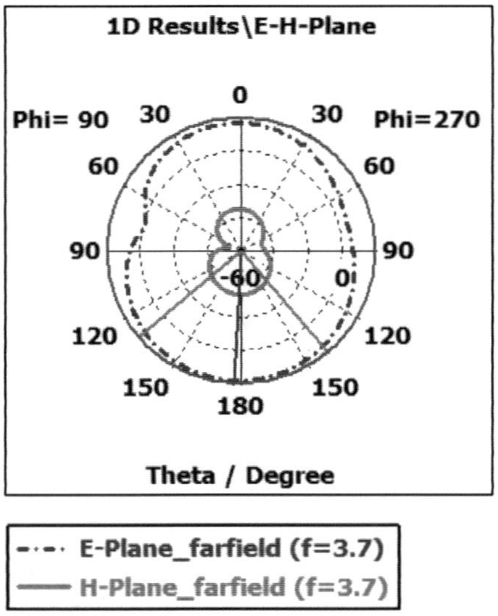

(a) (b)

Figure 5. Far-FieldE-Plane and H-Plane pattern;(a)at 3.5GHz(b) at 3.7GHz.

3. Filter design

The designed antenna offers the required radiation pattern, but it has much more than the required bandwidth. Thus, a filter is needed to limit the bandwidth to the required 400MHz.The proposed filter is to use a U-shaped slot etched on the microstrip line as shown in figure 6. This filter that was proposed in [14] has a compact design as it can be embedded into the microstrip line itself. This design was formerly used to achieve band-notched characteristics as in [12] and antenna isolation by filtering the coupled signal in[13]. The U-shaped slot has also shown a similar behavior in Frequency Selective Surface (FSS) applications [14]. The U-shaped slot resonates at a frequency f_r where its length equals multiple of the effective wavelength in the slot [14].

$$f_r = \frac{Nc}{2\times LS\times\sqrt{\varepsilon_{eff}}} \tag{8}$$

$$LS = 2Lu + d \tag{9}$$

Where N is an integer, c is the speed of light, Ls is the slot length, and d is the separation between the two legs of the slot. The FR4 substrate was used for the construction of the filter, where the filter dimensionsof 35mm*20mm*1.6 mm was used, and a feed line width (W_f = 3.1) was calculated from Equation 6 to achieve 50-Ohm impedance.

Using equations (8) and (9), the slot dimensions that realized a frequency of 3.6 GHz were found to be (Lu= 12.5 mm, Wu= 0.55 mm, and d= 1 mm) hence the total length is 26 mm.The obtained scattering parameters S_{11} and S_{21} for the proposed filter are shown in figure 7. A band-notch property of 15.6dB insertion loss at 3.598GHz has been achieved by the U-slot design.

The U-slot filter's characteristics, such as resonance frequency and band rejection depend on the dimensions of the U-slot. The slot length, slot width, and distance between the slot's legs are the parameters that control the characteristics of the filter. Simulated transfer characteristics for the U-slot filter are plotted in figure 8 for various slot lengths while keeping Wu=0.45mm and d=1mm unchanged. The frequency of resonance decreases as the slot length increases, which confirms equation (8).

The influence of the slot width is demonstrated in figure 9 when d=1mm and Lu=12.5mm. As can be seen, the width of the slot influences the width of the reject band, so that larger width results in wider bandwidth, while S_{11}gets slightly higher and S_{21}decreases slightly. Thus, the desired filter characteristics can be obtained by the proper choice of the filter dimensions guided by equations (8) and (9).

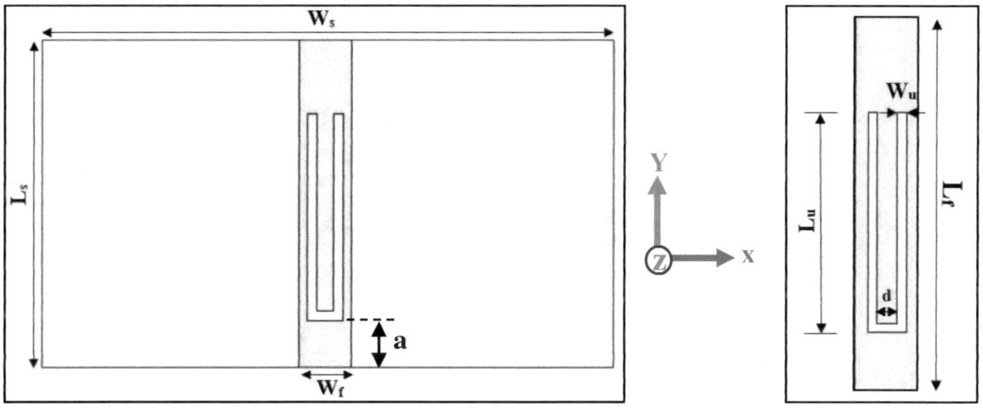

Figure 6. The geometry of the proposed band-stop filter using a u-slot etched on a microstrip line.

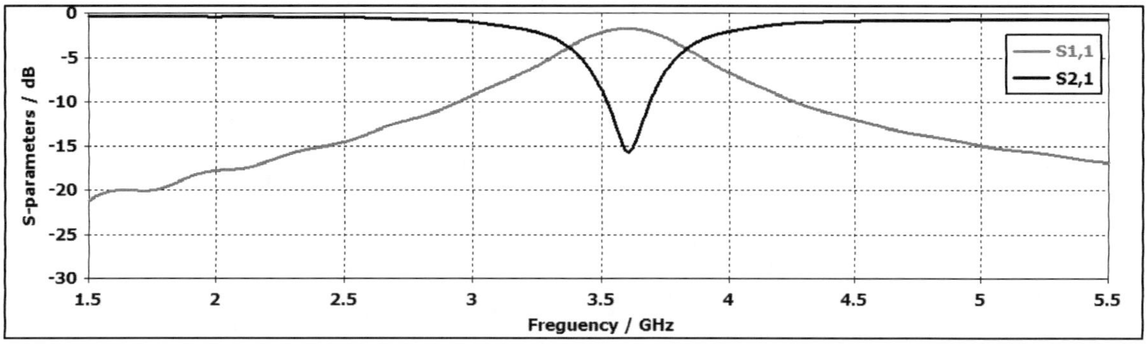

Figure7. Transfer characteristics of the U-slot filter.

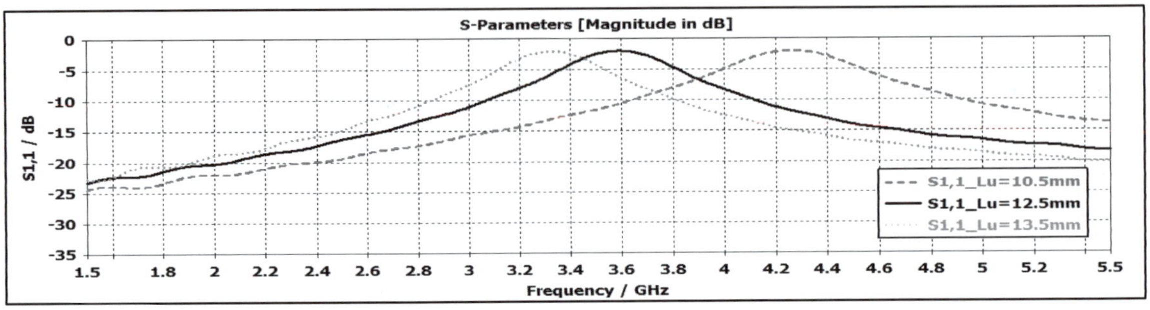

(a)

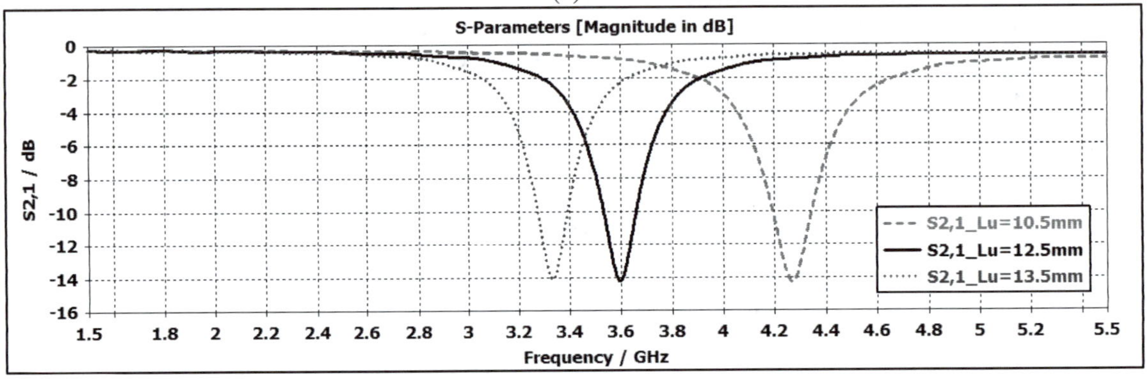

(b)

Figure 8. The frequency response of the U-shaped filter for various slot lengths at (Wu=0.45mm, d=1mm) ;(a) the reflection coefficient, (b) the transmission coefficient.

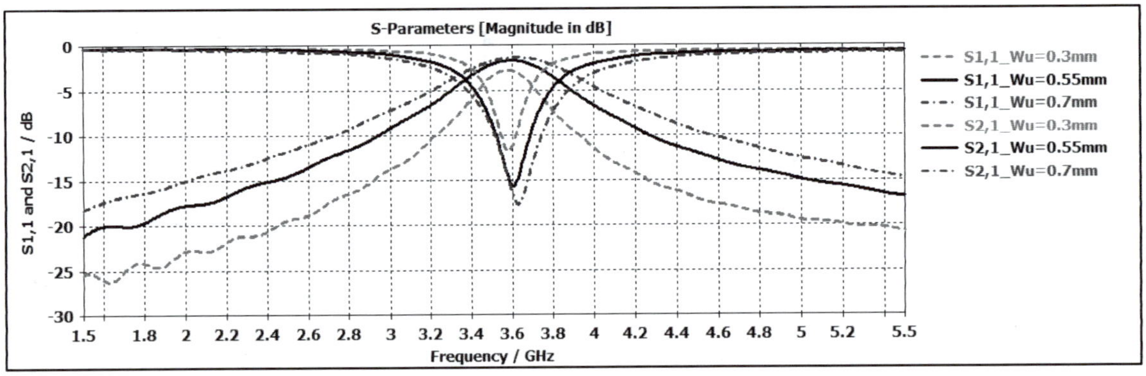

Figure9.S-parameters of the U-slot filter for various slot widths.

4. Design of the Filtering antenna

A U-slot resonator is placed on the feed line of the antenna designed in section 2 to limit its bandwidth to the desired 3.4–3.8 GHz range. The geometry of the proposed filtenna is shown in figure 10, and the optimized dimensions are listed in table 2. In this design, various values of the ground plane length L_g were considered to obtain the optimum values that achieve the required bandwidth (3.4-3.8GHz). Figure11illustrates the reflection coefficient S_{11} of the filtenna, where it covers a bandwidth of 412MHz (3.3952GHz to 3.8072GHz), and achieves a reflection coefficient S_{11} of -22.96 dB at 3.676 GHz. The figure also displays the frequency responses of the two-monopole antenna-1 and antenna-2 (without the U-shaped slot) that were designed in section 2, where much more than the desired 400 MHz bandwidth is shown. The introduction of the U-shaped slot has clearly improved the left part of the response, and a sharper response is also noticed at the right edge of the band.

IOP Conf. Series: Materials Science and Engineering 1152 (2021) 012011 doi:10.1088/1757-899X/1152/1/012011

The simulated peak gain of the proposed filtenna is displayed in figure 12. The peak gain of 3.1477 dBi is seen in the operating band. The figure also shows that the antenna achieved an efficiency higher than 73% across the desired band. The gain and efficiency drop to smaller values outside the band. Table 3 shows a comparison between various parameters of the three designed antennas. The filtenna has offered the required bandwidth with a small decrease in the gain.

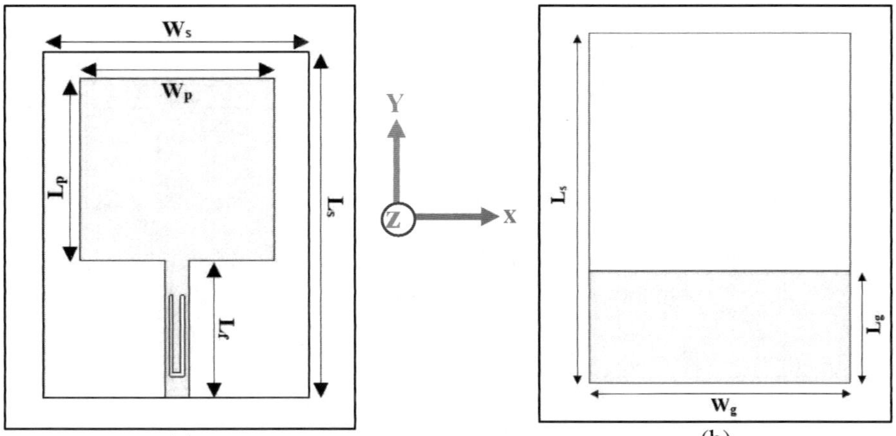

(a) (b)

Figure 10. The geometry of the proposed filtenna, (a) front view, (b) rare view.

Table 2. The dimensions of the proposed filtenna.

Parameter	Value in mm	Name
W_u	0.45	U-slotwidth
W_f	3.1	Feed linewidth
W_s	35	Substratewidth
W_p	25.6	Patch width
W_g	35	Ground plane width
L_u	10.9	Length of the U slot
L_f	19	Length of the feed line
L_g	18.3	Length of the ground plane
L_p	24	Length of the patch
a	2.2	Distance from the port to U-slot

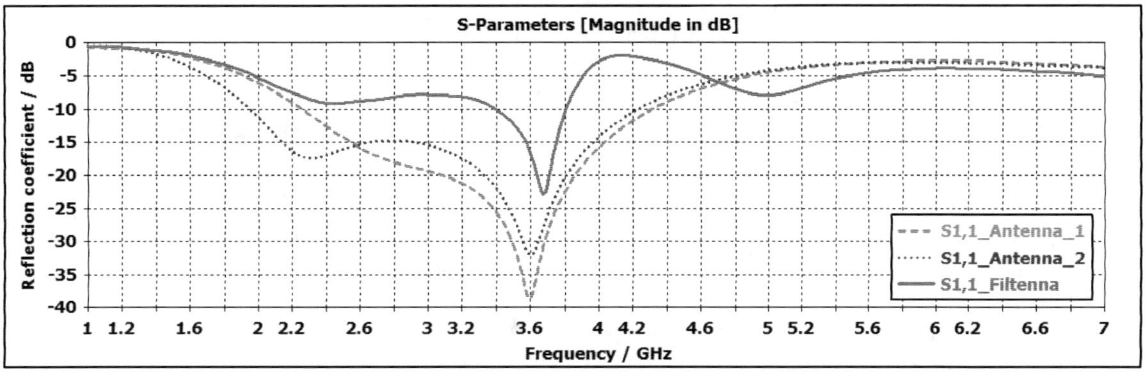

Figure 11. The reflection coefficient (S_{11}) of the proposed filtenna compared to antenna-1 and antenna-2(without the U-shaped slot).

Table 3. Comparison between the characteristics of antenna1-, antenna-2 and filtenna.

	Gain (dBi)	Bandwidth (GHz)	S_{11} (dB)
Antenna 1	3.1	2 .06(2.2613-4.3172)	-38.75
Antenna 2	2.72	2.282 (1.9523-4.2341)	-32
Filtenna	2.577	0.412 (3.3952 -3.8072)	-22.96

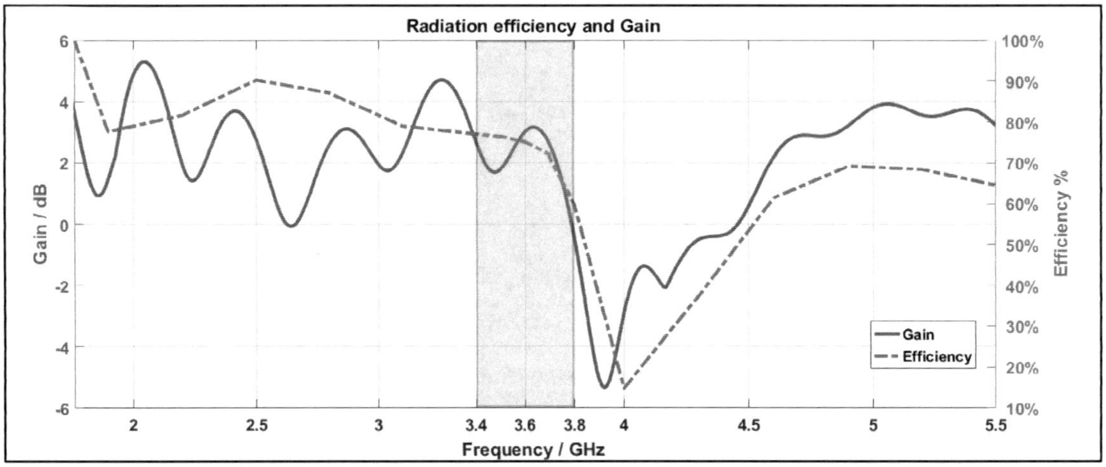

Figure 12. Gain and Radiation efficiency of the proposed filtenna.

The radiation patterns in the H-planes and E-planes are shown in figure 13. The antenna has the normal monopole pattern exhibiting the figure 8 shape in one plane and omnidirectional pattern in the other plane. The patterns are stable over the band of operation.

The surface current distribution for antenna-1 and the proposed filtenna are shown in figure 14 for the two frequencies; the center of the operating band (3.6 GHz) and the frequency of 4 GHz (outside the band). Comparing the results of antenna-2 that are shown in figures 14(a) and 14(c), (it can be seen that the current distribution on the radiating patch is basically similar at the two frequencies in the operating band (3.6 GHz) and outside the band (4 GHz). On the other hand, looking at the current distributions of the proposed filtenna, which are shown in figures 14(b) and 14-d, it is evident that outside the operation band (i.e., at 4 GHz), the density is limited on the radiating patch. This leads to small radiation efficiency and gain as confirmed by the results shown in figure 12, where the gain drops to -4 dB, and the efficiency drops to less than 20%. The influence of the U-shaped slot is to reduce the signal fed to the radiating patch drastically as can be seen in figure 14(d). Therefore, filtenna has fulfilled the requirement of low radiation/receiving properties outside the designated band of operation.

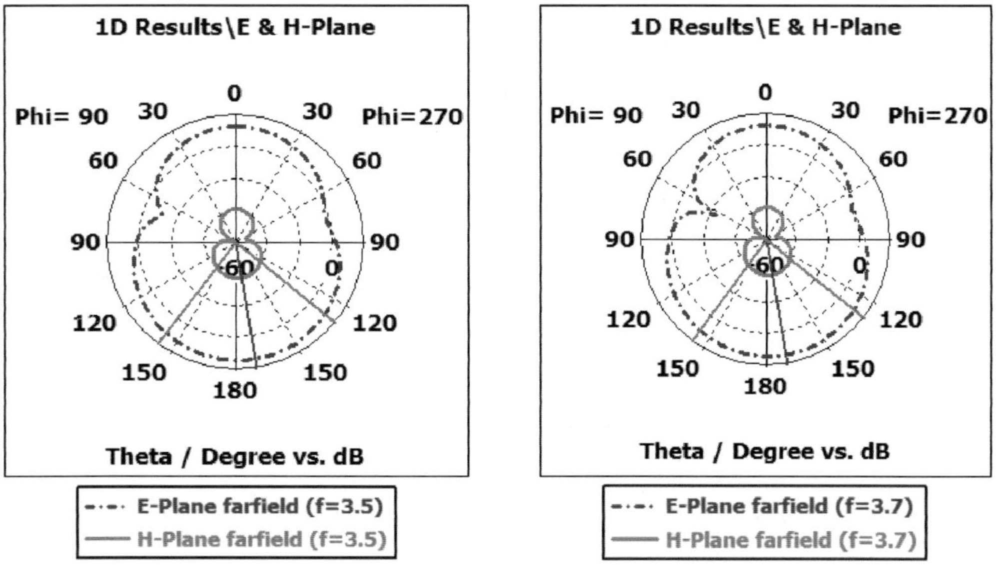

(a) (b)

Figure 13. The far-field patterns of the filtenna at various frequencies, (a) E/H-planes at 3.5GHz and (b) E/H-planes at 3.7GHz.

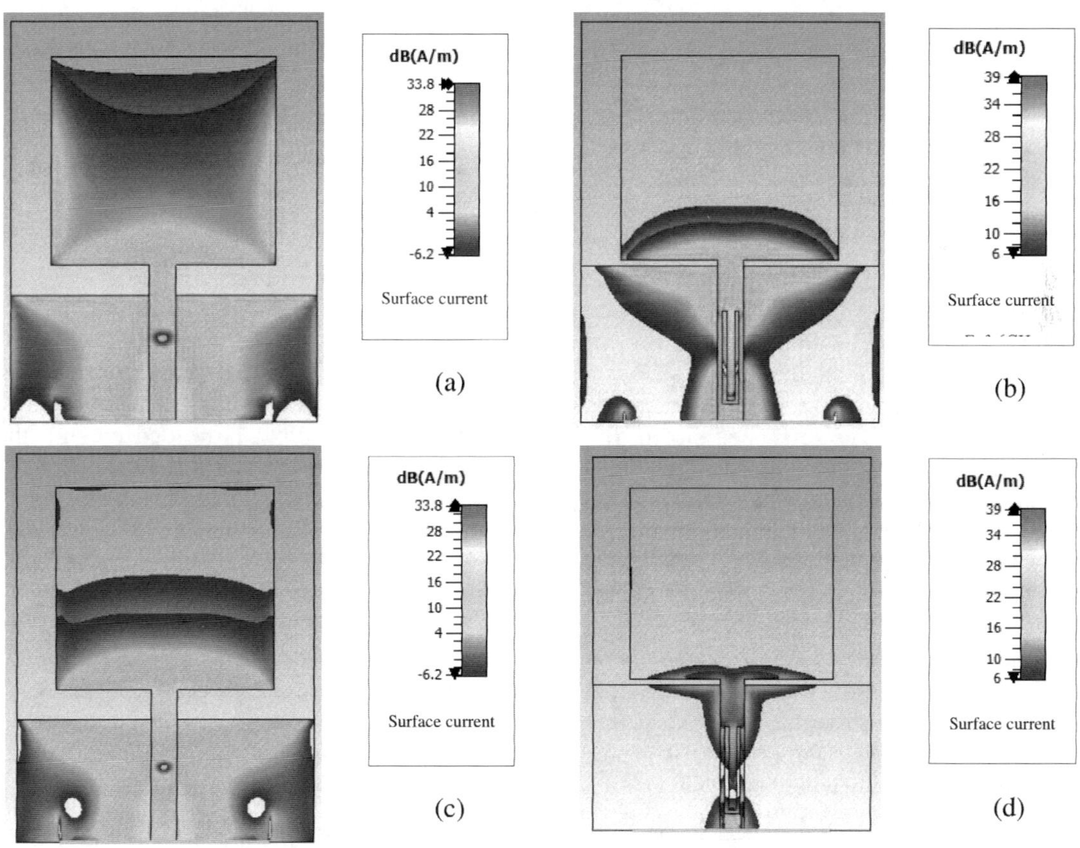

Figure 14. Simulated surface current distribution; (a) antenna MA1 at 3.6GHz, (b) Filtenna (proposed antenna) at 3.6GHz, (c) antenna MA1 at 4GHz , (d) Filtenna (proposed antenna) at 4GHz.

5. Comparison with other designs

The characteristics of the proposed filtenna are compared with those of other antennas presented in the literature, as listed in Table4.In [4]the last resonator element was replaced by a fan-monopole antenna with a compact size of 0.36 x $0.34\lambda_g^2$, but the bandwidth is less than 150MHz, and its achieved gain is 1.75dBi. In [6], the dipole antenna with a U-shaped microstrip-line and two I-shaped slots lead to a larger size compared to [4], but both antennas have a smaller size compared to the antenna proposed in this work. However, the proposed filtenna has a larger gain of 2.58 dBi, and a bandwidth of 412MHz satisfying the requirement of 400 MHz. Instead of filtering loops, short pins have been used in [16],but due to its large ground plane, the antenna is of considerable size (1.1 * 1.1 λg^2).A filter of the multi-stub method has been used in [17] in the feed line of a conventional inset-fed square microstrip patch antenna. However, this design needed twice the area occupied by the proposed filtenna. The proposed design shows various advantages compared to the other filtenna designs mentioned above, such as simple construction, compactness by using the feed line to achieve the filtration characteristic, and controllable bandwidth by easily adjusting the length of the U-slot.

Table 4. Comparison with other published designs of the proposed band-notch filtennas.

Ref.	Band frequency GHz	Dimension mm	ε_r / tanδ	S_{11}in dB	Resonant element
[4]	2.264-2.46	29*27	3.48 /0.0037	-21.9	Capacitive loaded loop (CCL)+ monopole
[6]	3.82-4.49	28*40	3.38/N.A.	-16.1	U-shaped and two I-shaped slots + Dipole
[16]	4.3-4.7	diameter= 99	2.65/N.A.	-22.6	Ring slot + Triangular patch
[17]	2.4-2.1	70*40	4.5 /0.022	-30	Multi-stub + Rectangular patch
This work	3.395-3.807	35*47	4.3/0.025	-22.96	U-slot + Monopole

6. Conclusions

The design and characterization of a filtering antenna (Filtenna) with a compact size of 35 x 47 mm^2 for the 3.4-3.8 GHz 5G applications has been demonstrated. The design consists of a conventional monopole antenna having a center frequency of 3.6 GHz, and a U-shaped slot embedded in the feed line of the monopole. The U-shaped slot provides the filtering property that is needed to limit the bandwidth to the desired value of 400MHz, and its length can be adjusted to control the bandwidth of the filtenna. Using CST Microwave Studio Suite, a comprehensive parametric analysis was performed to evaluate and optimize the characteristics of the proposed filtenna. The filtenna achieved a 412MHz band covering the frequency range (3.395-3.807) GHz. The filtenna showed a maximum gain of 2.577 dBi, a reflection coefficient of -22.96dB, and high rejection outside the operating band.

7. References

[1] Radio Spectrum Policy Group 2016 Strategic Roadmap towards 5G for Europe -Opinion on spectrum related aspects for next-generation wireless systems (5G) *Rspg16-032 Final* 6

[2] Ellinger F 2008*Radio frequency integrated circuits and technologies*(Berlin, Heidelberg : Springer Berlin Heidelberg) p 51.

[3] Lin C K and Chung S J 2011 A compact filtering microstrip antenna with quasi-elliptic broadside antenna gain response *IEEE Antennas Wirel. Propag. Lett.***10** 381–4

[4] Tang M C, Chen Y and Ziolkowski R W 2016 Experimentally validated, planar, wideband, electrically small, monopole filtennas based on capacitively loaded loop resonators *IEEE Trans. Antennas Propag.***64** 3353–60

[5] Ali A E T and Wang Y 2017 Integrated filtering planar dipole antenna using edge coupled feed *2016 Loughbrgh. Antennas Propag. Conf. LAPC 2016* 5–8

[6] Zhang Y, Zhang X Y and Pan Y M 2018 Low-Profile planar filtering dipole antenna with omnidirectional radiation pattern *IEEE Trans. Antennas Propag.***66** 1124–32

[7] Li D, Zhang Y, Zhang Y and Fan Y 2019 Dual-polarized filtering antenna using SRR without extra circuit *2019 IEEE Int. Symp. Antennas Propag. Usn. Radio Sci. Meet. APSURSI 2019 - Proc.* 1905–6

[8] Xi L 2019 A wideband planar filtering dipole antenna for 5G communication applications *Microw. Opt. Technol. Lett.***61** 2746–51

[9] Govindarajulu S R, Jenkel A, Hokayem R and Alwan E A 2020 Mutual coupling suppression in antenna arrays using meandered open stub filtering technique *IEEE Open J. Antennas Propag.***1** 379–86

[10] Awang Z 2014 *Microwave systems design*(Singapore:Springer Singapore) p 103.

[11] Sayidmarie K H and Fadhel Y A 2012 Design aspects of UWB printed elliptical monopole antenna with impedance matching *LAPC 2012 - 2012 Loughbrgh. Antennas Propag. Conf.* 1–4

[12] Sayidmarie K H and Najm T A 2013 Performance evaluation of band-notch techniques for printed dual band monopole antennas **3** 70–80

[13] Yahya L S, Sayidmarie K H, Elmegri F and Abd-Alhameed R A 2015 Crescent-shaped double-monopole antennas with reduced coupling for WLAN and WIMAX applications *2015 Internet Technol. Appl. ITA 2015 - Proc. 6th Int. Conf.* 393–8

[14] Al-Atrakchii M A, Sayidmarie K H and Abd-Alhameed R A 2020 Compact bandstop microstrip line filter using U-shaped slot *IETE J. Res.***0** 1–8

[15] M.A. Al-Atrakchii, K. H. Sayidmarie and R A A-A 2021 " A band pass frequency selective surface usingu-shaped slots" *Int. J. Microw. Opt. Technol.***Vol. 16** 36–43

[16] Wu T L, Pan Y M, Hu P F and Zheng S Y 2017 Design of a low profile and compact omnidirectional filtering patch antenna *IEEE Access***5** 1083–9

[17] Velidi V K, Kumar M, Subramanyam A V G, Sivareddy D, Kumar V S, Ramana D V., Sitaraman P V. and De A 2018 Uniplanar microstrip patch antenna with reduced harmonic radiation using multi-stub filter unit *2017 IEEE Int. Conf. Antenna Innov. Mod. Technol. Ground, Aircr. Satell. Appl. iAIM 2017* 1–5

Obtaining Performance Similar to Rectangular Planar Antenna Using Crossed Arrays with Smaller Number of Elements

Abdulrazaq Abdulhaq Khmees[1*] and Jafar Ramadhan Mohammed[1]

[1]College of Electronic Engineering, Ninevah University, Mosul-41002, Iraq

[*]E-mail: abdulrazaq.khmees2019@stu.uoninevah.edu.iq

Abstract. Conventional rectangular planar arrays with fully filled elements are practically very complicated especially for large arrays. They usually occupy a large space. Thus, any reduction in the space and the number of the array elements is highly desirable in many applications. In this paper, the authors present a novel technique for obtaining a performance similar to that of the conventional rectangular planar antenna using two orthogonal crossed arrays with a smaller number of elements. The proposed technique consists of redesigning the amplitude element excitations of only two crossed arrays using either deterministic or optimization algorithms. Simulation results are shown to prove the proposed idea.

1. Introduction

In certain applications such as massive MIMO wireless communication and the satellite systems, the weight of the used antenna array is need to be as small as possible and takes a small space. Thus, designing of such arrays with a fewer number of elements while maintaining a good radiation characteristic is highly desirable. Other advantages of such antennas with a fewer number of array elements include lower cost and great simplification in the array feeding network. Uniformly excited rectangular planar arrays are commonly used to generate pencil beam patterns with good directivity which can be made electronically steerable by properly controlling the progressive phases of the array elements [1]. The authors in[2], present a method for reducing the number of array elements based on the matrix pencil method (MPM). The method was applied to the non-uniform linear arrays to synthesize a pre-specified radiation patterns with limited number of elements. In[3], the authors present a narrowband MIMO imaging radar with two orthogonal linear T/R arrays where they were able to get an image resolution with a small number of array elements that is close enough to that of the conventional two-dimensional planar array that uses all of the array elements. In [4], the authors designed a MIMO cross array with azimuth-elevation combination beamforming capability by the use of DBF systems. They greatly reduced the number of the designed arrays with compared to the conventional two-dimensional rectangular array of one-way and two-way beamforming. Their proposed configuration were able to reduce the size of the designed array from (M x N) to only (M+N+4) elements. In[5], the authors analyzed the performance of the square planar arrays in term of their radiation patterns with that of the Mills cross array. They found that, although the square planar arrays have better efficiency, the Mills array was able to provide good directivity for the applications of a 5G mobile handset at 28 GHz.

In[6], the authors suggested to use the Bayesian compressive sensing to minimize the number of the array elements. They were able to achieve the required radiation patterns of both linear and planar arrays with minimum number of array elements. In [7], the authors use convex optimization to design the crossed array with minimum number of the array elements to achieve the required radiation

Content from this work may be used under the terms of the Creative Commons Attribution 3.0 licence. Any further distribution of this work must maintain attribution to the author(s) and the title of the work, journal citation and DOI.
Published under licence by IOP Publishing Ltd

pattern. However, in general, reducing the number of the array elements is associated with the losses in the directivity which is undesirable in many of the aforementioned applications.

Thus, instead of reducing or minimizing the number of the array elements, it is possible to select a certain number of the array elements to be controllable with required RF components such as variable attenuators, variable phase shifters, and other hardware to control the array radiation patterns according to the required goals. In[8–12], the authors suggested different optimization methods and configurations to control only the amplitude and the phase excitations of a number of the selected elements instead of all of the array elements. Thus, a great reduction in the array weight, cost, and feeding complexity was obtained without any loss in the directivity. The inverse fast Fourier transform (FFT) method has been also suggested, in [13]to synthesize antenna array patterns with controlled sidelobes and improved directivity.

In this paper, the authors demonstrate that by a simple configuration of two crossed arrays along with a proper choice of the element weightings, the radiation pattern of the crossed array which has a very small number of elements can be made equal to that of the conventional rectangular array with fully grid large number of elements. The element weightings of the proposed crossed array can be computed either deterministically through the use of triangular, Dolph, Taylor distributions or numerically through the use of genetic algorithm.

2. Principles of The Proposed Cross Array

Consider a fully filled conventional rectangular planar array with odd number of isotropic elements equal to$(2N+1)\times(2M+1)$ as shown in Figure 1. The array elements are assumed to be located in the x-y plane and the coordinate system's origin is set to be the array's geometric center. The array factor of this antenna can be expressed as[1]

$$AF(\theta,\emptyset)=\sum_{n=1}^{2N+1}\sum_{m=1}^{2M+1} w_{nm}\, e^{j[(m-1)kd_x\sin\theta\cos\phi]}e^{j[(n-1)kd_y\sin\theta\sin\phi]} \tag{1}$$

where w_{nm}, is the coefficients of the amplitude element excitation, d_x is the spacing between elements along the x-axis and d_y is the spacing between elements along the y-axis, $k=2\pi/\lambda$ and λ is the wavelength in free space.

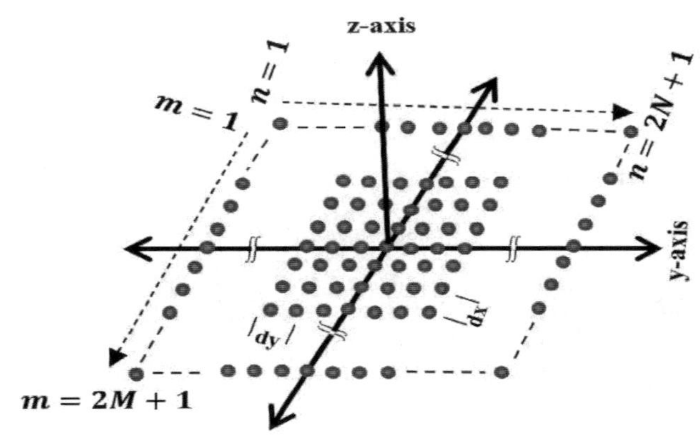

Figure. 1 Conventional rectangular array

For simplicity, we will assume a symmetric square planar array with total number of elements equal to $(2N+1)\times(2N+1)$. From equation (1), it can be seen that the synthesis of the fully filled rectangular array is complex and computationally intensive, especially when the number of elements is large. To simplify this complicated array, the authors suggest replacing this square planar array to only two orthogonal linear crossed arrays as shown in Figure 2. The total number of the elements in the crossed array is made to be $4(2N) + 1$. The patterns of the two crossed linear arrays with a total number of elements $4(2N) + 1$ have been combined to produce an effective pattern that is equivalent to that of the fully filled square planar with a total number of elements $(2N + 1)^2$. To achieve such matching between the radiation patterns of these two antenna arrays, the amplitude element excitation of the crossed arrays needs to be properly computed.

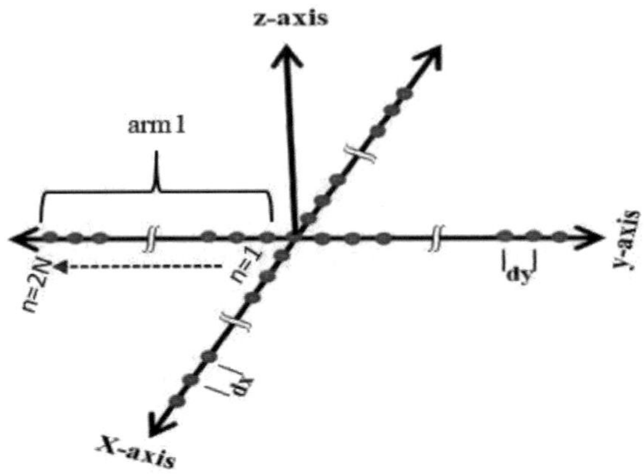

Figure. 2 Two orthogonal linear arrays

The array factor of the symmetric two orthogonal linear arrays can be written as: -

$$AF(\theta, \emptyset) = 2 \times w_0 + 2\sum_{n=1}^{n=2N} w_n \left[\cos\left(n(kd_x \sin\theta\cos\phi)\right) + cos\left(n\left(kd_y\, sin\,\theta \sin\emptyset\right)\right)\right] \quad (2)$$

If the element weightings $w_n$are all uniformly excited, then the resultant radiation pattern will have usually high sidelobe level[14]. Figure 3, show the radiation patterns in three-dimension at (N=2) for both conventional planar array with size $(2N + 1) \times (2N + 1)$and the proposed cross array with size $4(2N) + 1$and uniform excitations, i.e., $w_n = 1$ for all elements (This design is referred to as Design 1).

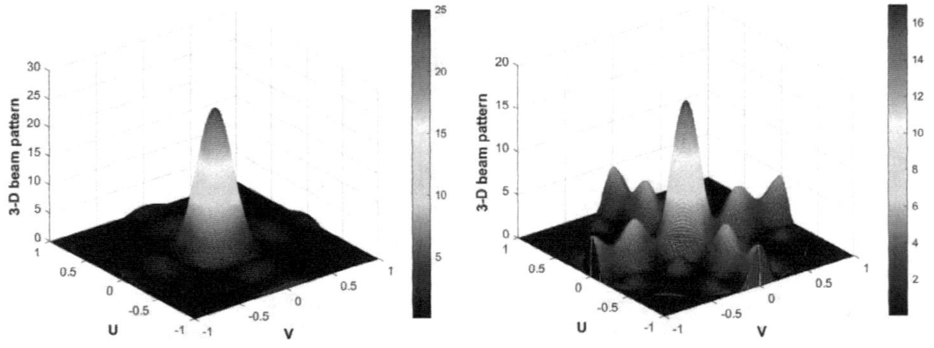

Figure 3. Three-dimension patterns of the uniform planar array 25 elements (left) and the proposed cross array 17 elements (right)

It can be seen from that the radiation pattern of the crossed array with uniform excitation has a relatively high sidelobe level. Thus, we need to redesign or recalculating the amplitude element excitations of the crossed array such that the sidelobes can be reduced.

IOP Conf. Series: Materials Science and Engineering 1152 (2021) 012012 doi:10.1088/1757-899X/1152/1/012012

One of the simplest techniques that can be used to reduce the side lobe level is by selecting the amplitude element excitations of the crossed array according to a specific taper, i.e., triangular taper (this design is referred to as Design 2) as given by the following equation

$$w_n = 2N + 1 - |n|, \qquad for \quad 0 \leq |n| \leq 2N \tag{3}$$

Where the weights of the elements at each arm from center to edge of crossed array having a tapered linear slope. Further, the above equation represents a straight line with a slope equal to 1. Moreover, it is also possible to apply other tapering such as Dolph (this design is referred to as Design 3) or Taylor (this design is referred to as Design 4) as will be shown in the next section. More important, the amplitude element excitations of the proposed cross array can be optimized using genetic algorithm to obtain lower sidelobe level (this design is referred to as Design 5).

3. Simulation Results

To demonstrate the usefulness of the proposed two orthogonal linear cross array, different examples are illustrated. In the first example, a small size arrays are considered where the amplitude element excitations of all the five designs of the proposed cross arrays and their corresponding radiation patterns are computed and compared. In the second example, a large size arrays are considered. In these two examples, antennas of an equally spaced linear arrays $d_x = d_y = \lambda / 2$ is considered, in addition, the amplitude-only control method is used to synthesis the excitation coefficients of the tested arrays. Thus, the phase element excitations are assumed to be zero.

For example 1 ,we assume a small square planar array with size equal to $(2N + 1) \times (2N + 1)$ =5 ×5and the coefficient weights of the element excitation are assumed to be uniform. The proposed cross array has a size equal to $4(2N) + 1 = 17$. The coefficient weights of this array are also assumed to be uniform (design 1). The coefficient weights of the element excitation are then redesigned using triangular taper (design 2), Dolph with required SLL=-20 dB (design 3), Taylor with SLL=-20 dB and $nbar$ =4 (design 4), and the genetic algorithm (design 5). Figure 4 and table I show the results of these five designs.

Furthermore, the dilution factor has been defined as the percentage ratio of the total number of elements in the proposed design, to the total number of the elements in the conventional square planar array. Thus, a smaller value of the dilution factor represents the better design.

$$\text{Dilution factor} = \frac{\text{Total element number in the proposed design}}{\text{Total element number in conventional square array}} \times 100\% \tag{4}$$

Accordingly, the Dilution factor of the proposed cross array under this case is $^{17}/_{25} = 68\ \%$.

IOP Conf. Series: Materials Science and Engineering 1152 (2021) 012012 doi:10.1088/1757-899X/1152/1/012012

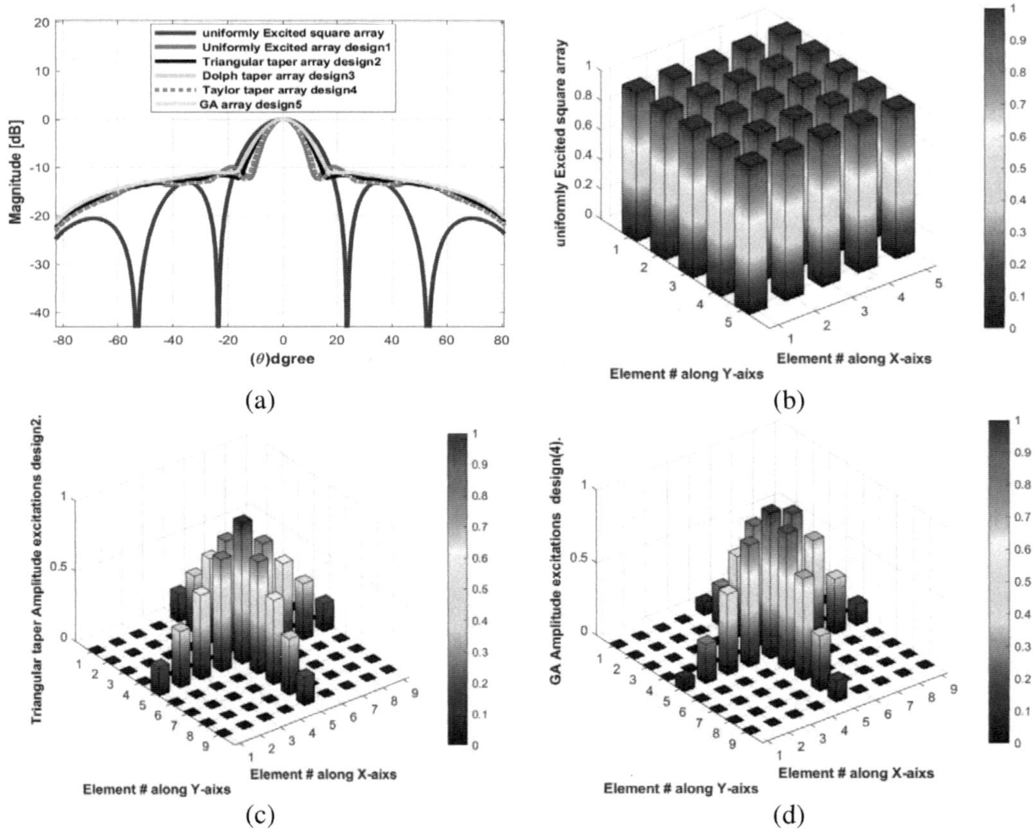

Figure 4. Results of the tested arrays for square array with size $(2N+1)\times(2N+1)$ =5 ×5 and crossed array with size $4(2N)+1=17$,(a)Radiation patterns of the designed arrays, (b) Amplitude excitations of the planar array, (c) Triangular taper (design2), (d) GA (design5).

Table I. Performance measures of the proposed designs with 17 elements and the conventional square array with 25 elements

The Method	Directivity [dB]	Peak SLL [dB]	FNBW [Deg.]	HPBW [Deg.]
uniformly Excited square array	15.5	-13.1	47	20.5
Uniformly Excited array design1	14.5	-9.9	24.6	12
Triangular taper array design2	15.1	-11.7	31.8	15.6
Dolph taper array design3	14.0	-11	27.3	13
Taylor taper array design4	14.07	-11	27.3	13
GA array design5	13.3	-11.3	31.3	16

From figure 4 and the table1, it can be seen that the level of the peak side lobe in the crossed array pattern has been slightly reduced. This reduction can be significantly reduced with larger array sizes as can be seen in the next example. Thus, the proposed crossed array method is found to be more suitable for the applications that require large array sizes such as massive MIMO.

For example 2, we assume a large square planar array with size equal to $(2N+1)\times(2N+1)$ =31 ×31and the coefficient weights of the element excitation are assumed to be uniform as previous. The proposed cross array has a size equal to $4(2N)+1=121$. The coefficient weights of proposed cross array are found using the five design methods as previous. Figure 5 and table II show the results of this case of large arrays. The dilution factor of this case is ${}^{121}/_{961} = 12.59\ \%$

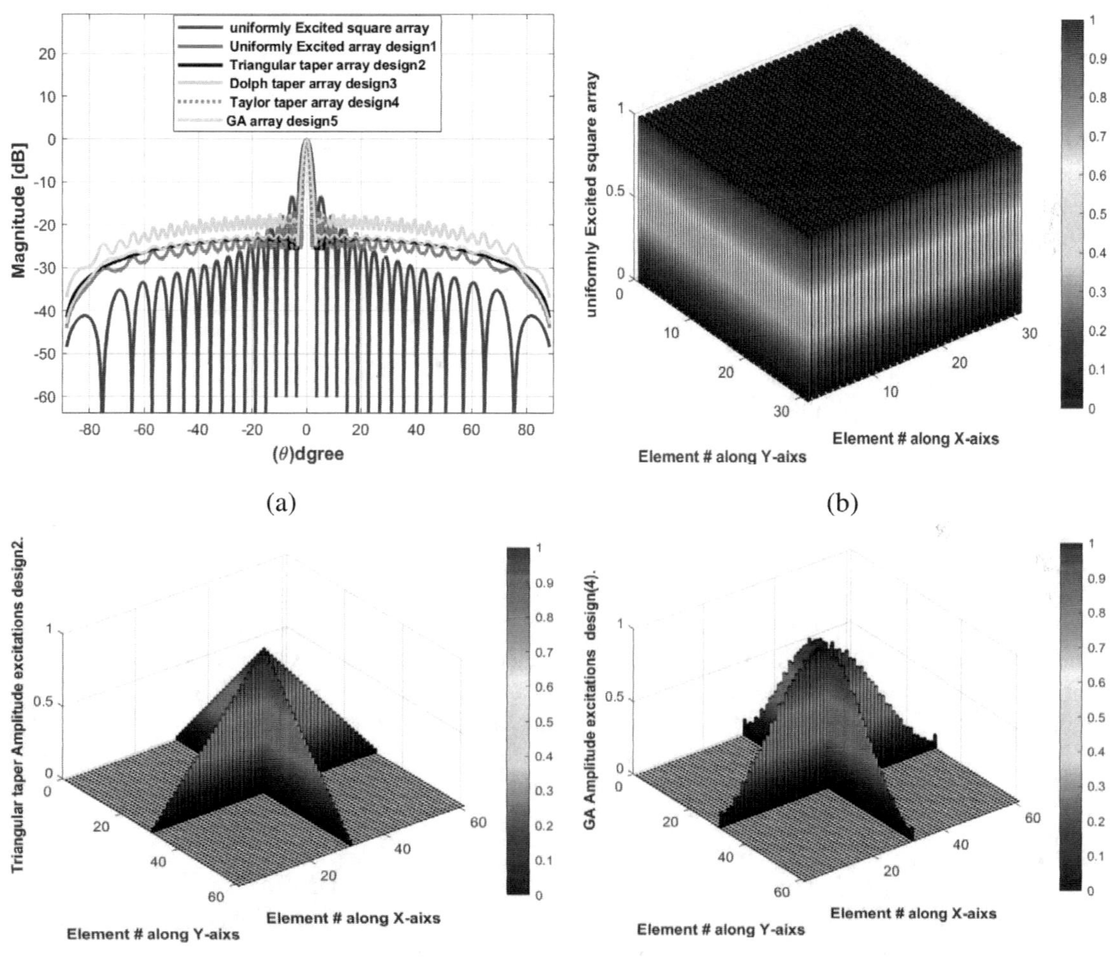

(a) (b)

(c) (d)

Figure5. Results of the tested arrays for square array with size $(2N+1)\times(2N+1) = 31\times31$ and crossed array with size $4(2N)+1=121$, (a) Radiation patterns of the designed arrays, (b) Amplitude excitations of the planar array, (c)Triangular taper (design2), (d)GA (design5).

Table II. Performance measures of the proposed designs with 121 elements and the conventional square array with 961 elements

The Method	Directivity [dB]	Peak SLL [dB]	FNBW [Deg.]	HPBW [Deg.]
uniformly Excited square array	30.9	-13.2	7.6	3.28
Uniformly Excited array design1	18.1	-12.3	3.58	1.6
Triangular taper array design2	18.7	-22.1	6.26	2.44
Dolph taper array design3	18.4	-17.5	4.02	1.8
Taylor taper array design4	17.7	-17.1	4.46	1.84
GA array design5	17.9	-22.5	6.4	2.42

From Figure 5 and table II, it can be seen that the results of the method that use the genetic algorithm are the best one among all other methods.

4. Conclusions

It has been shown from the presented results that the designed cross array with a total number of elements equal to $4(2N) + 1$ and adjusted amplitude excitations can be an alternative to the conventional rectangular planar array with a total number of elements equal to $(2N + 1) \times (2N + 1)$. For a small size array, for example1 N=2 (the total number of elements is 17), the Dilution factor was found to be 68%. This factor has been significantly reduced for a large array size, for example2 at N=15, to only 12.59 %. Thus, the number of the elements in the crossed arrays has been greatly reduced with compared to that of the conventional planar array. This reduction is come at the cost of relatively lower directivity. Nevertheless, the peak sidelobe level of the designed pattern of the crossed array is much lower than that of the conventional rectangular planar array which is another key advantage for the proposed array.

5. References

[1] Balanis C A 2016 *Antenna theory analysis and design* (Hoboken, New Jersey)
[2] Liu Y, Nie Z and Liu Q H 2008 Reducing the number of elements in a linear antenna array by the matrix pencil method *IEEE Trans. Antennas Propag.* **56** 2955–62
[3] Wang H and Su Y 2008 Narrowband MIMO radar imaging with two orthogonal linear T/R arrays *Int. Conf. Signal Process. Proceedings, ICSP* 2513–6
[4] Lu Y and Tang Y 2014 A study of cross array with MIMO DBF *IEEE Antennas Propag. Soc. AP-S Int. Symp.* 482–3
[5] M. J. Martínez Silva M S R P 2017 Radiation Pattern Analysis of Mills Cross Array of Patch Antennas for 5G Mobile Handset Applications *Int. J. Res. Electron. Electr. Eng.* **3** 1–6
[6] Zhang W, Li L and Li F 2011 Reducing the number of elements in linear and planar antenna arrays with sparseness constrained optimization *IEEE Trans. Antennas Propag.* **59** 3106–11
[7] Gu B, Chen Y, Jiang R and Liu X 2020 Optimization of sparse cross array synthesis via

perturbed convex optimization *Sensors (Switzerland)* **20** 1–17

[8] Mohammed J R and Sayidmarie K H 2014 Sidelobe cancellation for uniformly excited planar array antennas by controlling the side elements *IEEE Antennas Wirel. Propag. Lett.* **13** 987–90

[9] Mohammed J R 2018 A new antenna array pattern synthesis method with sidelobe control *J. Telecommun. Electron. Comput. Eng.* **10** 31–6

[10] Mohammed J R 2018 Element Selection for Optimized Multiwide Nulls in Almost Uniformly Excited Arrays *IEEE Antennas Wirel. Propag. Lett.* **17** 629–32

[11] Mohammed J R 2020 Phased Sub-arrays Pattern Synthesis Method with Deep Sidelobe Reduction and Narrow Beam Width Jafar

[12] Mohammed J R 2020 Simplified rectangular planar array with circular boundary for side lobe suppression *Prog. Electromagn. Res. M* **97** 57–68

[13] Mohammed J R 2021 Rectangular Grid Antennas with Various Boundary Square-Rings Array 10

[14] Bracewell R and Roberts J 1954 Aerial Smoothing in Radio Astronomy *Aust. J. Phys.* **7** 615

IOP Conf. Series: Materials Science and Engineering 1152 (2021) 012013 doi:10.1088/1757-899X/1152/1/012013

Comparative Study the Effect of Antenna Selection on Capacity and Energy Efficiency of 5G Systems

Sawsan N. Abdullah[1*], **Farhad E. Mahmood**[1] **and Y. E. Mohammed Ali**[1]

[1]Department of Electrical Engineering, College of Engineering, University of Mosul

*E-mail: sawsan.enp8@student.uomosul.edu.iq

Abstract. As a result of the rapid growth in wireless communications and the increase in the number of users of cellular phones around the world, massive multi-output, and multi-input (massive MIMO) systems have been used. The main drawback of such a system is the increase in the number of antennas and the amount of energy consumed. To mitigate that the number of active antennas can be reduced to save some energy. In this paper, we will study different algorithms for choosing the suitable antennas in MIMO systems, subject to keeping an acceptable level of bit rate and energy efficiency in the system. The study will also consider the amount of complexity for each algorithm based on the consumed time to perform each algorithm in MATLAB.

1. Introduction

The mobile systems has hegan for more than a few decades. Over the past two decades, the world has witnessed a gradual development of wireless communications towards the second, third and fourth generation wireless networks. Introducing digital modifications. Effective Frequency Reuse, packet-based Internet penetration.

Multi-output multi-input (MIMO) system technology is the most interesting type of wireless access to meet the 5G systems and beyond need [1]. Massive MIMO systems are an extension of the traditional MIMO technology that involves the use of hundreds or even thousands of antennas connected to the base station to improve spectral efficiency and throughput. The increase in capacity, throughput, and spectrum efficiency in large MIMO systems has made it a critical issue and has made a critical technology for emerging wireless standards [2].

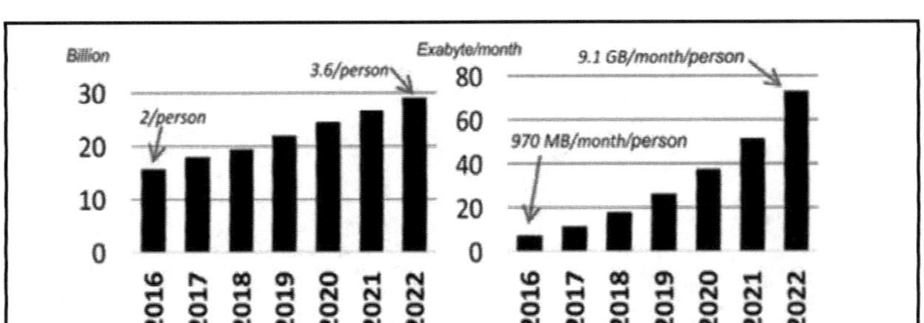

Content from this work may be used under the terms of the Creative Commons Attribution 3.0 licence. Any further distribution of this work must maintain attribution to the author(s) and the title of the work, journal citation and DOI.
Published under licence by IOP Publishing Ltd

Figure 1. Global mobile data traffic and growth in connected devices from 2017 to2022.

The main advantage of massive MIMO technology is that each user should only obtain a dedicated signal with the least interference or the least amount of interference potential from other stations [3]. Despite the significant advantages of massive MIMO, when it comes to power consumption, it is well known that BSs with a large number of active antenna elements suffer from an extremely high-power consumption since each antenna is connected to a separate power-demanding radio frequency (RF) chain. RF chains are responsible for 50%–80% of the total transceiving power consumption of communications systems [4].

One way to maintain the advantages offered by MIMO while reducing the energy consumption, cost, and hardware complexity is via using antenna selection (AS) techniques, However, unlike conventional MIMO systems where the AS is carried out with the main focus being on enhancing the performance, there has been a large number of researches on the selection of antennas in recent years [5-9]. In [5,6] the authors proposed an algorithm for selecting the receiving antenna in multi-input and multi-output systems. The researchers suggest two algorithms for selecting antennas, which are the Fast and Global The results showed that the proposed algorithms indicated that the capacity obtained is near ideal and close to the ideal case, but needs high memory due to the computational burden, i.e., it is complicated in terms of computation, and also showed that the global method is appropriate in the case of the number of selected antennas less than the number Antennas available. In [7] the researchers proposed an increase in the data rate through the use of Convex Optimization to determine the optimal antennas on a large scale in MIMO systems. The simulation results of the proposed method confirmed that it works well and its performance is close to the optimal selection algorithm and can reduce the cost by reducing the frequency chains And an increase in system efficiency. In [8], the researchers proposed a binary switching architecture for the solution that is simpler and sub-optimal but provides better signal quality, compared to a full switching network. To evaluate the proposed technique, compare the amplitude of the aggregate when using multiple dual switches with the full switching performance. It was found that the proposed dual switching gives a competitive performance close to the full switching. The results indicate that massive MIMO devices can be simplified by reducing the number of RF chains and simply selecting an antenna through binary switching architecture. The complexity is less and the loss in signal quality is less, but in return the number of RF is complete, that is, without reducing the RF, which is of a high cost.
As for the source [9], the authors selected the antennas and their effect on the channels, and the results showed that it is possible to reduce the complexity significantly in the devices and the energy consumption can be reduced without a significant deterioration in the performance of the system. In [10] the authors source a proposal for a sub-diagram and binding for the selection of antennas for an increase in capacity which can be achieved using dirty paper coding.
In this paper, four algorithms will be studied and compared and their effects on the selection of antennas in the fifth generation massive MIMO systems in terms of capacity and energy efficiency where the research paper is organized as follows on the system model in the Second Section, and in the Third Section the effect of the antenna selection in the fifth generation on the capacity will be discussed, while in the Fourth Section the study of the effect of the antenna selection on energy efficiency. After that, a summary of the results will be presented.

2. System Model

The base station is equipped with (Nt) from the antennas and the mobile station with (Nr) from the antennas, and the number of the selected antennas is(Ns). We assume that the channel is of the type of Rayleigh, and that the data is sent through the (dawn link) and the channel is similarly treated, which sends two experimental signals on The downlink in the system model, and the encoding of the transmitted signal used is of the MF type, and the signal is detected by the linear detectors of the

(MRC) type to clarify where Nr≤Nt (and the H)) represent the channel coefficient between the number of antennas Reported through the following matrix In this paperthe transmitted power Pt=0.1watt. The received signal can be given by the following equation:

$$y = Hx + n \tag{1}$$

Where $y \in^{c^{Nr\times 1}}$ represents the received signal by Nr and $x \in c^{Nt\times 1}$ represents the symbol vector for the transmitted information and $n \in c^{Nr\times 1}$ represents the AWGA noise vector in the receiver and $H \in^{c^{Nt\times Nr}}$: The channel matrix, which is a compound random variable, is formed by multiplying the number of transmitting antennas by the number of receiving antennas One equation can be illustrated as follows

$$\begin{bmatrix} y_1 \\ \vdots \\ \vdots \\ y_{Nr} \end{bmatrix} = \begin{bmatrix} h_{11} & \cdots & \cdots & h_{1,Nt} \\ \vdots & \ddots & \cdots & \vdots \\ \vdots & \cdots & \ddots & \vdots \\ h_{Nr,1} & \cdots & \cdots & h_{Nr\times Nt} \end{bmatrix} \begin{bmatrix} x_1 \\ \vdots \\ \vdots \\ x_{Nt} \end{bmatrix} + \begin{bmatrix} n_1 \\ \vdots \\ \vdots \\ n_{Nr} \end{bmatrix} \tag{2}$$

Table (1): assumed parameters.

Components	symbol	Assumed value
Number of transmitted antennas	Nt	16
Number of received antennas	Nr	2
Number of transmitted antenna selections	Ns	4
Number of reciver antenna selections	LR	
Signal to noise ratio	ϒ	6dB
Channel matrix	H	Complex number
Sub-channel matrix	Hsub	Complex number
Energy efficiency	EE	bit/joule
capacity	R	bit/sec/Hz

Where the base station is equipped with 16 antennas, considering that the base station in the fifth generation contains a large number of antennas and the mobile station is equipped with 4 antennas because most of the mobile devices in the fifth generation contain 5 antennas and thus becomes the channel array (16×4), unlike the fourth generation Which was the maximum of the channel array was made up of (4×2), (8×2), (2×2)……etc, as for the value of (SNR=6db) onlythis value was taken only for comparison between algorithms and also for power

3. Algorithms of selecting antennas

There are many algorithms in the literature to select antennas. It has been assumed consider four algorithms in our study: optimal algorithm antenna selection, random antenna selection (RAS), norm-based antenna selection (NBS), greedy algorithm antenna selection.

3.1 Random Antenna Selection (RAS)

This algorithm, as expressed via the flow chart in Figure 2, is considered one of the algorithms in which the antenna is randomly selected, and although it is the least time-consuming algorithm, it is also considered the least accurate and the least bit rate performing.

$$R^{dl} = |log_2(INr + \frac{\gamma dl}{Nt} HH^H)| \tag{3}$$

H^H: transpose Hermitia for channel matrix.
INr: matrix Identical.

IOP Conf. Series: Materials Science and Engineering 1152 (2021) 012013 doi:10.1088/1757-899X/1152/1/012013

R^{dl}: The capacity is represented in the downlink

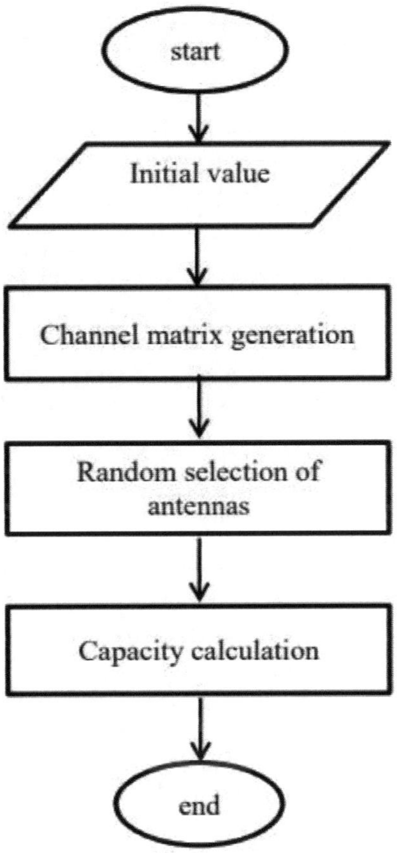

Figure 2. The flow chart of random select of antenna selection.

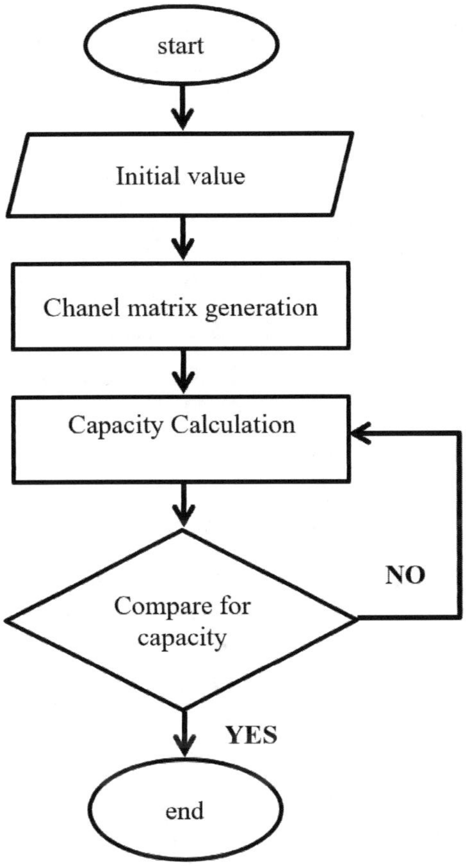

Figure 3. The flow chart of optimum antenna selection.

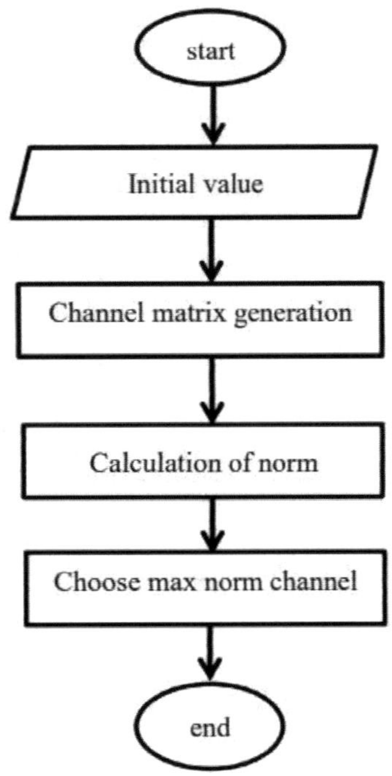

Figure 4. The flow chart of the norm Base algorithm.

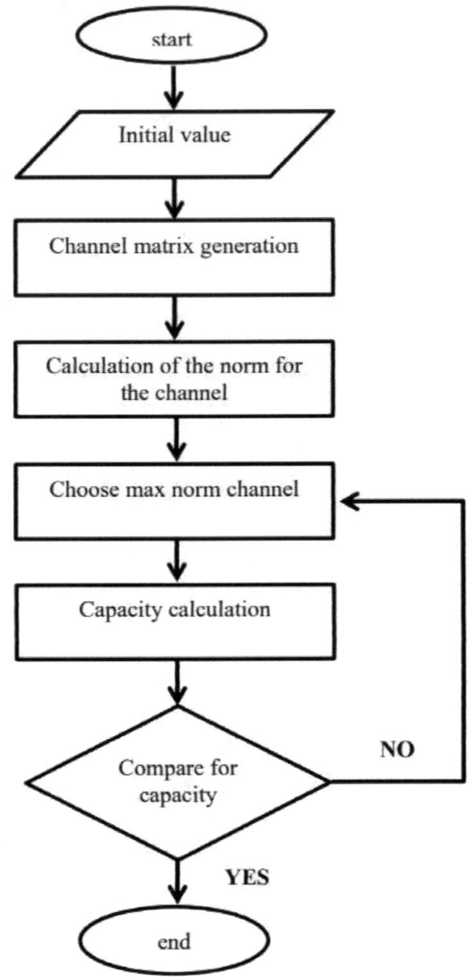

Figure 5. The flow chart of the Greedy algorithm.

3.2 Optimal algorithm antenna selection

It is considered the best option for choosing the antenna as explained in Figure 3,, but it has a very high complexity from a mathematical point of view as the bit rate formula is examined for a different group of antennas through the number of antennas in the reception and the number of antennas in the base station, and although it gives an ideal capacity, it takes a long period of time to choose an antenna. Suitable and for this reason, it has been used to compare it with other algorithms to obtain an algorithm of less complexity and close to the ideal case, and it can be expressed through the following relationship [12]

$$R^{dl} = max|log_2(INr + \frac{\gamma dl}{Nt} HH^H)| \quad (4)$$

3.3 Norm-based antenna selection (NBS)

A display of the Norm-based algorithm is shown in the flow chart in Figure 4. The number of antennas selected in the base station must be less than the number of antennas in the base station itself. Symbolizes the base station number of antennas selected NS such that NS<NT And the LR number of receiver antenna selection LR<NR, the full channel matrix $H \in C_{NR \times NT}$ is virtual as the MIMO

communications is happening over (LR × NS)-element subset channel matrix Hsub ∈ $C_{LR\times NS}$. In order to efficiently exploit the available resources, it is better to choose the antennas that have higher channel capacity. In general, increasing the channel gain has similar effect of having low noise, which leads to imporving the system performance. This motivates the NBS algorithm that selects the antennas based on the related to the higher channel matrix subset norm.

. Let Hsub ∈ CLR×NS denote the subset candidates of the full channel matrix H. The selected subset Hsub based on the NBAS criterion is found by solving the following optimization problem [11]:

$$H_{sub} = \text{arg}max_{HsubCH} \|H_{sub}\|^2 = \text{arg}max_{HsubCH} \sum_{N_t=1}^{L_T} \sum_{N_r=1}^{L_R} |H_{sub}\langle N_r, N_t\rangle|^2 \quad (5)$$

3.4 Greedy algorithm antenna selection

The following algorithm is proposed, which as shown in Figure 5, which is considered one of the algorithms of low complexity in mathematical terms, where the antenna is chosen in this algorithm based on the nature of the channel, that is, after the highest norm of the channel is calculated, these channels will be examined in terms of amplitude with the total rate If the capacity is greater than the total capacity, then the antenna will be turned off, otherwise the antenna will be energized. When comparing this algorithm with the ideal algorithm, we notice that although it gives capacity less than the ideal capacity, it is close to the ideal case and has less complexity in terms of mathematics and execution time.[13]

4. Energy Efficiency

The transmitter and receiver circuits usually include the so-called RF frequency chains, as shown in the figure above such as Power amplifier, Digital to anlage converter, Anlage to digital converter, Low noise amplifier, Mixer, these components may be active or inactive, depending on the operating mode of the phone [15].

The energy efficiency of the bit /joule unit can be given in MIMO systems. Energy efficiency can be obtained using the following formula

EE=bit rate/total power (6)

5. Results

In this section, it has been displied a comparison between the different algorithms in terms of capacity, energy efficiency, and in term of complexity.

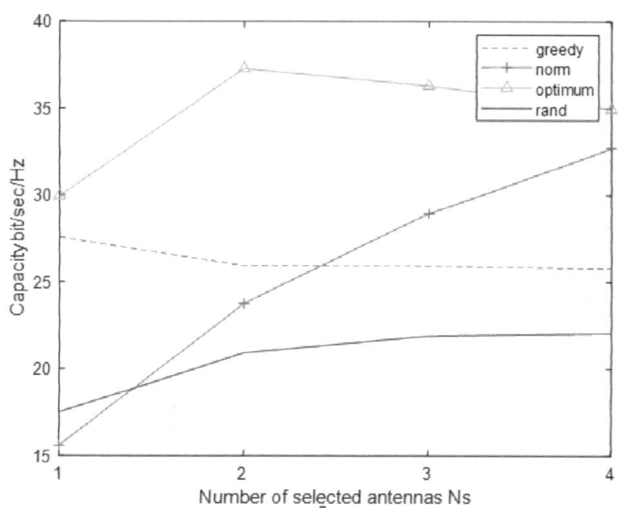

Figure 6. The bit rate vs the number of the selected antenna in the four algorithms.

In Figure 6. It has been shown that the optimum provides better performance in terms of bit rate with the number of selected antennas. Another insight from Figure 3 is that a comparison between Greedy & Norm, where the greedy can provide a better bitrate than the Norm at a low number of antennas, then it became harder for the greedy to stay better than the norm at a higher number of selected antennas. Lastly Random has the worst performance in terms of bit rate.

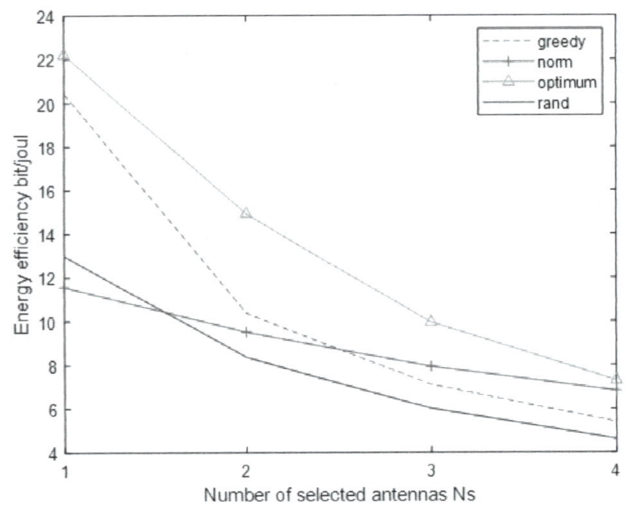

Figure 7. The energy efficiency vs the number of the selected antenna in the four algorithms.

Figure 7 shows the performance of the four algorithms in terms of energy efficiency. The results reinforce the idea of having higher energy efficiency when the low antenna selection. At higher antenna selection, the EE decreases gradually.

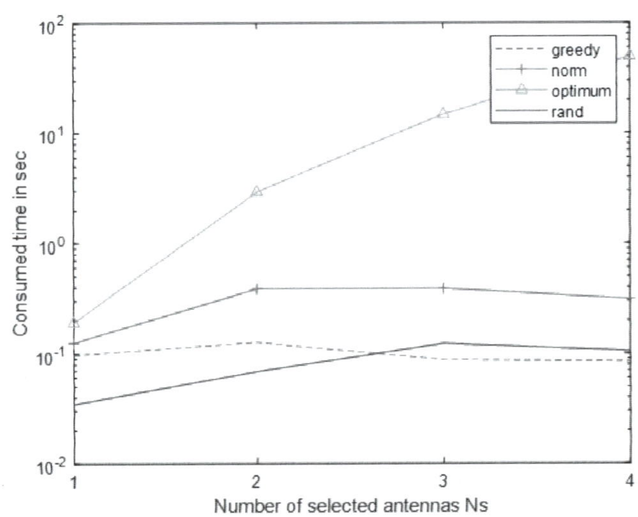

Figure 8. The consumed time with the number of antenna selection for the four algorithms.

From Figure 8, it is clear that the optimal algorithm consumes the highest time, which indicates that it has the highest complexity compare to other algorithms.

6. Conclusions

Through the comparison between the four algorithms under investigation, it was found that unlike the random and the norm algorithm, the optimal algorithm has the highest bitrate and energy efficiency with the cost of higher complexity. On the other hand, the Greedy algorithm can provide acceptable complexity with a valuable amount of bitrate and energy efficiency.

Acknowledgments

The authors would like to acknowledge the support provided by the College of Engineering, University of Mosul.

7. References

[1] Uthansakul, Peerapong, Nattaphat Promsuwanna, and Monthippa Uthansakul. "Performance of antenna selection in MIMO system using channel reciprocity with measured data." International Journal of Antennas and Propagation 2011

[2] Chataut, Robin, and Robert Akl. "Massive MIMO Systems for 5G and beyond Networks—Overview, Recent Trends, Challenges, and Future Research Direction." Sensors 20, no. 10 (2020): 2753.

[3] Alshammari, Ahmed, Saleh Albdran, and Mohammad Matin. "The effect of channel spatial correlation on capacity and energy efficiency of massive mimo systems." In 2017 IEEE 7th Annual Computing and Communication Workshop and Conference (CCWC), pp. 1-5. IEEE, 2017.

[4] L. M. Correia, D. Zeller, O. Blume, D. Ferling, Y. Jading, I. Godor, ´ G. Auer, and L. Van Der Perre, "Challenges and enabling technologies for energy aware mobile radio networks," IEEE Commun. Mag., vol. 48, no. 11, pp. 66–72, Nov. 2010.

[5] X. Gao, O. Edfors, F. Tufvesson, and E. G. Larsson, "Massive MIMO in real propagation environments: Do all antennas contribute equally?" IEEE Trans. Commun., vol. 63, no. 11, pp. 3917–3928, Nov. 2015.

[6] Uthansakul, Peerapong, Nattaphat Promsuwanna, and Monthippa Uthansakul. "Performance of antenna selection in MIMO system using channel reciprocity with measured data." International Journal of Antennas and Propagation 2011

[7] Mahboob, Saad, Rukhsana Ruby, and Victor CM Leung. "Transmit antenna selection for downlink transmission in a massively distributed antenna system using convex optimization." In 2012 Seventh International Conference on Broadband, Wireless Computing, Communication and Applications, pp. 228-233. IEEE, 2012

[8] Gao, Xiang, Ove Edfors, Fredrik Tufvesson, and Erik G. Larsson. "Multi-switch for antenna selection in massive MIMO." In 2015 IEEE Global Communications Conference (GLOBECOM), pp. 1-6. IEEE, 2015

[9] Gao, X., Edfors, O., Tufvesson, F., & Larsson, E. G. (2015). Massive MIMO in real propagation environments: Do all antennas contribute equally?. IEEE Transactions on Communications, 63(11), 3917-3928.

[10] GAO, Yuan; VINCK, Han; KAISER, Thomas. Massive MIMO antenna selection: Switching architectures, capacity bounds, and optimal antenna selection algorithms. IEEE Transactions on signal processing, 2017, 66.5: 1346-1360.

[11] Zhang, Peichang, Sheng Chen, and Lajos Hanzo. "Two-tier channel estimation aided near-capacity MIMO transceivers relying on norm-based joint transmit and receive antenna selection." IEEE Transactions on Wireless Communications 14, no. 1 (2014): 122-137.

[12] Gorokhov, Alexei, Dhananjay A. Gore, and Arogyaswami J. Paulraj. "Receive antenna selection for MIMO spatial multiplexing: theory and algorithms." IEEE Transactions on signal processing 51, no. 11 (2003): 2796-2807.

[13] Hamdi, Rami, Elmahdi Driouch, and Wessam Ajib. "Resource allocation in downlink large-scale MIMO systems." Ieee access 4 (2016): 8303-8316

[14] Alshammari, Ahmed, Saleh Albdran, and Mohammad Matin. "The effect of channel spatial correlation on capacity and energy efficiency of massive mimo systems." In 2017 IEEE 7th

Annual Computing and Communication Workshop and Conference (CCWC), pp. 1-5. IEEE, 2017.

[15] Mahmood, Farhad, Erik Perrins, and Lingjia Liu. "Energy-efficient wireless communications: From energy modeling to performance evaluation." IEEE Transactions on Vehicular Technology 68, no. 8 (2019): 7643-7654.

[16] Mahmoud, Farhad I., and Sami A. Mawjoud. "planning and design of a WCDMA network compatible with existing GSM system in Mosul city." *2008 5th International Multi-Conference on Systems, Signals and Devices*. IEEE, 2008.

Side Lobe Reduction Using Non-Uniform Linear Phased Arrays

Rasha Bashar Mohammed[1*] **and Jafar R. Mohammed**[1]

[1]Department of Communication Engineering, College of Electronics Engineering, Ninevah University, Mosul-41002, Iraq

[*]E-mail: rashakharoofa@yahoo.com

Abstract. Non-uniform amplitude excitation arrays are nowadays very desirable in the modern wireless communication systems. They have ability to provide required radiation characteristics such as low side lobes and good directivity. The uniform amplitude excitation arrays are good in providing good directivity and narrow beam width. However, these desired radiation features are come at the cost of relatively high side lobes. Thus, these types of arrays are not widely used in the wireless communication systems especially when these arrays are operated in a noisy and crowded environment that contains many interfering signals. Non-uniform amplitude excitations such as Dolph and Taylor are considered in this paper and their performances are investigated under various array parameters. Simulation results show that as the amplitude excitations of the array elements decaying at the edge elements more reduction in the sidelobe patterns can be obtained.

1. Introduction

The performance in terms of the sidelobe level of the linear uniform amplitude excitations arrays is not acceptable in many practical applications. Many approaches can be used to improve the radiation characteristics of the antenna arrays. One of the simplest approaches is the modification in the amplitude excitations of the array elements such that it has highest value at the centre of the array and then decaying gradually toward the array ends [1-7]. By this way the sidelobe level in the corresponding array radiation pattern can be significantly reduced. However, the main beam is found to be broadening which is not desirable due to its negative effect on the directivity or gain of the array. Other methods used phase excitations instead of the amplitudes to control the sidelobe level in which the phase shifters are needed [8-9]. These approaches that depend on the element phase excitations are more complex because these phases are also responsible for main beam scanning. Thus, a good control on both sidelobe reduction and main beam scanning is not easy to be implemented in practice. Much more complicated systems may be encountered when combining the amplitude and phase excitations of the array elements in one design.

Other researchers suggested controlling the locations of the array elements to achieve the required sidelobe level [10-13]. In this case, lowest sidelobe levels can be obtained by modifying the separation distances of the array elements either in analytical or numerical methods. Many different optimization algorithms were also used to design such arrays. Nevertheless, the Dolph distribution has many advantages that can be fully exploited in the work [14].

In this paper, the authors are focusing on studying the simplest analytical method that can be used to modify the amplitude excitations of the array elements while the phase excitations are all assumed constant. The Dolph distributions [15-16] are analyzed and studied and their corresponding radiation

Content from this work may be used under the terms of the Creative Commons Attribution 3.0 licence. Any further distribution of this work must maintain attribution to the author(s) and the title of the work, journal citation and DOI.
Published under licence by IOP Publishing Ltd

patterns are shown. Some comparisons are also presented to show the effectiveness of the design. Many design parameters are investigated and different array sizes such as small, medium and large are considered.

The remainder of this paper is organized as follows. In section 2, the background theory of the linear uniform amplitude excitation arrays is presented. Then, in section 3, the non-uniform amplitude excitation arrays are explained in details, while the simulation results and discussion are elaborated in section 4. Finally, the conclusion remarks of the paper are drawn in section 5.

2. Background Theory

Phased and Linear antenna arrays can be classified according to their element amplitude distribution where in general there are equal and unequal element excitations. The equal element excitations are commonly referred to as uniform arrays [17].

Uniform arrays are simple and their practical designs are very cheap, thus, they are commonly and widely used in practice. Here the term uniform is in the sense that the separation distance between any two adjacent elements is constant and equal and all the array elements have the same amplitude currents. As an example, consider the end fire array in which the amplitude currents are all constant. Its radiation direction is along the line of the array elements. Whereas, the phase of each element in the end fire array must be fed with a value equal to the difference to the separation distance of the adjacent elements.

Another practical case is the broadside array, which is the most commonly used antenna array in practice. In broadside array, a number of identical parallel elements are arranged along a line perpendicular to the array axis line. The broadside pattern can be obtained in a linear array with identical elements by exciting them with equal magnitude and phase. The array factor of such arrays is given by [15]

$$(AF)_{n=} \left[\frac{Sin\,N\,(kdcos\theta+\beta/2}{N\,(kdcos\theta+\frac{\beta}{2})}\right] \tag{1}$$

Where N represents number of elements, β progressive phase, k phase constant and d are the spacing between array elements. There are different array parameters such as Half Power Beam Width (HPBW), First Null Beam Width (FNBW), Side Lobe Level (SLL) and the Directivity (D) which their mathematical equations are given below [8].

$$HPBW = 2\left[\frac{\pi}{2} - \cos^{-1}\left(\frac{1.391\lambda}{\pi N d}\right)\right] \tag{2}$$

$$FNBW = 2\left[\frac{\pi}{2} - \cos^{-1}\left(\frac{\lambda}{N d}\right)\right] \tag{3}$$

$$D = 2N\left(\frac{d}{\lambda}\right) \tag{4}$$

3. Non-Uniform Amplitude Excitation Arrays

Non-uniform linear arrays are those in which the elements are fed with the currents of unequal magnitude. Some examples of the existing and well-known non-uniform arrays are Dolph-Tchebyscheff and Taylor arrays. Dolph arrays are considered as non-uniformly excited and equally spaced arrays in which the amplitudes of the antenna elements in the array are arranged according to the Dolph polynomial.

3.1 Dolph-Tchebyscheff Arrays

This type of array was originally introduced by the Dolph and it has many important and practical applications. The Dolph method is based on the properties of the Chebyshev polynomials, which gives the possibility of obtaining the maximum gain for a fixed or required level of the side lobes. The Chebyshev polynomial is defined as having equal ripples in the region bounded by x = ±1.

Additionally, the amplitude is varied between +1 and −1. The polynomial value outside this region rises exponentially [15, 16]. The polynomial can be computed according to the following equations

$$T_m(z) = \cos[m\cos^{-1}(z)] \qquad -1 \le z \le +1 \tag{5}$$

$$T_m(z) = \cosh[m\cosh^{-1}(z)] \qquad z < -1,\ z > +1 \tag{6}$$

Additionally, the half-power beam width and the directivity can be found by introducing a beam broadening factor, f, given approximately by [15]

$$f = 1 + 0.636\left\{\frac{2}{R_0} cosh\left[\sqrt{(Cosh^{-1}\ R_0)^2 - \pi^2}\right]\right\}^2 \tag{7}$$

While the directivity of this array type can be found as

$$D_0 = \frac{2R_0^2}{1+(R_0^2-1)\,f\,\frac{\lambda}{(L+d)}} \tag{8}$$

The simplest method to compute the half-power beam width of the Tschebyscheff arrays is by multiplying the half-power beam width of the uniform array by the beam broadening factor.

3.2 Taylor Arrays

Taylor distributions are another popular method in designing the arrays. It was first introduced by Taylor in 1955 and the design makes a compromise between beamwidth and sidelobe level which are the most wanted and desirable radiation features. In the theory, the side lobes are maintained at an equal with certain level. Since the side lobes are of equal level in such distributions and extend to infinity, this implies an infinite power [15]. In other words, the Taylor technique leads to a pattern whose first few side lobes (which they are closest to the main beam) are maintained at an equal and certain level; the remaining other side lobes decay monotonically. In fact, even the level of the closest side lobes exhibits a slight monotonic decay. This decay is a function of the space *u* over which these side lobes are required to be maintained at certain equal level. As this space increases, the rate of decay of the closest side lobes decreases [16]. For a large space of *u* (over which the closest side lobes are required to have an equal ripple), the rate of decay is negligible. Theoretically the normalized space factor that yields a pattern with equal-ripple side lobes is given by [15]

$$SF(\theta) = \frac{cosh\left[\sqrt{(\pi A)^2 - u^2}\right]}{\cosh(\pi A)} \tag{9}$$

The Taylor space factor is given by

$$Sf(u, A, \bar{n}) = \frac{\sin(u)}{u}\,\frac{\prod_{n=1}^{\bar{n}-1}\left[1-\left(\frac{u}{u_n}\right)^2\right]}{\prod_{n=1}^{\bar{n}-1}\left[1-\left(\frac{u}{n\pi}\right)^2\right]} \tag{10}$$

While u and u_n are obtained using the following equations

$$u_n = \pi v_n = \pi\,\frac{1}{\lambda}\,cos\theta_n \tag{11}$$

$$u = \pi v = \pi\frac{1}{\lambda}cos\theta \tag{12}$$

where θ_n represents the null locations. Additionally, the Taylor arrays have a parameter called scaling factor and it is represented by the symbol σ. This factor controls the spaces of the inner nulls so that

IOP Conf. Series: Materials Science and Engineering 1152 (2021) 012014 doi:10.1088/1757-899X/1152/1/012014

they blend smoothly with the outer ones. This factor also controls the width of the beam width in the Taylor pattern in which its beam width is usually greater than that of the Dolph-Tschebyscheff, and it can be given by

$$\sigma = \frac{\bar{n}}{\sqrt{A^2 + \left(\bar{n} - \frac{1}{2}\right)^2}} \tag{13}$$

Where the peak sidelobe level in decibels can be approximately calculated

$$SLL \cong -20\log(\cosh(\pi A)) \tag{14}$$

Finally, the half power beam width is given by

$$HPBW \approx 2\sin^{-1}\left\{\left[\frac{\lambda\sigma}{\pi l}\left[(cosh^{-1}R_0)^{-1} - \left(cosh^{-1}\frac{R_0}{\sqrt{2}}\right)^2\right]\right]^{1/2}\right\} \tag{15}$$

It is worth to mention the main applications of the non-uniform amplitude excitations arrays which can be summarized as:

- Non-Uniform amplitude excitations such as Dolph and Taylor are generally used to generate low side lobe level in which they are extensively used in the wireless communication systems to suppress the interfering signals.
- They compromise between the required or wanted sidelobe level and the desired half power beam width.
- Rectangular planar arrays which are used in radar applications can be created by combining many linear arrays parallel to each other [5].
- The Dolph and Taylor arrays can be extended to generate other types of beam pattern such as fan beam which is useful for coverage a wide space in one plane and generate a narrow beam in the orthogonal plane.
- These arrays have proved their effectiveness in the shipboard applications [15].

4. Simulation Results and Discussions

In this section, the results are obtained using the computer simulation and the theoretical approaches. In all considered arrays, the total number of the array elements is taken 10, 30, and 60 elements to illustrate the small, medium, and large array sizes. The array elements are distributed on the x-axis and on the linear form. The separation distance between any two successive elements is fixed and uniform at $d=\lambda/2$. For simplicity, the mutual coupling between the array elements and other scattering effects are neglected. For uniform amplitude excitation arrays, the amplitude of each element is chosen to be one and its phase is set to zero.

In the first example, the number of array elements is chosen to be N=10 elements. Figure 1 show the results of both Dolph and the Taylor array patterns. For comparison the uniform array pattern is also included in this figure. 'Figure 2' show the corresponding element amplitude excitations of these tested arrays. For Dolph and Taylor arrays, the required side lobe level is set in advanced during the design stage to be at -20 dB.

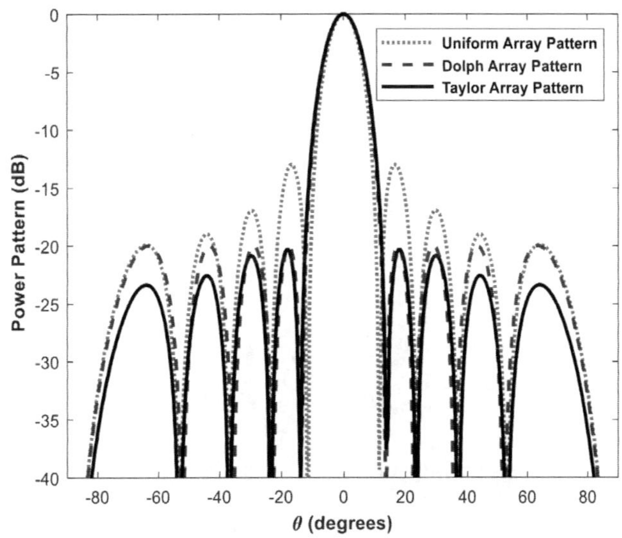

Figure 1. Radiation patterns for uniform, Dolph, and Taylor arrays and N=10 elements

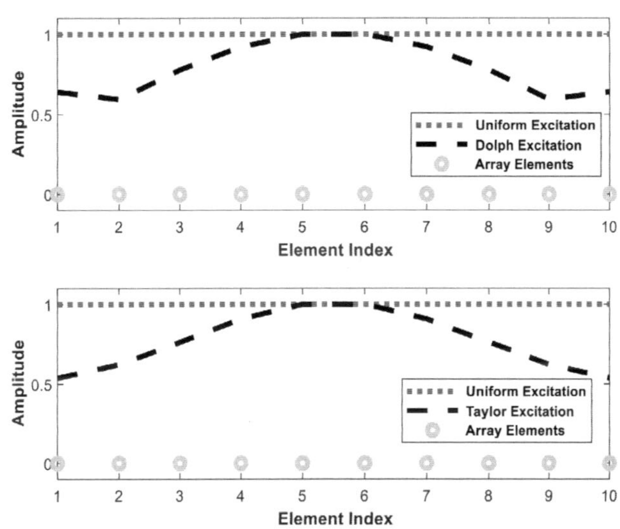

Figure 2. Amplitude excitations for the uniform, Dolph, and Taylor arrays and N=10 elements

In the second example, the number of the array elements is chosen to be N=30 elements. Figure 3 show the results of the uniform, Dolph and the Taylor array patterns, while Figure 4 show the corresponding element amplitude excitations of these tested arrays. In this case, the wanted side lobe level of the Dolph and Taylor arrays is chosen at -30 dB.

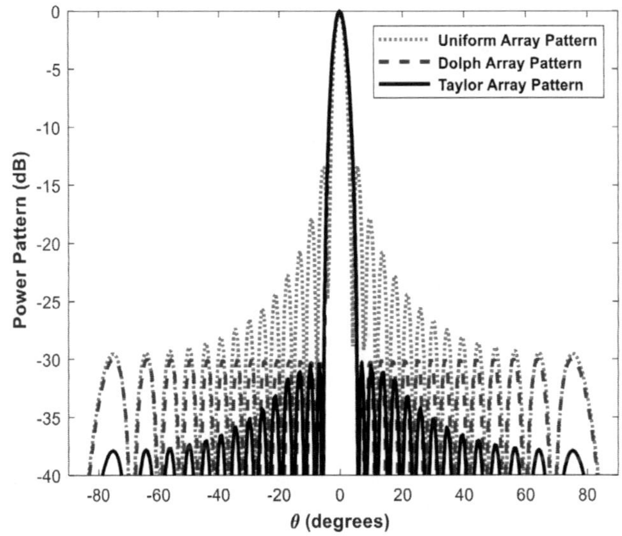

Figure 3. Radiation patterns for uniform, Dolph, and Taylor arrays and N=30 elements

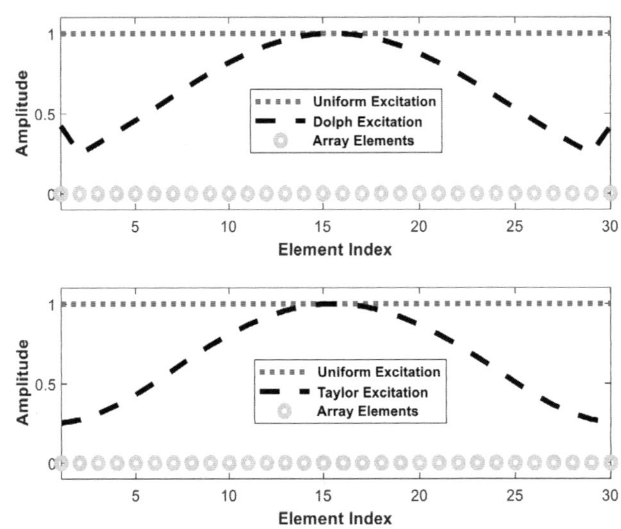

Figure 4. Amplitude excitations for the uniform, Dolph, and Taylor arrays and N=30 elements

In the third example, the number of the array elements is chosen to be N=60 elements. Figure 5 show the results of the uniform, Dolph and the Taylor array patterns, while Figure 6 show the corresponding element amplitude excitations of these tested arrays. In this case, the wanted side lobe level is chosen at -60 dB.

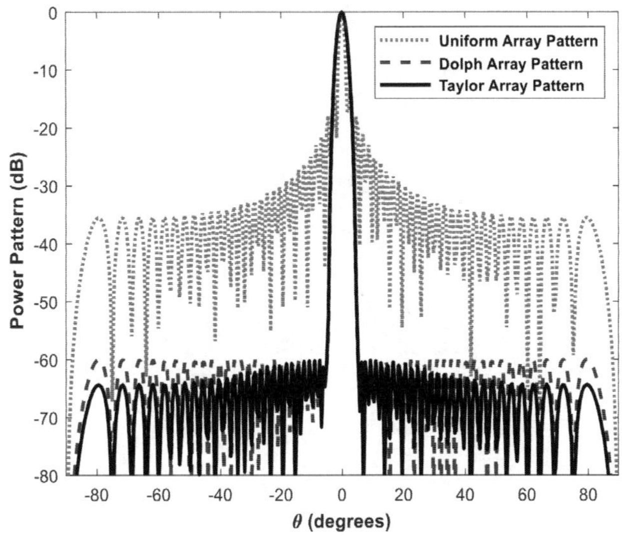

Figure 5. Radiation patterns for uniform, Dolph, and Taylor arrays and N=60 elements

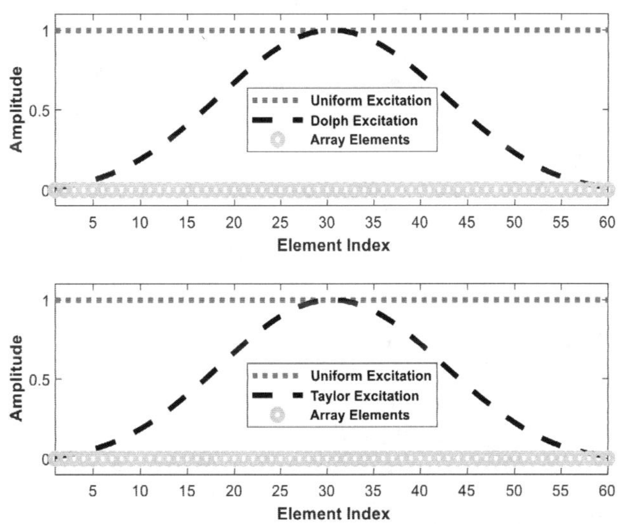

Figure 6. Amplitude excitations for the uniform, Dolph, and Taylor arrays and N=60 elements

From these figures, it can be seen that as the number of array elements increases the amplitude excitations of the Dolph and Taylor arrays become more identical to each other. Further, in all cases, the required side lobe levels have been effectively satisfied at the cost of getting wider beam widths and relatively lower directivities. In order to highlight these important features, some numerical calculations were carried out and presented in Table 1 below for a number of array elements equal to N=10, 16, and 30 elements respectively. In this table, the Dolph array is considered and the wanted side lobe level is chosen to be at -20 dB. Furthermore, two new important measures that compute the area under the main beam and the side lobe regions are introduced. By this way, we may have an interesting idea about how these areas may change which they are related to the directivity and the sidelobe. The area under the main beam region (AUMB) is computed by taking only the areas bounded between first null to null beam width, while the areas under the sidelobe regions (AUSL) are computed from areas that located outside the main beam region.

Table 1. Performance Measures of the uniform and Dolph Arrays

Features	Uniform Array			Non-Uniform Dolph Array		
	N=10	N=16	N=30	N=10	N=16	N=30
Directivity [dB]	10.01	12.05	14.78	9.84	11.87	14.42
HPBW[deg]	10.1	5.7520	3.0	11.20	6.520	3.52
Taper Efficiency	1	1	1	0.9622	0.9609	0.9207
Peak SLL [dB]	-13.23	-13.23	-13.23	-20	-20	-20
AUSL	0.038	0.041	0.093	0.0442	0.0531	0.068
AUMB	0.256	0.16	0.089	0.262	0.1796	0.0917

From this table, it can be seen that the area under the sidelobe region is generally decreasing with lower side lobe level. More important, the area under the main beam is found to be increased with reduced side lobe level. Thus, the directivity of the Dolph array is little lower than that of the uniform array.

5. Conclusions

From this study, it is found that there is a great need for making a compromise between the wanted reduction in the side lobe level and the obtained broadening in the main beam. Further it is found that the side lobe level is inversely proportional to the width of the main beam or the directivity of the array. Different array sizes are considered and studied to generate certain side lobe levels. The simulation results show that, in general, the uniform amplitude excitation arrays are able to provide higher directivity and narrower half power beam width among all other tested non-uniform excitation arrays. Nevertheless, the side lobe level of all designed Dolph and Taylor arrays are found to be satisfactory and satisfying the wanted low side lobe levels. Further, the performances of both Dolph and Taylor arrays were found almost same for large array sizes. The practical implementation of the non-uniform amplitude excitation arrays is also a great challenging issue where they usually require attenuators and the power dividers at each single element to realize the computed amplitudes.
This study can be further extended to include adaptive or optimization algorithms [18-20] that optimize between the side lobe level and the man beam width.

6. References

[1] Mohammed, J. R. and Sayidmarie, K. H. 2012 A New Technique for Obtaining Wide-Angular Nulling in the Sum and Difference Patterns of Monopulse Antenna, IEEE Antennas and Wireless Propagation Letters, vol.11, pp. 1242-1245.

[2] Mohammed, J. R. 2013 Design of Printed Yagi Antenna with Additional Driven Element for WLAN Applications, Progress in Electromagnetics Research C, Vol. 37, 67-81, ISSN:1937-8718.

[3] Mohammed, J. R. 2013 Phased Array Antenna with Ultra-Low Sidelobes, Electronics Letters, vol. 49, issue 17, pp. 1055-1056.

[4] Sayidmarie, K. H. and Mohammed, J. R. 2013 Performance of a Wide Angle and Wideband Nulling Method for Phased Arrays, Progress in Electromagnetics Research M, vol. 33, pp. 239-249.

[5] Mohammed, J.R, and Sayidmarie, K. H. 2014 Sidelobe Cancellation for Uniformly Excited Planar Array Antennas by Controlling the Side Elements, IEEE Antennas and Wireless Propagation Letters, vol.13, pp. 987 - 990.

[6] Mohammed, J. R. and Sayidmarie, K. H. 2014 A Null Steering Method by Controlling Two Elements, IET Microwaves, Antennas & Propagation, vol. 8, issue 15, pp.1348-1355.

[7] A., R., and K., A. 2016 Design and Analysis of Broadside Arrays of Uniformly Spaced Linear Elements. International Journal of Computer Applications, 156(6), 19–24. https://doi.org/10.5120/ijca2016912477

[8] Mohammed, J.R. 2017 Optimal Null Steering Method in Uniformly Excited Equally Spaced Linear Array by Optimizing Two Edge Elements, Electronics Letters, vol. 53, issue 11.

[9] Mohammed, J.R. 2018 Element Selection For Optimized Multi-Wide Nulls in Almost Uniformly Excited Arrays IEEE Antennas and Wireless Communication Letters. vol. 17, issue 4, PP. 629-632.

[10] K.P.D. 2017 Study of Broadside Linear Array Antenna with Different Spacing and Number of Elements. International Journal of Advanced Engineering Research and Science, 4(5), 190–194. https://doi.org/10.22161/ijaers.4.5.30.

[11] Mohammed, J.R. 2018 Thinning a Subset of Selected Elements for Null Steering Using Binary Genetic Algorithm", Progress in Electromagnetics Research M, vol. 67, pp. 147-157.

[12] Mohammed, J.R. 2019 Obtaining Wide Steered Nulls in Linear Array Patterns by Controlling the Locations of Two Edge Elements AEÜ International Journal of Electronics and Communications, vol. 101, pp. 145–151.

[13] Panduro, M. A., Mendez, A. L., Dominguez, R., & Romero, G. 2006 Design of non-uniform circular antenna arrays for side lobe reduction using the method of genetic algorithms. AEU - International Journal of Electronics and Communications, 60(10), 713–717. https://doi.org/10.1016/j.aeue.2006.03.006.

[14] Mohammed, J.R and Sayidmarie, K. H. 2017 Synthesizing Asymmetric SideLobe Pattern with Steered Nulling in Non-uniformly Excited Linear Arrays by Controlling Edge Elements International Journal of Antennas and Propagation, vol. 2017, Article ID 9293031, 8 pages.

[15] Balanis, C. A. 1996 Antenna Theory: Analysis and Design, 2nd Edition (2nd ed.). Wiley.

[16] Milligan, T. A. 2005 Modern Antenna Design. Wiley.

[17] Mohammed, J.R. 2017 Synthesizing Sum and Difference Patterns with Low Complexity Feeding Network By sharing Element Excitations International Journal of Antennas and Propagation, vol. 2017, Article ID 2563901, 7 pages.

[18] Mohammed, J.R and Sayidmarie, K. H. 2017 Performance Evaluation of the Adaptive Sidelobe Canceller with various Auxiliary Configurations AEÜ International Journal of Electronics and Communications, vol. 80, pp. 179–185.

[19] Mohammed, J.R. 2018 A New Antenna Array Pattern Synthesis Method with Sidelobe Control International Journal of Telecommunication, Electronics, and Computer Engineering, vol. 10. No.3, pp. 31-36.

[20] Mohammed, J.R. 2019 An Optimum Side Lobe Reduction Method with Weight Perturbation Journal of Computational Electronics, US Springer, vol.18, no.2, pp.705-711.

IOP Conf. Series: Materials Science and Engineering 1152 (2021) 012015 doi:10.1088/1757-899X/1152/1/012015

An Efficient Model for Low Voltage Distribution Smart Feeder Overcurrent Protection System Using Microcontroller

Akshay Kumar[1*] **and Viranjay M. Srivastava**[1]

[1]Department of Electronic Engineering, Howard College, University of KwaZulu-Natal, Durban - 4041, South Africa.

[*]E-mail: akshaylambha1996@gmail.com

Abstract. A Smart feeder is a growing energy distribution framework that is implanted with various types of energy distribution sources, as sustainable power sources, consolidated heat and power, and dispersed energy resources. The advantages of smart feeder arrangement are low transmission and dissemination cost, and, possibly, high dependability, high effectiveness, and low natural effect. Additionally, the administration of the electrical feeder system is getting increasingly problematic. The absence of data at the base feeder and the electricity administration network status have been distinguished as the significant obstruction to its successful monitoring and control of the electrical distribution system. This research work presents an advancement of the microcontroller-based smart feeder protection from overcurrent using fast using Solid State relay (SSR) to viable monitoring and control of the Feeder system. The simulation result of this work shows that the system is able to control and monitor the feeder protection system from overcurrent with the help of Microcontroller viable results that have been achieved with this system. In this research work, PhotoMOS relay cut-off time and overcurrent parameters have been calculated. The achieved results show significant results over the traditional EMR circuit protection. The PhotoMOS relay operating cut-off time 1.2, 1.3, and 1.4 ms with respect to the overcurrent values 0.8, 1.0, and 1.2 mA, which is significantly less than the EMR protection system. A 220 V electricity network was utilized to test the reliability and quality of the microcontroller-based PhotoMOS circuit. The proposed approach is demonstrated to be compelling in both grids associated and islanded modes.

1. Introduction

Smart feeder system performance and constant electricity distribution to the consumers are some of the significant problems nowadays. Therefore, fast safety and protection plans must be effective enough to conquer issues within a short period. Fault occurrence has to be detected rapidly by the distribution framework and promptly separated the forestall risks to the general public and utility staff [1, 2]. The sectional diagram of the smart feeder is shown in figure 1. Microcontroller integration in the relaying circuits improves the accessibility and reliable quality by constant self-monitoring of the overall distribution feeder framework [2-4]. Other than improving the presentation, microcontroller-based transfers likewise offer more noteworthy adaptability and the chance of including new capacities while simultaneously lessening the standard arrangement of hardware. Numerous favorable circumstances are conceivable, and the use of a microcontroller-based PhotoMOS relay is a received practice at present.

There is extensive existing research for microcontroller-based distribution system protection [6-9]. A microcontroller is an Integrated Circuit (IC) gadget utilized for controlling different parts of an electronic framework, ordinarily through a Micro-Processor Unit (MPU), memory, and a few

Content from this work may be used under the terms of the Creative Commons Attribution 3.0 licence. Any further distribution of this work must maintain attribution to the author(s) and the title of the work, journal citation and DOI.
Published under licence by IOP Publishing Ltd

peripherals. *Sabeel et. al.* [6] Feeder automation is the quickest developing pattern among utilities to improve distribution systems' dependability. One of the primary uses of feeder automation is programmed fault management. *Amare et. al.* [7], the future smart distribution feeder will comprise new parts and advancements with improved capacity whose fault conduct cannot be resolved with sureness. *Srivastava et. al.* [8] the smart protection system is essential to fulfilling the customer demand. *Patel et. al.* [9] the primary goals are to decrease the line losses, responsive power flows on the line, and to abstain from switching surge overvoltage because of turning on/off the capacitors. These devices are streamlined for implanted applications that require both preparing usefulness and nimble, responsive cooperation with digital, analog, or electromechanical parts [10, 11]. In the early 70s, a microcontroller was introduced for the first time, including 4 to 8 bit Micro-Controller Units (MCUs). After advanced developments, most microcontrollers being used today are inserted in other machine devices, for example, the auto sector, phones, machines, grid solutions, and peripherals for PC frameworks [12]. A microcontroller framework can be utilized for identifying issues in the feeder systems. The continuous observation of the various parameters and equipment types in the feeder framework to recognize electrical faults [13].

The microcontroller in smart grid technology is essential with any electric apparatus that stores, measures, show data, or computes a microcontroller chip inside it. The microcontroller device for monitoring and controlling the Smart distribution feeder is one of the most utilized ways to provide efficient and reliable electric power to users [14, 15]. Microcontroller additionally gives monitoring and protection to loads.

For the most part, they have some critical and non-critical loads. Critical loads ought to never be turned off. Non-critical loads can be turned on or off contingent upon Power utilization and distribution. In-home, we have an unbounded supply by distribution feeder. Be that as it may, power utilization is restricted by current suffering constraints of distribution wiring. At the point when the consumer turns more loads on, the current flowing into principal wire increments [14]. In the event that the consumer keeps on turning an ever-increasing number of loads, at that point, a cut-off is arrived at where the current surpasses the wiring limit causing the PhotoMOS relay to send a signal to the Circuit breaker to open the circuit [16]. As it may, switch all loads off if the overload current exceeds is not a smart idea. Here is our efficient model of LV distribution smart feeder overcurrent protection system that comes to play a crucial role and save the circuit and equipment from overcurrent [17].

The significant switching time of the system for the open and closed-circuit operations have been obtained using the Solid-State Relay device that resulted in switching time ratios (T_{OFF}/T_{ON}) of 0.03 and 0.06, respectively. The result shows the considerable gain in the operational switching time parameter using the solid-state relay device over the conventional Electromechanical (EM) relays [5, 11]. The obtained results also assure that the SSR is a useful relay circuit element. In this present model, the authors have proposed a protection framework consisting of an inlet with an AC 220 V transmission line as the source that is associated with a step-down transformer (220/6 V).

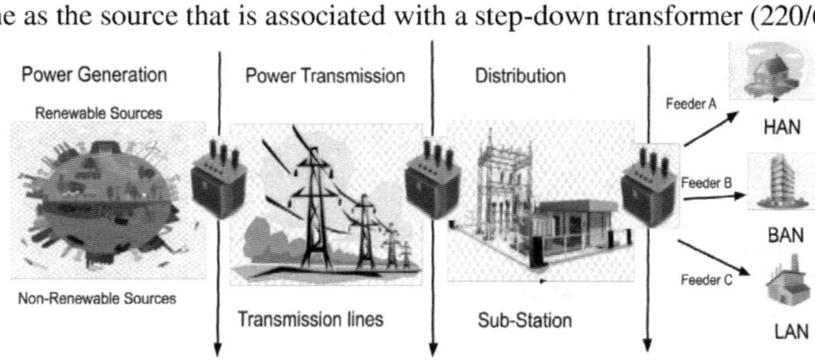

Figure 1. LV distribution smart feeder sectional diagram.

This paper has been organized as follows. Section 2 introduces the Smart Distribution Feeder structure construction. Section 3 presents the SSR interfacing with microcontroller. Section 4 presents the feeder protection system model using a fast-switching relay, Section 5 shows the protection system results. Finally, Section 6 concludes the work and recommends the future aspects.

2. System Description for LV Distribution Smart Feeder

The Low Voltage (LV) distribution smart feeder framework conveys power from the transmission network and carries it to consumers [18]. The smart feeder system, being an intermediary element for the distribution of power to the regional transformer. In the power system, electricity conveyance is considered to be the end-stage of power distribution to the consumers. With the power stations, it is necessary to decrease the transmission losses in overhead lines for that produced electricity has to be stepped up to a higher level so the power losses in produced electricity transmission will be less and make it increasingly productive [19]. To make transmission voltage progressively usable for the consumer's transmitted line voltage has to be stepped down. The transmission electrical cables will enter a dispersion substation where the voltage will be stepped down to low voltage levels, then it will be appropriated for use by industries, businesses, and residential consumers, as shown in figure 2. Power Reliability Assessment (PRA) to meet the consumer requirement with respect to energy, is the most significant parameter for suitable healing measures to give a continuous and safe power supply to end consumers.

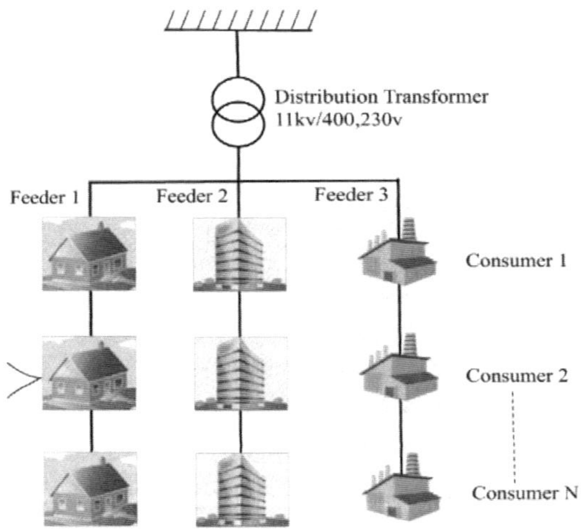

Figure 2. Smart feeder layout design.

Residential users regularly get the power from overhead or underground feeders exuding from utility-substations. The LV distribution smart feeder overcurrent protection and monitoring system is a key apparatus for power distribution utility to make the power system sound, solid, sheltered, and effective without contributing a lot. There are four types of LV distribution feeder systems used: Ring main, Meshed system, Radial, Parallel feeders, etc. [20]. In this proposed low voltage feeders exuding from a substation are observed by microcontrollers, relay circuits, and a circuit breaker, which will open when a fault is identified. Programmed microcontroller circuit Re-closers are also introduced to additionally isolate the feeder in this way, limiting the effect of faults.

3. Proposed Algorithm and Working for Interfacing Solid-State Relay (SSR) With Microcontroller

In this research work, the authors effectively interfaced Solid-State Relay (SSR) with a microcontroller utilizing a transistor. The transistors work as a switching circuit, conveying the necessary current requirements to the relaying circuit operation. Figure 3 shows the basic structure of a PhotoMOS relay device. Detailed working has been provided in the author's previous research work [11].

A microcontroller is a multipurpose, programmable, clock-driven register-based electronic device that peruses double guidelines. Since microcontrollers work on just binary codes, it's important to convert over our analog to digital signal utilizing Analog to Digital Converter (ADC) [21]. The transformation procedure of analog to digital sign and microcontrollers working appeared in figure 3. When pin PD7 of the PIC microcontroller goes high, the transistor BC130 turns-ON, and the current courses through the relay circuit. One diode d_1 associated with an arrangement is used to shield the transistor and the microcontroller during the relay operations.

Subsequent to getting this data-dependent on the programming given underneath in the relay data memory Read Only Memory / random Access Memory (ROM/RAM), the microcontroller takes the action of the tripping of the solid-state relay [22]. To calculate the relay tripping, the following values have been used.

$$T= \tau(s)((I \mid IP)^2-1)-K_d (\partial\theta \mid \partial t) \quad (1)$$

where $\tau(s)$ is the torque, I and I_p are the applied current and current value when Light Emitting Diode (LED) illuminate, respectively. The constants K_d and Θ are the damping factor and angle, respectively. Therefore, the setting angle will be:

$$TDS = \tau(s)/K_d((I \mid I_p)^2-1) (t) \quad (2)$$

Using this equation (2), trip time (t) can be calculated.

The proposed algorithm and work for interfacing Solid-State Relay (SSR) with microcontroller showed this algorithm.

```
==========================================================================

void main()
{
TRISB.F7 = 0;/Makes PD7 a output pin
do
{
PORTB.F7 = 1;//turns on ssr circuit
Delay ms(0.1);//0.1 ms delay
PORTB.F7 = 0;//turns off relay
Delay ms(0.1);/0.1ms delay
} while(1);
}

==========================================================================
```

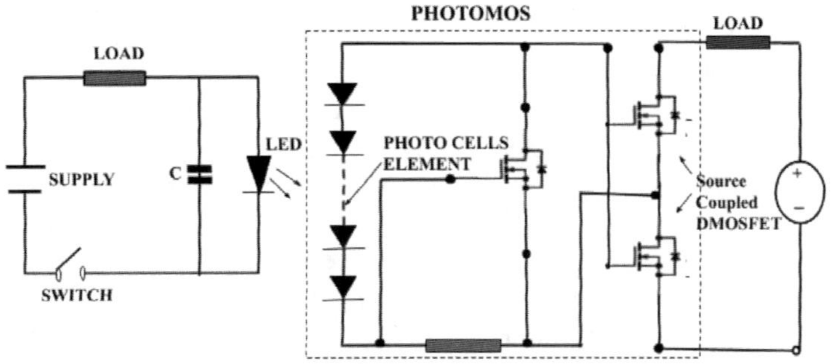

Figure 3. PhotoMOS relay circuit.

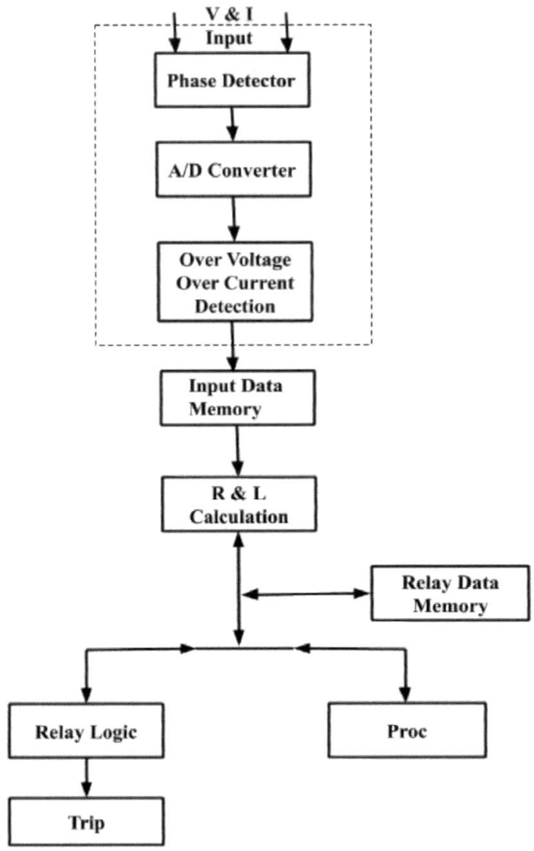

Figure 4. Program for relay interfacing and control.

The operation time of microcontroller-based relay circuits is given in the binary code very low so that the relay tripping operation will be fast, and our feeder system will be safe from the faults [23]. Hence the microcontroller based solid-state relay will be very in-line with the LV distribution feeder system.

4. Proposed Model of Microcontroller Based Monitoring and Protection Using SSR

In the power system, it is common for the feeders to experience operational contortion or faults inferable from the presentation to a few encompassing conditions. In a Smart feeder system, the microcontroller-based PhotoMOS relay circuits are liable for detecting the over-values (voltage, current, resistance, or reactance), list the fault parameters, and further convey the fault data to administrative control frameworks to distinguish the fault area [24].

In that circumstance, fusing a microcontroller-based PhotoMOS relay that gives numerous desirable advantages, for example, accurate tripping, less disturbance, and enhanced fault parameter to the resulting framework appear to be proper. A PhotoMOS (SSR) is an electronic switch that works on Photovoltaic impacts to switch High Voltage/Current using low power circuits utilizing. For fast switching, MOS innovation is applied to its control terminals [25]. To associate an SSR with a microcontroller, it needs a driving circuit shown in figure 5.

In figure 4, an effective LV distribution smart feeder protection system has been proposed using a Microcontroller-based solid-state relay circuit in operation [26]. The proposed model embodies the benefits of microcontroller-based solid-state relay circuits inside the general system that comprise other structure squares, such as step-down transformer (6 V), rectification assembly, sensing devices transduction, and an automation microcontroller to enable the operation of the circuit breaker.

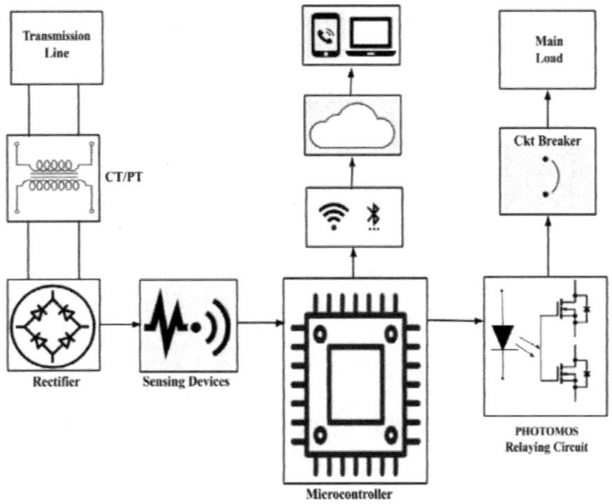

Figure 5. Proposed model of feeder protection system.

For example, with the assistance of a rectifying circuit, a bridge rectifier is associated in series to convert the feeder voltage from a 220 V AC supply to a 6 V DC. This DC supply connected with sensing devices to observe the circuit's abnormalities and provide information to microcontroller circuit units [27-31]. Any change in AC voltage can bring a power cut or a weak signal. Thereafter, such changes are logically evaluated.

5. Simulation Results

The simulation of microcontroller-based LV feeder protection system design and results have been shown in figure 6 and figure 7. The results of the overcurrent analysis of dynamic LV feeders utilizing microcontroller show that if the overcurrent value increases, the cut-off value of the SSR inversely decreases, as shown in figure 7. In this research work, relay, cut-off time, and overcurrent parameters have been calculated. The cut-Off time of the relay has been achieved using simulations and equations given above. With the calculation value of equation (1) to equation (3), authors have successfully calculated the relay trip circuit values. The achieved results show significant results over the traditional EMR circuit protection relay operating cut-off time 1.2, 1.3, and 1.4 ms with respect to the overcurrent values 0.8, 1.0, and 1.2 mA, which is significantly less compared to the EMR protection system.

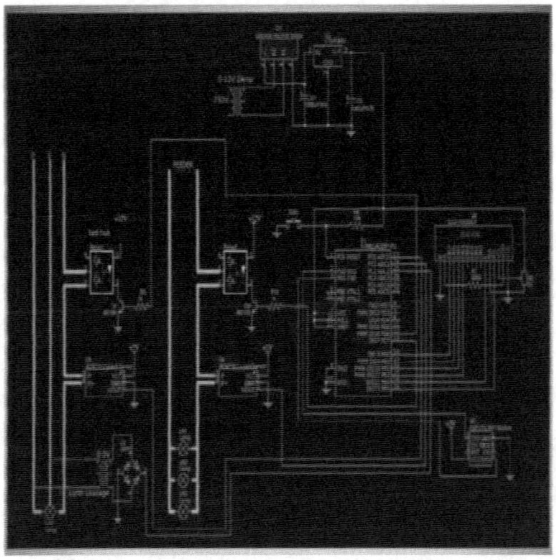

Figure 6. Microcontroller based feeder protection system simulation.

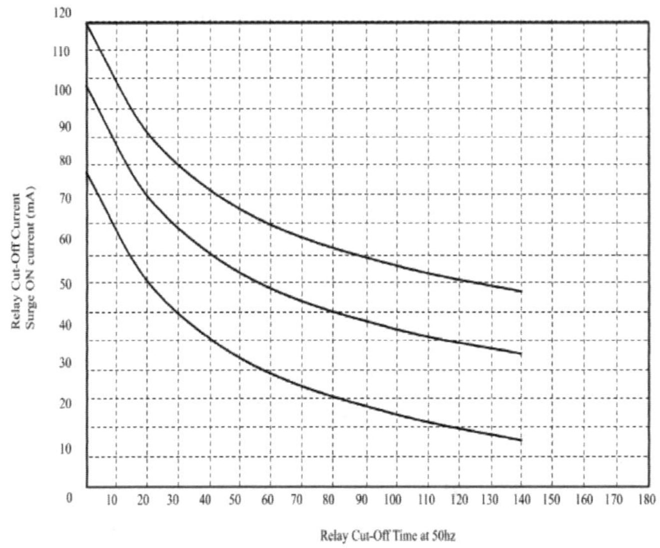

Figure 7. Relay Cut-off time characteristics.

It is commonly a decent practice to adjust down to the closest incentive to give extra headroom. Using a graph, for example, the one underneath will also help choose a cut-off value of our proposed model. Overall, to protect the LV feeder system, three tests have been conducted based on the results proposed circuit of the microcontroller-based system shows very positive outcomes for future research work. These characteristics show the cut-off time and overcurrent values.

6. Conclusion and Future Recommendation

This work furnishes that microcontroller-based LV feeders' performance using a solid-state relay is very significant for the electrical distribution system. Microcontroller and PhotoMOS relay play a crucial role in present-day power system security to detect and segregate various sorts of electric circuit faults. The choice of SSR and Microcontroller relies upon power rating, voltage, current rating, the impact of outside elements, etc. The design and implementation of the security of a disturbed electric system utilizing a microcontroller-based defensive system given. Because of the utilization of semiconductor, arc less fast switching of the system is conceivable with which the effectiveness, unwavering quality, and lifetime of the insurance unit increments.

An electric system's safety against overcurrent, under-voltage, and earth fault utilizes a robust microcontroller presently unveiled. However, it is suggested that different methodologies ought to be utilized to supplement this methodology. The assurance of the power network framework is a fundamental and unavoidable procedure to guarantee the LV Feeder's security since the faults are inescapable. The proposed model can further be utilized to monitor and control relaying circuits.

7. References

[1] Jones D and Kumm J J 2014 Future distribution feeder protection using directional overcurrent elements *IEEE Transactions on Industry Applications* **50(2)** 1385-1390.

[2] Eluvathingal A V and Swarup K S 2019 Protection scheme for smart distribution networks with inverter interfaced renewable power generating sources *Lecture Notes in Electrical Engineering* **580** 254-260.

[3] Chen L 2016 Overcurrent protection for distribution feeders with renewable generation," *International Journal of Electrical Power & Energy Systems* **84** 202-213.

[4] Baran M and El-Markabi I 2004 Adaptive over current protection for distribution feeders with distributed generators *IEEE PES Power Systems Conference and Exposition* (New York, USA) 715-719.

[5] Zhuang W, Dai J, and Liu Y 2019 Application of solid-state circuit breaker in selective relay protection of marine dc distribution system *IEEE Workshop on the Electronic Grid* (China) 1-6.

[6] Sabeel N, Alam A and Zaid M 2019 Feeder automation based strategy for reliability enhancement of radial distribution systems *International Conference on Power Electronics, Control and Automation* (India).

[7] Amare T and Helvik B E 2018 Dependability analysis of smart distribution grid architectures considering various failure modes *IEEE PES Innovative Smart Grid Technologies Conference* (Sarajevo) 1-6.

[8] Srivastava A and Parida S K 2019 Frequency and voltage data processing based feeder protection in medium voltage microgrid *IEEE PES Innovative Smart Grid Technologies Europe*, (Romania) 1-5.

[9] Patel Y, Panchal B, Parmar S, Shinde K and Prajapati P 2019 Power factor correction with microcontroller use solid state relay *Journal for Research* **5(2)** 681-685.

[10] Kacar S 2016 Analog circuit and microcontroller based rng application of a new easy realizable 4d chaotic system *Elsevier Optik* **127(20)** 9551-9561.

[11] Kumar A and Srivastava V M 2020 A novel feeder protection system using fast switching photomos relay *International Conference on Computing, Communication and Networking Technologies* (India) 1-6.

[12] Metering A S, Visalatchi S and Sandeep K K 2017 Smart energy metering and power theft control using Arduino & GSM *International Conference for Convergence in Technology* (India) 858-961.

[13] Marie T F B, Han, H and An B 2019 Microcontroller design for security system: implementation of a microcontroller based on STM32F103 microchip *International Journal of Embedded Systems* **11(5)** 541-550.

[14] Suteerawatananon S, Chompusri, Y, Charbkaew N and Bunyagul T 2018 Design of a low-cost microcontroller-based high impedance fault detector *International Conference on Electrical Engineering/Electronics, Computer, Telecommunications, and Information Technology* (Thailand) 552-555.

[15] Ali M A and Bendary F M 2013 Design of prototype non-directional overcurrent relay micro-controller-based *International Conference and Exhibition on Electricity Distribution*, (Stockholm) 1-4.

[16] Jhunjhunwala S, Pandey K and Kumar R 2018 A microcontroller based embedded system to provide complete self-protection (CSP) to any distribution transformer *International Conference on Power Energy, Environment and Intelligent Control* (India) 307-310.

[17] Chen C, Ku T and Lin C 2012 Design of phase identification system to support three-phase loading balance of distribution feeders *IEEE Transactions on Industry Applications* **48(1)** 191-198.

[18] Khadar A A, Hayath T M and Nagraj M S 2017 Design and implementation of ZigBee based smart grid system for power management *International Conference on Smart Technologies For Smart Nation* (India) 1339-1343.

[19] Rodriguez D F C, Osorio J D P and Ramos G 2018 Virtual relay design for feeder protection testing with online simulation *IEEE Transactions on Industry Applications* **54(1)** 143-149.

[20] Manditereza P T and Bansal R C 2018 Introducing a new type of protection zone for the smart grid incorporating distributed generation *IEEE Innovative Smart Grid Technologies* (Singapore) 86-90.

[21] The panasonic electric works Europe website 2020 Available: https://www.panasonic-electric-works.com/eu/photomos-relays.

[22] An W, Ma J J, Zhou H Y, Chen H S, Jun X and Jian X 2019 Application of an integrated protection and control system for smart distribution grid based on PTN and 4G LTE communication *International Conference on Smart Grid and Smart Cities*, (USA) 70-75.

[23] Mbunwe M J, Ogbuefi U C, Anyaka B O and Ayogu C C 2018 Protection of a disturbed electric network using solid state protection devices *World Congress on Engineering and Computer Science* (USA) 76-82.

[24] Mehta P and Makwana V H 2018 Radial feeder protection by definite time overcurrent relay *International Conference on Intelligent Systems and Signal Processing* (Singapore) 185-198.

[25] Hameed A A, Sultan A J and Bonneya M F 2020 Design and implementation of a new real-time overcurrent relay based on Arduino *International Conference on Materials Science and Engineering* 012005.

[26] Swathika O G and Hemamalini S 2015 Adaptive and intelligent controller for protection in radial distribution systems *Advanced Computer and Communication Engineering Technology* **362(4)** 195-209.

[27] Leeuw S and Srivastava V M 2021 Realization with fabrication of double-gate MOSFET based buck regulator *International Journal of Electrical and Electronic Engineering & Telecommunications* **10(1)** 66-75.

[28] Srivastava V M 2016 Schematic of boost regulator with DG MOSFET: A device modeling for energy transmission *13th Int. Conf. on Industrial and Commercial Use of Energy* (South Africa) 256-261.

[29] Shabani A and Mazlumi K 2020 Evaluation of a communication-assisted overcurrent protection scheme for photovoltaic-based dc microgrid *IEEE Transactions on Smart Grid* **11(1)** 429-439.

[30] Kavi M, Mishra Y and Vilathgamuwa M 2018 Morphological fault detector for adaptive overcurrent protection in distribution networks with increasing photovoltaic penetration *IEEE Transactions on Sustainable Energy* **9(3)** 1021-1029.

[31] Kumar A and Srivastava V M 2020 Proficient model of monitoring and controlling of low voltage distribution smart feeder protection system using IoT *International Journal of Advances Trends in Computer Science and Engineering* **9(5)** 9131-9137.

Antenna Azimuth Position Control Using Fractional Order PID Controller Based on Genetic Algorithm

Ali Mahmood[1*], Mohammed Almaged[1] and Abdulla I. Abdulla[1]

[1]Ninevah University, AL-majmoaa Street, Mosul 41002, Iraq

[*]E-mail: ali.mahmood@uoninevah.edu.iq

Abstract. This work shows the design procedure of fractional-order proportional–integral-derivative (FOPID) controllers for an antenna azimuth position system based on genetic algorithm optimization approach. FOPID controllers are considered as a special type of the classical PID controller in which the derivative and integral elements have orders of fractions between zero and one. Therefore, FOPID controllers comprise two additional variables (μ and λ) in comparison with the typical PID controller. The FOPID is designed to control the azimuth angle of the antenna. Genetic algorithm (GA) will be employed to find an optimal value for the five FOPID controller's gains using different type of fitness function including mean square error (MSE), integral square error (ISE) and integral time square error (ITSE). The achieved results exhibit an excellent steady and transient state performance of system, for instant fast settling and rise times as well as low present of peak overshoot.

1. Introduction

Antenna azimuth position control typically involves elevation control so that antennas, solar panels and sensors are accurately positioned. For instance, antennas are generally perpendicularly directed to an earth location while solar panels should be placed towards the sun for maximum power generation [1]. In general, the classical PID controller is considered as the leading type of feedback control in the industry. This is mainly because of their simple design and great ability to produce reasonable transient and steady state responses [2][3]. However, classical controllers reveal some downsides that may include sensitivity to variation in system parameter, performance decrease while system order increases and finally poor functioning with nonlinear systems [4], [5]. Therefore, FOPID controllers have lately gained great attention in the area of control system design in which using a non-integer PID controller is a successful way of enhancing the response of their traditional counterparts [6]. It has been concluded that FOPID controller has more freedom in the tuning process of the control parameters. In comparing with classical PID, fractional order PID reveals a better response characteristic at high order and nonlinear systems [7]. However, the design process of fractional-order PID contains some amount of complexities when compared with typical PID controllers. This is mainly due to the fact that FOPID has two additional integral tuning parameters that should be taken into consideration [8].
Lately, a significant amount of literature has been presented to discuss the tuning approaches of FOPID controller. These techniques are mainly classified as optimized, analytical and graphical approaches [9]. In optimization-based approach, the gain parameters of the controller are calculated to achieve some preferable time domain characteristics. Genetic algorithm is considered as one of the most commonly applied optimization techniques for tuning such controllers. This tuning approach is initially depending

Content from this work may be used under the terms of the Creative Commons Attribution 3.0 licence. Any further distribution of this work must maintain attribution to the author(s) and the title of the work, journal citation and DOI.
Published under licence by IOP Publishing Ltd

on estimating the fractional order calculus, then setting other gain parameters of the controller similarly [10].

2. Antenna Model

The model of the antenna azimuth control system that has been used is shown in figure 1. This system has five main subsystems. The potentiometers used to convert the azimuth angles (in deg.) to voltages while the power amplifier is used as the drive circuit for the motor. The motor and the load are presented in one system where its equation takes all the necessary parameters. Finally, the gearbox gives the gear ratio to increase the applied torque while reducing the speed. For the preamplifier used in figure 2, it will be replaced with the FOPID controller [11].

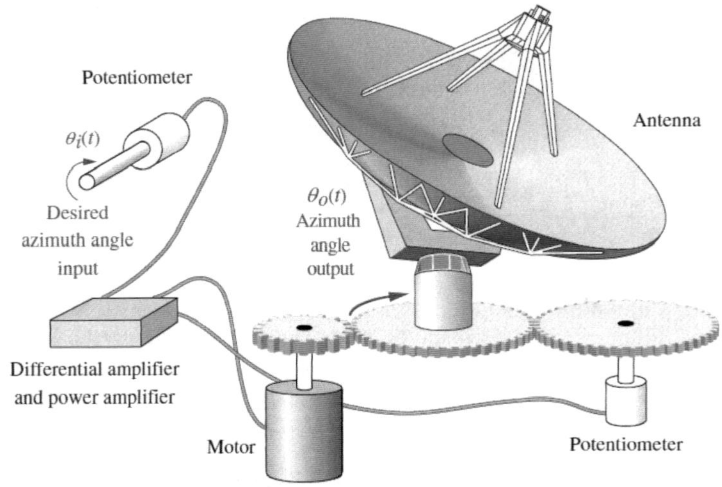

Figure 1. Layout of the antenna azimuth position control system [11].

In the antenna system, the schematic block diagram is represented in figure2 where it shows the physical subsystems with their parameters and internal connection.

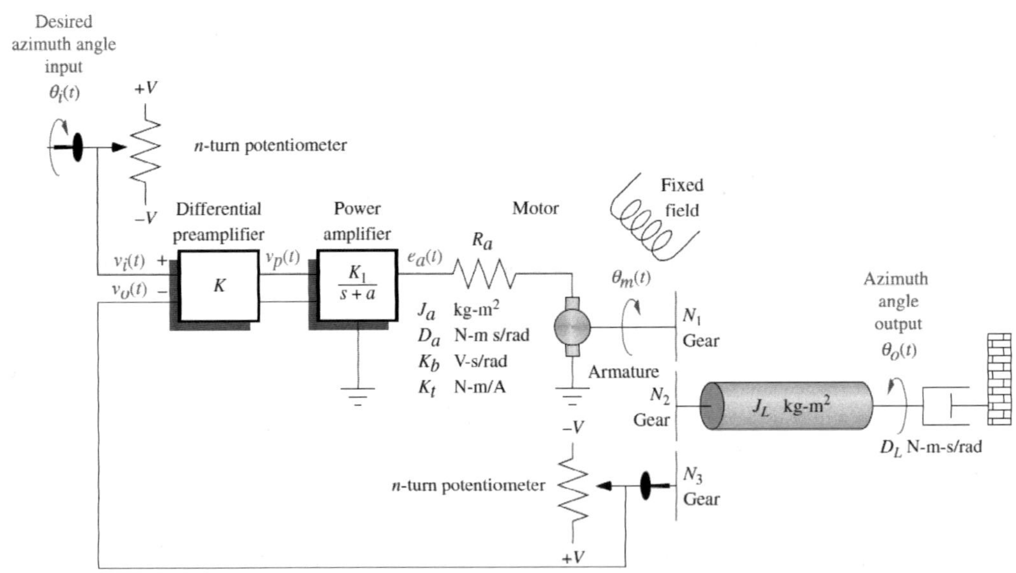

Figure 2. Schematic diagram of the antenna azimuth position control system [11].

Where:

J_a *and* J_L are motor inertial and load inertial constants respectively ($kg.m^2$)

D_a *and* D_L aremotor dampening and load dampening constants respectively (N-m s/rad)

R_a: Motor resistance (ohm)
$K_b$is back EMF constant (V.s/rad).
$Kt$is motor torque constant (N-m/A).
$N1$ and $N2$ are the gear teethes.

The block diagram shown in figure 3 represents the close loop system of the antenna azimuth position control system.

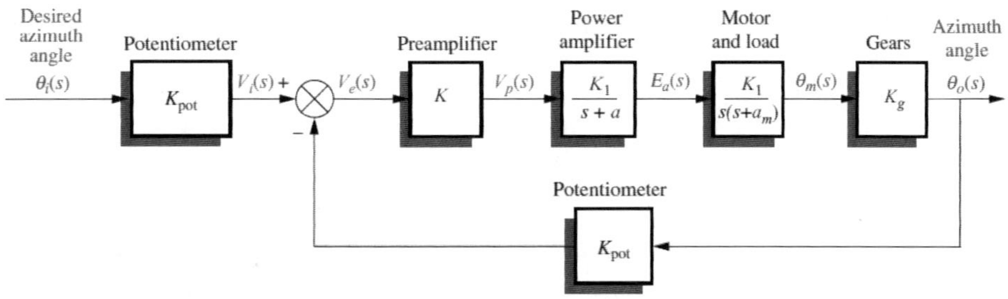

Figure 3. Block diagram of the antenna azimuth position control system [11].

By using equations (1-4), the transfer function of the motor and load can be represented in the simplified equation (5).

$$K_m = \frac{K_t}{J\,R_a} \quad (1)$$

$$a_m = \frac{D_m R_a + K_b K_t}{J R_a} \quad (2)$$

$$J = J_a + J_L(K_g)^2 \quad (3)$$

$$D_m = D_a + D_L(K_g)^2 \quad (4)$$

$$\frac{\phi m(s)}{E_a(s)} = \frac{K_m}{s(s+a_m)} \quad (5)$$

The gear ratio used is represented in equation (6) where it used to increase the speed supplied by the motor while increasing the provided torque. This is useful for the adjustment of the inertial components and the damping.

$$K_g = \frac{N_1}{N_2} \quad (6)$$

The overall parameters used for this system are represented in table 1.

Table 1. Schematic and block diagram parameters [11]

Schematic Parameter	Value	Block diagram parameters	Value
V	10	**K_{pot}**	0.318
n	10	**K1**	100
R_a	8	**a**	100
J_a	0.02	**K_g**	0.1
D_a	0.01	**K_m**	2.083
K_b	0.05	**a_m**	1.71
K_t	0.5		

N1	25
N2	250
N3	250
J_L	1
D_L	1

After substituting all the parameters in table 1, the overall equation of the azimuth of the antenna system is given by equation (7).

$$\frac{\theta_o(s)}{\theta_i(s)} = \frac{318}{s^3+101.7s^2+171\, s+318} \tag{7}$$

3. Fractional PID Controller Design

The fractional Order PID controller, also referred to as PI λ D μ , depends on the fractional calculus that includes two added parameters, which are (λ , and μ), with regards to the classical PID controller. These two additional parameters tend to increase the flexibility and robustness of the controller and enhance the time domain performance of the system. The main structure of the fractional order PID controller applied to closed loop systems is illustrated in figure 4.

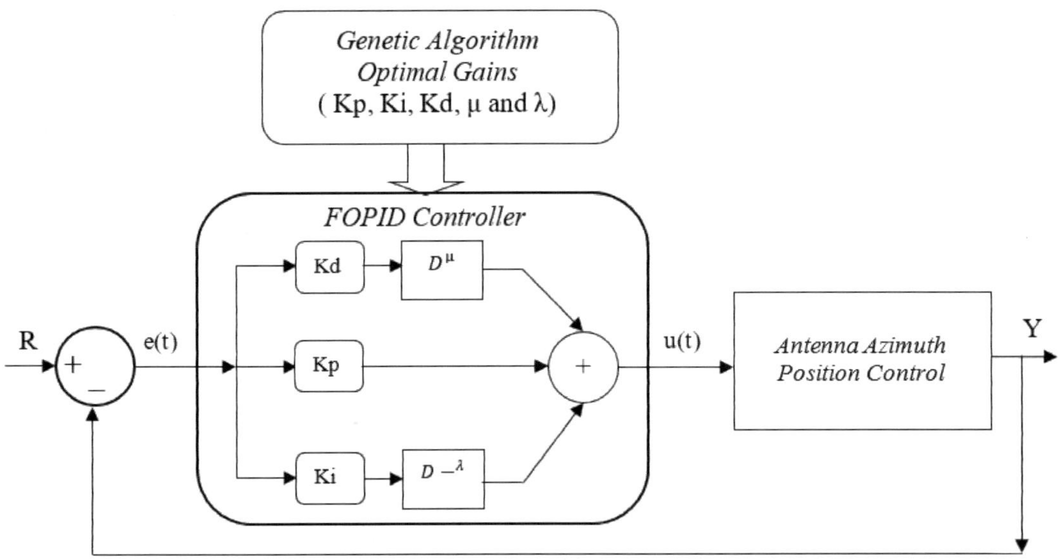

Figure 4. General FOPID architecture

The main equation of FOPID controller that defines the control action u(t) is given in equation (8).

$$u(t) = kp\, e(t) + ki\, D^{-\lambda}\, e(t) + kd\, D^{\mu}\, e(t) \tag{8}$$

The FOPID controller is designed depending on the fractional order theory. The fractional differential and integral can be written by using as ${}_aD_t^{\mu}$and ${}_aI_t^{\lambda}$. The symbol μ represents the differential order and λ is the integral order while a and t are the lower and upper bounds. There are mainly two kinds of fractional calculus. These are Riemann-Liouville and Grunwald-Letnikov calculus [12], [13]. Under zero initial condition, the Riemann-Liouville calculus can be defined in the S-domain as in equation (9)

$$\int_0^{\infty} e^{-st}\; {}_aD_t^x\, f(t)\, dt = S^x F(s) \tag{9}$$

IOP Conf. Series: Materials Science and Engineering 1152 (2021) 012016 doi:10.1088/1757-899X/1152/1/012016

By applying Riemann-Liouville calculus equation with Laplace transform to the control action equation, the controller output will be as represented in equation (10).

$$u(s) = \left(kp + \frac{ki}{s^{\lambda}} + kd\, S^{\mu}\right) e(s) \text{ Where } (\lambda \,\&\, \mu > 0) \qquad (10)$$

4. Genetic Algorithm

Genetic Algorithm (GA) is one of the most applied optimization search engines that imitates the chromosome natural development process. It starts with the representation of the chromosome structure thatis formed by all the elements that needs to be optimized to achieve the required response. Fractional order PID design procedure requires a selection of five gains values (Kp, Ki, Kd, μ and λ) in order to achieve an optimal output response. In general, GA comprises three main stages: Mutation, Crossover and Selection to enhance the initial population and hence produce a better response. The validity of the obtained solution is evaluated using a fitness function which is the measurement of the chromosome's quality. The objective functions employed in this particular design process are the Integral Time Square Error (ITSE), Integral Square Error (ISE) and Mean Square Error (MSE) that have the following equations [14]:

$$\text{ITSE} = \int_0^T t(e(t))^2 dt = \int_0^T t(r(t) - y(t))^2 dt \qquad (11)$$

$$\text{ISE} = \int_0^T (e(t))^2 dt = \int_0^T (r(t) - y(t))^2 dt \qquad (12)$$

$$\text{MSE} = \frac{1}{n} \sum_{i=1}^{n} (e(t))^2 \qquad (13)$$

The genetic algorithm process can be summarized in the flowchart shown in figure 5.

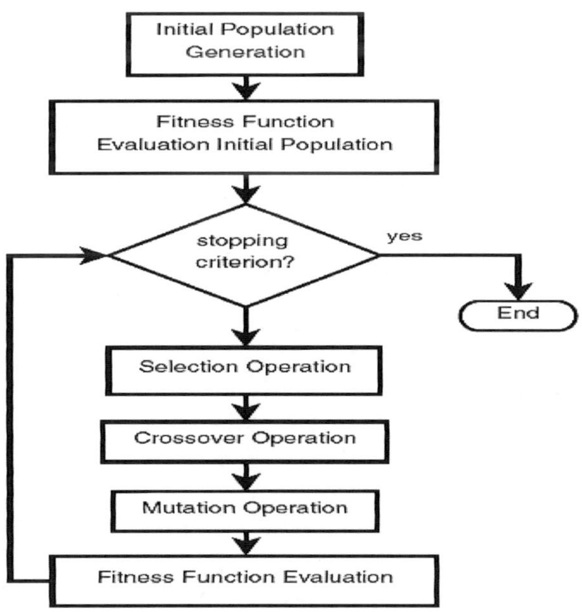

Figure 5. Genetic algorithm flowchart [13]

As per design requirements, maximum overshoot (Mp), Rise time (Tr) and Rise time (Ts) are required to be minimized, the fitness function selection was depending on that. ISE has the square function were used for dealing with the positive and negative values of the error and makes the system's performance faster while the integration is to eliminate the steady state error when times goes large. ITSE has an extra element over ISE, which is the time; this is for further improvement of the system's speed; however,

it could slightly increase the maximum overshoot. Finally, MSE has advantage over ISE and ITSE regarding the overshoot where it further decreases the maximum overshoot with no effect on the other specifications (Tr and Ts).
The main configuration of genetic algorithm tuning process for this particular design is implemented using a population size of 50 and 100 maximum number of generations with 0.05 and 0.2 selections and crossover probabilities respectively.

5. Simulation and Results

The first step in the simulation is to check the stability of the open loop system using Bode plot as shown in figure 6. As the gainmargin (Gm) and phase margin (Pm) are positive, the system is stable.

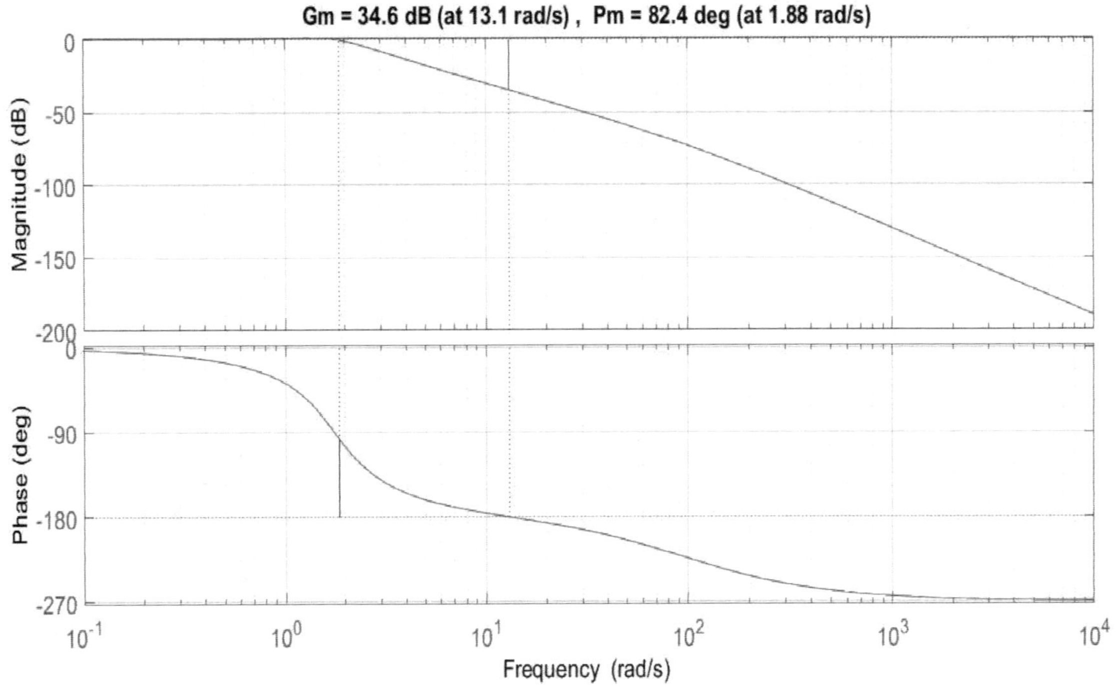

Figure 6. Stability analysis of the open loop system using Bode plot.

The parameters of GA-FOPID controllers are shown in figures 7, 8 and 9 for MSE, ISE and ITSE fitness functions respectively, while the step responses of GA-FOPID controllers for MSE, ISE and ITSE fitness functions are plotted in figure 10.

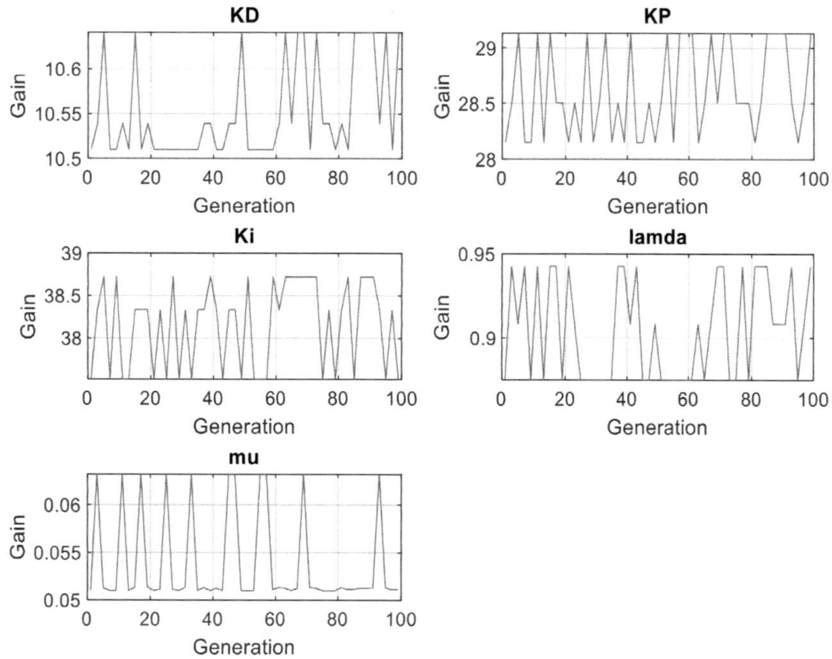

Figure 7. Generation's number of Kp, Ki, Kd, μ and λ value GA- FOPID controller for MSE fitness function.

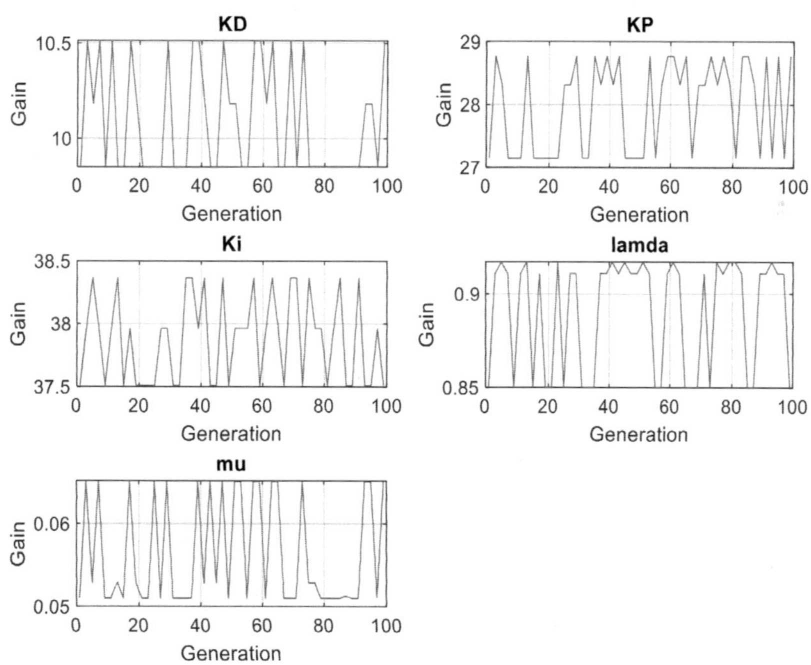

Figure 8. Generation's number of Kp, Ki, Kd, μ and λ value GA- FOPID controller for ISE fitness function.

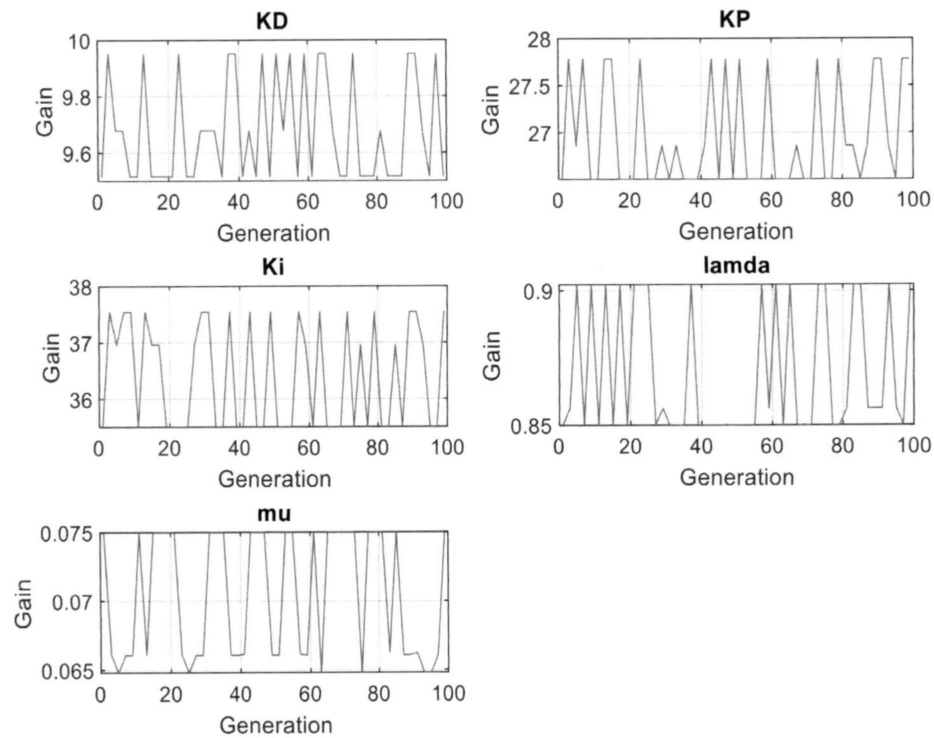

Figure 9. Generation's number of Kp, Ki, Kd, μ and λ value GA- FOPID controller for ITSE fitness function.

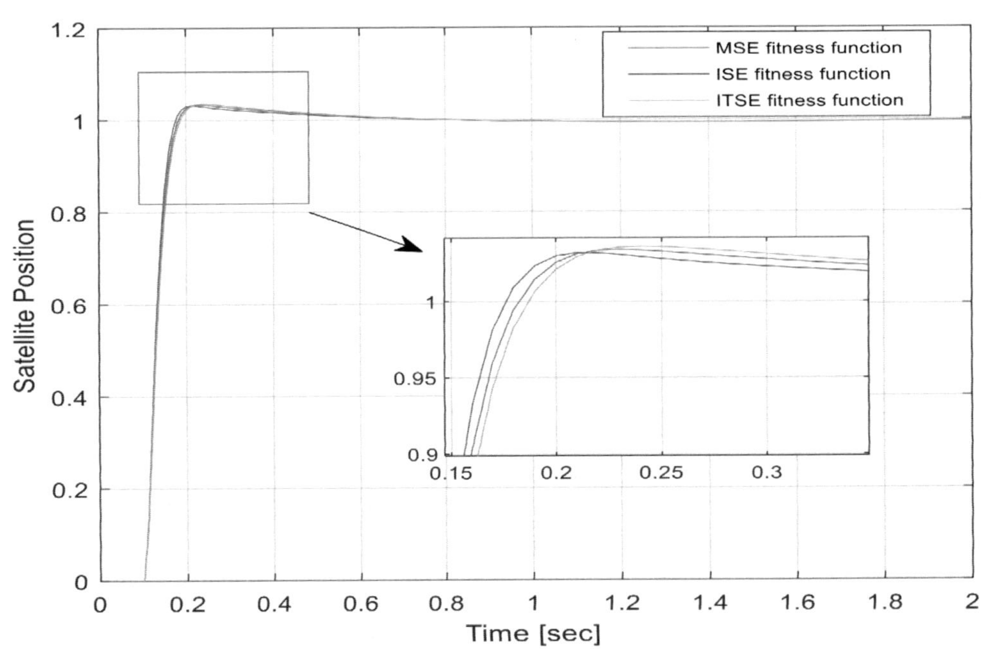

Figure 10. Antenna azimuth position response of GA-FOPID controller with MSE, ISE and ITSE fitness functions.

The Parameters of GA-FOPID controllers and the transient response specifications are summarized in table 2.

Table 2. Parameters of FOPID controller with the derivative order μ and integral order λ

GA-FOPID Controller	Kp	Ki	Kd	λ	μ	Ts	Tr	% Mp
MSE	28.151	37.51	10.51	0.875	0.0451	0.2534	0.0455	0.0455
ISE	27.151	37.51	9.851	0.843	0.0513	0.2651	0.0470	3.4458
ITSE	26.51	35.51	9.515	0.85	0.0752	0.2567	0.0457	3.4526

6. Conclusion

In this study, the antenna system has been modeled and analyzed in detail while its subsystems were described and transfer functions have been found for each of them. After that, this system is controlled by a FOPID controller with genetic algorithm optimization method. Three different fitness functions, MSE, ISE and ITSE, have been used. Results showed that FOPID controller with ISE and ITSE gives very similar results in term of %Mp and Tr; only Ts is further reduced with ITSE. However, the MSE fitness function gives better performance than ISE and ITSE in term of settling and rise times as well as low present of maximum peak overshoot.

7. References

[1] A. Zayed, "A New Discrete Self-tuning LQG Controller Applied to Geostationary Satellite System," *cetj.edu.ly*, Accessed: Dec. 16, 2020. [Online]. Available: http://www.cetj.edu.ly/magazine/number2/18.docx.

[2] A. Saleem, H. Soliman, S. Al-Ratrout, and M. Mesbah, "Design of a fractional order PID controller with application to an induction motor drive," *Turkish Journal of Electrical Engineering & Computer Sciences*, vol. 26, no. 5, pp. 2768-2778, 2018, doi: 10.3906/elk-1712-183.

[3] S. Ibrahim Khather, M. Almaged, and A. I. Abdullah, "Fractional order based on genetic algorithm PID controller for controlling the speed of DC motors," *International Journal of Engineering & Technology*, vol. 7, no. 4, pp. 5386-5392, 2018, doi: 10.14419/ijet.v7i4.25601.

[4] M. A. Rahimian and M. S. Tavazoei, "Improving integral square error performance with implementable fractional-order PI controllers," *Optimal Control Applications and Methods*, vol. 35, no. 3, pp. 303-323, 2014, doi: 10.1002/oca.2069.

[5] P. Cominos and N. Munro, "PID controllers: Recent tuning methods and design to specification," *IEE Proceedings-Control Theory and Applications*, vol. 149, no. 1, pp. 46–53, 2002, doi: 10.1049/ip-cta:20020103.

[6] I. Podlubny, L. Dorcak, and I. Kostial, "On fractional derivatives, fractional-order dynamic systems and PIλDμ-controllers," *Proceedings of the 36th IEEE Conference on Decision and Control*, vol. 5, pp. 4985-4990, 1997, doi: 10.1109/cdc.1997.649841.

[7] R. Duma, P. Dobra, and M. Trusca, "Embedded application of fractional order control," *Electronics Letters*, vol. 48, no. 24, pp. 1526-1528, 2012, doi: 10.1049/el.2012.1829.

[8] D. Valério and J. S. da Costa, "Tuning of fractional PID controllers with Ziegler-Nichols-type rules," *Signal Processing*, vol. 86, no. 10, pp. 2771-2784, 2006, doi: 10.1016/j.sigpro.2006.02.020.

[9] F. Merrikh-Bayat and M. Karimi-Ghartemani, "Method for designing PIλDμ stabilisers for minimum-phase fractional-order systems," *IET control theory & applications*, vol. 4, no. 1, pp. 61-70, 2010, doi: 10.1049/iet-cta.2008.0062.

[10] V. Singh and V. K. Garg, "Tuning of PID controller for speed control of DC motor using soft computing techniques - A review," *International Journal of Applied Engineering Research*, vol. 9, no. 9, pp. 1141-1148, 2014.

[11] Nise, Norman S. "Control systems engineering 6th edition." *New York* (2011).

[12] Y. Q. Chen, I. Petráš, and D. Xue, "Fractional order control - A tutorial," *American control conference. IEEE*, pp. 1397-1411, 2009, doi: 10.1109/ACC.2009.5160719.

[13] H. Zhao, W. Deng, X. Yang, X. Li, and C. Dong, "An optimized fractional order PID controller for suppressing vibration of AC motor," *Journal of Vibroengineering*, vol. 18, no. 4, pp. 2205-2220, 2016, doi: 10.21595/jve.2016.16652.

[14] M. Almaged, S. I. Khather, A. I. Abdulla, and M. R. Amjed, "Comparative Study of LQR, LQG and PI Controller Based on Genetic Algorithm Optimization for Buck Converters," *2019 11th International Conference on Electrical and Electronics Engineering (ELECO). IEEE*, pp. 1012–1017, 2019, doi: 10.23919/ELECO47770.2019.8990572.

Distributed Generation Embedding to Enhancing the Iraqi Grids Based on Optimization Technique

Ihsan Mousa Jawad[1*] **and Wafaa Said Majeed**[2]

[1]Department of Electrical Engineering, College of Engineering, University of Al-Mustansiriyah, Baghdad-IRAQ
[2]Department of Electrical Engineering, College of Engineering, University of Al-Mustansiriyah, Baghdad-IRAQ

[*]Email: wafaasaid2005@uomustansiriyah.edu.iq

Abstract. This research aims to analyze the impact of Distributed Generation (DG) on congestion line mitigation, power loss reduction, and voltage profile enhancement for Iraq's 400kV national grid system. The suggested Particle Swarm Optimization (PSO) and Exchange Market Algorithm (EMA) are implemented to specify the optimum size and location based on an analytic approach depended on voltage, active losses, and congestion lines as fitness. Then it is added to the super 400kV Iraqi grid. The results demonstrate the efficiency of the proposed approach in determining the optimum size and location, decreasing the congestion line, reducing power losses, reducing line flow, and improving the bus voltage profile. The results are obtained with proposed algorithms implementation by using programming language within MATLAB /R2018a environment.

1.Introduction

Energy demand has increased rapidly and is expected to continue to rise. However, power generation extension has been limited due to resources and limitations, contributing to higher transmission line tension and increased threats for overloaded lines. Distributed Generators, which they symbolize (DGs), are now more commonly used in electrical networks to meet rising demand.

Distributed Generation (compared with central power system) is a small source of electric power generation (typically range from less than a kW to hundreds of MW), which is located close to the load. It is not a part of the central power system referred to as the grid applies to large-scale production of power at concentrated installations. These installations are typically placed away from the users and linked to high voltage transmission lines (range from MW to GW). This concept has evolved to include connecting distributed generators on transmission networks as large-scale generators (hundred megawatts), such as gas turbine and Compound Heat Power CHP that connect to the customer side meter [1, 2].

One way to reduce losses and costs, voltage improvement, and congestion lines are by connecting the generators on the distribution network side to cover the demand for electrical energy from the load side [2]. Large-scale integration of electric power networks has often expanded the process and control complexities of the grid. Power grids are often exposed to unexpected contingencies (disturbances) such as dramatic

Content from this work may be used under the terms of the Creative Commons Attribution 3.0 licence. Any further distribution of this work must maintain attribution to the author(s) and the title of the work, journal citation and DOI.
Published under licence by IOP Publishing Ltd

alterations in load demand, short circuit faults, outages in one or more transmission lines, equipment outages, and generator failures [3-5].
The power grid's security relates to the threat's degree in its capability to manage unexpected disturbances (contingencies) without interruptions in customer service [6]. Any condition in the contingency set will contribute to the overloading of the transmission line or violates bus voltage limits during services. These contingencies must be quickly detected for a full evaluation. This critical approach to emergency detection is the emergency option that takes account of the generator or line interruption utilizing the maximum AC load flow software [7].
They were using the Genetic Algorithm (GA) to minimize losses and improve the voltage profile method. The DG unit's optimal rating is also optimized by looking at the Real Values of Power Flow Performance Indices for the most extreme emergencies [8]. DG units in distribution networks have specific impacts, like voltage level enhancement, energy quality, and loss reduction. The enormous scale of study papers will currently be collected from the DG machine for optimum sizing and optimization [9].
An approach used depends on finding a factor depending on the ratio of line load flow changes between two buses to determine congestion lines and manage them through distributed generations [10].
Mitigation congestion of transmission lines is implemented by adding DG to an appropriate location and optimal size based on PSO [11]. Optimal DG allocation and sizing based on Ant Colony Algorithms and Genetic Algorithms are implemented. Active power losses are evaluated as a function of power supplied by generators as different levels of generation are penetrated from DG [12]. An Exchange Market Algorithm (EMA) methodology is used to solve the optimal reactive control dispatch problems of an extremely non-linear power structure. In analyzing the power system, optimization issues are most critical and typically devised as an optimal power flow (OPF) problem [13]. In [14], research provides a strategy for optimizing the position and size of Distributed generation in the distribution systems based on improved sunflower optimization (ESFO) to mitigate system power loss. Based on the original sunflower optimization (SFO), a mutation technique is applied to update the best plant.
For configuring power distribution networks with the optimum allocation of several distributed generators, the paper suggests an improved equilibrium optimization algorithm (IEOA) combined with a proposed recycling technique. A multi-objective feature is created that considers total power loss and bus voltages enhancement concerning the device limits. Scan, refined genetic, and firefly [15].

2. Contingency screening

Maintaining security against possible threats is critical in the electric grid environment. The violation of its limitations of transmission lines represented part of the most specific threats. An index has been obtained to measure each line's series analysis if it reaches or exceeds the line limiter (MVA) to detect and prevent this issue by the device operator. Evaluation of the process result of the line limit violations called (contingency screening). This research applied the (*n*-1) method, which *n* means a line or generator outage. This index is called the Real Power Flow Performance Index (PI_{RPF}), where equation (1) as its general formula [8].

$$(PI_{RPF})^i = \sum_{\substack{all \\ branches}} \left(\frac{w_l}{2n}\right)\left(\frac{P_{flow\,l}^{(i)}}{P_l^{(max)}}\right)^{(2n)} + \sum_{j=1}^{N_{bus}} \left(\frac{w_j}{2n}\right)\left(\frac{\Delta V_j^{(i)}}{\Delta V_j^{(limit)}}\right)^{(2n)} \quad (1)$$

$$\Delta V_j^{(i)} = V_j^{(i)} - V_j^{(limit)} \quad (2)$$

$$V_j^{(limit)} = V_j^{(max)}, \quad \forall V_j^{(i)} \geq 1.0 \quad (3)$$

$$V_j^{(limit)} = V_j^{(min)} , \ \forall V_j^{(i)} < 1.0 \quad (4)$$

$$V_j^{(i)} = V_j^{(max)} , \ \forall V_j^{(i)} > V_{max} \quad (5)$$

$$V_j^{(i)} = V_j^{(min)} , \ \forall V_j^{(i)} < V_{min} \quad (6)$$

$$\Delta V_j^{(limit)} = \frac{V_j^{(max)} - V_j^{(min)}}{2} \quad (7)$$

Where, $i = 1 \ldots N_{lines}$, [$(PI_{RPF})^i$] denotes on the active power performance index for ith outage, [w_l] & [w_j] denote on weight factors for line l and for bus j respectively that chosen by the operator after taking into consideration the systems condition work, a[$P_{flow\,l}^{(i)}$] denotes on the line flow of the l th line for (i)th outage,a[$P_l^{(max)}$] denotes on the maximum rating of the lth line, N_{bus} denotes on total buses' number, and the term (2n) denotes the order of active power performance index, which considered as 2.

3. Finding Line Losses

The power losses in any line between buses i and j can written as the sum of power flows from equation (8) [16].

$$S_{Lij} = S_{ij} + S_{ji} \quad (8)$$

The summation of all line losses lead to the total losses of the network can be calculated using equation (9)

$$\text{Losses} = \sum_{K=1}^{N_{line}} S_L(K) \quad (9)$$

Where S_L is a loss of one branch, N_{line} denotes on the total lines, K acts specified line.

4. The Objective function and optimization

The GA determines the optimum DG site and size. The aims are used in GA-based optimization technology to assign the optimum DG site and size for reducing voltage deviation, active power loss, and performance index. The objective function represents by equation (10) as follows:

$$Minf = W_1 \sum_{i=1}^{N_{bus}} (V_{ref} - V_i)^2 + W_2 \sum_{j=1}^{N_{line}} P_{Lj} + W_3 \sum_{\substack{all\ congested \\ branches}} PI_{RPF} \quad (10)$$

Where V_{ref} means a reference voltage, V_i means voltage at ith bus, P_{Lj} means the active power losses at j th line, N_{bus} means total busses' number, N_{line} mean total lines' number, and W_1, W_2, and W_3 are mean the objectives weights for voltage deviation, active power losses, and active power performance index respectively.

5. The elements of (PSO)

The basic PSO elements can clarified as following: [17]

5.1. Particle s_{id}^k

It is The candidate's current location of the control variables where i=1, 2, 3, ... n, d = 1,2,3,......m, n represents control variables number, m is represent candidate particles number for each one of control variable. To illustrate that, let the vector of the control variables are [S1, S2, S3 ,...Sn], then, the set of particles of nth control variables Sn are (sn1, sn2, sn3, … , snD). Each particle represents a position in the search space solution.

5.2. Particle velocity v_{id}^{k}
Particle velocity is the current velocity of particles' motion in the swarm population at iteration *k*.

5.3. Individual best position ($pbest_{id}$ *)*
The best position that points to best fitness value for each particle is called the individual best position or (local best position) or (personal best position).

5.4. Global best position ($gbest_i$ *)*
Global best represents best position among all the individual better positions ever achieved, where $gbest_i^k$ represent the best position overall the individual best position (i.e., global position) for *i* th control variable at iteration *k*.

5.5. Updating
5.5.1. Velocity v_{id}^{k+1}
Updating the particle velocity can be according the following equation:

$$v_{id}^{k+1} = w\ v_{id}^{k} + c_1 \times rand_1 \times (pbest_{id}^{k} - s_{id}^{k}) + c_2 \times rand_2 \times (gbest_i^{k} - s_{id}^{k}) \tag{11}$$

where v_{id}^{k} is the current velocity of the particle at iteration *k*; v_{id}^{k+1} is the updating velocity of the particle at iteration *k*+1; s_{id}^{k} is the particle position at iteration *k*; ***w*** is the inertia weight; c_1 & c_2 are a randomly chosen number; $rand_1$ & $rand_2$ are a uniformly distributed random number between [0,1]; *k* is the iteration number.
If the velocity limits are violated by a particle, the algorithm sets its velocity equal to the violated limit.

5.5.2. Position
The current position can be updated as following equation:

$$s_{id}^{k+1} = v_{id}^{k} + v_{id}^{k+1} \tag{12}$$

If the position limits are violated by a particle, the algorithm sets its position to the violated limit.

5.6. Inertia weight
Selection of the weight factor must be in a way that making a balance between local and global searching as well as to make a faster convergence. The implementation of this shall be by choose a large value of the weight factor for the initial iterations and gradually decrease the weight factor in sequential iterations as in equation (13)

$$w = w_{max} - k \times \frac{(w_{max} - w_{min})}{K_{max.}} \tag{13}$$

where $w_{max} = 0.9$; $w_{min} = 0.4$; *k* represents currently iteration number; $K_{max.}$ represents maximum iteration number [17].

6. Principle of EMA
The capital market inspires EMA in which, depending on principles and their own experience and strategies, stockholders will take various decisions [18]. These choices can either stabilize or generate

volatility in the market. By splitting stocks and undertaking fewer risks, stockholders tend to maximize their benefits. Unlike the fluctuating style, citizens may achieve the maximum potential benefit in the prevailing balanced market situation by forecasting the current situation without considering the danger in their transactions. The possibility of swinging grades in the stock market can be either very advantageous or very dangerous to stockholders [19]. The EMA formulation implements the principles and fundamental requirements of the stock market are worked up for two states, namely states of equilibrium and fluctuations, as follows:

6.1. Balanced State

Stock market participants are grouped into three divisions according to their fitness functions:

- Superior stockholders are put in the first group (10 percent to 30 percent of total members). The participants of this community do not modify their stocks to retain their position in the market.
- Intermediate stockholders fall in the second group (20 percent to 50 percent of total members). By matching their stocks with stocks in the first type, these stockholders aim to attain a worldwide optimum.
- The remaining participants of the business are categorized into the third group with the lowest fitness feature value.

6.2. Fluctuating State

It divided to three categories as:

- Members try to maintain their place with the other owners and do not modify their stocks.
- Members in this group are seeking to boost their rank by altering their rank stocks.
- By adjusting the valuation of their stocks and a larger quest domain, members of this group continue to increase higher scores [20].

The Flowchart of procedures applying PSO and EMA to find the optimal DG size and location is shown in figure 1.

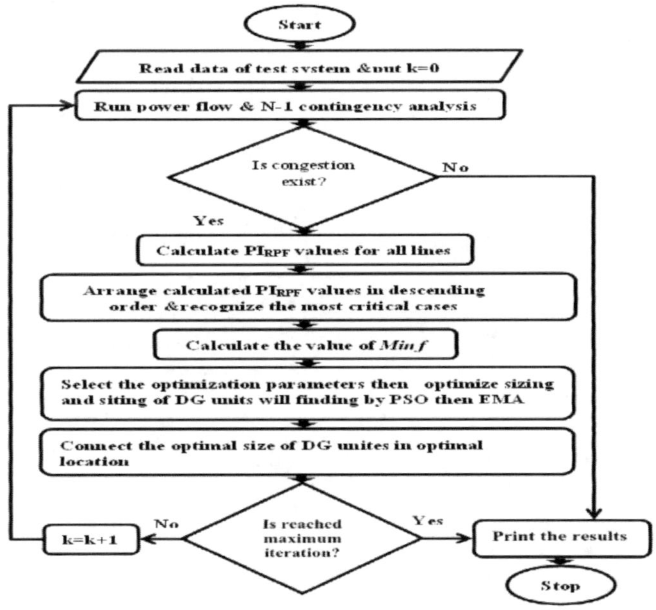

Figure 1. Flowchart of procedure applying PSO and EMA to find the **optimal** DG size and location.

7. The results and discussion

The proposed approach is applied to the Iraqi super grid 400kV where all data [21]. Iraqi Super Gird system 400 kV comprises (36) bus bars, (21) generator buses, (15) load buses, and (52) transmission lines. figure 7 represented the single line diagram of the 400 kV Iraqi Gird. The population size is 50, and iteration is 50, w_{max} equal 0.9, $w_{min.}$ equal 0.4, c2 and c1 equal 2. There are three emergencies in the network when the (*n*-1) method is applied. According to the arrangement in table 1, lines 30(15-20)(AMN4- KUTP) and line 31 (15-20)(AMN4- KUTP) represented the most dangerous lines. Lines number 30 and 31 are two lines connected between buses [15 (AMN4)] and [20 (KUTP)].

Table 1. Results of contingency ranking based on PI *power(first term)*, *PI voltage(second term)*, and PI_{RPF} from equation (1).

Ser. No.	Branch or unit No.	From bus	To Bus	*PI* power	*PI* voltage	PI_{RPF} total	Sum Congestion Line	Congestion line
1	30	15 (AMN4)	20 (KUTP)	1.2344	1.8025	3.0369	1	(31)(15-20) (AMN4-KUTP)
2	31	15 (AMN4)	20 (KUTP)	1.2344	1.8025	3.0369	1	(30)(15-20) (AMN4-KUTP)
3	20	11 (BGS4)	16 (BGC4)	1.1403	1.3025	2.4428	1	(25)(12-15) (BGE4-AMN4)

According to the research methodology with the proposed PSO and EMA, it is possible to mitigate the congestion lines, reduce the losses, improve voltage profile, and minimize load flow by connecting certain numbers of DG units in the network that leads to these goals. The initial generator size will be within a range (0-1000) MW as a group. Each group represents 1DG and has many generators. Table 2 and figures 2 and 3 are shown many things as the alleviation of congestion line from 3 lines to without line and the losses reduction, when applied PSO, from(46.22+j413.06) at the base case to (17.63+j153.73) (MVA), (14.54+j125.7) (MVA), (14.21+j122.6) (MVA) and (17.18+j149.34) (MVA). The losses percentage reduction are 61.85%, 68.54%, 69.26%, and 62.83%. The voltage profile is also improved where the voltage deviation changed from 0.01361 to 0.01335, 0.01322, 0.01316, and 0.01331 when connected 1DG, 2DGs, 3DGs and 4DGs, respectively. Table 3 and figure 2 shown the results by applying EMA. The losses are reduced to (21.86+j192.3) MVA at a percentage reduction of 52.7%. The voltage deviation reduced to 0.012944 when connected 4DGs, which illustrate the voltage improvement.

Table 2. Results of Grid with and without DG by PSO.

	PSO - Proposed method for Iraq 400kv grid							
Cases	P Loss (MW)	Q Loss (Mvar)	Congestion Lines	DG site bus	DG size (MW)	Loss Reduction (MW)	Loss Reduction %	Voltage Deviation (VD)
Without DG	46.22	413.06	3	—	—	—	—	0.01361
With 1DG	17.63	153.73	0	10	1000	28.59	61.85	0.01335
With 2DG	14.54	125.7	0	10 18	1000 455.6	31.68	68.54	0.01322
With 3DG	14.21	122.6	0	18 10 29	644.3 1000 195.1	32.01	69.26	0.01316
With 4DG	17.18	149.34	0	27 35 29 10	5.9 0 397.4 1000	29.04	62.83	0.01331

Table 3. Results of Grid with and without DG by EMA.

Cases	EMA approach for Iraqi 400kV grid							
	P Loss MW	Q Loss Mvar	Congestion Lines	DG site bus	DG size MW	Loss Reduction MW	Loss Reduction %	Voltage Deviation (VD)
Without DG	46.22	413.06	3	/	/	/	/	0.01361
Add 1DG	32.81	297.4	1	12	955	13.41	29.01	0.01341
Add 2DG	25.42	234.6	0	10 25	1000 344	20.8	45.002	0.01324
Add 3DG	20.82	201.3	0	15 17 34	720 1000 95	25.4	54.95	0.013012
Add 4DG	21.86	192.3	0	10 12 10 33	200 522 617 202	24.36	52.7	0.012944

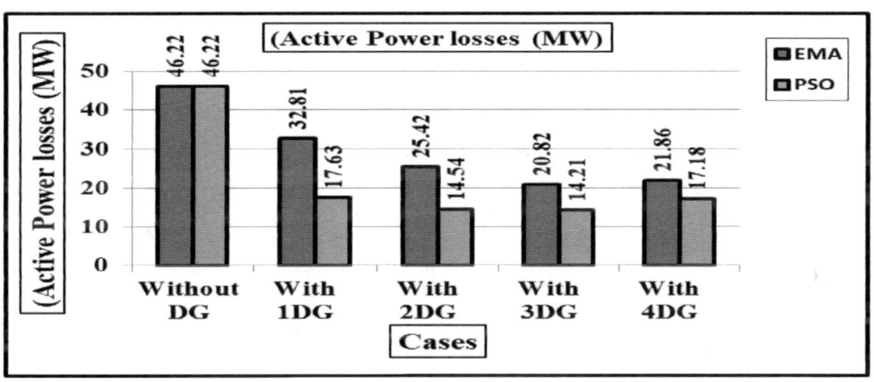

Figure 2. Illustrate the results of active power losses for Iraqi Grid 400kV with PSO and EMA.

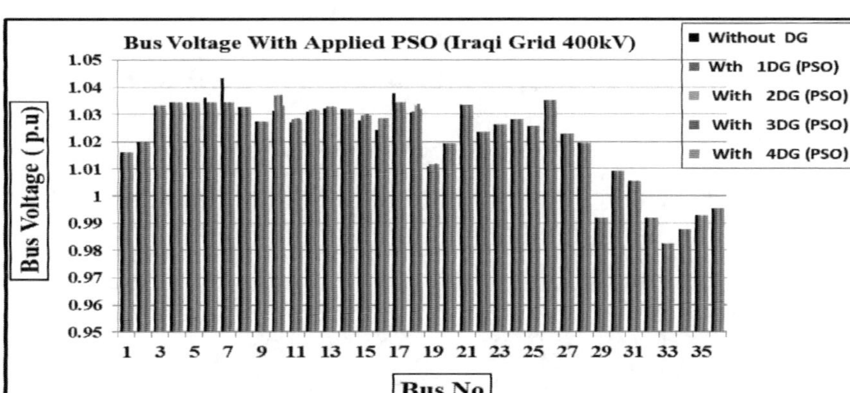

Figure 3. Bus voltage without and with DG by applying PSO.

Figure 4 shows the reduction of power flow when connected DGs to the network. The power flow percentage reduction is increased from 14.25% to 20.53% , and from 27.86% to 36.54% by applying EMA and PSO , respectively. The main reason for reducing the power flow is the inevitable result of the generating units' rescheduling and losses reduction result in reduction of the loads' drawn current. It leads to significant results, such as reducing the operation fuel cost and reducing harmful emissions.

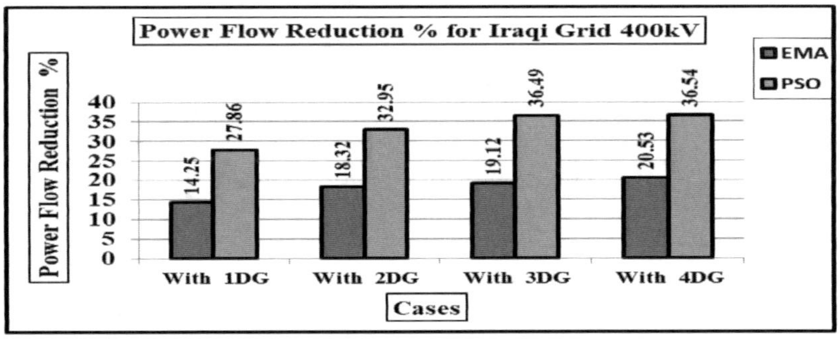

Figure 4. Percentage Power Flow Reduction for Iraqi Grid 400kV with DGs by PSO and EMA.

8. Loading 5% and 10%

For studying the grid's possibility for loading increasing with applying PSO and EMA, the loads were increased to 5% from the normal case (8803.9+j 3135.8) MVA to (9244.1+j3292.59) MVA. Also, increase the loads to 10%, which all become (9684.29 +j3449.38) MVA. That led to an increase in the congestion lines to 4 and 11 lines. The losses are increased and the voltage is dropped, as shown in table 4 and figures 5 and 6. After connecting the DG with ascending numbers to insert active power with applied PSO, we notice bus voltage improvement observed in buses 10, 11, 12, 13, 15, 16, 17, 18, and 19. The losses are reduced from (63.37 +j569.04) MVA to (16.44 +j143.18) MVA at 5% and from (84.95+j764.84) MVA to (15.85+j136.88) MVA at 10%. The congestion lines are reduced from 4 and 11 lines to without line at 5% and 10%, respectively. Also, applied EMA led to reduced losses to (25.54+j201.2) MVA and (27.85+j235.98) MVA at 5% and 10%, respectively, for four DGs.

Table 4. Results of loading the network with and without DG by PSO.

	Cases	Loading (MW)	Loading (Mvar)	Total Losses (MW)	Total Losses (Mvar)	Congestion line No. without outage	Congestion line No. with outage
	Base case without loading	8803.9	3135.8	46.19	413.08	Nil	3
Load increase 5% with PSO	Without DG	9244.1	3292.59	63.37	569.04	Nil	4
	With 1DG	8244.1	3292.59	25.32	223.28	Nil	0
	With 2DG	7510.4	3292.59	19.69	172.12	Nil	0
	With 3DG	6873.5	3292.59	15.61	135.1	Nil	0
	With 4DG	7348.03	3292.59	16.44	143.18	Nil	1
Load increase 10% with PSO	Without DG	9684.29	3449.38	84.95	764.84	1	11
	With 1DG	8684.29	3449.38	37.05	329.84	Nil	0
	With 2DG	7684.3	3449.38	23.32	204.65	Nil	0
	With 3DG	7375.9	3449.38	17.23	149.3	Nil	0
	With 4DG	7383.8	3449.38	15.85	136.88	Nil	0

Table 5. Results of loading the network with and without DG by EMA.

	Cases	Loading (MW)	Loading (Mvar)	Total Losses (MW)	Total Losses (Mvar)	Conges. line No. without outage	Conges. line No. with outage
	Base case without loading	8803.9	3135.8	46.22	413.06	Nil	3
Load increase 5% with EMA	Without DG	9244.1	3292.59	63.37	569.04	Nil	4
	With 1DG	8443.2	3292.59	29.62	312.5	Nil	1
	With 2DG	7919.6	3292.59	27.73	298.22	Nil	0
	With 3DG	702.9	3292.59	25.66	245.2	Nil	0
	With 4DG	7356.3	3292.59	25.45	201.2	Nil	0
Load increase 10% with EMA	Without DG	9684.3	3449.38	84.95	764.84	1	11
	With 1DG	8992.5	3449.38	41.1	397.2	Nil	3
	With 2DG	7823.6	3449.38	32.8	357.56	Nil	0
	With 3DG	7532.1	3449.38	29.33	243.3	Nil	0
	With 4DG	7462.6	3449.38	27.85	235.98	Nil	0

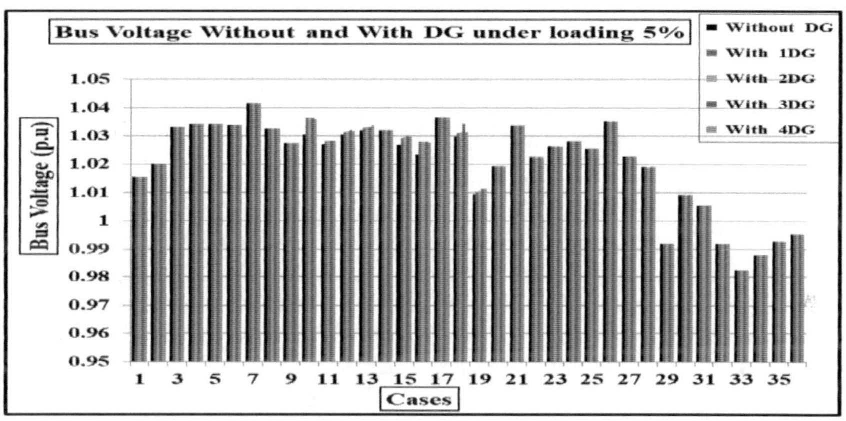

Figure 5. Bus voltage without and with DG Under Loading 5%.

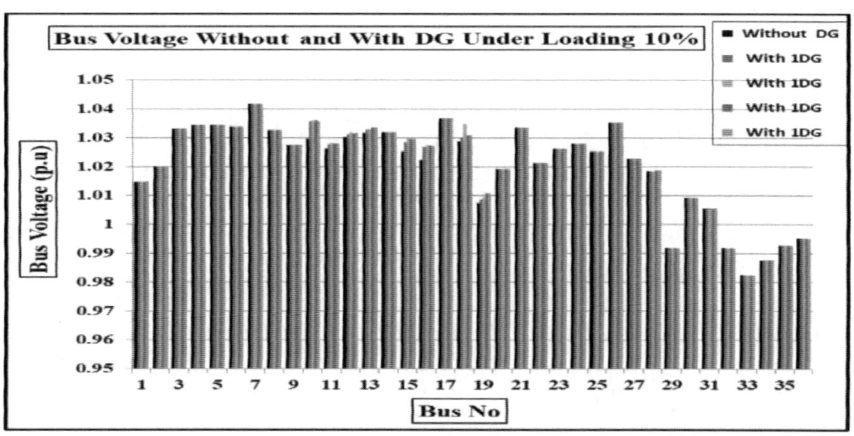

Figure 6. Bus voltage without and with DG under loading 10%.

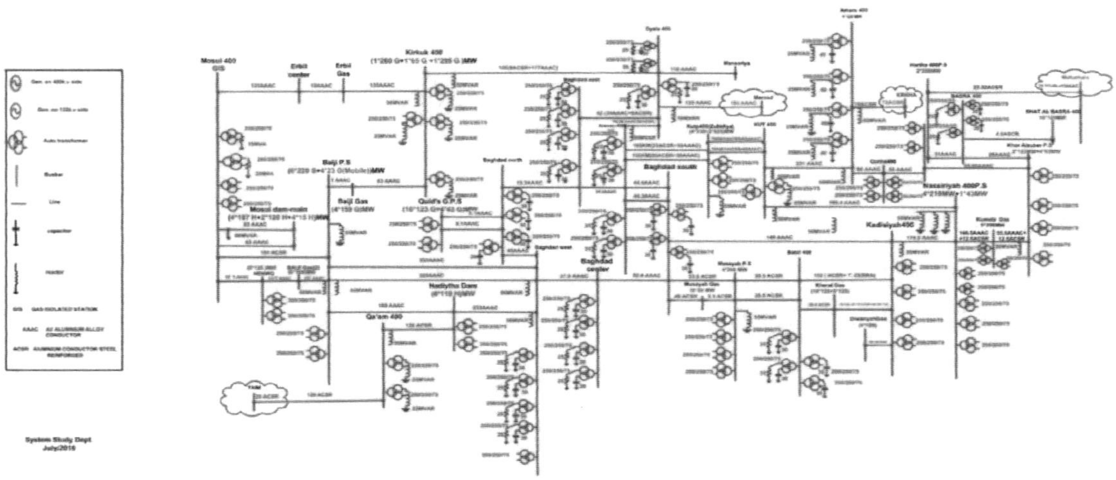

Figure 7. Single line diagram of the 400 kV Iraqi SHV Gird System.

9. Conclusion

In this paper, the proposed approach, which applies the contingency (n-1) line and generator outage for Iraqi high grid 400 kV (i.e., is carried out under single line outage and generator outage contingencies. Newton Raphson load flow within MATLAP programming language was applied. An optimization techniques PSO and EMA, have proven their effectiveness in obtaining important results, especially when connecting 1DG, 2DGs, 3DGs, and 4DGs to the network. This approach is applied to the Iraqi Super Grid 400kV to provide important solutions by connecting specific numbers of generating units in the optimal location and size to enhance its performance characteristic. Several important results were obtained that enhance the reliability of the network. The congestion lines are reduced from three lines to without line. The losses are reduced from(46.22+j413.06) at the base case to (17.63+j153.73) (MVA), (14.54+j125.7) (MVA), (14.21+j122.6) (MVA) and (17.18+j149.34) (MVA) with percentage reduction are 61.85%, 68.54%, 69.26%, and 62.83%. The percentage reduction of load flow increases from 27.86% to 32.95%, 36.49%, and 36.54%. In the same manner, when applying EMA. Also, the voltage profile is improved, as shown in the result of voltage deviation (VD) in table 2, table 3 and figure 3. These results have led to the enhancement and improvement of the network's performance, which is reflected in the improvement of other elements, such as reducing operation fuel costs and reducing environmental pollution emissions with the possibility of increasing its loading with additional loads. In comparison, these two algorithms found that relative preference to PSO.

10. References

[1] Dasan, S. B., Ramalakshmi, S. S., & Devi, R. K. (2009, December). “Optimal Siting and Sizing of Hybrid Distributed Generation Using EP”. In 2009 International Conference on Power Systems (pp. 1-6). IEEE.

[2] El-Khattam, W., & Salama, M. M. (2004). "Distributed Generation Technologies, Definitions and Benefits". Electric *power systems research*, **71(2)**, 119-128.

[3] P. S. R. Murty, 2017, " *Power System Analysis* ", **Second Edition**, Elsevier Ltd., USA.

[4] Leonard L. Grigspy, 2012, " *Power Systems*", **Third Edition**, CRC Press Tylor & Francis Group, London.

[5] P. Kundur, 1994, " *Power System Stability and Control*", Mc Graw-Hill Professional Publishing, New York.

[6] Ian Dobson and David E. Newman, 2017," Cascading Blackout Overall Structure and Some Implications for Sampling and Mitigation", *International Journal of Electrical Power & Energy Systems,* Vol. **86**, March, PP. 29-32.

[7] J.A refaee & H.Maghrabi,(1999). "Radial Basis function Network for Contingency Analysis of Bulk Power System", *IEEE Transaction on power system* Vol.**4** No.**2**.

[8] Singh, A. K., & Parida, S. K. (2013). "Congestion Management with Distributed Generation and Its Impact on Electricity Market. "*International Journal of Electrical Power & Energy Systems*, **48**, 39-47.

[9] El-Khattam W, Hegazy YG, Salama MMA.,(2005). "An Integrated Distributed Generation Optimization Model for Distribution" system planning. IEEE T Power Syst **20**: 1158-1165.

[10] Dehnavi, E., Aminifar, F., & Afsharnia, S. (2019). "Congestion Management Through Distributed Generations and Energy Storage Systems". *International Transactions on Electrical Energy Systems*, **29(6)**, e12018.

[11] Muthulakshmi, K., Sasiraja, R. M., & Kumar, V. S. (2016). The Phenomenal Alleviation of Transmission Congestion by Optimally Placed Multiple Distributed Generators Using PSO. *Circuits and Systems*, **7(8)**, 1677-1688.

[12] Zakaria, Y. Y., Swief, R. A., El-Amary, N. H., & Ibrahim, A. M. (2020, January). "Optimal Distributed Generation Allocation and Sizing Using Genetic and Ant Colony Algorithms". In *Journal of Physics*: Conference Series (Vol. **1447**, No. **1**, p. 012023). IOP Publishing.

[13] Rajan, A., & Malakar, T. (2016). Exchange Market Algorithm Based Optimum Reactive Power Dispatch. *Applied Soft Computing*, *43*, 320-336.

[14] Nguyen, T. T. (2021). “Enhanced Sunflower Optimization for Placement Distributed Generation in Distribution System”. International Journal of Electrical and Computer Engineering, **11(1)**, 107.

[15] Shaheen, A. M., Elsayed, A. M., El-Sehiemy, R. A., & Abdelaziz, A. Y. (2020). Equilibrium Optimization Algorithm for Network Reconfiguration and Distributed Generation Allocation in Power Systems. *Applied Soft Computing*, ***98***, 106867.

[16] Saadat, H. (1999). *Power System Analysis* (Vol. **2**). McGraw-Hill.

[17] Al-Bahrani, L., & Dumbrava, V. (2016). Optimal Power Flow Based on Particle Swarm Optimization. University Politehnica of Bucharest Scientific Bulletin Series C-Electrical Engineering and Computer Science, ***78*(3)**, 253-264.

[18] Statman, M. (1999). “Foreign Stocks in Behavioral Portfolios”. *Financial Analysts Journal*, ***55*(2)**, 12-16.

[19] Birnbaum, M. H. (2008). “New paradoxes of risky decision making”. *Psychological review*, ***115*(2)**, 463.

[20] Jafari, A., Khalili, T., Babaei, E., & Bidram, A. (2019). “A Hybrid Optimization Technique Using Exchange Market and Genetic Algorithms”. *IEEE Access*, ***8***, 2417-2427.

[21] Abdulsada, M. A., & Tuaimah, F. M. (2017). “Power System Static Security Assessment for Iraqi Super High Voltage Grid”. *International Journal of Applied Engineering Research*, **12(19)**, 8354-8365

A Rail line model with distributed parameters of track circuit

Ravshan Aliev[1*]

[1]Department of Information Systemsand Technologies, Tashkent State Transport University, Uzbekistan

[*]E-mail: AlievRavshan7@gmail.com

Abstract. Models of continuous jointlesstrack circuits with an adaptive receiver are presented, where the power supply is connected to the beginning or middle of the rail circuit, and also to the locomotive receiver. The aim of the work is to develop models of rail lines, which take into account the actual current distribution along the track line and which may be used to research adaptive jointlesstrack circuits. The electric schemes of the jointlesstrack circuits are given, equations, connecting currents and voltages at the ends of the rail line, coefficient of four-pole rail lines, algorithms for calculating track circuits. In conclusion, that the track circuit can operate at an insulation resistance four times lower than the permissible value, where the permissible insulation resistance is R_i = 1om*km. New approaches and tools for analyzing and synthesizing solutions are considered, with considering account real operating conditions and disturbing factors.

1. Introduction

The analysis of the operation of tonal jointless track circuits is devoted to [1–6, 10–12] and others.The issues of automation of the process of their design and reliability are considered in [7–9, 16-20]. On the problems of expertise and safety testing of railway automation and telemechanics systems are considered [18–20]. At the same time, there are practically no publications on the intellectualization of the solution of problems of the analysis and optimization of the parameters of track circuits in various operating modes.In connection with this actual seems to development of a hybrid expert system for analyzing the operation of a jointless tonal track circuit.

2. Methods

After a series of transformationsshows a generalized substitutionsscheme of a rail line for normal and shunt modes with a locomotive receiver for an unlimited jointlesstrack circuit [15], where shown: rail line RL, equivalent generator EG with internal resistanceZ_G, equivalent receiver with internal resistance Z_{er}, the first shunt with resistance $R_{sh1}$and the second shunt with resistance R_{sh2}, input resistances Z_{bi} and Z_{ei} of adjacent rail lines, input equivalent impedanceZ_{ei}^{I}, the directions of currents I, voltage U and lengths of sections l are shown.

Content from this work may be used under the terms of the Creative Commons Attribution 3.0 licence. Any further distribution of this work must maintain attribution to the author(s) and the title of the work, journal citation and DOI.
Published under licence by IOP Publishing Ltd

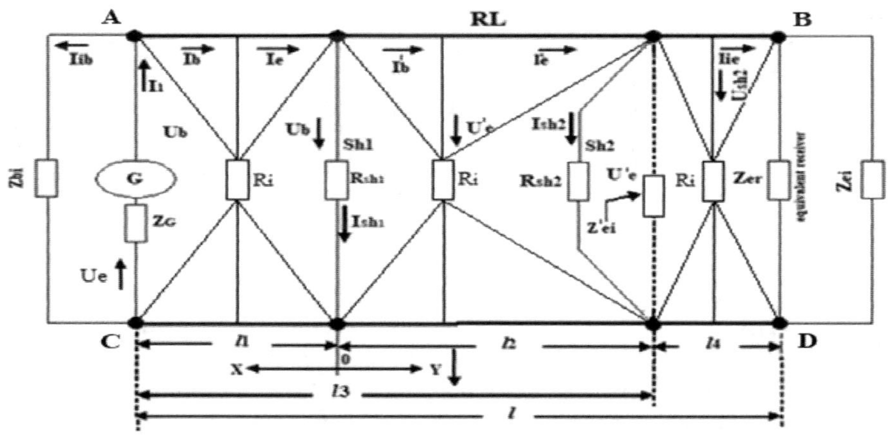

Figure 1. Generalized substitutionsschemeof an unlimited track circuit with locomotive receiver for normal and shunt modes

Voltage U_1 and current I_1 are related to voltage U_{sh2} and current I_{sh2} by equations:

$$U_1 = A_{shl}U_{sh2} + B_{shl}I_{sh2};$$
$$I_1 = C_{shl}U_{sh2} + D_{shl}I_{sh2}, \tag{1}$$

where

$$A_{shl} = c_1\left(ch\gamma l_1 + \left(\frac{Z_w}{R_{shl}}\right)sh\gamma l_1\right) + c_2(ch\gamma l_1 + sh\gamma l_1); \tag{2}$$

$$B_{shl} = Z_w(sh\gamma l_2 ch\gamma l_1 + c_3(ch\gamma l_1 + sh\gamma l_1)); \tag{3}$$

$$C_{shl} = (\frac{1}{Z_w})ch\gamma l_2(c_1 sh\gamma l_2\left(c_4\left(\frac{Z_w}{R_{shl}}\right) + c_5 ch\gamma l_2\right)) + c_2c_4; \tag{4}$$

$$D_{shl} = c_3(ch\gamma l_3 + sh\gamma l_3)c_4 + c_5 sh\gamma l_2. \tag{5}$$

where

$$c_1 = 1/th\gamma l_2 + (\frac{Z_w}{Z_{ei}^I})sh\gamma l_2;$$

$$c_2 = 1/cth\gamma l_2 + (\frac{Z_w}{Z_{ei}^I})ch\gamma l_2;$$

$$c_3 = 1/th\gamma l_2 + (\frac{Z_w}{R_{shl}})sh\gamma l_2;$$

$$c_4 = 1/th\gamma l_1 + (\frac{Z_w}{Z_{bi}})sh\gamma l_1;$$

$$c_5 = 1/cth\gamma l_1 + (\frac{Z_w}{Z_{bi}})ch\gamma l_1;$$

$$Z_{ei}^I = Z_w\frac{Z_{er}Z_{ei}ch\gamma l_3 R_i R_{sh1} + Z_w R_i R_{sh2}(Z_{er}+Z_{ei})sh\gamma l_3}{Z_{er}Z_{ei}sh\gamma l_3 R_i R_{sh1} + Z_w R_i R_{sh2}(Z_{er}+Z_{ei})ch\gamma l_3}. \tag{6}$$

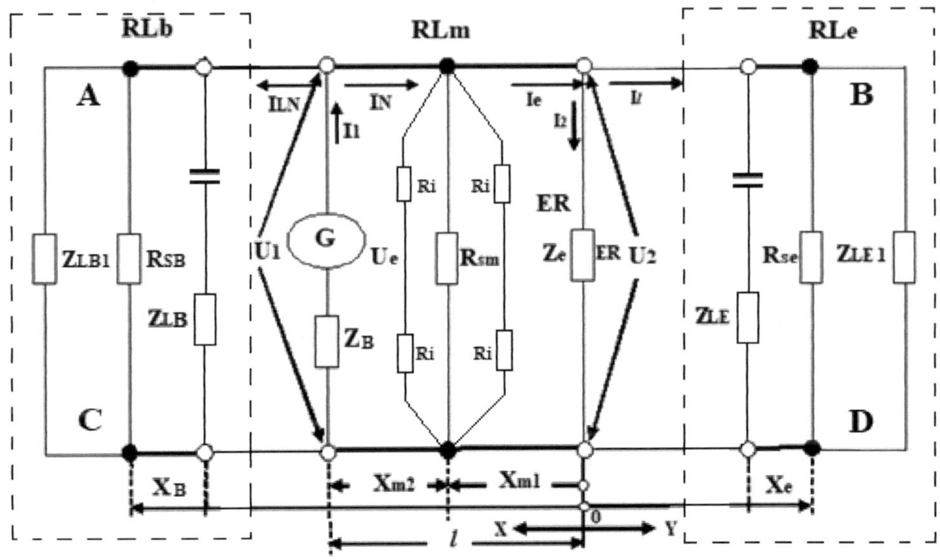

Figure 2. Generalized substitutionscheme of unlimitedtrack circuite with adaptive track receiver for normal and shunt modes

Figure 1 is converted to the figure 2 in order to represent the track circuit with the locomotive receiver in the form of the track circuit with the track receiver.

In figure 2 shows the rail line of the beginning R_{Lb}, the middle R_{Lm} and the end R_{Le} with shunts R_{SB}, R_{Sm}, R_{Se}, respectively. The impact on the rail line R_{Sm} of adjacent track circuits is taken into account by means of the resistances of the beginning Z_{LB} and the end Z_{LE}. (instead of Z_{ib} and Z^{I}_{ei} in figure 1). Instead of the second shunt, the resistance of which in figure 1 is designated R_{Sh2}, on figure 2 includes an equivalent ER receiver with a resistance Z_e. Through means of resistances Z_{LB1} and Z_{LE1} (instead of Z_{ei} in figure 1) the influence of rail lines is taken into account located behind the RL_b and RL_e shunts, respectively. On each of the rail lines RL_b, RL_m and RL_e, different insulation resistance and rail can be set, therefore, instead of the propagation constant γ, the designationsγ_N, γ_M and γ_E are provided, and instead of the wave impedance Z_w - designations Z_{wb}, Z_{wm} and Z_{we}. The dependences between the voltages and currents of the equivalent generator eG and the equivalent receiver ER are shown by equation (7). These equations are obtained from equation (1) with the replacement of the notation in figure1 to the designations in figure2:

$$U_1 = A_m U_2 + B_m I_2;$$
$$I_1 = C_m U_2 + D_m I_2, \tag{7}$$

where

$$A_m = c_1\left(sh\gamma_m x_{m2} + ch\gamma_m x_{m2} + \left(\frac{Z_w}{R_{sm}}\right) sh\gamma_m x_{m2}\right) + c_2(sh\gamma_m x_{m2} + ch\gamma_m x_{m2}); \tag{8}$$

$$B_m = Z_{wM}(x_{m1} ch\gamma_m x_{m2} + c_3 sh\gamma_m x_{m2}) + R_{se} Z_{LE1}(ch\gamma_m x_e); \tag{9}$$

$$C_m = \left(\frac{1}{Z_{WM}}\right) c_1 ch\gamma_m x_B \left(c_4\left(\frac{Z_{WM}}{R_{sm}}\right) + c_5 Z_B\right) + c_2 c_4 R_{se} x_e); \tag{10}$$

$$D_m = c_3 c_4 th\gamma_m x_e + c_5 cth\gamma_m x_e, \tag{11}$$

$$c_1 = 1/th\gamma_m x_{m1} + \left(\frac{Z_{wm}}{Z_{LE}}\right) sh\gamma_m x_{m1};$$

$$c_2 = 1/cth\gamma_m x_{m1} + (\frac{Z_{wm}}{Z_{LE}})ch\gamma_m x_{m1};$$
$$c_3 = 1/th\gamma_m x_{m1} + (\frac{Z_{wm}}{R_{SM}})sh\gamma_m x_{m1};$$
$$c_4 = 1/th\gamma_m x_{m2} + (\frac{Z_{wm}}{Z_{LE}})sh\gamma_m x_{m2};$$
$$c_5 = 1/cth\gamma_m x_{m2} + (\frac{Z_{wm}}{Z_{LE}})ch\gamma_m x_{m2}; \quad (12)$$

$$Z_{LE} = Z_{WE}\frac{R_{SE}Z_{LE1}(sh\gamma_e + ch\gamma_e)x_e + Z_{WE}(R_{se}+Z_{LE1})sh\gamma_e x_e}{R_{SE}Z_{LE1}(sh\gamma_e x_e - ch\gamma_m x_e) + Z_{WE}(R_{se}+Z_{LE1})ch\gamma_e x_e}; \quad (13)$$

$$Z_{LN} = Z_{WB}\frac{R_{SB}Z_{LB1}ch\gamma_b x_b + Z_{WB}(R_{sb}+Z_{LB1})sh\gamma_b x_b}{R_{SB}Z_{LB1}sh\gamma_b x_b + Z_{WB}(R_{sb}+Z_{LB1})ch\gamma_b x_b}. \quad (14)$$

On figure 3 shows the algorithm of the program for calculating a rail line with distributed parameters with an adaptive receiver powered from the end. Block-schemes of the algorithm have the following purpose:
Taking into account the proposed algorithm, programs were developed, which are an integral part of the track circuit model.

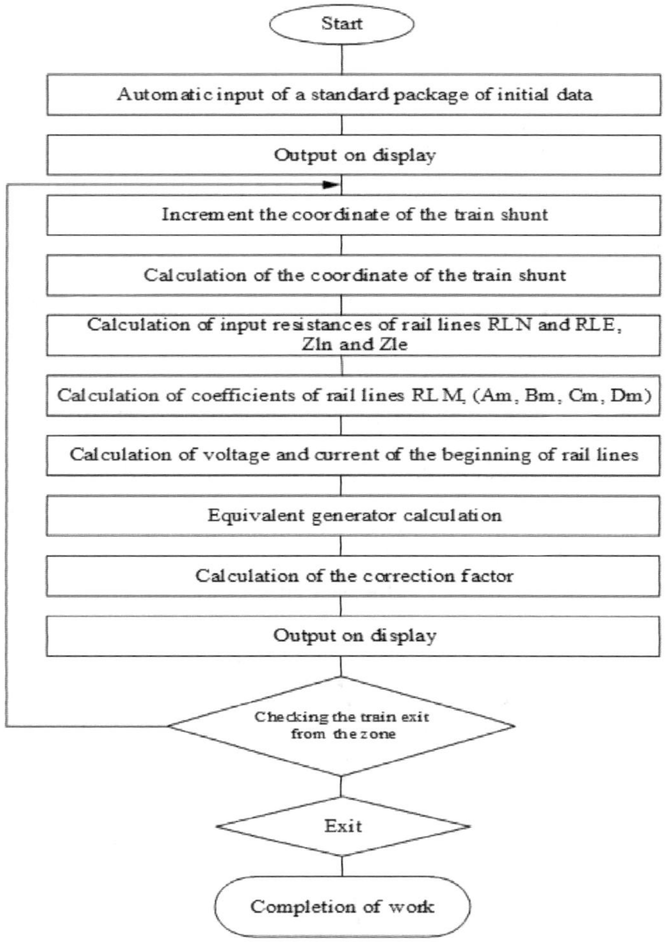

Figure 3. Algorithm work of the program for calculating the track circuits with adaptive receiver for rail line with distributed parameters.

3. Result and discussion

The model of a rail line with distributed parameters makes it possible to carry out studies of the ATC (adaptive track circuit) taking into account the real current distribution along the rail line. The disadvantage of the presented model is that cannot be used for jointlesstrack circuits powered from the middle. For the research of jointlesstrack circuits with two receivers at the ends of the rail line, a universal model is proposed.

3.1. Universal model rail line of the jointless track circuit with adaptive receiver

A typical running jointlesstrack circuit has one track generator and one or two track receivers. On the figure 4 shows the replacement scheme of the track circuit, which allows you to calculate and research a track circuit with an adaptive receiver, both with power from the end and from the middle. On the scheme shows: RLB start rail line, two rail lines middle $RLM^*$and RLM, rail line of end RLE, eG equivalent generator, an equivalent receiver of the beginning of ERe with resistances $Z_e^*$and an equivalent receiver of the end of ERe with resistancesZ_e, train shuntsR_{SB}, R_{SM}^*, R_{SM} and R_{SE}, input resistances Z_{LB1}, Z_{LE1}, Z_{LB}, Z_{LB}^*, Z_{LE}^* and Z_{LE}. The rail lines of the RLM and RLE of both schemes completely coincide, and the rail lines $RLM*$ and RLB on Fig.4 are a mirror image of the right side, i.e.rail lines RLM and RLE. By changing the values of insulation resistances, rails, shunts and their coordinates, as well as the parameters of equivalent receivers and a generator can be imitated any train situations and jointlesstrack circuit parameters.

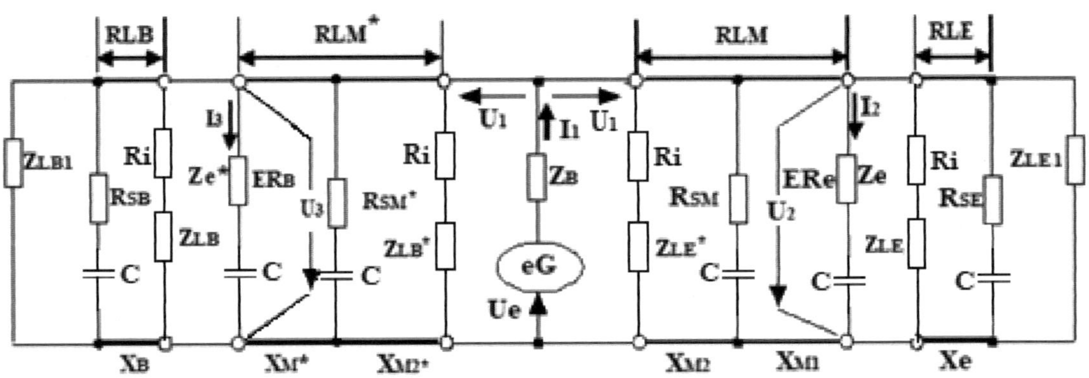

Figure 4. Universal replacement scheme of the track circuit with an adaptive receiver for normal and shunt modes

The dependences between the voltages and currents of the equivalent generator of the eG and the equivalent receiver of the ERe for the circuit in figure 4 can be written by analogy with equations (7-11) for the circuit in figure 2 (in the notation of the diagram in figure4):

$$U_1 = A_m U_2 + B_m I_2;$$

$$I_1 = C_m U_2 + D_m I_2, \tag{15}$$

where

$$A_m = \tau_1 \left(ch\gamma_m X_{m2} + \left(\frac{Z_{wm}}{R_{Sm}}\right) sh\gamma_m X_{m2} \right) + \tau_2 sh\gamma_m X_{m2}; \tag{16}$$

$$B_m = Z_{wm}(sh\gamma_m X_{m1} ch\gamma_m X_{m2} + \tau_3 sh\gamma_m X_{m2}; \tag{17}$$

$$C_m = (\frac{1}{Z_{wm}})(\tau_1 \left(\tau_4 \left(\frac{Z_{wm}}{R_{Sm}}\right) + \tau_5 \right) + \tau_2 \tau_4; \tag{18}$$

$$D_m = \tau_3\tau_4 + \tau_5 sh\gamma_m X_{m1}; \tag{19}$$

$$\tau_1 = th\gamma_m X_{m1} + \left(\frac{Z_{wm}}{Z_{LE}}\right) cth\gamma_m X_{m1} + R_i(Z_{LE} ch\gamma_e x_e + R_{SE} sh\gamma_e x_e);$$

$$\tau_2 = ctsh\gamma_m X_{m1} + \left(\frac{Z_{wm}}{Z_{LE}}\right) th\gamma_m X_{m1} + R_i(Z_{LE} ch\gamma_e x_e + R_{SE} sh\gamma_e x_e);$$

$$\tau_3 = th\gamma_m X_{m1} + \left(\frac{Z_{wm}}{R_{Sm}}\right) cth\gamma_m X_{m1} + R_i(Z_{LE} ch\gamma_e x_e + R_{SE} sh\gamma_e x_e);$$

$$\tau_4 = th\gamma_m X_{m2} + \left(\frac{Z_{wm}}{Z_{LB*}}\right) cth\gamma_m X_{m2} + R_{SM*} ch\gamma_b x_b + Z_{LB*} sh\gamma_b x_b;$$

$$\tau_5 = cth\gamma_m X_{m2} + (\frac{Z_{wm}}{Z_{LB*}}) th\gamma_m X_{m2} + R_{SM*} ch\gamma_b x_b + Z_{LB*} sh\gamma_b x_b, \tag{20}$$

$$Z_{LE} = Z_{WE} \frac{R_{SE} Z_{LE1} ch\gamma_e x_e + Z_{WE}(R_{se} + Z_{LE1}) sh\gamma_e x_e}{R_{SE} Z_{LE1} sh\gamma_e x_e + Z_{WE}(R_{se} + Z_{LE1} ch\gamma_e x_e}; \tag{21}$$

$$Z_{LB*} = (Z_{LB} A_m + B_{m*})/(Z_{LB} C_{m*} + D_{m*}). \tag{22}$$

The dependences between the voltages and currents of the equivalent generator of the eG and the equivalent receiver of the ERb for the scheme in figure4 can be written by analogy with equations (15 and 21):

$$U_1 = A_{m*} U_3 + B_{m*} I_3;$$
$$I_1 = C_{m*} U_3 + D_{m*} I_3, \tag{23}$$

where

$$A_{m*} = \tau_{1*}\left(ch\gamma_{m*} X_{m2*} + \left(\frac{Z_{wm*}}{R_{Sm*}}\right) sh\gamma_{m*} X_{m2}\right) + \tau_{2*} sh\gamma_{m*} X_{m2*}; \tag{24}$$

$$B_{m*} = Z_{wm*}(sh\gamma_{m*} X_{m2*} ch\gamma_{m*} X_{m1*} + \tau_{3*} sh\gamma_{m*} X_{m1*}; \tag{25}$$

$$C_{m*} = (\frac{1}{Z_{wm*}})(\tau_{1*}\left(\tau_{4*}\left(\frac{Z_{wm*}}{R_{Sm*}}\right) + \tau_{5*}\right) + \tau_{2*}\tau_{4*}; \tag{26}$$

$$D_{m*} = \tau_{3*}\tau_{4*} + \tau_{5*} sh\gamma_{m*} X_{m1*}; \tag{27}$$

$$\tau_{1*} = ch\gamma_{m*} X_{m1*} + (\frac{Z_{wm*}}{Z_{LE}}) sh\gamma_{m*} X_{m1*};$$

$$\tau_{2*} = sh\gamma_{m*} X_{m1*} + (\frac{Z_{wm*}}{Z_{LE}}) ch\gamma_{m*} X_{m1*};$$

$$\tau_{3*} = ch\gamma_{m*} X_{m1*} + (\frac{Z_{wm*}}{R_{Sm*}}) sh\gamma_{m*} X_{m1*};$$

$$\tau_{4*} = ch\gamma_{m*} X_{m2*} + (\frac{Z_{wm*}}{Z_{LE*}}) sh\gamma_{m*} X_{m2*};$$

$$\tau_{5*} = sh\gamma_{m*} X_{m2*} + (\frac{Z_{wm*}}{Z_{LE*}}) ch\gamma_{m*} X_{m2*}, \tag{28}$$

$$Z_{LB} = Z_{WN} \frac{R_{SB} Z_{LN1} ch\gamma_b x_b + Z_{WB}(R_{sb} + Z_{LB1}) sh\gamma_b x_b}{R_{SB} Z_{LB1} sh\gamma_b x_b + Z_{WB}(R_{sb} + Z_{LB1} ch\gamma_b x_b}; \tag{29}$$

$$Z_{LE*} = (Z_{LE} A_m + B_{m*})/(Z_{LE} C_m + D_m). \tag{30}$$

In equation (20), which refer to the right side of the circuit in figure 4, you need to enter the value $Z_{LE*}$from equation (22), which belongs to the left side of the scheme, and in equation (28) which refer to the left side of the scheme, you need to enter the value $Z_{LE*}$from equation (22), which belongs to the right side of the scheme. This excludes the possibility of presenting in explicit as the values of the input resistances Z_{LB*} and Z_{LE*}. The solution of the system of equations is possible only by a numerical method.

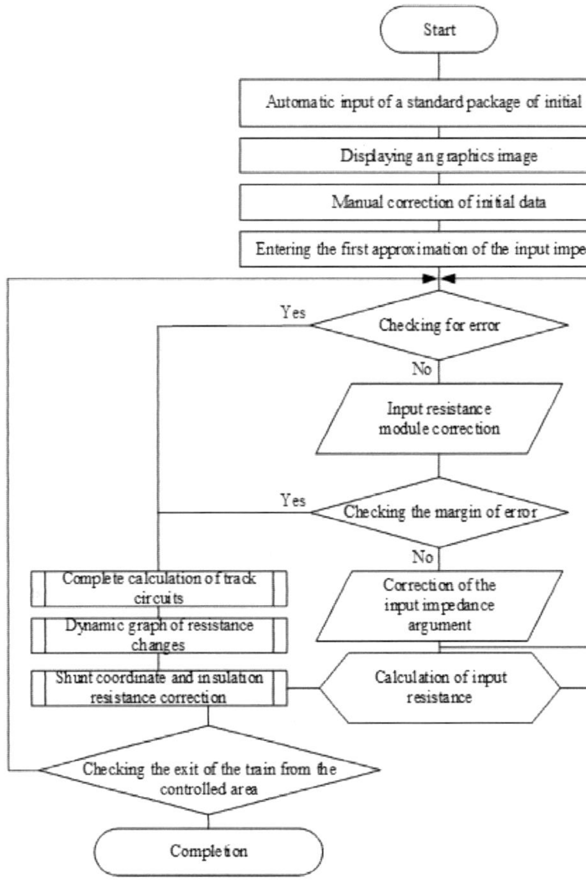

Figure 5. Algorithm of the universal track circuit calculation programs

In figure 5 shows the algorithm for the operation of the universal program for calculating the rail line. The model of a rail line with distributed parameters makes it possible to carry research adaptive track circuities into account the real current distribution along the rail line in the presence of one or two track receivers.

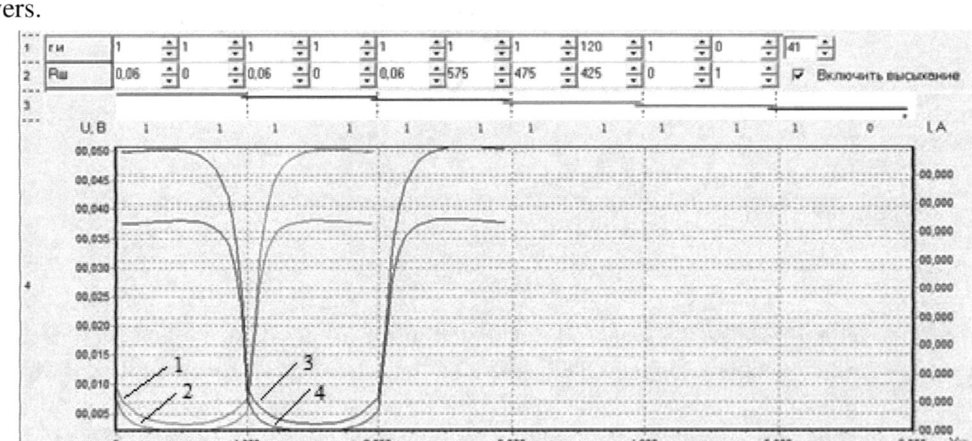

Figure 6. Graphs of the influence of insulation resistance on the voltage of the track receiver

Where, on 1 graph shows the first track circuit, determines the voltage U at insulation resistance $R_i = 1$ om*km and $R_{sh1} = 0{,}06$ om, 2 graph at insulation resistance $R_{sh1} = 0$ om, 3 graph for the second track circuit, i.e. in our case, an adjacent jointlesstrackcircuit, with an insulation resistance $R_i = 1$ om*km and $R_{sh2} = 0{,}06$ om,4 graph, respectively, $R_{sh2} = 0$ om consequently, adaptive join less track circuits define the voltage and fix free track section.

4. Conclusion

The proposed models of the rail lines make it possible to carry out studies of rail circuits with an adaptive receiver in conditions of significant longitudinal asymmetry of insulation resistance under the intense influence of atmospheric influences.

For the analysis of jointless track circuits with an adaptive receiver have to consider of condition (the presence of shunts, the value of the insulation resistance of each radar line, the dynamics and range of variation of this resistance, etc.) several jointless track circuits at the same time. Their mutual influence on each other.

In conditions, when monitoring condition of each jointless tone track circuit depends on the state of the rail lines of other sections, model needs to be investigated, which would represent a collection of rail lines. The research of such track circuits differs significantly from the study of track circuits with relay action receivers.

5. References

[1] Aliev, R.M. and Aliev, M.M. 2020 Methods for calculating the coefficients of the rail fourpoleof sensors control tonal frequency," *Chemical Technology, Control and Management*: Iss. 5, Article 13. pp 68-71.DOI: https://doi.org/10.34920/2020.5-6.68-71

[2] Shamanov V. I. 2019 Pomekhynaappparatururelsovykhtsepeyiavtomaticheskoylokomotivmoysignalizatsii. Sredstvazashchity [Noise disturbance in track circuit equipment and automatic cab signaling. Interference protection facilities]. Uchebnoyeposobiye [Teaching aid] *Moscow, Uchebno-metodicheskiytsentrpoobrazovaniyunazheleznodorozhnomtransporte* [Training and Methodology Centre for Railway Transport] Publ., 303 p.

[3] Leushin V. B. 2009 Osobennostystrukturrelsovykhtsepeyavtoblokirovky [Design specificitiesof track circuits of automatic block system]. *Uchebnoyeposobiye [Teaching aid]. Samara, SamGUPS* [Samara State Transport University] Publ., 136 p.

[4] Kravtsov Yu. A., Arkhipov E. V., Antonov A. A. &Bakin M. E. 2013 Rabotosposobnostputevykhpriemnikovrelsovykhtsepeytonalnoy chastity pry vozdeistviisetevogotokaelektropoezda "Sapasan" [Operating capacity of audio frequency track circuit receivers during line current effect of the electric train "Sapsan"]. *VestniktransportaPovolzhya*[Bulletin of the Volga region transport], **3**, 14–19.

[5] VatulinYa. S., Dergachev A. I. &Chuyan S. N. 2014 Diagnostirovaniyerelsovykhsoyedineniy. Sistemyavtomatizirovannogoproektirovaniyanatransporte [Diagnosis of rail bonds.Computer-aided design systems for transport]. Tezisydokladov V Mezhdunarodnoynauch-no-prakticheskoykonferentsiistudentov, aspirantovimolodykhuchenykh. Federalnoyeagenstvozheleznodorozhnogotransporta [Brief outline reports of the fifth International Research and Training Conference for students, postgraduates and young scholars. Federal Railway Transport Agency]. Saint Petersburg, PGUPS [St. Petersburg State Transport University] Publ., pp. 9–15.

[6] Theeg G. & Vlasenko S. 2020 *Railway Signalling& Interlocking*. 3ed Edition. Germany, Leverkusen, PMC Media House GmbH, 552 p.

[7] Aliev, R. M. &Aliev, M.M. &Tokhirov, E. T. 2020 *Methodology for Determining the Optimal Values of Resistance at the Ends of the Jointless Track Circuit with Considering Twofold Shunting* International Journal of Emerging Trends in Engineering Research, IJETER 8 (9), 5048-5052

[8] Shamanov V. 2018 Formation of Interference from Power Circuits to Apparatus of Automation and Remote Control.*Proceedings of 16th IEEE East-West Design & Test Symposium (EWDTS`2018).* Russia, Kazan, pp. 140–146. DOI: 10.1109/EWDTS.2018.8524676.

[9] Zasov V. 2018 Stability Verification of Statistical Methods for Signal Separation. *Proceed-ings of 16th IEEE East-West Design & Test Symposium* (EWDTS`2018). Russia, Kazan, pp. 803–807. DOI: 10.1109/EWDTS.2018.8524683.Automation on Transport. No 2, Vol. 6, June 2020221Electronic simulation

[10] Sedykh D., Gordon M., Zuyev D. &Skorokhodov A. 2018 Analysis of the Amplitude and Phase-Manipulated Signals of Automation Devices via Bluetooth Technology. *Proceedings of 16th IEEE East-West Design & Test Symposium* (EWDTS`2018). Russia, Kazan, pp. 703–710. DOI: 10.1109/EWDTS.2018.8524605.

[11] Efanov D. V., Osadchy G. V., Khóroshev V. V. &Shestovitskiy D. A.2019 Diagnosticsof Audio-Frequency Track Circuits in Continuous Monitoring Systems for Remote ControlDevices: Some Aspects. *Proceedings of 17th IEEE East-West Design & Test Symposium* (EWDTS`2019). Batumi, Georgia, pp. 162–170. DOI: 10.1109/EWDTS.2019.8884416.

[12] N.E. Fedorov2006 *Relay and microelectronic systems of interval regulation of train traffic*: textbook.allowance.Part 1 - Samara: SamGAPS, - 163 p.

[13] Vasilenko M.N., Denisov B.P.2006Kultin V.B., Rastegaev S.N., Calculation of parameters and performance check of jointless tonal rail circuits // Izvestiya PGUPS.- No 2. - P. 104–112.

[14] VS Dmitriev, VA Minin 1992 *Systems of automatic blocking with rail circuits of a tone frequency* /- Moscow: Transport - 182 p.

[15] Yu. I. Polevoy 2006 *Relative track circuits: textbook.manual for universities* /- Samara: SamGAPS- 80 p.

[16] M.B.Sokolov, SA Veselkov, VB Kultin 2005 Modeling of the equipment of tonal track circuits of the third generation *Collection of articles of the scientific and technical international conference "Science Week 2005"*.- St. Petersburg: PGUPS - S. 163-165.

[17] Denisov B. P., Kultin V. B., Rastegaev S. N. &Vorobey N. Yu. 2010 Metodykontrolyakorrektnostypostroeniyaskhemzameshcheniyatonalnykhrelsovykhtsepey v ARM-TRTs [Testing methods for the design validity of equivalent circuits for audio frequency track circuits in ARM-TRTs (Automated Work Station-Audio Frequency Track Circuits]. *IzvestiyaPeterburgskogouniversitetaputeysoobshcheniya* [Proceedings of Petersburg State Transport University]. Saint Petersburg, PGUPS [St. Petersburg State Transport Univer-sity] Publ., iss. 4 (25), pp. 110–119.

[18] Vorobey N. Yu. 2011 Avtomatizatsiyapostroyeniyaskhemzameshcheniyatonalnykhrelsovykhtsepey s ispolzovaniyemformalizovannogoopisaniya [Design automation of equivalent circuits for audio frequency track circuits involving formal description]. *IzvestiyaPeterburgskogouniversitetaputeysoobshcheniya [Proceedings of Petersburg State TransportUniversity*]. Saint Petersburg, PGUPS [St. Petersburg State Transport University] Publ., iss. 3 (28), pp. 47–53.

[19] Rastegaev S. N. &Vorobey N. Yu. 2011 Avtomatizatsiyaformirovaniyaskhemzameshcheni-yadlyarascheta TRTs [Automation of equivalent circuit formation for TRTs calculation]. Avtomatika, svyaz, informatika: nauchno-populyarniyproizvodstvenno-tekhnicheskiyzhurnal[Automation, communication, information technology: educational industrial and engineeringjournal]. *Moscow, OAO "Russian Railways*" Publ., no. 4, pp. 12–13. (In Russian)

[20] Bulavskiy P. E., Markov D. S., Sokolov V. B. &Konstantinova T. Yu. 2015 Formalizatsiyaalgoritmicheskogoopisaniya system obespecheniyazhiznennogotsiklazheleznodorozh-noyavtomatikyitelemekhaniky [Algorithmic description characterization of the lifecycle supporting systems of railway automation and remote control]. *Avtomatikanatransporte* [Transport automation], vol. 1, No. 4, pp. 418–432.

[21] Markov D. S., Nasedkin O. A., Vasiliev D. A. &Butuzov M. A. 2018 Ponyatiyniy apparat ekspertizyiispytaniyzheleznodorozhnoynabezopasnost [Conceptual framework for se-curity examination and testing of railway automation]. *Avtomatikanatransporte* [Transport automation], vol. 4, no. 1, pp. 30-45.

An adaptive search direction algorithm for the modified projection minimization optimization

Rana Z. Al-Kawaz[1] and Abbas Y. Al-Bayati[2*]

[1]Department of Mathematics, College of Basic Education, University of Telafer, Tall'Afar, Mosul, Iraq.
[2]University of Telafer, Tall'Afar, Mosul, Iraq.

*E-mails:profabbasalbayati@yahoo.com

Abstract. In this paper, we proposed a double-search direction algorithm with a free of derivatives based on a different parameter update (η_k) which gives us an update on the double search trend. We used anewly updated formula for the projection parameter within the formula for the proposed search direction. When comparing our numerical results for the proposed algorithm with some standard published algorithmswe obtain efficient numerical results. The proposed algorithm, especially, isused to solve large-scale nonlinear problems by combining two search directions in one search direction. Also, to demonstrate the general convergence of the proposed new algorithm under some circumstances.The numerical performance of the new proposed algorithm on some nonlinear test functions proved the efficiency of this algorithm.

1. Introduction

Optimization issues took a lot of space in different disciplines such as engineering, geosciences, physics, mechanics, medicine, and so on, but the difficulty of these issues was that they were of the non-linear and complex type. At present, researchers have been able to overcome this problem through numerical optimization methodsof derivative-free to solve such problems as in [1-6].The most prominent and widespread use of derivative-free methods for solving a system of nonlinear equations, which we can represent through:

$$F(x) = 0 \tag{1}$$

Knowing that the function$F: R^n \rightarrow R^n$is a nonlinear function. The premise of this function R^nrefers to the real space of the dimension-n measured by the Euclidean standard $\|.\|$.Researchers have given methods to solve this system in equation (1) through derivative-free methods as in[7-10]. Iterative methods start with a starting pointx_0 and then continue to update the points through the:

$$x_{k+1} = x_k + \alpha_k d_k \tag{2}$$

But these iterations equation (2) dependent on the step $\alpha_k$and the search direction $d_k$may continue without reaching the goal, so there is a method for updating the points based on the quadratic form for the length of the step, known as:

$$x_{k+1} = x_k + \alpha_k e_k + \alpha_k^2 c_k \tag{3}$$

The first idea of the double-search direction equation (3) suggested in [11] where $\alpha_k > 0$ is the step-size, x_{k+1} is the new point of 'equation (3)', x_k is the previous point while e_k and c_k are the first and the second search directions, respectively.The search direction $d_k$in equation (2) is obtained from:

$$F(x_k) + J(x_k)d_k = 0 \tag{4}$$

$where\, J(x_k)$ is the Jacobian matrix that equal to $F'(x_k)$ or an approximation of it.One of the good characteristics of Newton's method is the speed in reaching the optimal solution and its fast convergence, but it requires the Jacobian matrix to be computed. It is best to use a suitable Jacobian matrix $J(x_k)$approximation such thattransformation:

$$J(x_k) \cong \eta_k I \tag{5}$$

Content from this work may be used under the terms of the Creative Commons Attribution 3.0 licence. Any further distribution of this work must maintain attribution to the author(s) and the title of the work, journal citation and DOI.
Published under licence by IOP Publishing Ltd

Where I is the identity matrix and $f(x)$is a modular function that defines:

$$f(x) = \frac{1}{2}\|F(x)\|^2 \tag{6}$$

Note that the problem of the equation in (1) is equivalent to the optimization problem:

$$min\, f(x), \quad x \in R^n \tag{7}$$

When the double-search direction method is observed in equation (3), the iterative information is used in multiple steps, and the curves are searched to generate new repeat points.Researchers continue to develop a special type of search trend, for example, Petrović and Stanimirović [12] deal with a double-direction to solve unconstrained optimization issues. The double-step length system transformation is suggested in [13] to boost the numerical efficiency and global convergence properties of double-direction methods. In [14], Halilu and Waziri recently suggested an Improved matrix-free procedure for solving nonlinear equation structures by the approach of double-step length.Ibrahim's [15] suggested method can be interpreted as an extension using the projection technique of the three-term modified Polak-Ribiére-Polyak and the three-term Hestenes-Stiefel conjugate gradient method.Researchers continued to develop as in [16- 21]. A prerequisite of line search is to properly reduce function values, that is, to evaluate function values:

$$\|F(x_{k+1})\| \leq \|F(x_k)\| \tag{8}$$

The projection approach relies on the use of a monotone case F to accelerate equation (2) and change the new point using repetition. As in:

$$z_k = x_k + \alpha_k d_k \tag{9}$$

The hyperplane, as an original iterative, is:

$$H_k = \{x \in R^n | F(z_k)^T(x_k - z_k) = 0\} \tag{10}$$

To start using the projection technique, we use the new point x_{k+1} as given in the [22, 23] to be the projection of x_k onto the hyperplane H_k. So, can be evaluated as:

$$x_{k+1} = P_\Omega[x_k - \delta_k F(z_k)] \tag{11a}$$

$$\delta_k = \frac{F(z_k)^T(x_k - z_k)}{\|F(z_k)\|^2} \tag{11b}$$

We organized the article in the order: Section 2, deals with the two new modified algorithms (MS-RA and MD-RA). Section 3 deals with introducing some new theorems that prove the convergence of the newly proposed algorithms (MS-RA and MD-RA). Section 4, concerns the numerical results which demonstrate the efficiency of the newly proposed algorithms when compared to some standard algorithms. Finally, Section 5 deals with general conclusions.

2. Two New Modified Algorithms (MS-RA and MD-RA).

In this section, we suggest reducing the two search directions equation (3) into a single one to locate the study direction by depending on the projection technique. This is made possible by having the two directions be similar, i.e. $e_k = c_k$. We say that the unique search path of the $e_k$and c_k in equation (3) is defined as:

$$e_k = c_k = -\eta_k^{-1} F(x_k) \tag{12}$$

Put equation (13) into equation (3), we get:

$$x_{k+1} = x_k + \alpha_k(-\eta_k^{-1}(1 + \alpha_k))F(x_k) \tag{13}$$

From equation (13), we can conclude that the double-direction will become:

$$d_k = -\eta_k^{-1}(1 + \alpha_k)F(x_k) \tag{14}$$

To get acceleration in the new algorithm we will impose the parameterη_{k+1} used in equation (14) as:

$$\eta_{k+1} = e^{\left(\frac{\|s_k\|^2-\|y_k\|^2}{e^{\|F(x_{k+1})\|^2}}\right)} \tag{15}$$

where $y_k = F_{k+1} - F_k$ and the difference between the two-point is $s_k = x_{k+1} - x_k$ .Therefore, through the two equations (13) and equation (14), we can get:

$$x_{k+1} = x_k + \alpha_k d_k \tag{16}$$

Sun and Liu and Wang[24] presented a multi-parameterized conjugated gradient (SLW) algorithm for solving monotone equations with convex constraints based on the projection update formula that:

$$x_{k+1} = P_\Omega\left[x_k - \vartheta\bar{\delta}_k[\tau F(x_k) + F(z_k)]\right] \tag{17a}$$

$$\bar{\delta_k} = \frac{F(z_k)^T(x_k-z_k)}{\|[\tau F(x_k)+F(z_k)]\|^2} \tag{17b}$$

Where ϑ and τ any positive value. In this article, we used an improved projection method equation (17) and the new parameter η_{k+1} in equation (15) for two new algorithms to be designed in the points as the difference between the new algorithm and the algorithm (SLW):

1- Integration of the new single-search direction equation (12) with the newly updated parameter equation(15) in the new point equation (16) to obtain (MS-RA).

2- Merging the new double-search direction equation (14) with the newly updated parameter equation (15) in the new point equation (16) to obtain (MD-RA).

We present the steps of each of these algorithms in-depth to explain the concept of the numerical algorithms used in this research.

*2.1. **New Modified Single Search Direction Algorithm (MS-RA)***

Step 0: Given $x_0 \in \Omega$,η_0, r, σ, μ, τ $\in(0,1)$, $\alpha > 0$, $\varepsilon > 0$, set k=0.
Step1: Compute $F_k = F(x_k)$ and test If $\|F_k\| \le \varepsilon$ yes, then stop; otherwise, continue to the next step.
Step 2: Compute search direction d_k 'equation (12)'.
Step 3: Set z_k from equation (9) and compute step length α_k using this line-search:

$$-F(x_k + \mu\, r_k^m d_k)^T d_k \ge \sigma\, \mu\, r_k^m \|\tau F(x_k) + F(x_k + \mu\, r_k^m d_k)\|\|d_k\|^2 \tag{18}$$

Step 4: If $z_k \in \Omega$ and $\|F(z_k)\| \le \varepsilon$ stop, else compute x_{k+1} from 'equation (16)'.
Step 5: Determine η_{k+1} using 'equation (15)'.
Step 6: Set k=k+1, and go to 2.

*2.2. **New Modified Double Search Direction Algorithm (MD-RA)***

Step 0: Given $x_0 \in \Omega$,η_0, r, σ, μ, τ $\in(0,1)$, $\alpha > 0$, $\varepsilon > 0$, set k=0.
Step 1: Compute $F_k = F(x_k)$ and test If $\|F_k\| \le \varepsilon$ yes, then stop; otherwise, continue to the next step.
Step 2: Compute search direction d_k equation (14).
Step 3: Set z_k from (13) and compute step length α_k using this line-search from equation (18).
Step 4: If $z_k \in \Omega$ and $\|F(z_k)\| \le \varepsilon$ stop, else compute x_{k+1} from equation (16).
Step 5: Determine η_{k+1} using equation (15).
Step 6: Set k=k+1, and go to 2.

3. Convergence Analysis.

In the previous section, we proposed two new algorithms (MS-RA and MD-RA) depending on the parameter η_{k+1} in equation (15). Now in this section, we will present several theorems to prove the global convergenceproperty of the two new proposed algorithms, but before that, we must give the basic assumptions some amount of attention which is

*3.1. **Assumption A***

- Suppose there is a set level defined by:

$$\Omega = \{x|\|F(x)\| \le \|F(x_0)\|\}$$

- There is x^*that belongs to R^n, where $F(x^*) = 0$ is true.

- Let the F function be continuously and differentiablein some neighborhood N, i.e. N of x^* contained in Ω.
- On N, i.e., there is a positive definite Jacobian of F restricted, i.e. there are a positive constant M > m > 0 are such that:

$$\|F'(x)\| \leq M, \ \forall x \in N. \tag{19}$$

and

$$m\|d\|^2 \leq d^T F'(x)d, \ \forall x \in N, d \in R^n . \tag{20}$$

Assumption A means that the special solution of equation (1) in N stands for x^*. Since $F'(x_k)$ is approximated by $\eta_k I$ along the search directionss_k.

3.2. *Assumption B*

If we consider it, $\eta_k\, I$ is a good approximation of $F'(x_k)$, which implies that:

$$\|(F'(x_k) - \eta_k\, I)d_k\| \leq \varepsilon \| F(x_k)\| \tag{21}$$

where $\varepsilon \in (0,1)$ is a small quantityfor more information on equation (21) see [25].

3.3. Theorem (Descent Direction)

Suppose Assumption B holds and the new algorithms (MS-RA) and (MD-RA) produces $\{x_k\}$. Then, d_k in equation (12) and equation (14) in the direction of the descent of $f(x_k)$ at x_ki.e.

$$\nabla f(x_k)^T d_k < 0 \tag{22}$$

Proof:

We will divide the proof into two parts, each part concerned with an algorithm to change the search direction in each of them as follows:

Part 1: when dealing with the (MS-RA) algorithm, we will need a search direction from the equation (12), as:

$$\nabla f(x_k)^T d_k = F(x_k)^T F'(x_k) d_k$$

$$= F(x_k)^T [(F'(x_k) - \eta_k I)d_k - F(x_k)]$$

$$\nabla f(x_k)^T d_k = F(x_k)^T (F'(x_k) - \eta_k\, I)d_k - \|F(x_k)\|^2 \tag{23}$$

by Cauchy-Schwarz inequality, we have:

$$\nabla f(x_k)^T d_k \leq \| F(x_k)\| \|(F'(x_k) - \eta_k\, I)d_k\| - \|F(x_k)\|^2 \tag{24}$$

If equation (21) satisfy then,

$$\nabla f(x_k)^T d_k \leq \varepsilon \|F(x_k)\|^2 - \|F(x_k)\|^2 \leq -(1 - \varepsilon)\|F(x_k)\|^2 \tag{25}$$

Hence for $\varepsilon \in (0,1)$, this proves of part 1 is true.

Part 2: when dealing with the (MD-RA) algorithm, we will need a search direction from equation (12) and equation (14), as:

$$\nabla f(x_k)^T d_k = F(x_k)^T F'(x_k) d_k$$

$$= F(x_k)^T [(F'(x_k) - \eta_k I)d_k - (1 + \alpha_k)F(x_k)]$$

$$\nabla f(x_k)^T d_k = F(x_k)^T (F'(x_k) - \eta_k\, I)d_k - (1 + \alpha_k)\|F(x_k)\|^2 \tag{26}$$

by Cauchy-Schwarz inequality, we have:

$$\nabla f(x_k)^T d_k \leq \| F(x_k)\| \|(F'(x_k) - \eta_k\, I)d_k\| - (1 + \alpha_k)\|F(x_k)\|^2$$

If equation (21) satisfy then,

$$\nabla f(x_k)^T d_k \leq \varepsilon \|F(x_k)\|^2 - (1 + \alpha_k)\|F(x_k)\|^2 \leq -(1 - \varepsilon + \alpha_k)\|F(x_k)\|^2$$

Hence this proves part 2. This means that the newly proposed algorithms (MS-RA) and (MD-RA) have descent search directions.

3.4. ***Remark***

From the Theorem (descent direction), we can deduce that the norm functions $f(x_k)$ is a decline for d_k, which implies that:

$$\|F(x_{k+1})\| \leq \|F(x_k)\| \leq \ldots \leq \|F(x_0)\|$$

This implies that $x_k \in \Omega$.

3.5. ***Lemma (Bounded*** η_{k+1}***)***

Suppose Assumption A holds and $\{x_k\}$ is algorithm-generated (MS-RA) and $\{x_k\}$ (MD-RA). So, a constant $M > m > 0$exists, such that for all k:

$$\frac{\|s_k\|^2 - \|y_k\|^2}{e^{\|F(x_{k+1})\|^2}} \leq \frac{(1-M^2)D}{B} \tag{27}$$

Proof:

We get From Assumption A:

$$y_k^T s_k \geq m\,\|s_k\|^2 \tag{28}$$

from [26], we have:

$$M^2 y_k^T s_k \geq m\,\|y_k\|^2 \tag{29}$$

then,

$$[m\,\|s_k\|^2 - m\,\|y_k\|^2] \leq [y_k^T s_k - M^2 y_k^T s_k]$$

$$[\|s_k\|^2 - \|y_k\|^2] \leq \frac{[1-M^2]}{m} y_k^T s_k$$

from the above Remark we have:

$$m\,\|s_k\| \leq \|F_{k+1}\| \leq \|F_{k+1} - F_k\| \leq M\,\|s_k\| \tag{30}$$

Let $e^{m^2\|s_k\|^2} \cong C$ and $\|s_k\|^2 \cong D$. Hence,

$$\frac{\|s_k\|^2 - \|y_k\|^2}{e^{\|F(x_{k+1})\|^2}} \leq \frac{\frac{[1-M^2]}{m} y_k^T s_k}{e^{\|F(x_{k+1})\|^2}} \leq \frac{[M^2-1](-m\|s_k\|^2)}{m\, e^{m^2\|s_k\|^2}} \leq \frac{[1-M^2]\,D}{C}$$

The inequality (27) is true. Using (27), η_{k+1} is generated by the update of formula (15) and we can deduce that $\eta_{k+1}I$ inherit the positive definiteness of $\eta_k I$.

3.6. ***Lemma (Bounded*** $\mathbf{d}_k$***)***

Suppose the assumptions A and B are retained and the algorithm $\{x_k\}$ is generated (MS-RA) and $\{x_k\}$ (MD-RA). Then a b>0 constant exists such that $\forall k > 0$,

$$\|d_k\| \leq b_i \tag{31}$$

where i=1,2.

Proof:

In this lemma, we will present two elements, each of which relies on the search direction resulting from an algorithm, while in:

Part 1: We have equations (12,15), and Assumption A:

$$\|d_k\| = \left\| -e^{\frac{-(\|s_k\|^2 - \|y_k\|^2)}{e^{\|F(x_{k+1})\|^2}}} F(x_k) \right\| \tag{32}$$

and using the result of Lemma (Bounded η_{k+1}),

$$\|d_k\| \leq e^{\frac{[1-M^2]\,D}{C}} \|F(x_k)\| \tag{33}$$

$$\|d_k\| \leq [B_1 \|F(x_0)\|] \tag{34}$$

$$\|d_k\| \leq b_1 \tag{35}$$

where B_1>0 and $b_1 = B_1\|F(x_0)\|$.

Part 2: From equations (14,15), and Assumption A we have:

$$\|d_k\| = \left\| -(1+\alpha_k)\, e^{\frac{-(\|s_k\|^2 - \|y_k\|^2)}{e^{\|F(x_{k+1})\|^2}}} F(x_k) \right\| \tag{36}$$

and using the same procedure used in the first part,

$$\|d_k\| \leq (1+\alpha_k) e^{\frac{[1-M^2]\,D}{C}} \|F(x_k)\| \tag{37}$$

$$\|d_k\| \leq [(\|F(x_0)\| + \alpha_k \|F(x_k)\|) e^{\frac{[1-M^2]\,D}{C}}] \tag{38}$$

$$\|d_k\| \leq [(\|F(x_0)\| + B_2) e^{\frac{[1-M^2]\,D}{C}}] \tag{39}$$

$$\|d_k\| \leq b_2 \tag{40}$$

where B>0 and $b_2 = (\|F(x_0)\| + B_2) e^{\frac{[1-M^2]\,D}{C}}$.

The theorem below deals with the global property of convergence. To show that there is an accumulation point of x_k under a few appropriate conditions, which is a solution to the problem equation (1).

3.7. ***Theorem (Global Convergence).***
Supposing Assumption B holds, the algorithm $\{x_k\}$ (MS-RA) is generated and the algorithm (MS-RA) is generated (MD-RA). Assume for more$\forall k > 0$,

$$\alpha_k \geq \zeta \frac{|F(x_k)^T d_k|}{\|d_k\|^2} \tag{41}$$

where ζ is some positive constant. Then

$$\lim_{k\to\infty} \|F(x_k)\| = 0 \tag{42}$$

Proof:
We have from equation (31), and (Descent Direction Theorem)

$$\lim_{k\to\infty} \|s_k\| = 0 \tag{43}$$

and the bounded of $\|d_k\|$, we have:

$$\lim_{k\to\infty} \alpha_k \|d_k\|^2 = 0 \tag{44}$$

From equation (41) and (44) it follows that:

$$\lim_{k\to\infty} |F(x_k)^T d_k| = 0 \tag{45}$$

We will take two parts according to the two new algorithms at this point of the proof, as in:

Part 1: According to the (MS-RA) algorithm and from equation (12), we have:

$$F(x_k)^T d_k = -\eta_k^{-1} \|F(x_k)\|^2$$

$$-\eta_k F(x_k)^T d_k = \|F(x_k)\|^2$$
$$|\eta_k||F(x_k)^T d_k| = \|F(x_k)\|^2$$

and as in the theorem, we imposed:

$$|\eta_k|\frac{1}{c}\alpha_k\|d_k\|^2 \geq \|F(x_k)\|^2 \quad (46)$$

While

$$|\eta_k| = \left| e^{\frac{\|s_{k-1}\|^2-\|y_{k-1}\|^2}{e^{\|F(x_k)\|^2}}} \right| \leq e^{\frac{[1-M^2]D}{C}} \leq \rho$$

so, from the equation (46), then

$$0 \leftarrow \frac{\rho}{c}\alpha_k\|d_k\|^2 \geq \|F(x_k)\|^2 \geq 0 \quad (47)$$

Consequently, the equation (42) is valid and the proof for Part 1 is complete.

Part 2: According to the (MD-RA) algorithm and using equation (14), we have:

$$F(x_k)^T d_k = -\eta_k^{-1}(1+\alpha_k)\|F(x_k)\|^2 \quad (48)$$
$$-\eta_k F(x_k)^T d_k = (1+\alpha_k)\|F(x_k)\|^2$$
$$\|F(x_k)\|^2 = \|-\eta_k F(x_k)^T d_k\| - \|\alpha_k F(x_k)\|^2 \quad (49)$$
$$\|F(x_k)\|^2 \leq |\eta_k||F(x_k)^T d_k| \quad (50)$$

while $|\eta_k| \leq \delta$ as in part 1, so from the equation (50), then

$$0 \leftarrow \delta|F(x_k)^T d_k| \geq \|F(x_k)\|^2 \geq 0 \quad (51)$$

The equation (42) is then valid and the evidence for Part 2 is complete.

4. Numerical Performance.

Our computational results for comparisons between the two recent proposed algorithms (MS-RA) and (MD-RA) and the two standard algorithms (HWY) [26] and (SLW) [24] that are devoid of the derivative for solving such nonlinear test problems will be presented in this section. In our implementation of all three algorithms in a laptop calculator, we used the Matlab R2018b program with its Corei5 requirements. As for the instruments used in the two algorithms, such that:

Table1: Numerical parameters used

Parameters	Value of new algorithms	Value of old algorithms
η_0	0.3	0.5
μ	0.5	0.5 , 0.9
r	0.9	0.9
σ	0.02	0.01
$w_1 = w_2$		10^{-4}
τ	0.1	0.1
ϑ	1.7	1.7

and

$$\|F(x_k)\| < 10^{-8}$$

The results of several non-derivative functions are found by the program through several initial points indicated in the table below:

Table2: The startingpoints

Variables	Starting point

x_1	$(1,1,1,..,1)^T$
x_2	$(0.2,0.2,0.2,..,0.2)^T$
x_3	$(20,20,20,..,20)^T$
x_4	$(\text{rand},\text{rand},\text{rand},..,\text{rand})^T$

These algorithms we implemented within dimensions n (1000, 2000, 5000, 7000, 12000). All such algorithms are recognized by their performance in:

Table3: Tools of comparison

Variables	Starting point
Iter	The number of iterations
Eval-F	The number of functions evaluations
Time	TimeCPU in the second

The test problems $F(x) = (f_1, f_2, f_3, \dots, f_n)^T$ where $x = (x_1, x_2, x_3, \dots, x_n)^T$, for $i = 1,2,\dots,n$ and $\Omega = R_+^n$ are from [27,28] and listed as follows:

Problem 1: $F_i(x) = x_i - \sin x_i$

Problem 2: $F_i(x) = e^{x_i} - 1$

Problem 3: $F_i(x) = \sqrt{c}\,(x_1 - 1), i = 2,3,..,n-1$; $F_n(x) = \frac{1}{4n}\sum_{j=1}^{n} x_j^2 - 1/4, c = 1 * 10^{-5}$.

Problem 4: $F_i(x) = \ln(|x_i| + 1) - \frac{x_i}{n}$

Problem 5: $F_i(x) = \min(\min(|x_i|, x_i^2), \max(|x_i|, x_i^3))$

Problem 6: $F_1(x) = x_1 - e^{\frac{\cos(x_1+x_2)}{n+1}}$

$F_i(x) = x_i - e^{\frac{\cos(x_{i+1}+x_i+x_{i-1})}{n+1}}, \text{ for } i = 2,3,..,n-1$

$F_n(x) = x_n - e^{\frac{\cos(x_{n-1}+x_n)}{n+1}}$

Problem 7: $F_i(x) = \frac{i}{n} e^{x_i} - 1$

Problem 8: $F_1(x) = e^{x_1} - 1$; $F_i(x) = e^{x_i} - x_{i-1} - 1$

Problem 9: $F_i(x) = \sum_{i=1}^{n} |x_i|^i$

Problem 10: $F_i(x) = \sum_{i=1}^{n} |x_i|$

Problem 11: $F_i(x) = \max_{i=1,..,n} |x_i|$

Problem 12: $F_i(x) = \sum_{i=1}^{n} |x_i|\, e^{-\sum_{i=1}^{n} \sin(x_i^2)}$

Problem 13: $F_i(x) = \sum_{i=1}^{n} |x_i|^{i+1}$

Figures (1, 2, and 3) are used to equate (HWY) and (SLW) with (MS-RA) and (MD-RA) algorithms when flipping the search path from 1 to 2 using the Dolan and Mor'e style [29]. Two initial points, the first 1, 2, 3, and 4, were compared. The figures are around the initial point 1 since they work better in such a way that:

IOP Conf. Series: Materials Science and Engineering 1152 (2021) 012019 doi:10.1088/1757-899X/1152/1/012019

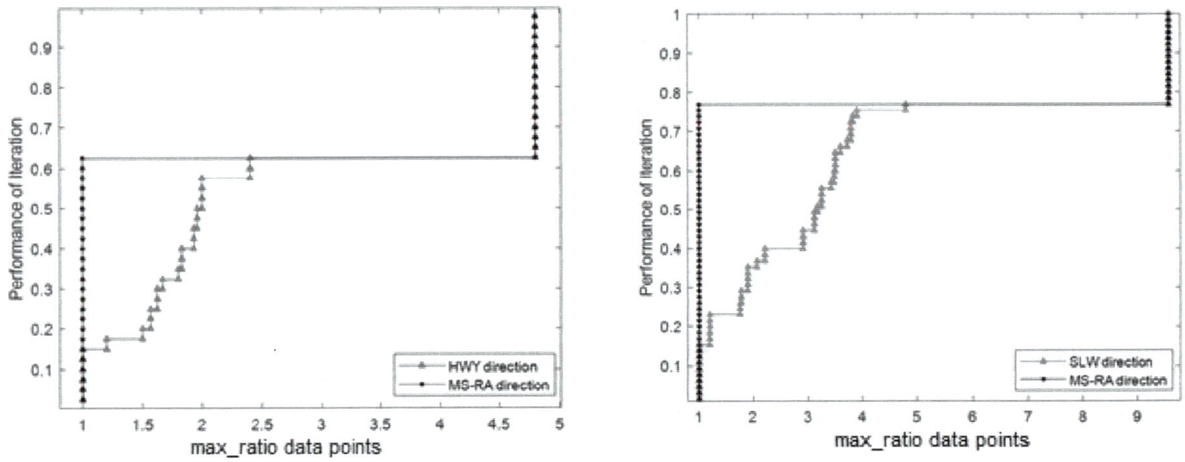

(a) (b)

Figure 1. Iterations efficiency for the (MS-RA)with((a) HWY and (b) SLW) algorithms.

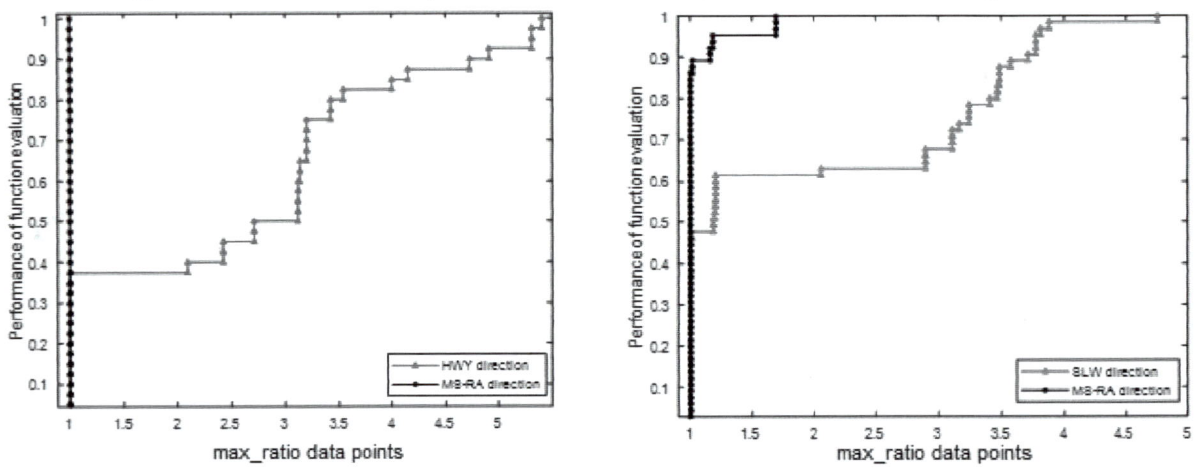

(a) (b)

Figure 2. Function evaluation efficiency for the (MS-RA)with((a) HWY and (b) SLW) algorithms.

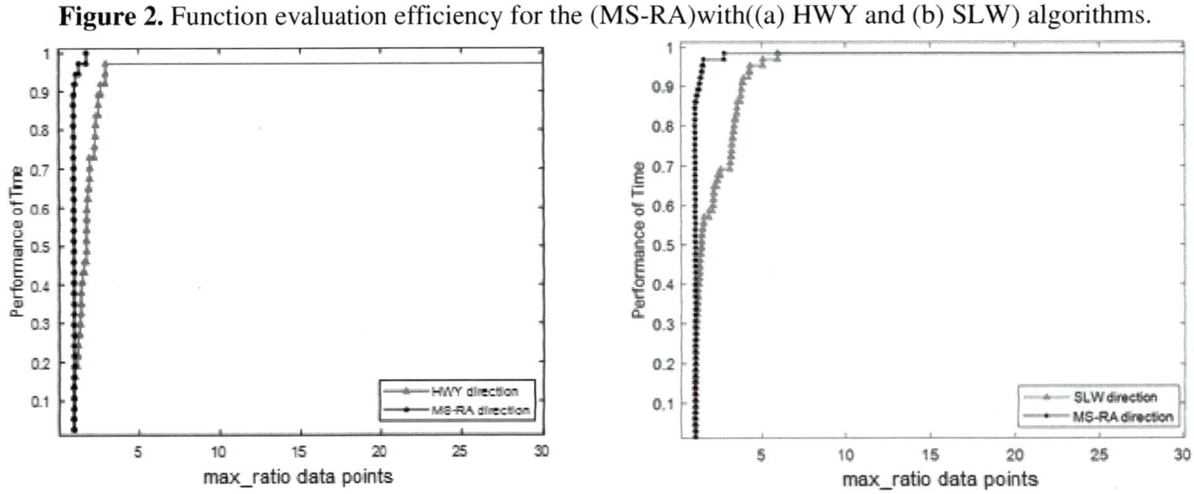

(a) (b)

Figure 3. Time efficiency for the (MS-RA)with((a) HWY and (b) SLW) algorithms.

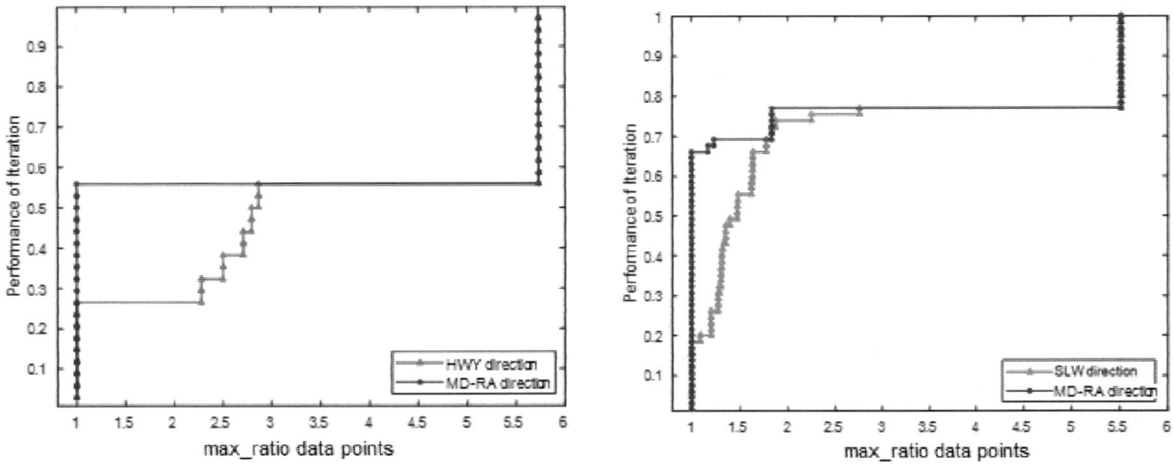

(a) (b)

Figure 4. Iteration efficiency for the (MD-RA)with((a) HWY and (b) SLW) algorithms.

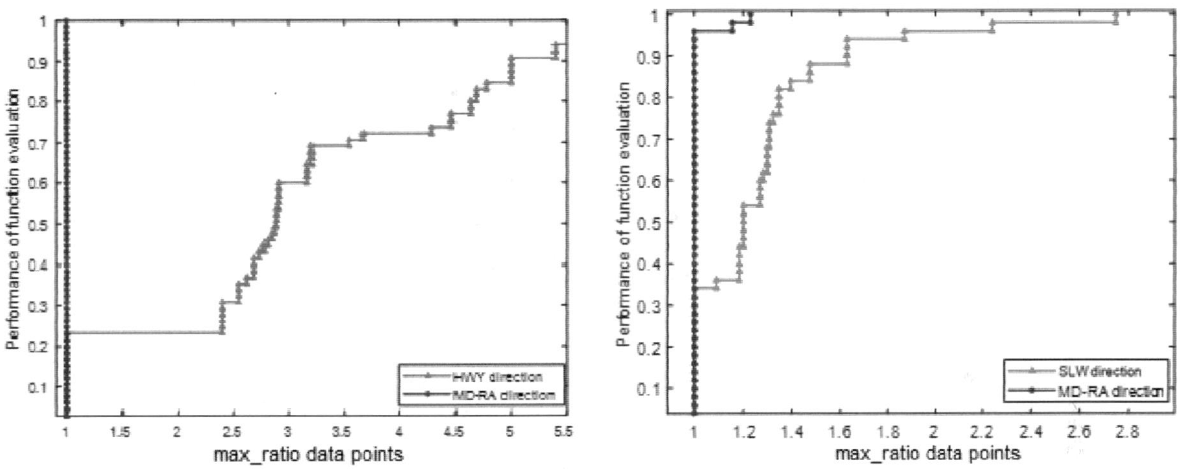

(a) (b)

Figure 5. Function evaluation efficiency for the (MD-RA)with((a) HWY and (b) SLW) algorithms.

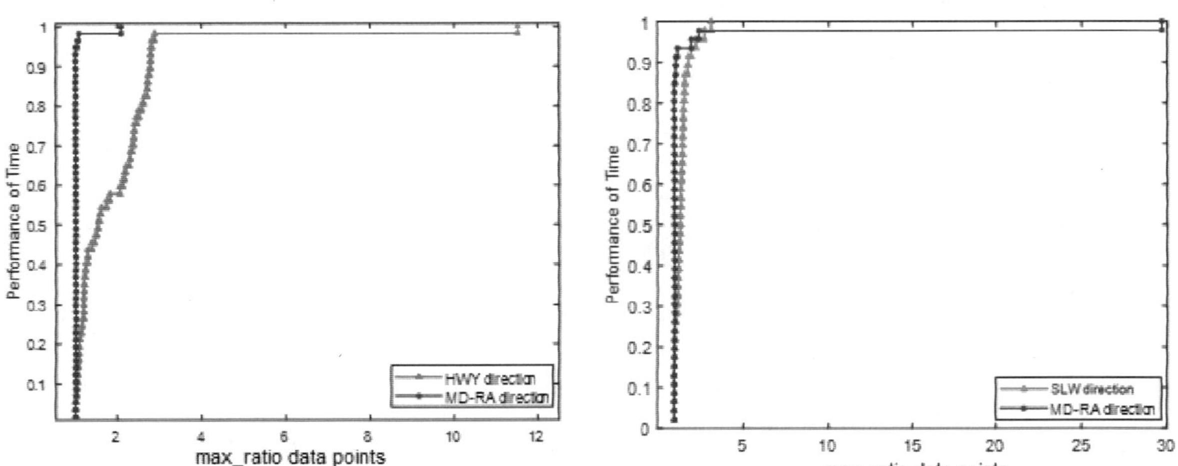

(a) (b)

Figure 6. Time efficiency for the (MD-RA)with((a) HWY and (b) SLW) algorithms.

The first part describes the output of the first new algorithm (MS-RA) with a repeated point with one search direction equation (12), where we show the figures (1), (2), and (3) which we compared with the output of the (HWY and SLW) algorithm also with one search direction, and the comparison was rendered in three ways (the number of iterations, the number of times the function evaluations is calculated, time spent in implementation).
As for the second part, represented by the figures (4), (5), and (6), the second new algorithm was used, which replaced one search direction 'equation (12)' with two search directions in the previous iterative point (14), and the contrast was made with the first new algorithm in the same normalization steps.

5. Conclusions.

The results in the six figures within the numerical performance section show how efficiently the two new algorithms (MS-RA) and (MD-RA) are when we compare their performance with the previous standard algorithms (HWY and SLW). The direction of research in the two new algorithms gave us faster results with fewer iterations when we used the first point, no matter how much we increased the dimensions of the variables. Also, the two algorithms converge to reach the optimum point for solving nonlinear functions.

6. References.

[1] Ahmed H I Al-Kawaz R Z and Al-Bayati A Y 2019 Spectral Three-Term Constrained Conjugate Gradient Algorithm for Function Minimizations *Journal of Applied Mathematics* **2019**
[2] Hamed E T Al-Kawaz R Z and Al-Bayati A Y 2020 New Investigation for the Liu-Story Scaled Conjugate Gradient Method for Nonlinear Optimization *Journal of Mathematics* **2020**
[3] Al-Kawaz R Z Al-Bayati A Y and Jameel M S 2020 Interaction between un updated FR-CG algorithms with an optimal Cuckoo algorithm *Indonesian Journal of Electrical Engineering and Computer Science* **19** 1497-1504
[4] Al-Arbo A A and Al-Kawaz R Z 2020 Implementation of a combined new optimal cuckoo algorithm with a Gray wolf algorithm to solve unconstrained optimization nonlinear problems *Indonesian Journal of Electrical Engineering and Computer Science* **19** 1582-1589
[5] Al-Arbo A A and Al-Kawaz R Z 2021 A fast spectral conjugate gradient method for solving nonlinear optimization problems *.Indonesian Journal of Electrical Engineering and Computer Science* **21** 429-439
[6] Al-Kawaz R Z and Al-Bayati A Y 2021 Show off the efficiency of the Dai-Liao method in merging technology for monotonous non-linear problems *Indonesian Journal of Electrical Engineering and Computer Sciences.* **21** 505-515
[7] Abubakar A B Kumam P and Mohammad H 2020 A note on the spectral gradient projection method for nonlinear monotone equations with applications *Computational and Applied Mathematics* **39** 1-35
[8] Halilu A S and Waziri M Y 2018 An improved derivative-free method via a double direction approach for solving systems of nonlinear equations *Journal of the Ramanujan Mathematical Society* **33** 75-89
[9] Abdullahi H Halilu A S and Waziri M Y 2018 A Modified Conjugate Gradient Method via a Double Direction Approach for solving large-scale Symmetric Nonlinear Systems *Journal of Numerical Mathematics and Stochastics* **1** 32-44
[10] Halilu A S Waziri M Y and Musa Y B 2020 Inexact Double Step Length Method For Solving Systems Of Nonlinear Equations *Statistics, Optimization & Information Computing* **8** 165-174
[11] Đuranović-Miličić N 2008 A multi-step curve search algorithm in nonlinear optimization *Yugoslav Journal of Operations Research* **18** 47-52
[12] Petrović M J and Stanimirović P S 2014 Accelerated double direction method for solving unconstrained optimization problems *Mathematical Problems in Engineering* **2014**

[13] Halilu A S and Waziri M Y 2017 A transformed double steplength method for solving large-scale systems of nonlinear equations *Journal of Numerical Mathematics and Stochastics* **9** 20-23.
[14] Halilu A S and Waziri M Y 2017 Enhanced matrix-free method via a double step length approach for solving systems of nonlinear equations *International journal of applied mathematical research* **6** 147-156
[15] Ibrahim A H Kumam P and Kumam W 2020 A family of derivative-free conjugate gradient methods for constrained nonlinear equations and image restoration *IEEE Access* **8** 162714-162729
[16] Broyden C G 1965 A class of methods for solving nonlinear simultaneous equations *Mathematics of computation* **19** 577-593
[17] Waziri M Y Leong W J Hassan M A and Monsi M 2010 A new Newton's method with a diagonal Jacobian approximation for systems of nonlinear equations *Journal of Mathematics and Statistics* **6** 246-252
[18] Halilu A S Dauda M K Waziri M Y and Mamat M 2019 A Derivative-Free Decent Method Via Acceleration Parameter for Solving Systems of Nonlinear Equations *Open Journal of Science and Technology* **2** 1-4
[19] Waziri M Y and Sabi'u J 2015 A derivative-free conjugate gradient method and its global convergence for solving symmetric nonlinear equations *International Journal of Mathematics and Mathematical Sciences* **2015**
[20] Bouaricha A and Schnabel R B 1998 Tensor methods for large sparse systems of nonlinear equations *Mathematical Programming* **82** 377-400
[21] Zhou W and Shen D 2014 An inexact PRP conjugate gradient method for symmetric nonlinear equations *Numerical Functional Analysis and Optimization* **35** 370-388
[22] Ortega J M and Rheinboldt W C 2000 Iterative solution of nonlinear equations in several variables *Society for Industrial and Applied Mathematics* **30**
[23] Solodov M V and Svaiter B F 1998 A globally convergence, inexact Newton method for systems of monotone equations *In Reformulation: Non-smooth Piecewise Smooth Semi-smooth and Smoothing Methods* 355-369
[24] Sun M Liu J and Wang Y 2020 Two Improved Conjugate Gradient Methods with Application in Compressive Sensing and Motion Control *Mathematical Problems in Engineering* **2020**
[25] Yuan G and Lu X 2008 A new backtracking inexact BFGS method for symmetric nonlinear equations *Computers & Mathematics with Applications* **55** 116-129
[26] Halilu A S Waziri M Y and Yusuf I 2020 Efficient matrix-free direction method with line search for solving a large-scale system of nonlinear equations *Yugoslav Journal of Operations Research*
[27] Yang X S 2010 *Engineering optimization: an introduction with metaheuristic applications* (John Wiley & Sons)
[28] Jamil M and Yang X S 2013 A literature survey of benchmark functions for global optimization problems *International Journal of Mathematical Modelling and Numerical Optimisation* **4** 150-194
[29] Dolan E D and Moré J J 2002 Benchmarking optimization software with performance profiles Mathematical Programming **91** 201-213

Acknowledgments.
The research is supported by the College of Computer Sciences and Mathematics, University of Mosul, and College of Basic Education, University of Telafer, Mosul, Republic of Iraq. The authors declare that there are no conflicts of interest regarding this work.

Authors'Background

Your Name	Title*	Research Field
Rana Z. Al-Kawaz	Ph.D.Candidate	Numerical Optimization
Abbas Y. Al-Bayati	Full Professor	Computer Oriented Numerical Optimization

Realization of Dependable Digital Systems for Safety-Critical Computer Systems using FPGAs

F N kassab bashi[1], S S Khairullah[1*] and C R Elks[2]

[1]College of Engineering, Computer Engineering Department, University of Mosul, Mosul, Iraq.
[2]College of Engineering, Virginia Commonwealth University, Richmond, USA.

[*]E-mail: shawkat.sabah@uomosul.edu.iq

Abstract. Digital systems built based on field-programmable-gate-arrays (FPGAs) are increasingly being used in safety-critical computer systems. These systems typically require high levels of integrity, reliability, and performance to ensure the correct operation of the critical application in which the digital systems are embedded. To achieve the safety and functional objectives, many hardware architectural scenarios based on fault tolerance techniques were proposed. This research paper presents a new dependable digital system (DDS) that consists of fault tolerant blocks (FTBs), spare functional blocks (SFBs), synchronization units, and input output units. Markov-based chain have been used to model the correct operation of the proposed hardware system that is targeted to be realized on FPGA-based technology. This system was designed to be reliable against transient faults (TF), permanent faults (PF), and hardware common cause failures (CCF). The reliability analysis of the proposed system was accomplished with Markov modeling techniques which could express the regenerative behavior of the digital system. Certain states in the system represent system failure, while others represent fault-free behavior in the presence of faults. Finally, based on the MathWorks Simulink simulation results of modeling the DDS digital system, the system does meet its functionality and reliability requirements.

1. Introduction

Field-Programmable Gate Arrays (FPGAs) are integrated circuits (ICs) that are capable of implementing custom digital designs based on the interaction between hardware and software components that could be storage elements, reconfigurable computing blocks, switch boxes, multiplexers and decoders. These embedded elements have the ability of being modified and reconfigured at runtime to execute the task of a new application [1]. In fact, many studies have indicated that FPGA's is a viable technology for safety-critical applications such as space and avionic control systems, industrial automated systems, and nuclear power industry. The key characteristics of this technology that are beneficial for safety critical systems are; custom configurability, partitioning of functionality, true concurrency, and high performance [3].

In contemporary safety related digital systems, that use the Static random-access memory (SRAM) technology the progressive reduction in device feature size and increase in clock rates are making them more susceptible to the effects of ionizing radiation, electromagnetic interactions, and high temperatures. Safety-critical digital systems are expected to operate correctly in the presence faults, diverse failure modes that stem from external threats that could be error, fault, or attack. These faults can cause the system to generate wrong data values and go into an unsafe state in which the error must

Content from this work may be used under the terms of the Creative Commons Attribution 3.0 licence. Any further distribution of this work must maintain attribution to the author(s) and the title of the work, journal citation and DOI.
Published under licence by IOP Publishing Ltd

be detected and tolerated. These faults are often classified as a single transient fault, a single permanent fault, or multiple sequential permanent faults. Transient-faults are temporary changes in the data value and can be caused by radiations. For example, Single-Event-Upsets (SEUs) are the change of state in the contents of a memory cell from a logic '1' to logical '0', or from a logic '0' to a '1'. On the other hand, Multiple-Bit-Upsets (MBUs) are examples of multiple faults affecting on the system concurrently due to the continual shrinking of the dimensions in the SRAM-based transistor [13].
Furthermore, permanent faults or hard-errors caused by system wear-out can cause changing the contents of the data permanently [2]. In recent years, numerous research works have been developed to solve the issue of soft-errors and calculate its reliability [9] [10] [11], and other works have tried to tolerate the effects of both soft-errors and hard-errors in the FPGA-based digital technology [4] [6] [14]. However, proposing a comprehensive dependable digital system that is hierarchal, deterministic, and scalable is still essential in the research field. In this paper, we aim to design a deterministic fault-tolerant digital system (DDS) that is reliable against different failure modes that could be transient faults (TFs), permanent faults (PFs), or hardware common cause failures (CCFs). MathWorks Simulink tool was used in modeling the hardware components of the proposed system and verifying the functional and safety requirements. Finally, the dependability and reliability analysis of the system was achieved by using Markov-based chains and injecting transient faults into the functional blocks was performed at different levels of the proposed system.

2. Related works

Several studies on FPGA design and fault tolerance have recently been published. However, the manufacturers are still looking for further reliable chips that are low in cost (delay, complexity, hardware, area, and so on) and have a high yield elaboration of safety-critical applications contra various types of faults like transient- faults, intermittent-faults, and permanent-faults. On the other hand, from the user's perspective, fault tolerance becomes a real requirement, especially for certain applications where the time needed for the repair process (downtime) has a critical impact on the application [5]. A brief presentation of hardware fault tolerance methods is discussed in this section.
A new self-healing hardware architecture inspired by the concepts of biology was proposed in [1] and [3], these architectures merge between cellular-based biological concepts, operational schematics of international standard (IEC 61131-3), and traditional fault-tolerance techniques. In [1] Three lines of defence contra discovered faults was suggested by Khairullah; the first one is a hybrid unit, which is responsible for tolerating the discovered transient faults. The second one is a local healing layer, which is responsible for tolerating permanent faults and the third one is a global healing layer, which is responsible for tolerating additional permanent faults, and fault management of the critical service layer. In [3] Khairullah and et al. proposed an architecture that marshaled in two levels; the first one is a critical functions layer, which is used to provide the meant service of the implementation. And the second one is a healing layer, which regularly manages the application's proper execution and generates health syndromes in the case of failure appears in the function layer.
A horse shifting allocation method was proposed in [5], in which the apportionment of the (spare CLBs) in 2-dimensional arrays was introduced, allowing the SRAM part of the FPGA to be modified and have the ability to shift in on-chip data horizontally and vertically. When trouble is detected, it can be easily avoided by transferring user data to a nearby spare instead of the defective CLB, causing the defective CLB to become unusable. In this method, each CLB block size 3 X 3 has a maximum of two CLBs.
Miculka et al. (2013) [4], have present a fault-tolerant method in FPGA that have the ability to mitigate transient fault and sundry permanent fault that occur sequentially. To mitigate transient fault, they use partial-dynamic-reconfiguration (PDR) and for permanent faults mitigation, they use a specific FT architecture that occupies fewer resources than the already used and excludes the FPGAs faulty part. In this technique, they based on the stored precompiled configurations in external memory and they used the relocation technique to reduce the space that wonted for a partial bit stream.
A Master-Slave TMR inspired technique was proposed in [6] for fault tolerance in SRAM-based FPGA, single and double faults affecting configurable logic blocks (CLBs) are being taken up. This

method depends on triple-modular redundancy (TMR) in conjunction with the master-slave technique, as well as partial reconfiguration for permanent fault toleration. In the suggested MSU architecture, there are two kinds of CLBs, CLB-Master, and CLB-Slave. If a fault appears in CLB-Slave, a partial reconfiguration on the MSU will exchange the defective CLB-S with one of the three CLB-M, and then defective CLB-S is ignored.

An adaptable and resilient hardware architecture was proposed in [7], this architecture was designed by a classical active redundancy model, TMR with spares in SRAM-based FPGA. In this method the replacement of a triple-redundant model failed modules with spare modules will occur, a developed controller in the proposed architecture gives them a possibility to adapt between Triple-Modular-redundancy with one spare, or Triple-Modular-redundancy with two spares.

3. Overview of the proposed fault-tolerant digital system

The presented digital system below is designed to achieve high levels of reliability with respect to different fault models such as transient, permanent, and common cause hardware faults. Another objective of the design of this system is to be scalable, efficient, and deterministic. In the proposed Dependable Digital System (DDS) shown in Figure 1, four levels of the organization are discussed. Level (0) contains the function block (FB) operations, which are capable of executing four arithmetic operations, seven logic operations, and two shift operations. In level (1) a Triple-Modular-Redundancy (TMR) is embedded, which is responsible for tolerating transient faults. Furthermore, levels (2) and (3) use the concept of Reallocation Manager (RM) and take advantage of using duplication with compare (DWC) circuit to detect and correct faulty function blocks. Two spare function blocks (SFBs) were used to replace the faulty functional blocks effected by two serial permanent faults or two simultaneous common cause faults (CCF). Moreover, in order to make easier to program from a programmability point of view, we adopted a Function Block Programming model similar to the international standard-"IEC 61131-3 Programmable Logic Controllers".

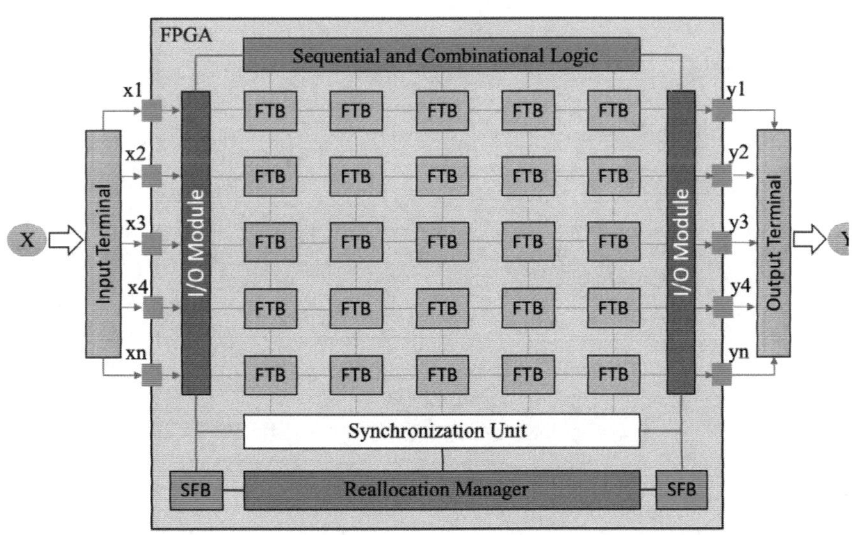

Figure 1. The proposed fault-tolerant digital system

Figure 1 illustrates the basic Functional components of our fault tolerance digital system. These are the I/O Terminal, I/O Module, Fault Tolerance Blocks (FTBs), Two Spare Function Blocks (SFBs), Sequential and Combinational Logic, Synchronization unit, and Reallocation Manager (RM). The valid execution of each function block (FB) is observed using two voting circuits and one DWC circuit comprised of XOR gates (See Figure 2(a)). Once a fault test is performed by detecting transient faults in the functional blocks and is observed in the input register of any FB, these faults will be tolerated by Triple-Modular-Redundancy (TMR) embedded in the same FB. Furthermore, if a detection of permanent faults is observed by passive FT method; duplication with a comparison in one or two FBs, at different times or even at the same time, the Reallocation Manager (RM) (See Figure 2(b)), is responsible for ignoring this FB and establish a new connection with the new spare (SFB) as it is described in the flowchart in Figure 2(b). The RM takes the address of fault FB and makes rerouting to reach the input of spare FTFB. This strategy requires eight clock cycles to generate the resulted output date through the spare FB.

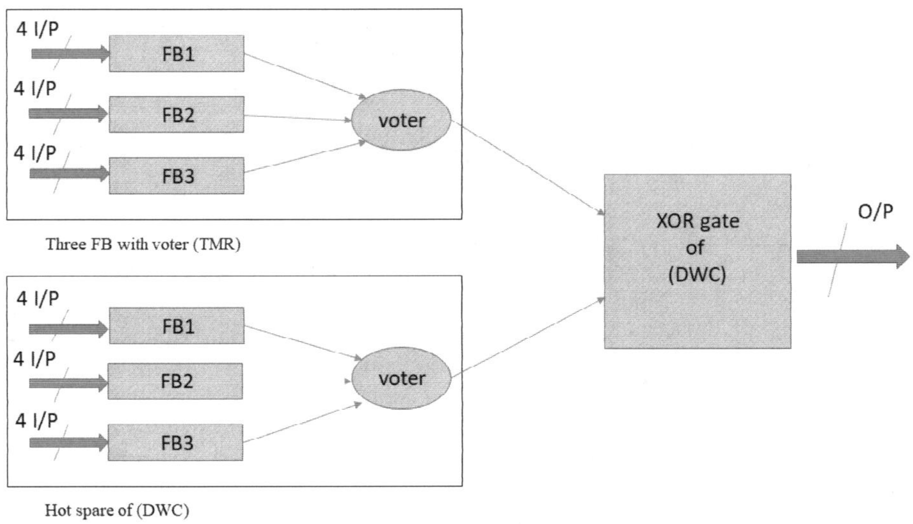

Figure 2(a). Scheme of PF detection and correction

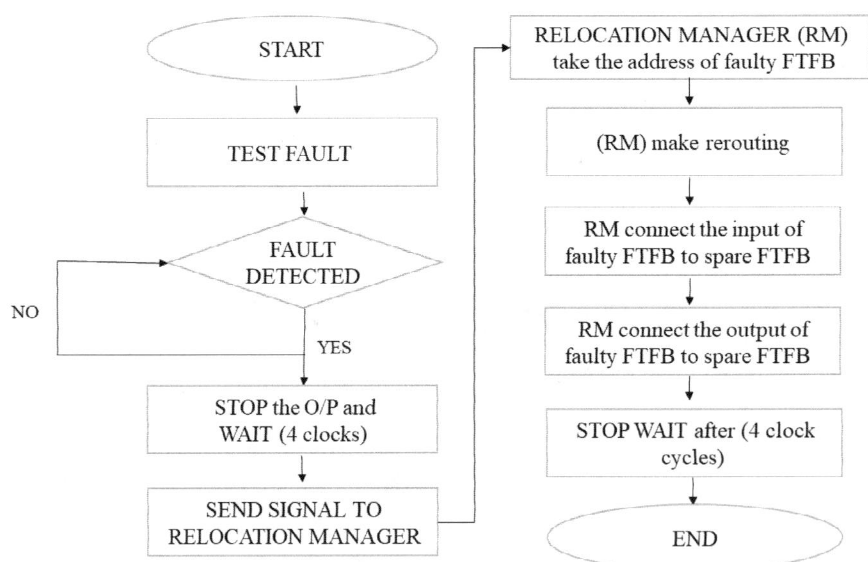

Figure 2(b). Flow chart of PF and CCF

In our proposed architecture, we have library of functional blocks organized in a two dimensional-array that consists of 25 Fault Tolerance Function Block (FTFB) as is shown in Figure 1, highlighted with yellow colour. Every five FTFB are connected in parallel as one container in such a way that a group of containers will serially be connected. The internal structure of each FTB consists of four stages which are input, operation, voting, and output. These four stages are controlled by a state machine-based control flow. Consequently, we need 4 clocks to activate one FTFB and compute the resulted data output as shown in Figure 3.

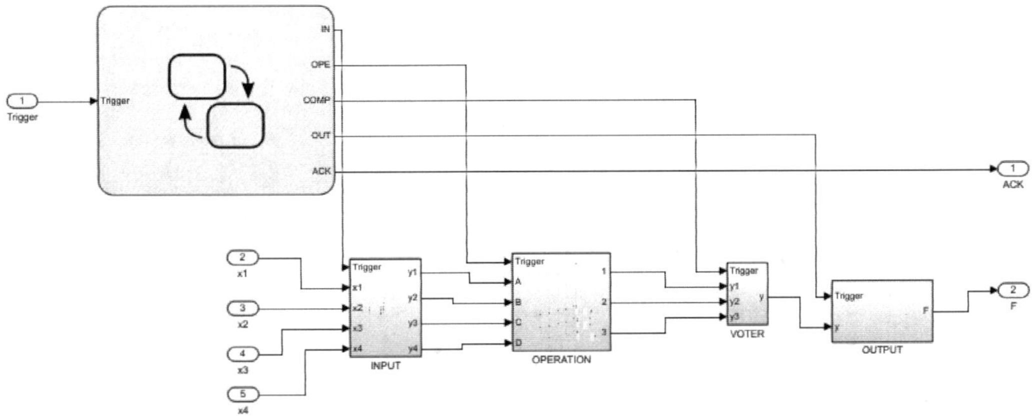

Figure 3. Block diagram of state machine-based control flow

Figure 4 (a) shows a block diagram of one row or container of the proposed DDS digital system that contains different hardware components like AND gates, inverters, multiplexers, OR gates, XOR gates, FTFB, and SPARE units. One container is capable of detecting an attack of two sequential common cause faults (CCFs) that can occur in one of the following pairs of functional blocks: (FTFB1, FTFB2), (FTFB1, FTFB3), (FTFB1, FTFB4), (FTFB1, FTFB5), (FTFB2, FTFB3), (FTFB2, FTFB4), (FTFB2, FTFB5), (FTFB3, FTFB4), (FTFB3, FTFB5), (FTFB4, FTFB5) at one time. This failure mode is repaired based on two spare functional blocks SPARE1 and SPARE2. Figure 4 (b) shows the hardware implementation for Figure 4 (a) using MathWorks Simulink Software design tool.

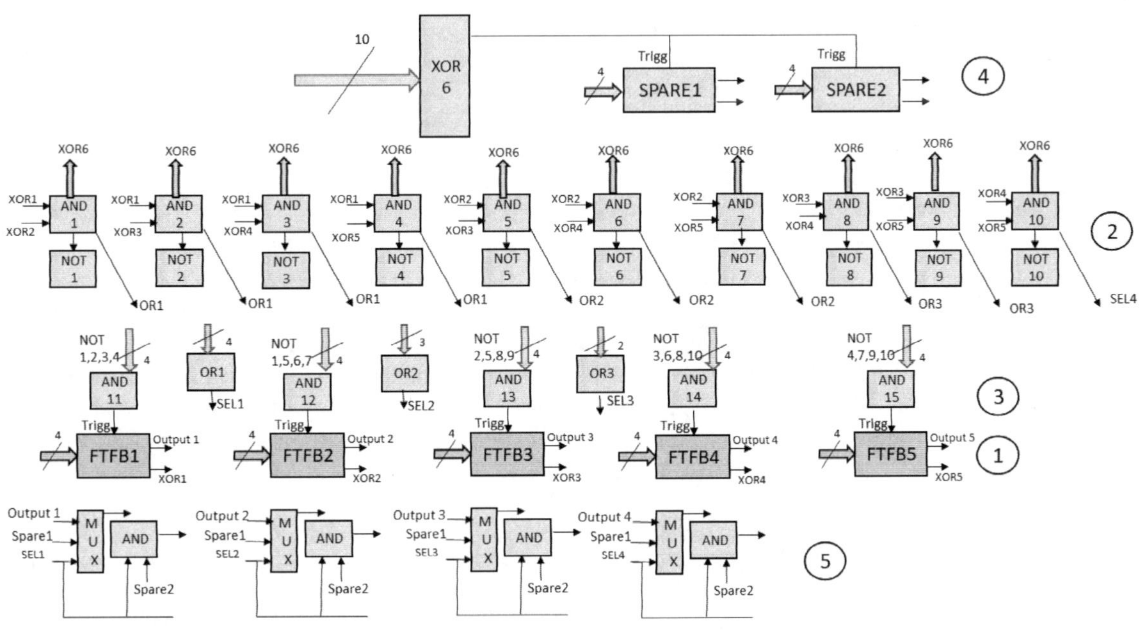

Figure 4(a). Block diagram of CCF detection and correction

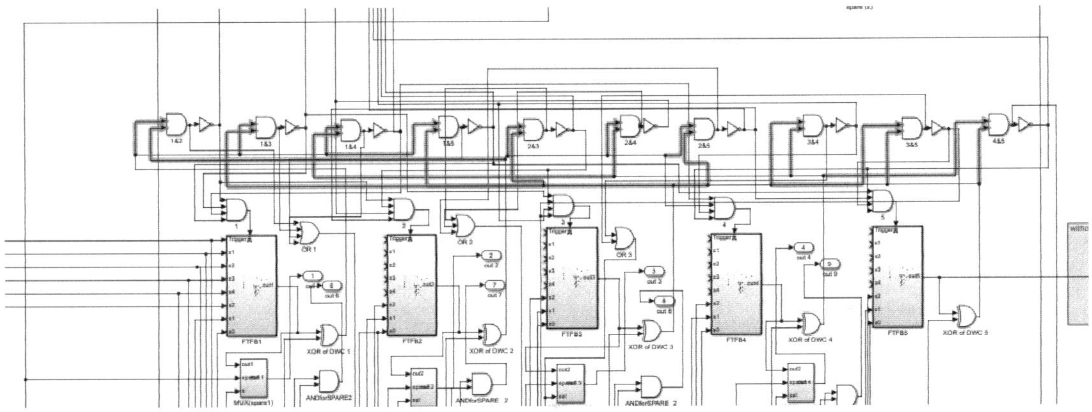

Figure 4(b). Hardware implementation of CCF detection and correction

Referring to Figure 4(a), (b), if the operation of DDS flow successfully without any fault detection, 4 clock cycles are sufficient to activate every container, and 4*5=20 clock cycles are sufficient to activate all the containers as it is described in Table 1. In another word, in the 1st clock, the input of the five FTFBs in the first container will be activated, in the 2nd clock, a special operation is executed depending on the selected value of the multiplexer input lines, in the 3rd clock, a majority voter will choose the correct output that will be flow to the output port, and finally, in the 4th clock, DWC will test the accuracy of permanent fault or common cause fault, if there is no fault the output will flow normally.

In one scenario we suppose that we have an effect in the fifth FTFB of the second container and the fifth FTFB in the fourth container, so in the first four clocks the operation will flow normally, so we consume five outputs and no delay appear. In the 5th clock, the input of the five FBs in the second container will enter, in the 8th clock the device will test the accuracy of permanent fault (PF) or common cost fault (CCF), the PF or CCF appearance in the fifth FTFB will cause to enter the system in a wait state by the usage of synchronization unit. At the same time, a specific signal will activate the Reallocation Manager, the RM will change the routing of the system, by ignoring the faulty FB and activate the spare one, by connecting the inputs of faulty FTFB to the Spare FTFB, and the output of spare FTFB instead of the output of the faulty FTFB. This scenario will lead to recovering the proposed faulty component and continue to perform its task despite error occurrence at the expense of systems delay which will increase depending on wait time (we suppose 4 clock cycles) to complete the FB exchange.

In the 13th clock cycle, if no fault is detected, the data input of all five FTFBs in the third container will be measured concurrently, so the output for each FTFB will flow and generate data normally at the end of the 16th clock. In the 20th clock, another detection of Pf will appear in the fifth FTFB of the fourth container, so the same scenario of container two will repeat and the output will flow in the 24th clock. In the 25th clock, the input of the five FTFBs in the fifth container will enter, no fault is detected so the output will flow normally in the 28th clock. If we compare the delay of PF tolerance with CCF tolerance, we can see that CCF (for two parallel FTFB) detection and correction will take four clocks less than PF detection and correction, that because the latest case appears serially.

Table 1. Number of clocks and their actions

No. Clock	FB active	Container No.	No. of I/P data	No. memories	No. control data	No. of O/P data
4	FB11	1	20	20	5	5
	FB12	1				
	FB13	1				
	FB14	1				
	FB15	1				

8	FB21	2	40	40	10	10
	FB22	2				
	FB23	2				
	FB24	2				
	FB25	2				
12	FB31	3	60	60	15	15
	FB32	3				
	FB33	3				
	FB34	3				
	FB35	3				
16	FB41	4	80	80	20	20
	FB42	4				
	FB43	4				
	FB44	4				
	FB45	4				
20	FB51	5	100	100	25	25
	FB52	5				
	FB53	5				
	FB54	5				
	FB55	5				

As it can be seen in Figure 5, we assume that one operation needs to obtain the value of the following Boolean logic expression ((NOT 7) AND 12) OR (7 AND 7), So firstly we must obtain the value of NOT 7, then the result must AND with 12, this case will take 8 clocks because the two operations are happening serially. On the other side, to obtain the result of 7 AND 7 will take 4 clocks to appear, we will need another 4 clocks to obtain the final value, so the total number of clocks will be (12 clocks). If a fault appears in one of these operations (NOT, AND1, AND2, and OR), we will need additional 4 clocks to replace the faulty FB with the spare one, 'Figure 5' shows that we have a fault occurrence appears at AND1 operation.

Figure 5. Timing diagram of fault detection in AND gate FTFB

4. Markov Mathematical Analysis

Combinatorial mathematics approaches are used to analyze the reliability of safety-critical applications. Hazard Tree Analysis, Reliability Block Diagrams, Fault Tree Methods, and Probabilistic Risk Analysis are some of these methods. These standard reliability analysis approaches are based on stateless and time-invariant mathematics. When encountering systems with regenerative behaviors, repair, and reconfiguration features, it's important to use the more efficient Markov technique to model them [1]. We can define a Markov process that is a stochastic process whose action is solely determined by the system's current state. The order in which the device arrived at its current state has no impact on how it will behave in the future [12].

4.1 Reliability Analysis of the Proposed hardware system

To calculate the proposed hardware DDS model's reliability, a five-state Markov model was created (See Figure 6). As shown in table 2, this model has three operating states, one failing secure state, and one failing insecure state. Each of the states is in one of four situations: completely operational, fail-operational, fail-secure, or fail-insecure.

Table 2. Case and its description of markov modelled system depend on Figure 3.

case	Description
A	completely operational System (all the FBs are operated properly and two spares are available)
B	First Fail-Operational (one FB is failed by permanent fault and detected by DWC, so one spare is used for repair)
D	Second Fail-Operational (second FB is failed by permanent fault and detected by DWC, so the second spare is used for repair, so no spare remains)
FS	Fail- secure Operational (third FB is failed by permanent fault and detected by DWC, it cannot repair because no spares remain)
FI	Fail- insecure (one, two or three FBs are failed permanently without detection)

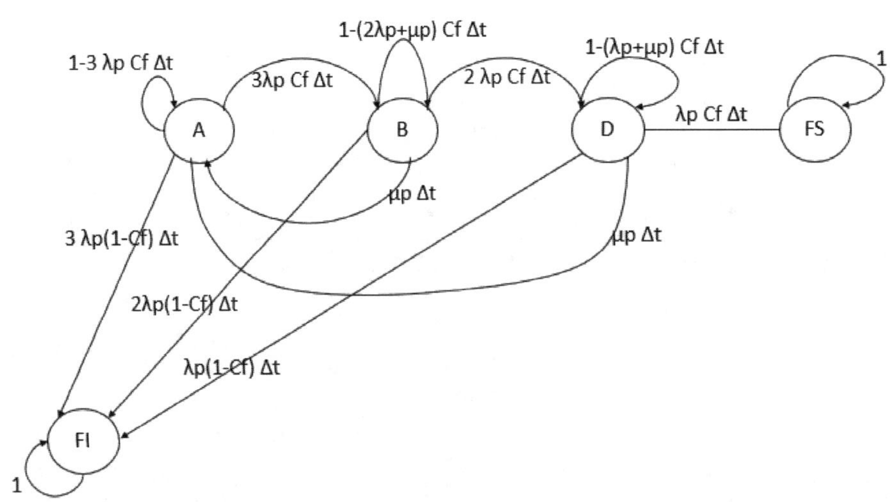

Figure 6. Transition state diagram for the Dependable Digital System (DDS)

The DWC system for (level 2) DDS was modelled using the Markov model that depends on Repair fault operation (μp) and Fault coverage (Cf). State-A represents a case where all the FBs are operating properly and the system is completely operational. In state-B one FB is failed by permanent fault and detected by DWC, so the system is still operational but one module has failed, the transition from A to B happened at 3λp. Furthermore, the system could return to its original state-A at repair rate (μp) by using the first spare. In state-D, a second FB is failed by permanent fault and detected by DWC and the second spare is used for repair, so no spare remains. The transition from (B) to (D) happened at 2λp, furthermore; the system could return to its original state-B at repair rate (μp). In state-FS, a third FB is failed by permanent fault and detected by DWC, but it cannot repair because no spares remain. The transition from (D) to (FS) happened at λp. State-FI represents our DDS where one, two, or three FBs are failed permanently without any detection.

The designed system's Markov model can be written in matrix form as:

$Psystem\ (t+\Delta t) = Tsystem * Psystem\ (t)$
Where;
T system: represents the state transition matrix,
P system (t): denotes the likelihood of being in the corresponding state at time t.
$Psystem\ (t+\Delta t)$: is the possibility about being in the identical state at (t + Δt) time.

Furthermore; we expect that every functional cell follows the exponential failure rule and has a fix failure rate λ at all times when we calculate the reliability of the Markov model. So, we can compute the system's probability of failure at a time (t + Δt) as follows:

$$P\ (t + \Delta t) = 1 - e^{-\lambda \Delta t} = \lambda\, \Delta t$$

$$PA\ (t + \Delta t) = (1 - 3\lambda p\ Cf\ \Delta t)\ PA\ (t) + \mu p\ \Delta t\ PB\ (t) + \mu p\ \Delta t\ PD\ (t)$$
$$PB\ (t + \Delta t) = [1 - (2\lambda P + \mu p)\ Cf\ \Delta t]\ PB\ (t)$$
$$PD\ (t + \Delta t) = [1 - (\lambda P + \mu p)\ Cf\ \Delta t]\ PD\ (t)$$
$$PFS\ (t + \Delta t) = \lambda P\ Cf\ \Delta t\ PD\ (t) + PFS\ (t)$$
$$PFI\ (t + \Delta t) = 3\lambda p\ (1 - Cf)\ \Delta t)\ PA\ (t) + 2\lambda P\ (1 - Cf)\ \Delta t\ PB\ (t) + \lambda P\ (1 - Cf)\ \Delta t\ PD\ (t) + PFI\ (t)$$

The reliability can be computed as follows:
R (t) = 1- PFI (t) = PFS (t) + PD (t) + PB (t) + PA (t)

Where: -

$$\frac{PA\ (t + \Delta t) - PA\ (t)}{\Delta t} = -3\lambda p Cf\ PA\ (t) + \mu p\ PB\ (t) + \mu p\ PD\ (t)$$
$$\frac{PB\ (t + \Delta t) - PB\ (t)}{\Delta t} = -(2\lambda P + \mu p)\ Cf\ PB\ (t)$$
$$\frac{PD\ (t + \Delta t) - PD\ (t)}{\Delta t} = -(\lambda P + \mu p)\ Cf\ PD\ (t)$$
$$\frac{PFS\ (t + \Delta t) - PFS\ (t)}{\Delta t} = \lambda P\ Cf\ PD\ (t)$$
$$\frac{PFI\ (t + \Delta t) - PFI\ (t)}{\Delta t} = 3\lambda p\ (1 - Cf)\ PA\ (t) + 2\lambda P\ (1 - Cf)\ PB\ (t) + \lambda P\ (1 - Cf)\ PD\ (t)$$

$$P\text{-system}\ (t + \Delta t) = \begin{vmatrix} PA\ (t + \Delta t) \\ PB\ (t + \Delta t) \\ PD\ (t + \Delta t) \\ PFS\ (t + \Delta t) \\ PFI\ (t + \Delta t) \end{vmatrix}$$

A Markov model's two-dimensional state transition matrix would look like this:
P-system (t + Δt) =

$$\begin{vmatrix} -3\lambda p Cf & \mu p & \mu p & 0 & 0 \\ 0 & -(2\lambda P + \mu p) Cf & 0 & 0 & 0 \\ 0 & 0 & -(\lambda P + \mu p) Cf & 0 & 0 \\ 0 & 0 & \lambda P\ Cf & 0 & 0 \\ 3\lambda p (1 - Cf) & 2\lambda P (1 - Cf) & \lambda P (1 - Cf) & 0 & 0 \end{vmatrix}$$

$$\text{P-system}(t) = \begin{vmatrix} PA(t) \\ PB(t) \\ PD(t) \\ PFS(t) \\ PFI(t) \end{vmatrix}$$

Allowing the timing interval Δt to fall to zero through (algebraic manipulation) outcome to the following differential equations: -

$$\frac{d\,PA(t)}{dt} = -3\lambda p\, Cf\, PA(t) + \mu p\, PB(t) + \mu p\, PD(t)$$

$$\frac{d\,PB(t)}{dt} = -(2\lambda P + \mu p)\, Cf\, PB(t)$$

$$\frac{d\,PD(t)}{dt} = -(\lambda P + \mu p)\, Cf\, PD(t)$$

$$\frac{d\,PFS(t)}{dt} = \lambda P\, Cf\, PD(t)$$

$$\frac{d\,PFI(t)}{dt} = 3\lambda p (1 - Cf)\, PA(t) + 2\lambda P (1 - Cf)\, PB(t) + \lambda P (1 - Cf)\, PD(t)$$

Using Laplace Transform, we obtain the following equations

$S\,PA(S) - PA(0) = -3\lambda p\, Cf\, PA(S) + \mu p\, PB(S) + \mu p\, PD(S)$
$S\,PB(S) - PB(0) = -(2\lambda P + \mu p)\, Cf\, PB(S)$
$S\,PD(S) - PD(0) = -(\lambda P + \mu p)\, Cf\, PD(S)$
$S\,PFS(S) - PFS(0) = \lambda P\, Cf\, PD(S)$
$S\,PFI(S) - PFI(0) = 3\lambda p (1 - Cf)\, PA(S) + 2\lambda P (1 - Cf)\, PB(S) + \lambda P (1 - Cf)\, PD(S)$

We suppose that the system begins in a perfect state at time t=0 in the analysis, so: -
PA (0) = 1, PB (0) = 0, PD (0) = 0, PFS (0) = 0, PFI (0) = 0.
So, the Reliability is the probability of being in either State A, State B, State D or FS State
R (t) = PA (t) + PB (t) + PD (t) + PFS (t).

5. Conclusion

In this research paper, a novel FPGA based fault tolerant architecture for safety critical systems is proposed and developed. The important aspects of the proposed architecture are it is adaptive resilience to a wide class of faults while maintaining good performance and uninterrupted service. The basic fault detection and tolerance patterns build upon, duplication with comparison, and triple modular redundancy, but extend these concepts with respect to the system level organization of fault tolerance and fault resilience. The proposed architecture system shows significant promise in meeting high levels of fault resilience, according to the simulation results in MathWorks Simulink software when modeling the DDS digital system verified against its specifications. To increase the dependability levels of the proposed system, future works will include design and implementation of more robust digital systems inspired by biological concepts. In addition, MathWorks Simulink Design verifier tool will be used to check the right execution of our proposed design as verification & validation methods (V&V).

6. References

[1] Khairullah S 2018 Toward biologically-inspired self-healing, resilient architectures for digital instrumentation and control systems and embedded devices (Virginia: Virginia Commonwealth University: Theses Diss) pp 1-144.

[2] Di Carlo S, Gambardella G, Prinetto P, Rolfo D, Trotta P and Vallero A 2014 A novel methodology to increase fault tolerance in autonomous FPGA-based systems *2014 IEEE 20th International On-Line Testing Symposium (IOLTS)* (Spain, Platja d'Aro: IEEE) pp 87–92.

[3] Khairullah S and Elks C 2020 Self-repairing hardware architecture for safety-critical cyber-physical-systems *IET Cyber-Phys. Syst. Theory Appl.* **vol 5(1)** (England Wales: IET Digital Library), pp 92–99.

[4] Miculka L, Straka M, and Kotasek Z 2013 Methodology for fault tolerant system design based on FPGA into limited redundant Area *2013 Euromicro Conference on Digital System Design* (USA, CA, Los Alamitos: IEEE) pp 227–234.

[5] Kshirsagar R and Sharma S 2012 Fault tolerance in FPGA through horse shifting *2012 Fifth International Conference on Emerging Trends in Engineering and Technology* (Japan, Himeji: IEEE) pp 228–232.

[6] Lahrach F, Doumar A, Châtelet E, and Abdaoui A 2010 master-slave TMR inspired technique for fault tolerance of SRAM-based FPGA *2010 IEEE Computer Society Annual Symposium on VLSI* (Greece, Lixouri: IEEE) pp 58–62.

[7] Farias M, Nedjah N and de Caravlho P 2020 Active redundant hardware architecture for increased reliability in FPGA-based nuclear reactors critical systems *2020 23rd Euromicro Conference on Digital System Design (DSD)* (Slovenia, Kranj: IEEE) pp 608–615.

[8] Chmiel M, Kulisz J, Czerwinski R, Krzyzyk A, Rosol M, and Smolarek P 2016 An IEC 61131-3-based PLC implemented by means of an FPGA *Microprocess. Microsyst.* **vol 44** (Netherlands: Elsevier) pp 28–37.

[9] Luo P and Zhang J 2011 SEU mitigation strategies for SRAM-based FPGA **Vol 8196 81960N-1** (China: Beijing) pp 1–10.

[10] Zheng M, Wang Z, and Li L 2015 DAO: Dual module redundancy with AND/OR logic voter for FPGA hardening 2015 *First International Conference on Reliability Systems Engineering (ICRSE)* (China, Beijing: IEEE) pp 1–5.

[11] Afzaal U, and Lee J 2018 A self-checking TMR voter for increased reliability consensus voting in FPGAs *IEEE Trans. Nucl. Sci* **vol 65(5),** IEEE pp 1133–1139.

[12] Dubrova E 2012 *Fault-Tolerant Design* (New York: Springer), pp 1–194.

[13] Shaker M, Hussien A, Alkady G, Amer H, and Adly I 2019 Mitigating the effect of multiple event upsets in FPGA-based automotive applications *8th Mediterranean Conference on Embedded Computing (MECO)* (Montenegro, Budva: IEEE) pp 1–4.

[14] Bolchini C, Miele A, and Sandionigi C 2013 Autonomous fault-tolerant systems onto SRAM-based FPGA platforms *Journal of Electronic Testing* **vol 29** (New York: Springer), pp 779–793.

IOP Conf. Series: Materials Science and Engineering 1152 (2021) 012021 doi:10.1088/1757-899X/1152/1/012021

Performance Optimization on GPGPU & Multicore CPU Using Roofline Model

Noor M. Allayla[1*] **and Shefa A. Dawwd**[1]

[1]Computer Engineering Department, College of Engineering, University of Mosul, Mosul, Iraq.

*E-mail: noor.mowafeq@uomosul.edu.iq

Abstract. The roofline model introduced in this paper to evaluate the best optimized platform for training the neural network that used to recognize handwritten digits under multicore CPU and general-purpose GPU (GPGPU) as hardware environment. The pattern parallel training technique for MNIST dataset is applied. The parallel network training of MNIST using different data layout of multicore CPU and GPGPU is presented. Different bottlenecks have been explained by applying the roofline model. The most suitable platform is selected according to layouts and constrains either for memory or computation bounds. The computational intensity of all rooflines is moved toward right, then the performance is increased. As a result of optimization, and with the diversity of the available data size, core number, operational strength, the most suitable hardware platform is selected.

1. Introduction

Applying parallelism on the systems becomes widespread to increasing the performance of computer architectures. The Graphics Processor Unit (GPU) is suitable to be used in a different type of applications, while other applications are more suitable to be implemented on a multicore processor. So, which one is a more suitable in a specific application? One of the methods to choose is a roofline model.

The roofline model is a performance model used in recent years to understand and analyse the performance and bottlenecks in varying platform [1]. The roofline model is a very simplified model, so it is executed very fast. The roofline model could be applied on the CPU, GPU and the memory architectures [2]. This gives a multiple option for computing on varied platforms. Applying the performance on specific application depends on difficulty of the factors that need to be taken when using a multicore. Since the exclusiveness properties of the processor architecture vary for each processor [3], the roofline model gives the performance valuation of a specific application depending on the postulate that performance is bounded by the bandwidth between the processor and the memory or by the maximum computational performance of the specific architecture. The roofline model firstly designed to rating the CPU execution, but can easily applied on the GPU [4]. Some works use the roofline are presented: Yu Jung Lo and others, measured sustained PCIe (peripheral component interconnect express) throughput GPU, based on architectural specifications. The achieved results from the properties of the GPU architecture aided with the calculations using roofline model, guided them to select the appropriated platform [5]. Aleksandar Ilic and others, analyses and applied the roofline model, and proposes a new approach to provide a more sight performance modelling of modern architectures [4]. Nugteren and others, trying to improve performance before implementation using algorithm classification to predict performance prior the implementation by modifying the roofline model to enable architectural option through performance forecasting prior to the development of architecture specific code by the boat hull model. For six example algorithms using a GPU, this model proved performance predict without requiring code [6]. Choi and others, also applying the roofline model on the time, energy, and power, the roofline model applied on the features of an algorithm such as locality and concurrency with time and energy costs. The difference in an algorithm optimization for time versus an algorithm optimization for energy is explained, and the time-energy trade-offs is achieved [7].

Content from this work may be used under the terms of the Creative Commons Attribution 3.0 licence. Any further distribution of this work must maintain attribution to the author(s) and the title of the work, journal citation and DOI.
Published under licence by IOP Publishing Ltd

The roofline model gives a perfect performance estimation of the neural network training process under CPU and GPU environments. A neural network is applied to exceeded problems inclusive extraction the feature, noise reduction, classifications and image matching, it is effective to exceed the problems where the mathematical solution is difficult to apply.
Typically, large data pattern is desired to train the neural networks. Whenever increasing the size of the neural network pattern, the time required to train the network is increased exponentially [8] [9]. In this research, we used the multicore Central Processing Units (CPUs), and GPU with shared memory as a cost-effective way to increasing the performance. Then we introduce the roofline model to show the CPU, GPU, and the memory bottlenecks and trying to improve the performance regard to each designed model.

2. The Roofline Model

The visual representation of the roofline model concludes two roofs: firstly, memory roof, limited by the bandwidth of external memory, the second is a computational roof, limited by the maximum computational performance of the processor [10]. The roofline rates the kernel of a specific application as a data transfer from one or more memories to a processor where it can be used for the computation. The results are written back to the memories. The movements of data between the memory and the processor are bounded by the characteristics of the interconnection between them. The memory traffic with the processor has the greatest latency, and it is a form of bottleneck. For example, when data is transfer to the processor from the memory, the caches make sure this data is kept locally if it is re-used (unless there is a great amount of data used that could cause the data to be overwritten in any layer of the cache hierarchy). The maximum computational performance consists of single precision floating-point operations (such as compares, add, subtract or multiply) [11].

2.1 Computational Roofline

For a specific architecture, the computational roofline model is achieved by applying equation:

$$C_{roof} = \#cores * C_f * C_{ops} \tag{1}$$

The computational roofline is represented by C_{roof}. The C_{roof} can be calculated by multiplying the number of cores (#cores) by the clock frequency of the cores (C_f), and the floating-point number of the operations that can be performed per core, per clock cycle (C_{ops}). The C_{ops} take multiple function units per core; it is calculated as shown 'in equation (2)' by summing the number of the function units and their respected width:

$$C_{ops} = \sum_{0}^{q-1} FU\ width \tag{2}$$

The number of function units per core represented by q [12]. In this paper, we used different platforms, after applying the above equations. The results are show in table 1:

Table 1. Computational Roofline of Platforms.

The Platform	#cores	C_f	C_{OPS}	C_{roof}
CPU3210	2	2.5	4	20
CPU4200	2	2.5	4	20
CPU2670	4	2.2	8	70.4
GeForce610	48	0.81	2	77.76
GeForce740	384	0.98	2	752.64

2.2 Memory Bandwidth

The memory bandwidth B_{roof} can be calculated using equation:

$$B_{roof} = M_f * M_{dt} * M_{width} * \#channels \quad (3)$$

Where M_f is the memory frequency at which a memory unit moved the data, M_{dt} is the size of data moved per clock cycle, M_{width} specify the width in bytes of a single memory moved and #channels is the number of memory channels [12]. The roofline model (Figure 1) can be obtained by using equations (1) to (3). The B_{roof} of our platforms according to platforms specification is shown in table 2; at this point the celling point is calculated as C_{roof} / B_{roof}:

Table 2. The Celling Points of Platforms.

The Platform	B_{roof}	C_{roof} / B_{roof}
CPU3210	25.6	0.78
CPU4200	25.6	0.78
CPU2670	21.3	3.3
GeForce610	14.4	5.4
GeForce740	80	9.4

2.3 Operational Intensity

When the computations are performed a large amount of data are required to move from the memory, the computational performance is expressed in floating point operations per second (flops/sec). The operational intensity is expressed in floating point operations per byte (flops/byte). Increasing the operational intensity leads to bounding the kernel by the computational roofline of a specific architecture, which performs a large number of floating-point operations per byte moved from the memory. While decreasing operational intensity leads to bounding the kernel by the memory bandwidth. The operational intensity can be controlled by the software programmer.

2.4 Computational Performance

After estimating the operational intensity of a specific kernel, the application could be drawn as a vertical line on the roofline, as shown in Figure 1.

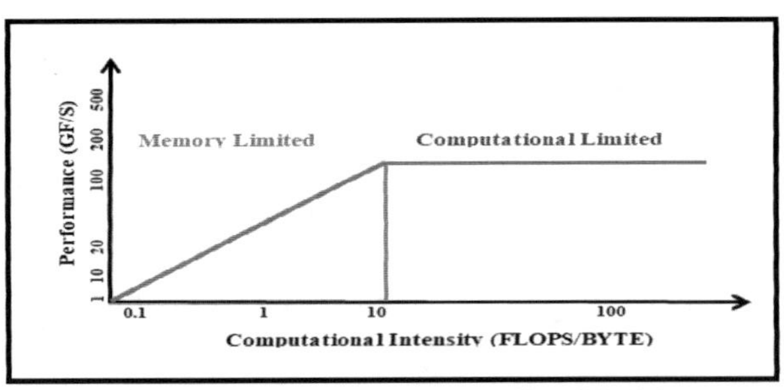

Figure 1. Application on Roofline Model.

It is possible to achieve the maximum performance of a specific application on the specific architecture using this model. For example, when a specific kernel is considered, it can be deduced when the kernel optimally uses the resources of that architecture, it could change to hit the memory bandwidth roof. When conclude the time required to executing the kernel, it can be drawn as a point on the roofline graph and the computational performance of the kernel can be extracted by calculating the number of floating-point operations [10].

3. The Case Study and Mapping

The MNIST (Mixed National Institute of Standards Technology) handwritten digit database (showed in figure 2) is applied on the neural network. The MNIST sample can be founded in [5].

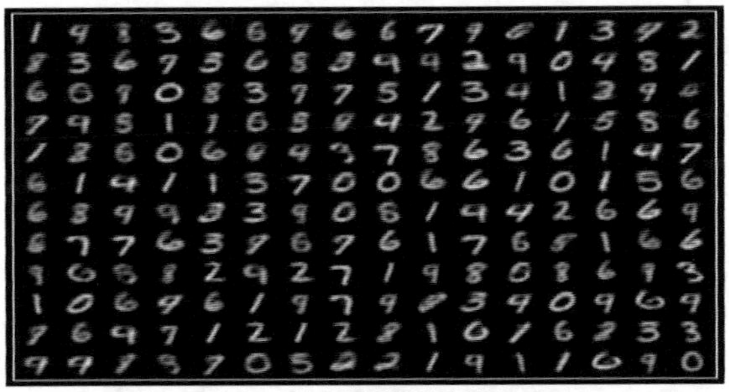

Figure 2. The MNIST Database.

It contains 60,000 patterns; each one represents a digital image used to train the classifier. 10,000 patterns are used for testing. All images are black and white. The size is normalized and centred in a fixed-size image where the centre of the intensity lies at the centre of the image of 28 × 28 pixels. The dimensions of each sample vector are 28 * 28 = 784, where each element is binary one or zero [11]. The reason behind choosing this database is: it is a standard and simple database for fast-testing theories and algorithms. Also testing neural networks for actual issue in the real world. The MNIST database have already been pre-processed including segmentation and normalization, so it could be a good start for us with minimal efforts on pre-processing.
The PTT (Pattern Parallel Training) technique is implemented for parallel training on MNIST database as a 60000 pattern for training and as 10000 patterns as a network tester. A network of 784 neurons in the input layer, a 350 neuron in the hidden layer (this number of neurons is selected by trial and error), and 10 neurons in the output layer is designed. A 28 × 28 pixels pattern is applied on the input layer for the network and the output from 0 to 9 is produced. The neural network structure is shown in Figure 3.

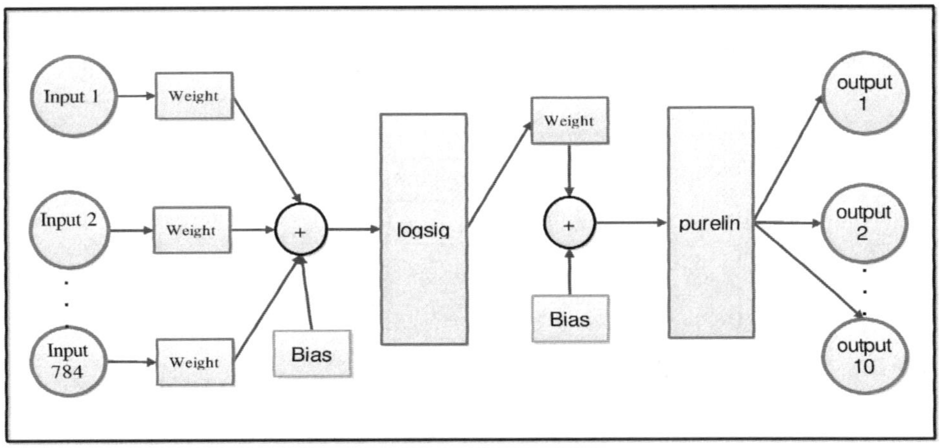

Figure 3. The ANN Architecture.

4. Results and Discussion

The neural network is trained on MNIST using different specification platforms as shown in table 3. Then different bottlenecks have been explained by applying the roofline model as presented later.

Table 3. Platform Specifications.

The Platform	RAM	The Freq.	Core NO.	GPU Type	GPU NO.	Memory of GPU
i7-2670QM	8GB	2.2GHZ	7	Intel® HD Graphics3000	-	-
i5-3210M	6GB	2.5GHZ	5	GeForce610	48	1GB
i5-4200M	6GB	2.5GHZ	5	GeForce740	384	1TB

4.1 CPU Bottlenecks

By determining the operation intensity of the training program which equals to 1.7869, the result on the first platform that contains five cores, is shown in Figure 4. The CPU roof is 20, then increasing the CPU cores to seven cores by using another platform, the roof becomes 70.4.

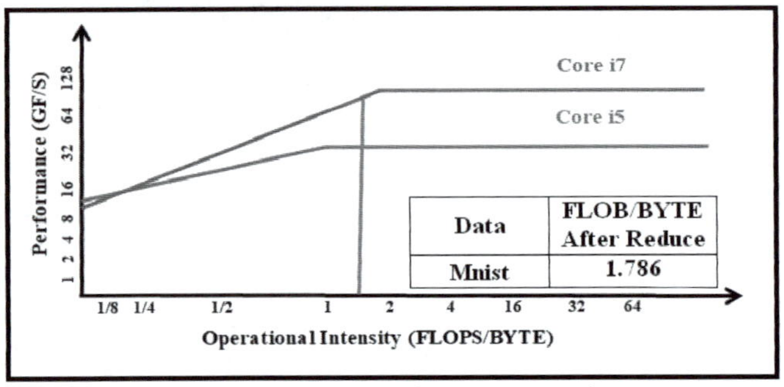

Figure 4. CPU Roof Representation.

When increasing the cores number, the line of application going toward CPU roof, and approaching from the celling point, while it was overtaken the celling point. This means that the result is improved by increasing the cores number.

4.2 Memory Bottlenecks

After applying the training on multicore CPU platforms, the memory roof 0.7514 is achieved. Then when trying to decrease the operation (by simplifying some operations), this leads the memory roof to be equals to 1.7869. Figures 5, shows the memory roof of two different platforms (5 cores CPU and 7 cores CPU respectively).

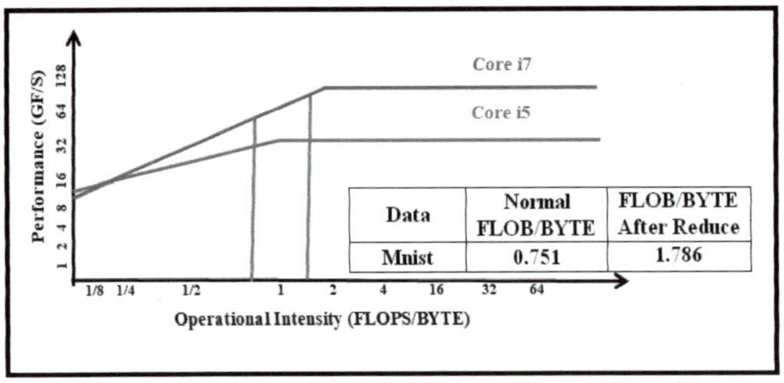

Figure 5. Memory Roof of Core i5 and Core i7.

From Figure 5, one can see that, approaching the celling point is achieved in the normal operation intensity when core i5 is used, while approaching the celling point at core i7 is achieved when reducing the operation intensity.

After applying the roofline calculation of different GPU platforms, a memory roof at 1.8869 is achieved. Then when trying to reduce the operation intensity, the memory roof equals to 6.9184. Figures 6, shows the memory roof with two different platforms: 48 GPU core and 384 GPU core.

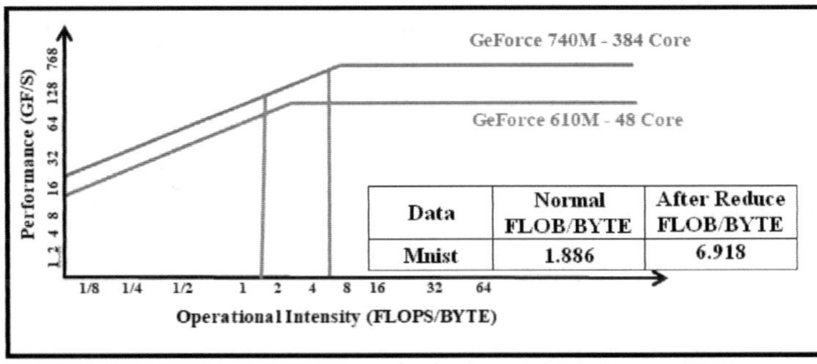

Figure 6. Memory Roof of 48 and 384 Cores GPU.

As shown in the Figure 6, approached to the celling point in both GPU types are achieved when the operation intensity is reducing.

4.3 GPU Bottlenecks

After applying the training on the neural network of data MNIST on two different platforms, GPU roof at 77.76 in 48 cores GPU is achieved. Then, when applying the training on 384 core GPU, GPU roof at 752.4 is achieved, as shown in Figure 7.

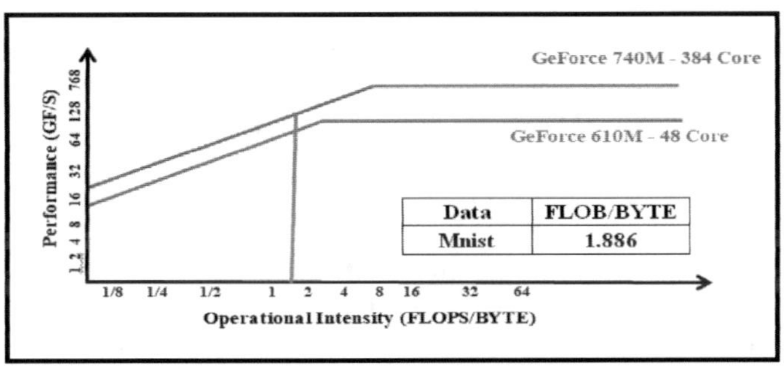

Figure 7. GPU Roof Representation.

The Figure 7; show that approaching the celling point at the GeForce 610 is closer than that of GeForce 740.

5. Training Time

The processing time of different platforms that used in training before and after roofline optimization is show in Figure 8.

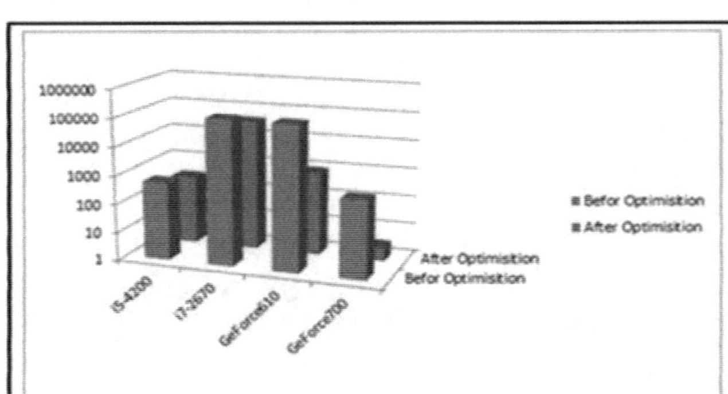

Figure 8. Training Time Before and After Roofline Optimization.

The benefit of using the optimizations is clearly shown due to reducing the training time in all types of the platforms.

6. Conclusion

The neural network based handwritten digit recognition system on MNIST database is presented in this work. The parallel training using different types of platforms is achieved. Then the roofline model is applied to investigate bottlenecks of both CPU and GPGPU for the performance optimization purposes. The platform of best specification is selected according to layouts and constrains either for memory or computation bounds. Suitable results are achieved when the data size, operational intensity and the core numbers are suitably identified.

7. References

[1] Nugteren, C. and Corporaal, H., 2012, February. The boat hull model: adapting the roofline model to enable performance prediction for parallel computing. In Proceedings of the 17th ACM SIGPLAN symposium on Principles and Practice of Parallel Programming (pp. 291-292).

[2] Cabezas, V.C. and Püschel, M., 2014, October. Extending the roofline model: Bottleneck analysis with microarchitectural constraints. In 2014 IEEE International Symposium on Workload Characterization (IISWC) (pp. 222-231). IEEE.

[3] Denoyelle, N., Goglin, B., Ilic, A., Jeannot, E. and Sousa, L., 2017, November. Modeling large compute nodes with heterogeneous memories with cache-aware roofline model. In International Workshop on Performance Modeling, Benchmarking and Simulation of High Performance Computer Systems (pp. 91-113). Springer, Cham.

[4] Ilic, A., Pratas, F. and Sousa, L., 2013. Cache-aware roofline model: Upgrading the loft. IEEE Computer Architecture Letters, 13(1), pp.21-24.

[5] LeCun, Y., Bottou, L., Bengio, Y. and Haffner, P., 1998. Gradient-based learning applied to document recognition. Proceedings of the IEEE, 86(11), pp.2278-2324.

[6] Nugteren, C. and Corporaal, H., 2012, February. The boat hull model: adapting the roofline model to enable performance prediction for parallel computing. In Proceedings of the 17th ACM SIGPLAN symposium on Principles and Practice of Parallel Programming (pp. 291-292).

[7] Choi, J.W., Bedard, D., Fowler, R. and Vuduc, R., 2013, May. A roofline model of energy. In 2013 IEEE 27th International Symposium on Parallel and Distributed Processing (pp. 661-672). IEEE. [8] Wang, Y.E., Wei, G.Y. and Brooks, D., 2019. Benchmarking tpu, gpu, and cpu platforms for deep learning. arXiv preprint arXiv:1907.10701.

[9] Alzantot, M., Wang, Y., Ren, Z. and Srivastava, M.B., 2017, June. Rstensorflow: Gpu enabled tensorflow for deep learning on commodity android devices. In Proceedings of the 1st International Workshop on Deep Learning for Mobile Systems and Applications (pp. 7-12).

[10] Cabezas, V.C. and Püschel, M., 2014, October. Extending the roofline model: Bottleneck analysis with microarchitectural constraints. In 2014 IEEE International Symposium on Workload Characterization (IISWC) (pp. 222-231). IEEE.

[11] Lorenzo, J.A., Pichel, J.C., Pena, T.F., Suárez, M. and Rivera, F.F., 2011. Study of Performance Issues on a SMP-NUMA System using the Roofline Model. In Proceedings of the International Conference on Parallel and Distributed Processing Techniques and Applications (PDPTA) (p. 1). The Steering Committee of The World Congress in Computer Science, Computer Engineering and Applied Computing (WorldComp).

[12] van Doormaal, R.P.M., Corporaal, H. and Peemen, I.M., 2012. Parallel Training of Large Scale Neural Networks: Performance Analysis & Prediction.

Power consumption measurements of WSN based on Arduino

U T Salim[1], S L Qaddoori[2], A Al-Saegh[1*] and Q I Ali[1]

[1]Computer Engineering Department, College of Engineering, University of Mosul, Mosul, Iraq.
[2]Electronic Engineering Department, College of Electronics Engineering, University of Ninevah, Mosul, Iraq.

*E-mail: ali.alsaegh@uomosul.edu.iq

Abstract. To deal with the deficiencies and limitations for existing wireless sensor network (WSN) nodes, such as low generality, low degree of variability and poor flexibility, Arduino WSN node's development benefits are integrated with Bluetooth wireless communication characteristics. This paper introduces four simple point-to-point WSN power measurement methods. Each WSN node is constructed from Arduino Mega 2560 microcontroller, Bluetooth module, and different sensors. The paper's goal is to characterize the power consumption at different power-saving modes. So, numerous power measurement methodologies are presented to measure the power consumption of the WSN node, in real-time, for different power saving modes such as (idle, sleep, power-down, etc.). These methodologies also determine the power consumption at communication states. Then the results obtained from these states show that the idle mode consumes more power than the other modes. In addition, the transmitted state consumes the major power regardless of the data rate used during the communication between the two ends.

1. Introduction

Vast number of projects on wireless sensor networks (WSNs) were carried out duo to the effectiveness of this field of research on the daily life [1,2]. Among the applications of WSNs are environmental data collection[3], robotically handling agriculture irrigation, gathering parameters of the patient's state, identifying mechanical faults, military arena [4] and many others. The main purpose of WSNs is detecting periodically actions within a specific motivating area, process them, and then routing them as digital data to the base station.

Normally, the WSNs comprises of a big number of small sensor nodes that are spatially distributed with limited resources: battery power, capacity of computation, memory and radio range [5]. However, there exists a variety of challenges which may encountered and greatly affect the lifespan of the network; e.g. distance to the base station, distance among nodes, energy consumption[6,7], data processing capability, and may be some others. One of those crucial and dominant issues is the amount of consumed energy which is encountered during routing data among the nodes as well as during the actions of sensing phase [5]. A lot of researches have dealt with energy saving procedures, some of them attempted to achieve that throughout investment different protocols, which comprise MAC protocols, routing protocols, data transmission models, communication architectures, and special protocols, etc. [4]. Other researches attempted to manage energy saving in system level design and the architecture of nodes,

Content from this work may be used under the terms of the Creative Commons Attribution 3.0 licence. Any further distribution of this work must maintain attribution to the author(s) and the title of the work, journal citation and DOI.
Published under licence by IOP Publishing Ltd

whereas the main essential components of a standard node are: Radio (RF) communications, computing unit (MCU/FPGA and Memory), sensor/actuator unit, and power management unit[8].

In order to construct a sophisticated system with low power consumption, a lot of WSN architectures must communicate with very low duty cycle. Each of the nodes within the WSN can be enforced to work with low power, this is achieved by keeping the controller and communication interfaces in a low-power (sleep state) and beside that all the sensors and other peripheral devices are switched off. A node wakes up for a short time periodically, senses if an action is required, circulates information to other nodes within the network then it sleeps again [8]. However, such a system requires some manipulation with the implemented duty cycle in real-time applications, but still these applications will still consume higher power than others.

This paper presents a sophisticated way to measure power consumption of different situations using a number of wireless sensors connected to the Arduino kit. Then, comparison results of those situations are presented and discussed in order to show the power consumption at each state (idle, sleep, send, receive) of WSN operation.

The rest of this paper is introduced as follows: section 2 discusses some of the previous related works. Section 3 describes the used practical electronic circuits, and the way of measuring its power consumption is declared in section 4. Section 5 presents the practical obtained results and gives detail discussion for comparing different work situations. Then the comparison between Arduino Mega and ESP32 is declared in section 6. Finally, section 7 concludes the paper.

2. Related Works

To manipulate the power consumption problem and hence prolong the lifespan, many researches have inspected a variety of aspects in measuring and saving power of WSNs.

Due to the different activities of a person, the locations of sensor nodes placed on the body will always change. Thus, using a static routing algorithm causes packet and energy losses during transmission. Khan et al. [9]used a dynamic routing algorithm to address the problem of node movement in wireless body area sensor networks. They used the information of outstanding energies of nodes, hop count to sink distance, and throughput when nodes select the next hop node to forward data.

Elhoseny et al.[10] proposed a self-clustering method using genetic algorithm for improving power consumption in heterogeneous Wireless Sensor Networks. Comparing to other five methods, their proposed one was able to extend the network lifespan. They got improvement of 33.8% and 13%based on the first-node-die and the last-node-die are 33.8% and 13%, respectively, with respect to the second-best performance

Nair et al. [11] proposed a wireless node architecture in order to decrease power consumption in internet of things based wireless sensor networks (IOTWSN). The architecture is based on Bluetooth Low Energy (BLE) wireless standard and also a hybrid topology that connects the nodes to each other. They proved that BLE requires low operating power, able to fast connect, attains data throughput similar to traditional wireless techniques, able to deal with real-time monitoring of data.

Lin et al. [12]presented the Adaptive Transmission Power Control (ATPC) in wireless sensor networks. They showed that those topologies which are based on static power transmission are not suitable for the physical world, instead the adaptable transmission power control has the capability of power maintaining. In ATPC each node figures out a model for each of its neighbours, this model describes the relationship between transmission power and link quality. Then upon that, the individual link quality of a node is dynamically maintained over time.

Deng et al. [13] suggested several variables in sound source localization that affect the path loss exponent. Sound source localization is normally used in battlefield environments where low power consumption is required. They proved that those variables have the ability to reduce power consumption and also reduce localization error.

Pal and Nasipuri[14] proposed Power control and Routing (PCOR) protocol that enforces the sensor nodes in WSN to adjust their energy consumption based on the existing energy resources that permits continuous network operation as long as possible. They achieved that by reducing the overhearing at those nodes suffering from low-energy resources.

Liu et al. [15] used data fusion method in order to decrease the amount of data to be transmitted and hence the amount of consumed energy in WSN. They built their fusion algorithm on hybrid delay-aware clustering (HDC) to achieve a trade-off between the network delay caused by the fusion scheme and the energy consumption.

3. Hardware Tools Overview

This section presents some details about the Arduino board and sensors used in our experiments.

3.1. Arduino Mega 2560 Board

The Arduino Mega 2560 is a microcontroller ATmega2560 based board [16,17]. It is an open-source platform allowing the end user to have access to the product design and implementation. The board simply comprises of a microcontroller with all electrical circuits that are necessary for the operation of the microcontroller. The end user can easily upload/download code between the board and a computer throughout a USB cable. Table 1 summarizes the features of the Arduino Mega 2560 board used in our experiments.

The Arduino software comprises of two parts: Arduino Boot-Loader and Arduino IDE. Arduino Boot-Loader is a small program residing within the microcontroller which makes the controller distinct and grants it the power of combination to the Arduino IDE and the Arduino board. The Arduino IDE contains a compiler, serial monitor, etc. Furthermore, Arduino is an open source hardware and software platform which uses Atmel microcontroller as core hardware component and C or C++ as core software language [17].

Table 1. Features of Arduino Mega 2560.

Feature	Type
Hardware serial ports	4 UARTs
Operating Voltage	5V
Input Voltage (recommended)	7-12V
Input Voltage (limits)	6-20V
Digital I/O Pins	54 (of which 14 provide PWM output)
Analog Input Pins	16
DC Current per I/O Pin	40 mA
DC Current for 3.3V Pin	50 mA
Flash Memory	256 KB of which 8 KB used by bootloader
SRAM	8 KB
EEPROM	4 KB
Clock Speed	16 MHz

3.2. Bluetooth

The HC-05 Bluetooth is a serial communication module that can serves in two working modes: the automatic connection work mode and the order response work mode. One of these modes can be activated via alternating the input voltage to pin (PIO11). When the module works in the automatic connection mode, it has three roles to work on: Master, Slave and Loopback; and the module will transmit and receive data according to the last set role. However, when the module works in the order response mode, the user will have the ability to send AT command in order to set the control parameters. Bluetooth module setup is described in Table 2.

3.3. Sensors

Each sensor differs in its function and has various features so the sensors are uneven in power consumption. In this paper, the sensors have been selected based on the availability in the Lab and ease of connection with the board. Besides, they consume the power from the board without using an external source and give its sensing value instantly that will be used in the communication operations. The used Sensors are: ultrasonic sensor, photoresistor, PIR motion sensor, and temperature and humidity sensor.

Table 2. Bluetooth module setup.

Command	Response	Function
Master and Slave module		
AT	OK	Test
AT+ORGL	OK	Restore default status
AT+RMAAD	OK	Delete all authenticated devices in the pair list
AT+UART= 9600,0,0	+UART:9600,0,0 OK	Set/ Inquire- serial parameter
Master module		
AT+NAME= BTMASTER	+NAME:BTMASTER OK	Set/ inquire device's name
AT+ROLE=1	OK	Set/ inquire module role
AT+CMODE=0	OK	Set/ Inquire - connection mode
AT+LINK= FCA8,9A,31F7	OK	Connect device
AT+BIND= FCA8,9A,31F7	OK	Set/Inquire - bind Bluetooth address
Slave module		
AT+NAME=BTSLAVE	+NAME:BTSLAVE OK	Set/ inquire device's name
AT+ROLE=0	OK	Set/ inquire module role
AT+CMODE=1	OK	Set/ Inquire - connection mode
AT+ADDR?	+ADDR:FCA8:9A:31F 7 OK	Get module Bluetooth address

4. Experiment Setup

In order to emulate the actual wireless sensor networks, an experiment is implemented using a practical model that is consisting of two electronic circuits. The circuit shown in Figure 1 (its block diagram is in Figure 3) is used for transmitting sensor signals and the other circuit, shown in Figure 2, is for receiving those signals (its block diagram is in Figure 4). The circuits are drawn with the help of the open-source program Fritzing. The experimental model comprises of two Arduino Mega boards that are communicating with each other via two Bluetooth models in the master/slave means. Five different sensors are used for transmitting and receiving signals between the two boards; hence power consumption can be evaluated upon the exchanging of those signals.

The operation at transmitter and receiver sides are declared as flowcharts in Figure 5 and Figure 6. The transmitter checks an interrupt signal for reading the voltage values of sensors if that interrupt is raised, otherwise it goes sleep. All sensor values are assembled is one packet and sent via the Bluetooth. The packet is received at the other side via a Bluetooth as well. A corresponding LED for each of the sensors is lighted depending on the received values of sensors.

4.1. Power Management Modes

The microcontroller on Arduino board supports different modes of operation, these modes make an efficient use of power resources and hence prolong the lifespan. Those modes are activated by software, a brief description for each of them is given here. The idle mode put the CPU in halt state while the timer/counters, SRAM, SPI port, and interrupt system stay in operative state. The power down mode keeps the register contents but freezes the oscillator, deactivating all other chip functions until the following interrupt or hardware reset. The Power-Save mode allows the asynchronous timer to stay working in order to give the user the ability to maintain a timer base while the rest of the device is sleeping [22]. The ADC Noise Reduction (ADCR) mode also put the CPU in halt state and stops the input/output circuitry as well, and leave the asynchronous timer and ADC to be running; this minimizes switching noise during ADC conversions [23]. The standby mode permits almost no time start-up of the device together with very low power consumption, this is achieved by putting all integrated circuits in sleep mode except the crystal/resonator oscillator which stay in running. In Extended Standby mode, both the asynchronous timer and the main oscillator remain to run [22]. Table 3 shows the active clock domains and wake up sources in the different sleep modes.

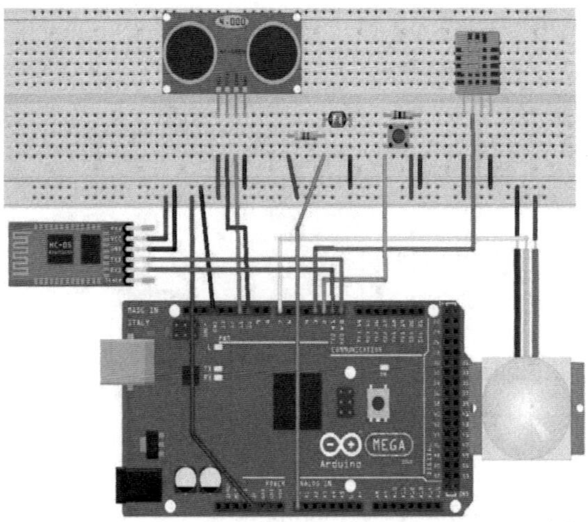

Figure 1. Transmitter circuit. Bluetooth as a master with several sensors are connected to the Arduino board. These sensors produce signals to be transmitted to the receiver side.

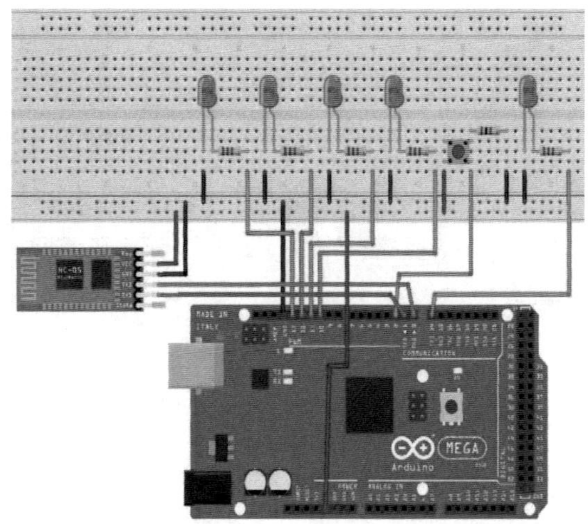

Figure 2. Receiver circuit. Bluetooth as a slave with several actuator indicators are connected to the Arduino board. These actuators give indications about the received signals from the transmitter side.

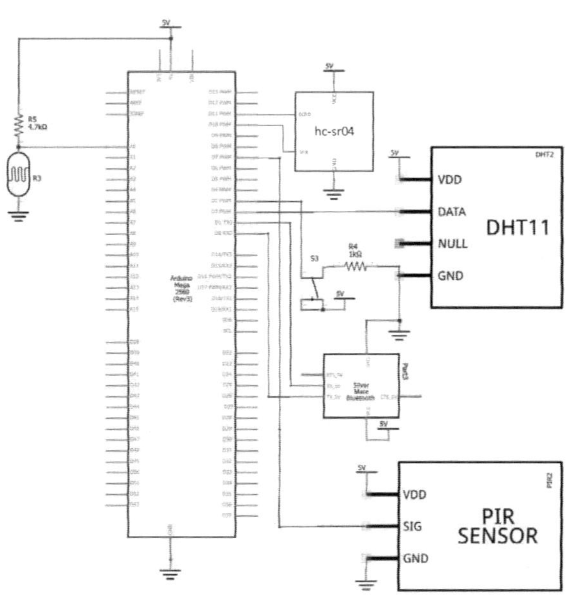

Figure 3. Block diagram of the transmitter circuit.

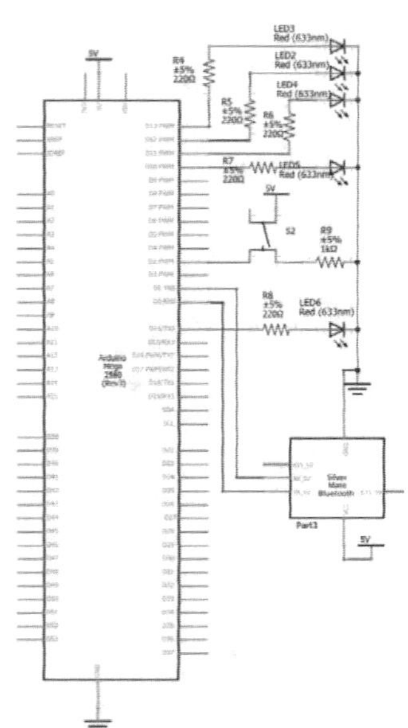

Figure 4. Block diagram of the receiver circuit.

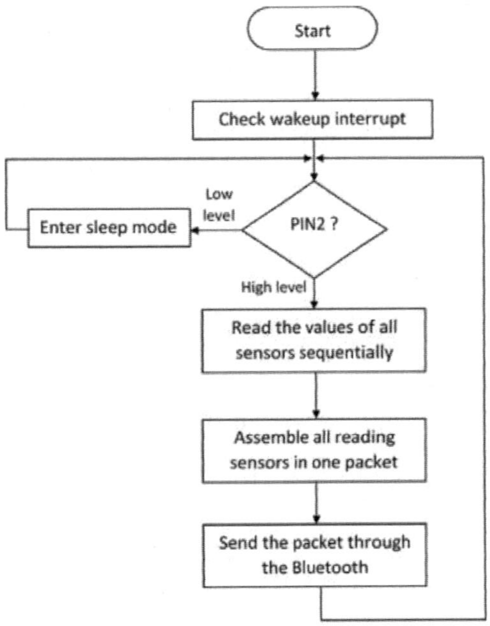

Figure 5. Transmitting side flowchart.

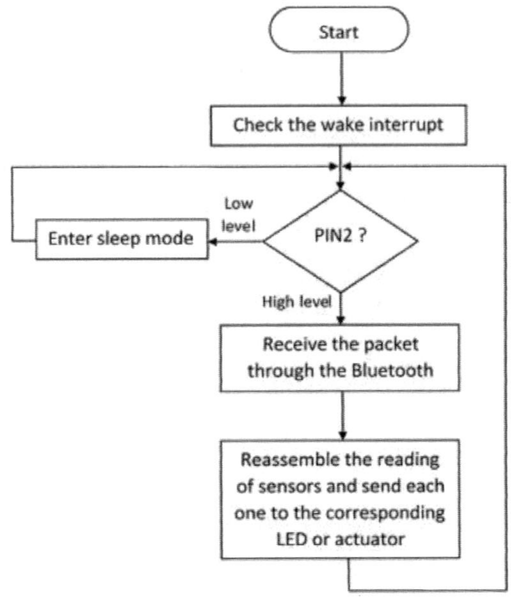

Figure 6. Receiving side flowchart.

Table 3. Sleep modes of the microcontroller on Arduino board.

	Active clock domains					oscillators		Wake up sources						
Sleep mode	Clkcpu	Clk flash	Clk io	Clkadc	Clkasy	Main clock source enable	Timer osc enable	Int7:0 and pin change	Twi address match	Timer2	Spm\eeprom ready	adc	Wdt interrupt	Other I/O
Idle			x	x	x	x	x^b	x	x	x	x	x	x	x
ADCNRM				x	x	x	x^b	x^c	x	x^b	x	x	x	
Power-Down								x^c	x				x	
Power-Save					x		x^b	x^c	x	x			x	
Standby[a]						x		x^c	x				x	
Extended Standby					x^b	x	x^b	x^c	x	x			x	

[a] Only recommended with external crystal or resonator selected as clock source.
[b] If Timer/Counter2 is running in asynchronous mode.
[c] For INT7:4, only level interrupt.

4.2. Power Measurement Methodologies

Current and voltage measurements are the most practical methods used to properly calculate the power consumption. Basically, the power value can be calculated using one of the formulas:

$$P = V \times I \tag{1}$$

$$P = \frac{V^2}{R} \tag{2}$$

$$P = I^2 \times R \tag{3}$$

For a known voltage value, the current value can be measured by connecting a personal computer (PC) and Arduino Mega board either through a USB-meter or a Digital Multi-Meter (DMM) as described in Figure 7 and Figure 8. In Figure 7, the electric current value is measured directly through a USB-meter and the measured output is displayed on the USB-meter monitor.

While in Figure 8, the circuit has been splited and the DMM is inserted with the probe which will be put into the current socket and the selector will be pointed on the current choice, then the electric current is passed through the DMM. Where it is measured and displayed on the DMM monitor. Experimental methodologies measurements are presented as illustrated in Figure 9 and Figure 10. In Figure 9, instead of measuring the electric current, the voltage will be measured over 1-ohm resistance using a DMM by putting the probe into the voltage socket and the selector is pointed on the voltage indicator, then the measured output value is displayed on the DMM monitor. Furthermore, Figure 10 is same as Figure 9 but it is used Arduino Uno board instead of the DMM, where the measured output is displayed on a personal computer (PC) through serial monitor Arduino IDE.

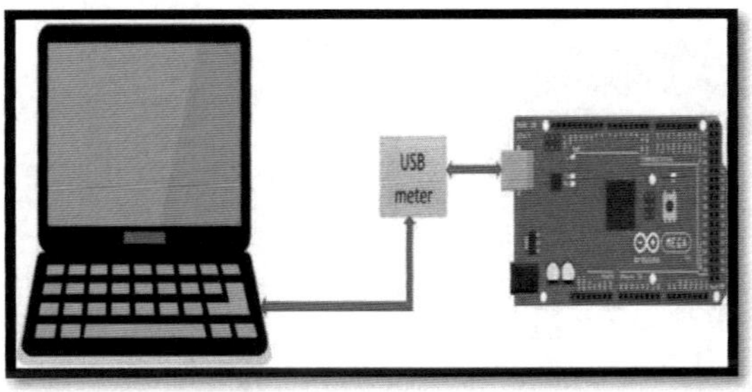

Figure 7. The first method (measure the current using a USB meter)

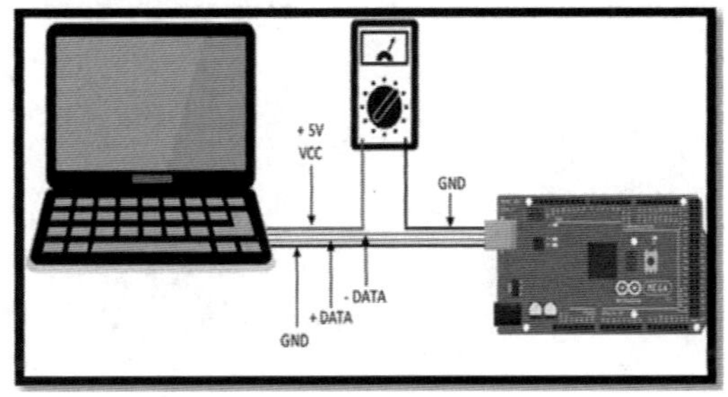

Figure 8. The Second method (measure the current using DMM)

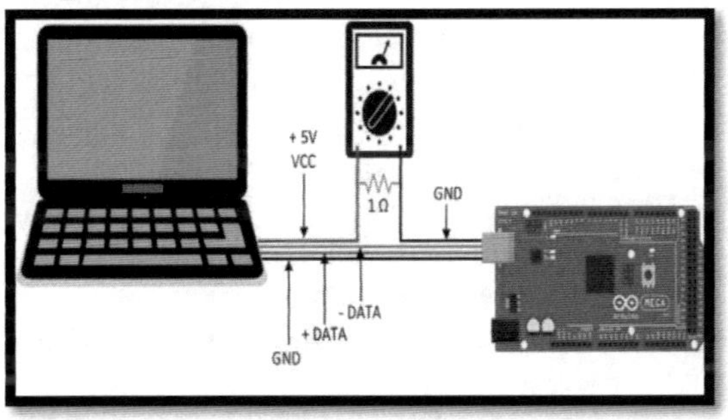

Figure 9. The Third method (measure the voltage using DMM)

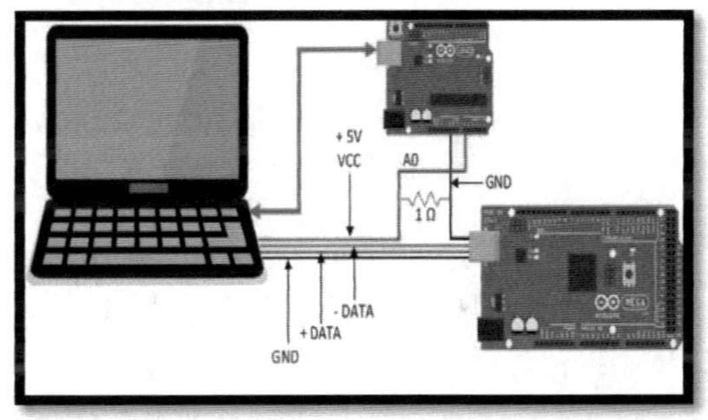

Figure 10. The Fourth method (measure the voltage using Arduino-Uno)

5. Results and Discussion

Power consumption is tested with respect to the state and role of the device. As, the measuring method checks the impact of sleeping modes on power consumption regardless of the data rate type.

Where, at data rate (9600), the experimental results in Figure 11 show that there is a discrepancy among the measured values using the DMM, USB meter, and Arduino Uno board for the same case. The advantages and disadvantages for each power measuring method are displayed in Table 4.

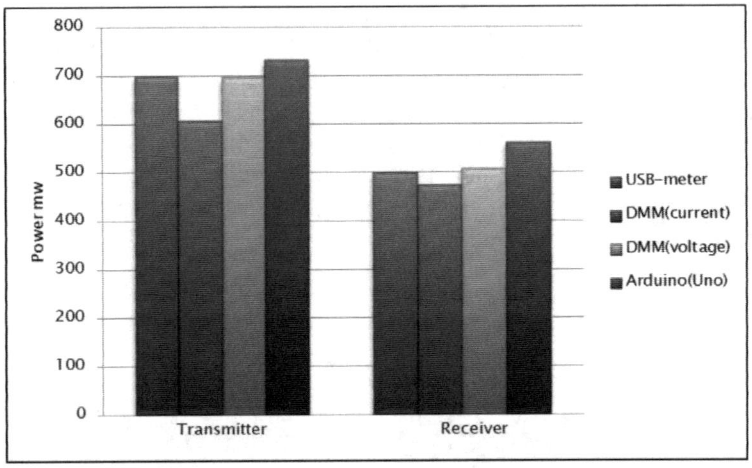

Figure 11. Transmitter and receiver power consumption at data rate 9600 for different power measuring methods

Table 4. Comparison between the presented power measuring methods.

Method	Advantages	Disadvantages
Measure the current using a USB meter	• USB voltage and current can be easily measured. • The USD operation is not influenced, so the current can be measured from any USB device.	• Accurate measurements of current can be obtained when at high current levels but not at low current levels. • It has an internal current shunt resistance of 0.05 ohm, and thus the voltage drop at 2.5A is 0.125V. • Related to the shape of the device the measurement ranges are 3~7.5V and 0~2.5A of the voltage and current, respectively.
Measure the current using DMM	• The DMM often has diverse current ranges that are implemented using different shunt resistors to change the value of R.	• It damages the PC USB port if the current draw is too high • The resistance (RA) of the current meter (i.e. Ammeter) must be much smaller compared to the resistance (R) of the test element to have an accurate current measurement.
Measure the voltage using DMM	• The DMM often has different voltage ranges that are implemented using different shunt resistors to change the value of R. Some ranges also have a gain stage on the measured voltage.	• The resistance (RV) of the Voltmeter must be much higher compared to the resistance (R) of the test element to have accurate voltage measurement.
Measure the voltage using Arduino-Uno	• It is a simple and economic tool that can be used instead of the DMM.	• Its accuracy depends on ADC conversion and is determined by the number of bits and voltage reference.

IOP Conf. Series: Materials Science and Engineering 1152 (2021) 012022 doi:10.1088/1757-899X/1152/1/012022

Figures 12 and 13 display the monitoring voltage at Transmitter and receiver sides respectively on the serial plotter using Arduino Uno during 5 mins. Each voltage level is multiplied by (5/1024) to find the real voltage value where the value (5) is a source voltage and the value (1024) is ADC resolution (10 bits).

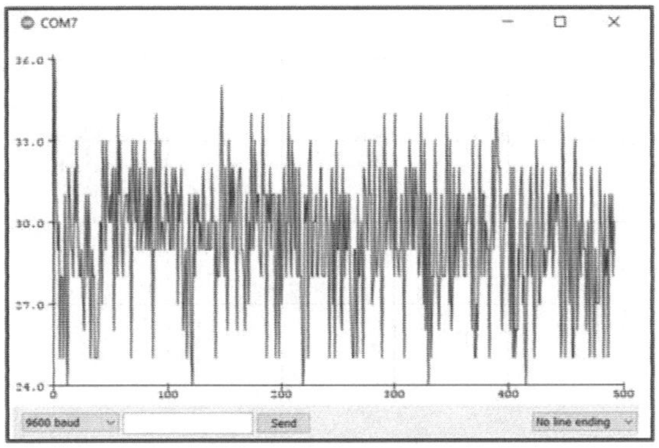

Figure 12. Monitoring voltage at Transmitter side using Arduino Uno during 5 mins.

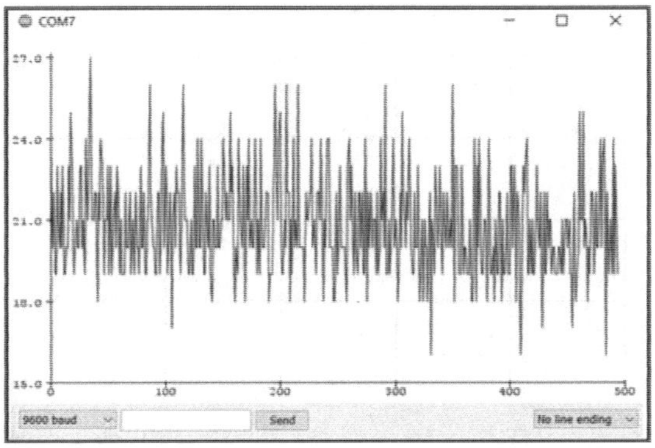

Figure 13. Monitoring voltage at receiver side using Arduino Uno during 5 mins.

Figure 14 illustrates the effect of power consumption for different power saving modes. Where, the power consumption of idle mode is larger than other modes, while the PWR_DOWN mode consumes less power, depended on the details in Table 3. This note will be confirmed regardless of the data rate type.

Besides, the different operating modes of the microcontroller unit (MCU) permit power saving based on the application's requirements. Selecting one of the sleep modes causes shutting down unused modules in the in the MCU as stated below for each of the used modes in the experiments result:

Idle mode stops the CPU but allows the SPI, ADC, USART, analog comparator, watchdog,2-wire serial interface, timer/counters, and the interrupt system to continue operating. This sleep mode fundamentally halts the CPU clock and FLASH clock, while permitting the other clocks to work. External and internal interrupts wake up the MCU.

ADC Noise Reduction mode (ADCNRM) stops the CPU but allows the ADC, the external interrupts, timer/counter2, 2-wire serial interface address match, and the watchdog to continue operating (if enabled). This sleep mode principally halts CPU clock, I/O clock, and FLASH clock, while allowing the other clocks to run.

Power-down mode stops the external oscillator, while the 2-wire serial interface, the external interrupts, and the watchdog continue operating (if enabled). Only an external reset, a brown-out reset, a watchdog reset, 2-wire serial interface address match, an external level interrupt on INT7:4, an external interrupt on INT3:0, or a pin change interrupt can wake up the MCU. This sleep mode essentially halts all generated clocks, permitting operation of asynchronous modules only.

Power-Save mode is identical to Power-Down, with one exception: if timer/counter2 is enabled, it will keep running during sleep. This allows the user to maintain a timer base while the rest of the device is sleeping.

Standby mode is identical to the Power-Down mode with the exception that the oscillator is remained running while the rest of the device is sleeping. This permits very fast start-up combined with little power consumption.

Extended Standby mode is identical to the Power-Save mode with the exception that the oscillator is kept running.

The power consumption of different components in presented WSN node are shown in Figure 15. It is indicated that the power consumed by the Board only compared with the power consumed at transmitting and receiving can be insignificant.

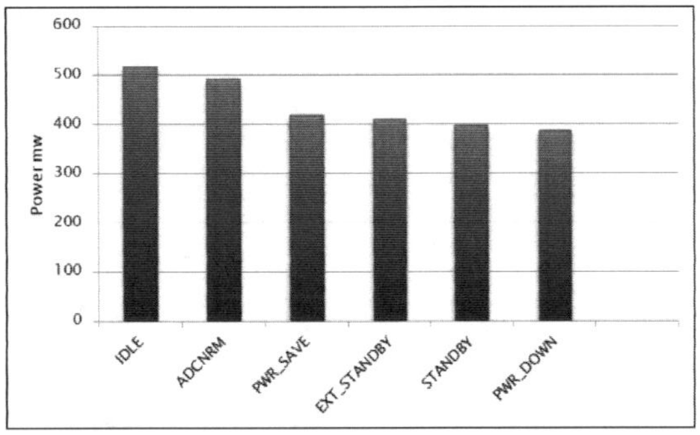

Figure 14. Power Consumption at Transmitter Side With Different Sleep Modes at Data Rate 9600.

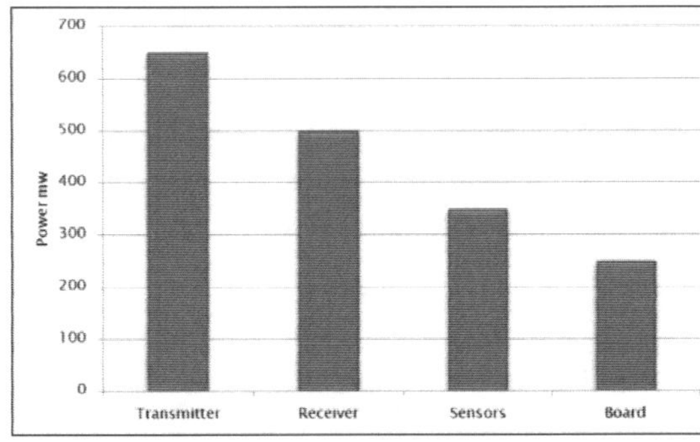

Figure 15. Typical power consumption of different components.

6. The Arduino Mega compared to the ESP32

These two microcontrollers are totally dissimilar. Where they have diverse hardware architecture such as the processing capability, the memory size, the communications features, the number of GPIOs that they expose, and much more. So, the company that develops and designs the ESP32, has introduced a vast effort in writing software to cover the hardware hole between the Arduino Mega and the ESP32. As a result, the ESP32 operates with the Arduino IDE by the ESP32-Arduino Core installation and the combination between these two microcontrollers is noteworthy [24].

Obviously, the ESP32 has unique features that are not exist in the ArduinoMega, but the Arduino Megais a much better choice because It's a humbler device, more robust, meeker to program, more forgiving, in addition, to mistakes and problems in the wiring, and easier to set up. In fact, the ESP32 is a better economic option than Arduino Mega, which means that can be gotten a more powerful board for a lesser cost [24]. In the future, the existing Arduino skills can be used to operate with the ESP32 that can be treated as a supercharged ArduinoMega.

7. Conclusion

This paper demonstrates several power measurement methods for a WSN node based on the Arduino development board equipped with the Bluetooth module, where both are very prevalent in small-scale research as well as wireless automation solution. These simple, and flexible power measurement methodologies are chosen to calculate how much power is spent at each power-saving mode such as (idle, sleep, power-down, etc.) and communication states. From the testing results, it can be noted that these measurement methods diverse in their accuracy and ease of use and can be used according to the user's needs. The idle mode consumes power larger than the other modes because most components on the board are turned on, and the major power is consumed at transmitted state regardless of the data rate used during the communication between the two ends. Towards the end, a better function can be attained when integrating the Arduino kit with the Bluetooth module. As the experimental results display the WSN nodes operate stead and the lower power consumption performance is prominent.

8. References

[1] Member A S, Karpov E F, Karpova E, Suchkov A, Mironov S, Karelin A, Member A B and Spirjakin D 2015 Compact Low Power Wireless Gas Sensor Node with Thermo Compensation for Ubiquitous Deployment *IEEE Transactions on Industrial Informatics* **11(6)** 1660-1670

[2] Yun W-K and Yoo S-J 2021 Q-Learning-Based Data-Aggregation-Aware Energy-Efficient Routing Protocol for Wireless Sensor Networks *IEEE Access***9** 10737-10750

[3] Javaid N, Cheema S, Akbar M, Alrajeh N, Alabed M S and Guizani N 2017 Balanced Energy Consumption Based Adaptive Routing for IoT Enabling Underwater WSNs *IEEE Access***5** 10040–10051
[4] Peng J and Chen Y 2015 A low energy consumption WSN node *Int. J. Embed. Syst.***7** 318–323
[5] Laouid A, Dahmani A, Bounceur A, Euler R, Lalem F and Tari A 2017 A distributed multi-path routing algorithm to balance energy consumption in wireless sensor networks *Ad Hoc Networks***64** 53–64
[6] Joshi P, Gavel S and Raghuvanshi A S 2021 Estimating Energy Consumption for Various Sensor Node Distributions in Wireless Sensor Networks *Nanoelectron. Circuits Commun. Syst.***692** 289–299
[7] Ali Q I 2018 GVANET project: an efficient deployment of a self-powered, reliable and secured VANET infrastructure *IET Wirel. Sens. Syst.***8** 313–322
[8] Popovici E, Magno M and Marinkovic S 2013 Power management techniques for Wireless Sensor Networks: A review *Proc. 2013 5th IEEE Int. Work. Adv. Sensors Interfaces, IWASI 2013* 194–198
[9] Khan R A, Xin Q and Roshan N 2021 RK-Energy Efficient Routing Protocol for Wireless Body Area Sensor Networks *Wirel. Pers. Commun.***116** 709–721
[10] Elhoseny M, Yuan X, Yu Z, Mao C and El-minir H K 2014 Balancing Energy Consumption in Heterogeneous Wireless Sensor Networks using Genetic Algorithm **7798** 1–4
[11] Nair K, Kulkarni J, Warde M, Dave Z, Rawalgaonkar V, Gore G and Joshi J 2016 Optimizing power consumption in iot based wireless sensor networks using Bluetooth Low Energy *Proceedings of the 2015 International Conference on Green Computing and Internet of Things, ICGCIoT 2015 (IEEE)* 589–593
[12] Lin S 2016 ATPC : Adaptive Transmission Power Control for Wireless Sensor Networks *ACM Trans. Sens. Networks***12** 1–31
[13] Deng F, Guan S, Yue X, Gu X, Chen J, Member S, Lv J and Li J 2017 Energy-Based Sound Source Localization with Low Power Consumption in Wireless Sensor Networks *IEEE Transactions on Industrial Electronics* **64(6)**4894-4902
[14] Pal A and Nasipuri A 2019 Joint power control and routing for rechargeable wireless sensor networks *IEEE Access***7** 123992–124007
[15] Liu X, Zhu R, Anjum A, Wang J, Zhang H and Ma M 2020 Intelligent data fusion algorithm based on hybrid delay-aware adaptive clustering in wireless sensor networks *Futur. Gener. Comput. Syst.***104** 1–14
[16] Kumar N S, Vuayalakshmi B, Prarthana R J and Shankar A 2017 IOT based smart garbage alert system using Arduino UNO *IEEE Reg. 10 Annu. Int. Conf. Proceedings/TENCON* 1028–1034
[17] Badamasi Y A 2014 The working principle of an Arduino *2014 11th International Conference on Electronics, Computer and Computation (ICECCO)* 1–4
[18] Li S E, Li G, Yu J, Liu C, Cheng B, Wang J and Li K 2018 Kalman filter-based tracking of moving objects using linear ultrasonic sensor array for road vehicles *Mech. Syst. Signal Process.***98** 173–189
[19] Qi Y, Soh C B, Gunawan E and Low K S 2015 Ambulatory Measurement of Three-Dimensional Foot Displacement during Treadmill Walking Using Wearable Wireless Ultrasonic Sensor Network *IEEE J. Biomed. Heal. Informatics***19** 446–452
[20] Amri I, Dian Atmajati E, Salam R A, Yuliza E, Munir M M and Khairurrijal 2017 Potentiometer a simple light dependent resistor-based digital *Proceeding - 2016 Int. Semin. Sensors, Instrumentation, Meas. Metrol. ISSIMM 2016* 24–27
[21] Yang W, Qiao S, Song Q, Liu Z and Yang J 2015 The design and implementation of wireless temperature and humidity control system based on nRF905 *Proc. 2015 10th IEEE Conf. Ind. Electron. Appl. ICIEA 2015* 753–756
[22] Berthier F, Beigne E, Vivet P and Sentieys O 2015 Power gain estimation of an event-driven wake-up controller dedicated to WSN's microcontroller *Conf. Proc. - 13th IEEE Int. NEW*

Circuits Syst. Conf. NEWCAS 2015 1–4

[23] Wannenburg J and Malekian R 2015 Body Sensor Network for Mobile Health Monitoring, a Diagnosis and Anticipating System *IEEE Sens. J.* **15** 6839–6852

[24] Maier A, Sharp A and Vagapov Y 2017 Comparative analysis and practical implementation of the ESP32 microcontroller module for the internet of things Proceeding - *2017 Internet Technologies and Applications ITA* 143-148

Model of Integrated of Car Brake-Active Suspension of Quarter Car Model By Using LQR Control Under Periodic Road Excitation

Muchammad Harly[1], Marji[1], Yoto[1] and Paryono[1]

[1]Faculty Of Engineering, Universitas Negeri Malang, Indonersia

*Email: Muchammadharly.ft@um.ac.id

Abstract. The most important technology in a vehicle is the stability of the direction of the vehicle and the comfort of the passengers. For stability, direct yaw control is used via ABS / TCS / DSC, while for convenience, use the active suspension. These two solutions also contribute to one another. In this research paper is produced and explained about the use of LQR control to control the response of the sprung mass to achieve high damping ratios in isolating the excitation force from nonlinear road surfaces. The simulation results prove that the damping ratio controlled by LQR was better than compared by using PID.

1. Introduction

The final suspension is a research object that is very essential and is being carried out by world researchers because it is one of the important factors of vehicle road handling comfort and stability. The results of previous research on the suspension consisted of 3 important concentrations, namely the first plant model consisting of quarter, half and full suspension. Second, in terms of control, research has been carried out using passive, semi-active and active. and third of many control methods that use PID control, Sliding mode, Observer etc. for that we will review the previous research, before explaining the results of our study The ideal target control response is minimal rolling and maximum grip-wheel ground surface. S.F. Youness, E.C. Lobusov [1] published a research on suspension control ¼ car model that is no-network and networking, with PID and LQR control approaches, the results of the performance of suspense control control concluded that LQR is better than PID both no-network and networking. Prabu Krishnasamy, Jancirani Jayaraj, Dennie John [2] concluded from the results of his research on pneumatic active suspension with PID control compared to passive suspension, all model car models with LabView software. The results of PID control response for pneumatic active suspension are better than passive suspension. Narinder Singh1, Himanshu Chhabra and Karansher Bhangal [3] studied the 2 DOF ¼ car model by comparing 2 control H∞ and μ- synthesis, the results of this study concluded that μ synthesis is better than H∞ from the robust side. Zhi-Jun Fu, 1 Bin Li, 2 Xiao-Bin Ning, and Wei-Dong Xie [4] show the results of research on ¼ car using optimal control approximate dynamic programming (ADP) instead of LQR. Optimal control applies single adaptive critic NN Hamilton-jacobi-bellman solution. The results of the close loop stability analysis for both controls concluded that the optimal control is better than LQR. Michiel Haemers1, Stijn Derammelaere3, Clara-Mihaela Ionescu1, Kurt Stockman, Jasper De Viaene, Florian Verbelen1 [5] presents research results from the full-car state space model to meet the dynamic characteristics of the ISO 8608 suspension. State feedback control is used. with parameter optimization using a genetic algorithm (GA). The results of the study explain that this control system is better than those that are not tuned with GA. Rosmazi Rosli, Musa Mailah, Gigih Priyandoko

Content from this work may be used under the terms of the Creative Commons Attribution 3.0 licence. Any further distribution of this work must maintain attribution to the author(s) and the title of the work, journal citation and DOI.
Published under licence by IOP Publishing Ltd

[6], and Thanh-Phong Pham, Olivier Sename , Luc Dugard [7] promoted the results of active suspense control by comparing 2 control methods, namely hybrid active force control (AFC) -iterative learning (IL) with PID. The results of the study concluded that AFC-IL is better than PID. Qinghua Meng Chih-Chiang Chen Pan-Wang Zong-Yao Sun Bingji Li1 [8] described the results of his investigative research on the differences in the use of 2 active suspension controls, namely active suspension homogenous controller (ASHC) and SMC for passive suspension. The results showed that ASHC was better than SMC. Ammar A. Aldair and William W.J. Wang [9] was also proposing and Evolutionary Algorithm Based Fractional Order PID + μ Controller Design for a Full Vehicle Nonlinear Active Suspension System with successfully. Abdolvahab Agharkakli, Ghobad Shafiei Sabet, Armin Barouz [10] proposed a Simulation and Analysis of Passive and Active Suspension System Using Quarter Car Model for Different Road Profile. The result active suspension has superior performance than passive. Tongchit Suthisripok, Chanwit Wongrattanapornkul Somchai Poonyaniran, And Adirak Kanchanaharuthai [11] studied a Disturbance Observer Based Control for Active Suspension Systems was better performance compared with PID control. Sorin Marcu, Dinel Popa, Nicolae–Doru Stănesc, Nicolae Pandrea [12] has designed and investigated a Model for the study of active suspensions. Mathematical Modeling and Control of Active Suspension System for a Quarter Car Railway [13], investigated from Mathematical Modeling side and Control of active suspension system for a quarter car railway to control the vertical damper ratio under uncertainty railway contour. Aizuddin Fahmi Mohd Riduan, Noreffendy Tamaldin, Ajat Sudrajat, and Fauzi [14] also made a Review on active suspension system used by BMW car. Nouby M. Ghazaly, Ahmad O. Moaaz [15] studied the future development and analysis of vehicle active suspension system. Pavan Kumar.G [16] Muchammad harly, IN sutantra, mauridhi HP [17][18][19] also studied active suspension which is integrated with DSC control and active steering which can adapt the driver's behavior Pinandhito, M.R., Indriawati, K., Harly, M. [20] studied an experimental and computational investigation of Active fault tolerant control design in regenerative antilock braking system of electric vehicle with sensor fault and at the end Diarnise, S., Ya'umar,, Indriawati, K., Harly, M.[21] electric vehicle with actuator fault . From all the studies above, there are still problems that because the variable disturbance factor only involves axial excitation, without considering longitudinal excitation, which affects the physical phenomena in the field, so this study (ECU-HASB) involves both.

2. Proposed Turning, Braking/Traction on Random Road Full Car Suspension Model

2.1 Axial excitation

The DoF, it includes unsprung mass, namely X_w, which is the motion of the wheels and sprung mass X_b, which is the motion of the vehicle chassis / body as shown in figure 1.

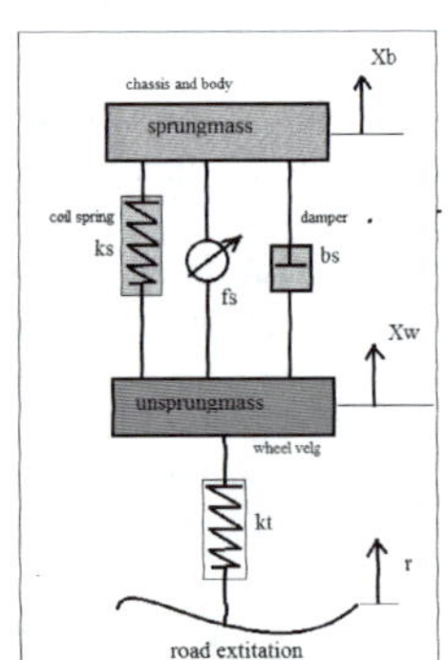

Figure 1. Model 3 DoF without Brake Vertical Force

If it is assumed

$$X_1 = X_b \ , \ X_2 = X_w \ , X_3 = \dot{X}_b \ , \ X_4 = \dot{X}_w$$

The motion equation of the sprung as can be represented by the equation (1)

$$m_b.\ddot{X}_b = k_s(X_w - X_b) + b_s(\dot{X}_w - \dot{X}_b) + f_s \quad (1)$$

The equation for the movement of un-sprung mass can be represented by the equation (2)

$$m_w.\ddot{X}_w = k_t(r - X_w) - k_s(X_w - X_b) - b_s(\dot{X}_w - \dot{X}_b) - f_s \quad (2)$$

Equations (1) and (2) can be represented in the form of state space 3 as follows

$$\dot{X} = A.X + BU + EW \quad (3)$$

Where U is the controlled force (f_s) and the highway excitation container (r) with the state space matrix constants as follows:

$$A = \begin{Bmatrix} 0 & 0 & 1 & 0 \\ 0 & 0 & 0 & 1 \\ -\frac{k_s}{m_b} & \frac{k_s}{m_b} & -\frac{b_s}{m_b} & \frac{b_s}{m_b} \\ \frac{k_s}{m_w} & -\frac{(k_s + k_t)}{m_w} & \frac{b_s}{m_w} & \frac{-b_s}{m_w} \end{Bmatrix}$$

$$B^T = \begin{Bmatrix} 0 & 0 & \frac{1}{m_b} & \frac{1}{m_w} \end{Bmatrix}$$

$$E^T = \begin{Bmatrix} 0 & 0 & 0 & \frac{k_t}{m_w} \end{Bmatrix}$$

2.2 The implication of Vertical and longitudinal brake force excitation as shown in figure 2 make superposition to vertical motion of road excitation. Formula is presented below.

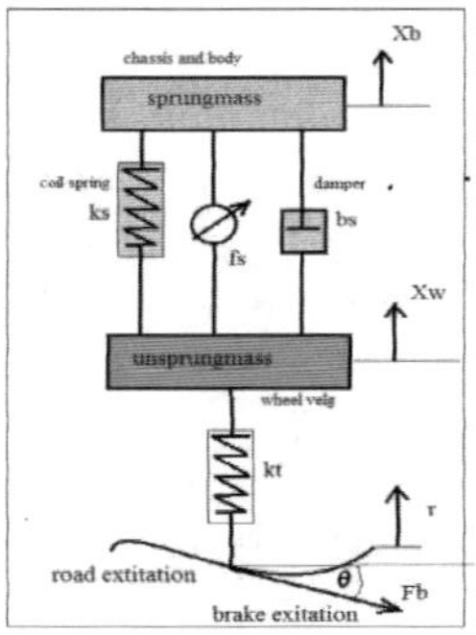

Figure 2. Model 3 DoF active suspension with brake force vertical superposition

The equation for the movement of the sprung mas can be represented by the equation (4)

$$m_b.\ddot{X}_b = k_s(X_w - X_b) + b_s(\dot{X}_w - \dot{X}_b) + f_s \quad (4)$$

IOP Conf. Series: Materials Science and Engineering 1152 (2021) 012023 doi:10.1088/1757-899X/1152/1/012023

The motion equation of the un-sprung mass which is excited by the brake force F_b, shown in figure 2 implicate to add vertical extra force to the un-sprung mass. It can be represented by the equation (5)

$$m_w.\ddot{X}_w = k_t(r - X_w) + F_b \sin\theta - k_s(X_w - X_b) - b_s(\dot{X}_w - \dot{X}_b) - f_s \quad (5)$$

$$m_w.\ddot{X}_w = k_t(r - X_w) + F_b \sin\theta - k_s(X_w - X_b) - b_s(\dot{X}_w - \dot{X}_b) - f_s \quad (5)$$

If the change of road-tire geometry is periodic, the equation becomes

$$m_w.\ddot{X}_w = k_t(r - X_w) + F_b \sin\omega.t - k_s(X_w - X_b) - b_s(\dot{X}_w - \dot{X}_b) - f_s \quad (6)$$

Equations (4) and (5) can be represented in the form of state space as follows

$$\dot{X} = A.X + BU + EW + DZ \quad (7)$$

Where U is the controlled force (f_s) and W is the highway excitation (r) and Z is the brake force (F_b.sin $\boldsymbol{\omega}$.t). With the state space matrix constants as follows

$$A = \begin{Bmatrix} 0 & 0 & 1 & 0 \\ 0 & 0 & 0 & 1 \\ -\frac{k_s}{m_b} & \frac{k_s}{m_b} & -\frac{b_s}{m_b} & \frac{b_s}{m_b} \\ \frac{k_s}{m_w} & -\frac{(k_s + k_t)}{m_w} & \frac{b_s}{m_w} & \frac{-b_s}{m_w} \end{Bmatrix}$$

$$B^T = \begin{Bmatrix} 0 & 0 & \frac{1}{m_b} & \frac{1}{m_w} \end{Bmatrix}$$

$$E^T = \begin{Bmatrix} 0 & 0 & 0 & \frac{k_t}{m_w} \end{Bmatrix}$$

$$D^T = \begin{Bmatrix} 0 & 0 & 0 & \frac{\sin\omega.t}{m_w} \end{Bmatrix}$$

2.3 Axial and longitudinal drive force excitation

If the acceleration sensor variable is (a) and variable, then the speed of the sprung mass X_b and un-sprung mass X_w and the body level sensor is X_b (0) as the initial condition, it can be formulated by the equations (8), (9), (10):

$$\dot{X}_b = \int a_b \, dt \quad (8)$$

$$\dot{X}_w = \int a_w \, dt \quad (9)$$

$$X_{b(0)} = \iint a_b \, dtdt \quad (10)$$

3. Proposed Suspension LQR Control

To test the LQR control, all the quarter car modeling equations above are compiled into the state space equation $\dot{X} = A.X + BU + EW + DZ$ with A, B, C, D which have been determined according to the vehicle suspension parameters as depicted in figure 3. The first step is to define the function object as follows:

$$J = \frac{1}{2}\int_{to}^{tf}\{(x_{(t)i}{}^{T}.Q.x_{(t)i}) + (u_{(t)j}{}^{T}.R.u_{(t)j})\}\,dt$$

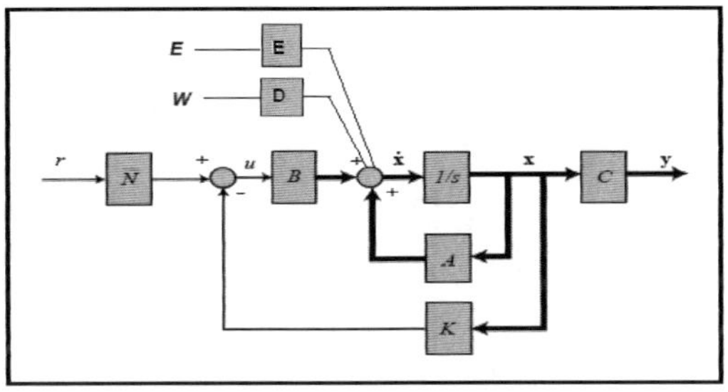

Figure 3. LQR control design suspension state space

Then second determined the values of Q, R as control state and excitation state from A, B, C, D and Q, R has been obtained P with Riccati equestion to get the value of K.

$$P.A_{ii} + A_{ii}{}^{T}.T\,P - P.B_{ij}.R^{-1} - B_{ij}{}^{T}.P + Q = 0$$

$$K = (R^{-1}\,B^{T}\,P)$$

Then third from here it can be calculated the value of control vector U. Where the control vector U can be formulated as follows

$$U_i = (-K.x_{(t)i})$$

4. Simulation result and discussion

The simulation has been setup with the excitation frequency of the highway geometry as follows 1 rad / sec, 2 rad / sec and 4 rad / sec and constant amplitude of 0.5 meters. In this research, the results of the LQR control are compared to the damper ratio with PID controls with the same plant. Then the most important thing is that the brake excitation force which is projected in a vertical direction is also included while this has not been done by other researchers.

The results of the real time response of the vertical displacement for a street frequency of 1 rad / sec are to produce a damping ratio of 8 as shown in the figure 4 while by using PID reach 7.3 only. Mean it can isolate the sprung mass from road vibration by reducing one eighth compared road amplitude.

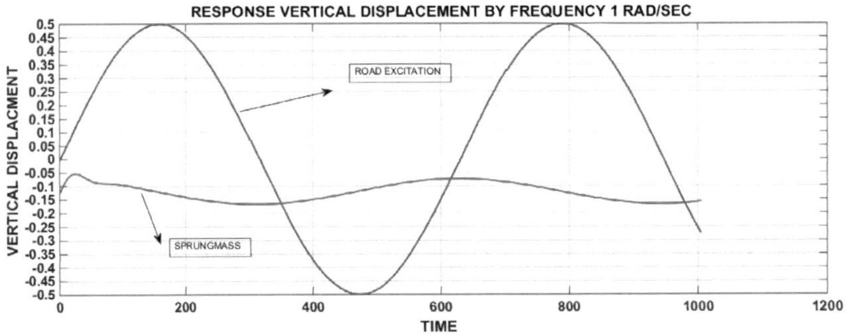

Figure 4. Damper Ratio Response by Frequency 1 rad/sec

Then for the second simulation, the road geimetry frequency and vehicle speed have been set so that it becomes 2 rad / sec while maintaining the 0.5 meter amplitude. The result of the damping ratio is 6 as shown in the figure 5 while by using PID reach 5.7 only. Mean it can isolate the sprung mass from road vibration by reducing one sixth compared road amplitude.

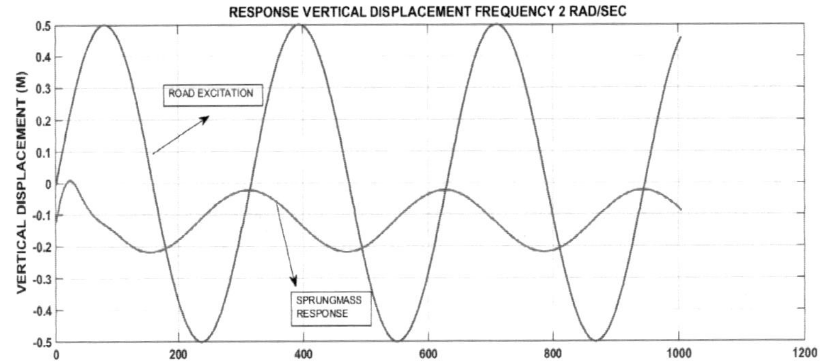

Figure 5. Damper Ratio Response by Frequency 2 rad/sec

In the third experiment, it was tried to increase the frequency geometry and vehicle speed which produced 4 rad / sec with a performance damping ratio of 2 to 3 as shown in the figure 6 while by using PID reach 2 only. Mean it can isolate the sprung mass from road vibration by reducing half to one third compared road amplitude

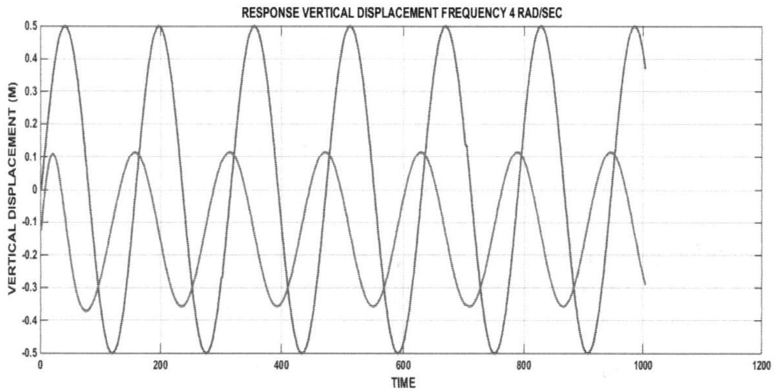

Figure 6. Damper Ratio Response by Frequency 4 rad/sec

5. Conclusion

From the results of the discussion in the chapter above, it can be concluded as follow:

1) In average the performance LQR control suspension in research were better than compare by using PID control
2) Incorporating the longitudinal brake force excitation into a vertical excitation superposition to the vertical force of the road geometry is very significant

6. Reference

[1] S.F. Youness, E.C. Lobusov, 2018, Networked Control for Active Suspension System, 13th International Symposium “Intelligent Systems” (INTELS’18)

[2] Prabu Krishnasamy – Jancirani Jayaraj – Dennie John, 2013, Strojniški vestnik ,2013, Experimental Investigation on Road Vehicle Active SuspensionJournal, of Mechanical Engineering 59(2013)10, 620-625 © 2013 Journal of Mechanical Engineering. All rights reserved. DOI:10.5545/sv-jme.2012.925

[3] Narinder Singh, Himanshu Chhabra and Karansher Bhangal, 2016, Robust Control of Vehicle Active Suspension System, International Journal of Control and Automation Vol. 9, No. 4 (2016), pp. 149-160 http://dx.doi.org/10.14257/ijca.2016.9.4.15

[4] Zhi-Jun Fu, Bin Li, Xiao-Bin Ning, and Wei-Dong Xie, Online Adaptive Optimal Control of Vehicle Active Suspension Systems Using Single-Network Approximate Dynamic Programming, Hindawi Mathematical Problems in Engineering Volume 2017, Article ID 4575926, 9 pages https://doi.org/10.1155/2017/4575926

[5] Michiel Haemers, Stijn Derammelaere, Clara-Mihaela Ionescu, Kurt Stockman, Jasper De Viaene, Florian Verbelen, (2018), Proportional-Integral State-Feedback Controller Optimization for a Full-Car Active Suspension Setup using a Genetic Algorithm, Preprints of the 3rd IFAC Conference on Advances in Proportional- Integral-Derivative Control, Ghent, Belgium, May 9-11, 2018

[6] Rosmazi Rosli, Musa Mailah, Gigih Priyandoko, 2014, Active Suspension System For Passenger Vehicle Using Active Force Control With Iterative Learning Algorithm, Wseas Transactions On Systems And Control

[7] Thanh-Phong Pham, Olivier Sename , Luc Dugard, 2015, Design and Experimental Validation of an H1 Observer for Vehicle Damper Force Estimation, Univ. Grenoble Alpes, CNRS, Grenoble INP>, GIPSA-lab, 38000 Grenoble, France. >Institute of Engineering Univ. Grenoble Alpes (e-mail: thanh-phong. pham2, olivier.sename, luc.dugard@ gipsa-lab.grenoble-inp.fr).

[8] Qinghua Meng, Chih-Chiang Chen, Pan Wang3 Zong -Yao Sun4 Bingji Li, 2018, Study on vehicle active suspension system control method based on homogeneous domination approach, wileyonlinelibrary.com/journal/asjc, 2019 Chinese Automatic Control Society and John Wiley & Sons Australia, Ltd.

[9] Ammar A. Aldair and William W.J. Wang,2018, Evolutionary Algorithm Based Fractional Order PID + μ Controller Design for a Full Vehicle Nonlinear Active Suspension System, AVEC 10

[10] Abdolvahab Agharkakli, Ghobad Shafiei Sabet, Armin Barouz,2012, Simulation and Analysis of Passive and Active Suspension System Using Quarter Car Model for Different Road Profile, International Journal of Engineering Trends and Technology- Volume3Issue5- 2012

[11] Tongchit Suthisripok, Chanwit Wongrattanapornkul Somchai Poonyaniran, And Adirak Kanchanaharuthai, 2018, Disturbance Observer Based Control for Active, Suspension Systems, International Journal of Innovative Computing, Information and Control ICIC International c 2018 ISSN 1349-4198 Volume 14, Number 6, December 2018

[12] Sorin Marcu, Dinel Popa, Nicolae–Doru Stănesc, Nicolae Pandrea,2017, Model for the study of active suspensions, IOP Conference Series: Materials Science and EngineeringIOP Conf. Series: Materials Science and Engineering 252 (2017) 012032 doi:10.1088/1757-899X/252/1/012032

[13] Ahmed, M. I., Hazlina, M. Y., and Rashid, M. M.,2014, Mathematical Modeling and Control of Active Suspension System for a Quarter Car Railway, 2016, Special Issue: The 3rd International Conference on Mathematical Applications in Engineering 2014 (ICMAE'14), Journal homepage: http://einspem.upm.edu.my/journal

[14] Aizuddin Fahmi Mohd Riduan, Noreffendy Tamaldin, Ajat Sudrajat, and Fauzi, 2018, Review on active suspension system, SHS Web of Conferences 49, 02008 (2018) https://doi.org/10.1051/shsconf/20184902008 ICES 2018

[15] Nouby M. Ghazaly, Ahmad O. Moaaz, 2014, The Future Development and Analysis of Vehicle Active Suspension System, IOSR Journal of Mechanical and Civil Engineering (IOSR-JMCE) e-ISSN: 2278-1684,p-ISSN: 2320-334X, Volume 11, Issue 5 Ver. V (Sep- Oct. 2014), PP 19-25 www.iosrjournals.org

[16] Pavan Kumar.G, 2015, Semi-Active Suspension for an Automobile, International Journal of Engineering Research & Technology (IJERT) ISSN: 2278-0181 IJERTV4IS070431 www.ijert.org (This work is licensed under a Creative Commons Attribution 4.0 International License.) Vol. 4 Issue 07, July-2015

[17] M.Harly, IN sutantra, 2012, Hybrid selforganizing map and locally recurrent neural network-based adaptive back through for improving integrated vehicle stability control, International Journal of Modeling and Simulation, https://www.tandfonline.com/doi/abs/10.2316/Journal.205.2012.2.205-5301

[18] M. Harly, IN sutantra, 2009, New Method of Neuron Design Based on Discrete Z-Function To Adapt The Change Of Integrated Vehicle Stability Control Order, International Journal of Computational Intelligence And Applications, https://www.worldscientific.com/doi/10.1142/S146902680900259X

[19] M.Harly, IN sutantra, 2012, Vehicle integrated stability control using Hybrid Fuzzy C-Mean Clustering- Adaptive Back Propagation scheme, IASTED Austria

[20] Pinandhito, M.R., Indriawati, K., Harly, M., 2019, Active fault tolerant control design in regenerative antilock braking system of electric vehicle with sensor fault, AIP Conference https://doi.org/10.1063/1.5095276

[21] Diarnise, S., Ya'umar, Indriawati, K., Harly, M., 2019, Design active fault tolerant control System for regenerative anti-lock braking system of electric vehicle with fault at actuator, AIP Conference https://doi.org/10.1063/1.509

IOP Conf. Series: Materials Science and Engineering 1152 (2021) 012024 doi:10.1088/1757-899X/1152/1/012024

Integrated Artificial Intelligent with RS and GIS for Long-Term Drought Prediction

Bashar Muneer Yahya[1*], Dursun Zafer Seker[2] and Basman Younus Hameed[1]

[1]Remote Sensing Center, Mosul University, Mosul, Iraq.
[2] Istanbul technical university, Civil Engineering Faculty, Istanbul, Turkey.

[*]E-mail: bashar1974@uomosul.edu.iq

Abstract. The accurate evaluation of drought plays an urgent role in the optimal solution for future development. The present research aimed to predict and assess drought in Nineveh governorate northwestern Iraq by integrated a fuzzy logic procedure with Geographical Information System (GIS) environment. Primary meteorological data for six stations were used in this study. Vegetation cover changes were detected using remote sensing analysis in which there was a decrease in the vegetation cover's area by 17.7% from 592.3 km^2 in the year 1992 to 487.46 km^2 in the year 2016. Standardized Precipitation Index (SPI) as a proxy for drought conditions was simulated by developing the model of Adaptive Neuro-Fuzzy Inference System (ANFIS) as an artificial intelligent approach where its performance reached to 82% when evaluating the Nash-Sutcliffe coefficient and Root Mean Square Error (RMSE)as performance criteria. However, the long-term meteorological drought was an assessment where the results show that the drought level will intensify in a future by 10.1% with more successive droughts levels. Those results were clarified through developing spatial distribution map for long-term drought simulation by employing the (GIS) as a database to develop future plans for the study area.

1. Introduction

All life components are negatively affected via a natural phenomenon referred to as drought. Yet, such phenomenon can be monitored easily, in which proactive solutions might be set for reducing its harms when identifying its predicted values for future. Many life sectors like environment, cultural, economy etc. are affected by drought factors that have a harsh impact on these sectors in several respects, especially if its effects last for a long period [1,2]. Drought can be characterized through its duration, severity, and spatial impact on a specific area, yet due to the various climate characteristics, there is a difficulty in determining when drought will start and where or when it will adversely affect a certain area [3]. In Iraq, the drought's impact has lately become evident via many observations, such as the reduction in agricultural land and vegetation cover which in the past produced various economic crops [4]. Previous reports and researches indicating that all the parts of Iraq are suffering from drought, reaching advanced phases in certain parts. These studies are supported by the International Organization for Migration [5] which reported that drought was the main reason for the migration of 4263 families consisting of 25,578 individuals between 2007 and

Content from this work may be used under the terms of the Creative Commons Attribution 3.0 licence. Any further distribution of this work must maintain attribution to the author(s) and the title of the work, journal citation and DOI.
Published under licence by IOP Publishing Ltd

2009 in many regions of Iraq like Salah Adin and the Nineveh governorates. The development of applications for solving problems associated with forecasting climate variables or drought indices is usually complicated as a result of many variables that controlled in one way or another for the issue under consideration. Artificial Intelligence (AI) might be specified as anapplication in computer science simulating the human's capability to make logical decisions [6, 7]. AI help in solving difficult problems facing humans with cost-efficiency and reliability. Various tasks might be provided by AI such as speech recognition, decision-making, predict variables, images recognition, and so on. AI includes six main procedures, Evolutionary Computation (EC), Machine Learning (ML), Computational creativity, Fuzzy systems, Chaos theory and Probabilistic methods [8, 9, 10].

Based on fuzzy logic methods that developed by [11]. An ANFIS model was developed for simulating the long- term drought changes in the study area. It is well known that both fuzzy logic and artificial neural networks as two artificial intelligent approaches have advantages and disadvantages recognized by many applied studies according to many abilities [12, 13, 14, 15]. Artificial Neural Networks (ANNs) were developed as mathematical models simulating the nervous system in human mind for the purpose of resolving real world issues requiring complicated calculations [16, 17]. The output of the (ANNs) is a combination of input variables and fuzzy roles that already designed according to experience [18, 19]. There are various techniques of Remote Sensing (RS) used for detecting the vegetation changes (for instance, remote sensing indices, supervised classification, image differencing, correlation or rations [20] where these methods proved their efficiency and accuracy throughout the obtained results in many vegetation changes situations like desertification, fire and vegetation diseases. The Normalized Difference Vegetation Index (NDVI) can be defined as a quantified vegetation remote-sensing index used for measuring the levels of the difference between near-infrared and red spectrum observed or reflected from vegetation cover. Mostly, it is utilized for discriminating vegetated areas from non-vegetated ones, also it might be utilized for evaluating vegetation healthy status [21] this index was utilized originally via [22].

Standardized Precipitation Index(SPI) is the majorly utilized drought indices for estimating drought. In the literature reviews, there are many summaries of drought analysis using several drought indices and statistical methodologies as in the studies of [23] where drought was analyzed according to its magnitude and level in different areas using different procedures. When precipitation is lower than or fluctuates over the average value in any region, it has negative impacts on soil moisture, groundwater, and natural reservoir storage. This led Mckee, Doesken and Kleist to realize the need for a simple and flexible index that shows how low precipitation impacts the soil [24]. SPI is used in this study to assess drought in the study area because of its simplicity, its wide range of drought classification, its effective indication of soil moisture, its effectiveness in the predicting of short-term drought, and its spatial invariance in interpretation.For spatial management and data manipulation Graphic Information Systems (GIS) environmentis increasingly being regarded as a useful tool for many application like drought detection and monitoring as an example [25]. Today, GIS was used along with various up-to-date technologies such as AI and RS for various drought researches in which GIS has the ability for translating the results related to such up-to-date technologies into identified formulas enable classifying the drought as well as its extension over time in various forms which might be grasped easily [26, 27].

The present research has three main objectives:

1- Detect the effects of climate change and expansion drought area in the study area through vegetation cover analysis using remote sensing analysis and drought analysis using SPI index.

2- Develop ANFIS model based on four meteorological data as (temperature, wind speed, evaporation and rainfall) as input data that used to predict SPI values.

3- The average mean of the predicted SPI that's gained from the ANFIS model was calculated and the final predicted SPI was used for assessing the meteorological drought. These results were integrated inside the GIS environment as separate layers to create spatial distribution maps for drought assessments.

2. Methods and Materials

2.1 Study area

Nineveh is one of the Iraqi provinces that located in northwestern of Iraq between longitude of (41° 30' – 44° 30') and latitude of (35° 00' – 37° 00') and has a dry and semi-arid climate. Figure 1, shows that the province has 37323 km^2 as total area, representing approximately 8.6% of Iraq's total area. There are two considerations for monitoring, prediction, and assessment drought in the study area. Firstly, is the economic importance of this province for grain crops production such as wheat and barley, so protecting this area is an economic priority. Secondly are the recent environmental changes represented by sand and dust storms, the decline of vegetation cover, the fluctuation of rainfall, and the rise in temperature. These changes have negatively affected the population of the province environmentally and psychologically as stated in many local and international reports.

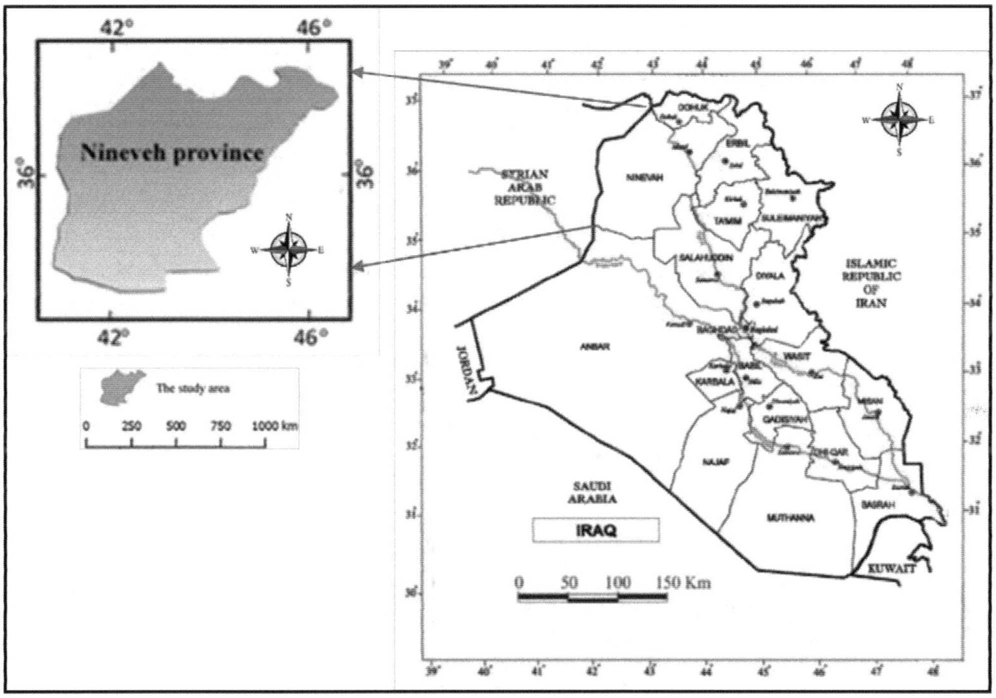

Figure 1. Genearal view of the study area.

2.2 Materials

There are a total of six meteorological stations in the study area (Tel-Afar, Mosul, Hadher, Ba shiqah, Sinjar and Rabiaa) as illustrated in figure 2, showing a total of 4 weather variables records (temperature, rainfall, evaporation and wind speed) belonging to such meteorological stations between (1970 and 2017) have been utilized in this work as 540 samples regarding each one of the variables [28]. For obtaining realistic results when using Fuzzy logic as analysis procedure continues time series must be used to training and testing the designed artificial intelligence models [15, 29]. For remote sensing analysis including supervised classification as well as NDVI methods, multi-date satellite imagery was used, these imageries belong to

Landsat 7 satellite with path 170, row 35 captured on April 25th 1992 and May 16th 2016 captured via enhanced thematic mapper (ETM+) sensor onboard Landsat-7 satellite.

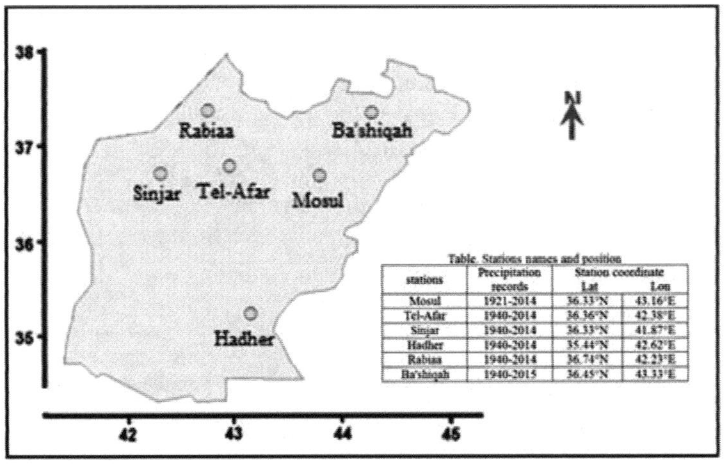

Table. Stations names and position

stations	Precipitation records	Station coordinate Lat	Lon
Mosul	1921-2014	36.33°N	43.16°E
Tel-Afar	1940-2014	36.36°N	42.38°E
Sinjar	1940-2014	36.33°N	41.87°E
Hadher	1940-2014	35.44°N	42.62°E
Rabiaa	1940-2014	36.74°N	42.23°E
Ba'shiqah	1940-2015	36.45°N	43.33°E

Figure 2. Meteorological stations distribution in the study area.

2.3 Methods

The Fuzzy Logic or multi-valued logic system was developed for simulating the thinking of the human nature regarding a problem under-study, in which the answer is not essentially no or yes (classic logic), the answer might go to another phase such as uncertain, vague and indecisive [30]. A fuzzy inference system includes several sections that are represent the structures of these system shown in Figure 3.

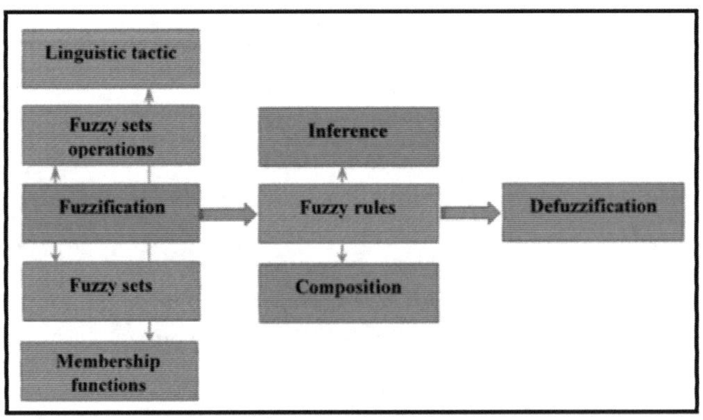

Figure 3. Categories of the Fuzzy Logic Control system.

Fuzzy sets Membership Functions (MF) can be used in a huge vary of domains in which facts are ambiguous, indistinct or uncertain such as in bioinformatics. The best approach for the MF to be defined is mathematically expressing it and clearly specifying its parameters. As indicated in the example submitted via [31,32]. Equation (1) is applied to carry the requirements of designed the Fuzzy logic system and via utilizing max and min rules, the alternative expression with regard to equation (1) might be stated as shown in the equation (2).

$$MF(x; a, b, c, d) = \begin{cases} 0 \, x \leq a \\ \frac{x-a}{m-a} \; a < x \leq m \\ 0.3 \; m \leq x \leq n \\ \frac{x-n}{b-n} n \leq x \leq b \\ 0.8 \; b \leq x \leq c \\ \frac{d-x}{d-c} \; c \leq x \leq d \end{cases} \tag{1}$$

$$MF(x; a, b, c) = \max\left(\min\left(\frac{x-a}{b-a}, \frac{c-x}{c-b}, 0.3, \frac{x-n}{b-n}, 0.8, \frac{d-x}{d-c}\right), 0\right) \tag{2}$$

Where, a is the lowest possible value, m, n, b and c are the middle possible value and d is the max possible value in the judgement interval. If the lowest possible value= middle possible value= max possible value then the fuzzy numbers functions get a crisp number.

The model of NFIS is specified via ANN and Fuzzy Logic model consists of Non-linear Auto-regressive Network with Exogenous inputs (NARX) created for predicting SPI based on the predicting weather the variable records with regard to 6 meteorological stations. Also, NARX is one of the significant procedures to predict and model the time series tacking into consideration, the non-linearity and complexity of time series patterns as indicated in the equation 4.

$$y(k+1)=f_{ANN}=(y(k), y(k-1),\ldots y(k-n+1),u(k),u(k-1),\ldots u(k-m+1)+\epsilon(k) \tag{3}$$

In which, y(k+1) represents the projected output of the model, fANN represents the nonlinear function indicating the behaviour of the system, y(k), u(k) and ε(k) were output, input as well as the vectors of the approximation error at k, n and m time instances the order of u(k) & y(k). ANFIS can be defined as a combination of NN and Fuzzy Logic created via [8]. Majorly, it was specified as a representation of the graphical network regarding the Sugeno-type fuzzy systems which are connected with capabilities of the neural learning. In addition, the network consists of nodes with certain functions collected in layers. Furthermore, the ANFIS has the ability of constructing a network realization of IF / THEN rules which might be indicated by [12] as can be seen in 'Figure 4' showing the develop ANFIS's flow chart.

NDVI is used for finding vegetation cover changes in the area of the study. It might be mathematically represented via equation (4) in the following way:

$$\text{NDVI} = (\text{NIR} - \text{RED}) / (\text{NIR} + \text{RED}) \tag{4}$$

In which,NIR represents the reflection in near infrared spectrum and RED represents the reflection in red range of spectrum. In this formula, the results are generating a value that is in the range from -1 to +1, in which values between -1 and 0 are representing the non-vegetated regions like the urban land or abandoned land and, while the values more than 0 to +1 are representing the current vegetation in area under-study. Put differently, high yield of the NDVI value in the case where the high reflectance in NIR channel and the low reflectance in red channel and vice versa have been identified.

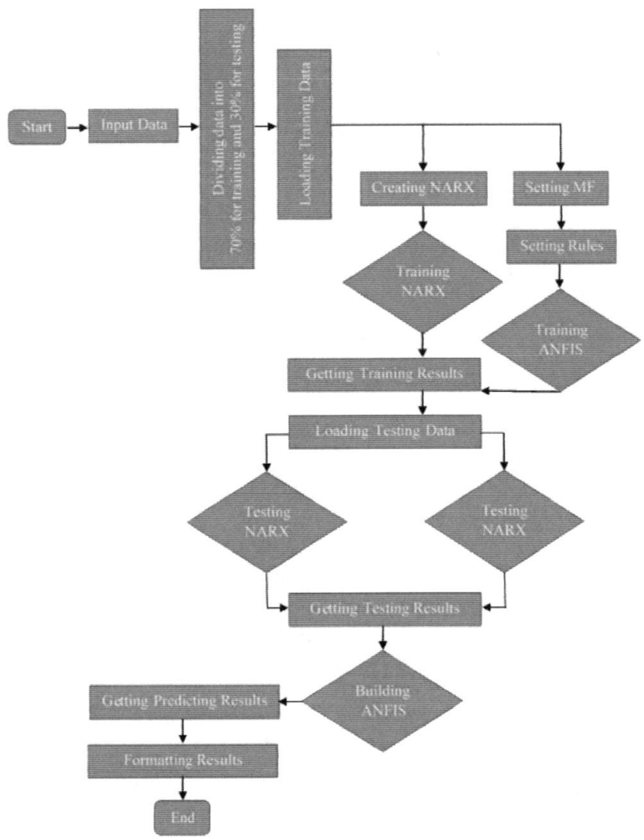

Figure 4. Developed flow-chart structure of ANFIS.

The use of the NDVI is significant, particularly in three sectors, first in evaluating, differentiating and demonstrating the global or local vegetation cover specified via green bio-mass of farming, forests along with the rangeland vegetation cover. Second, the NDVI might be utilized to indicate the drought offering an early warning process. Thirdly, the NDVI might be utilized to indicate the earth surface moisture. The image's continuous grey tone is converted via the density slicing into density interval series or each corresponding to certain digital ranges. The method of density slicing might be utilized for showing the areas bounded via polygons. The approach is excellent in separating the gray levels which might be little to viewer.

SPI is acting as one of the drought monitoring indicators with regard to early warning, which allow assessing the severity and extent of the drought. In SPI, the major statistical analysis is based on how the data of the time series have been converted to normal distributions by means of Gamma Distribution Probability Function where the arithmetic average value equals 0 [33]. In such index, the positive values are indicating that there has been an elevation in the precipitations from yearly average values, which has been specified as wet period. The differences between positive values as well as yearly average related to precipitation is representing a precipitation surplus, whereas the negative values indicating the fact that there were deficits from precipitation's yearly average, which has been specified as period of the drought. Also, deficiency is representing differences between negative values as well as yearly average regarding the precipitations in investigated year. A study conducted by [24] developed rules used to classify the levels of the dryness, in which the values ≤-1 are indicating the conditions of the drought, while the values ≥1 are indicating the conditions of the wetness.

3. Results

For the purpose of achieving the aim of the present research, as well as matching the goals as it has been stated in the section of the introduction, results that have been produced from the variety of the utilized approaches in the present research have been given below:

3.1. Remote Sensing Analysis

In this section, the levels of the NDVI have been calculated, where it has been discovered that the levels of the vegetation have been in the range between -0.548 and 0.280and between -0.10 and 0.642 for the images (16-5-2016) and (25-4-1992), respectively as has been illustrated in Figure 5.

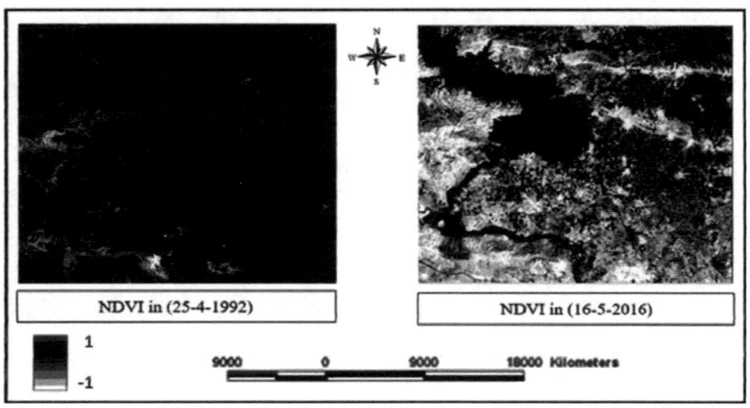

Figure 5. The analysis of the NDVI for part of the area of the study.

Images that have been resulted from the application of this approach were at the grey levels, which makes it infeasible for human eyes to be capable of identifying those levels and easily separating them for the purpose of making the result of the NDVI clearer and for the estimation of the vegetation changes 'impact, the approach of the density slicing has been utilized for the classification of the level of the NDVI range, that has been split to three classes based upon the available vegetation in the area of the study. Those levels are a non-vegetation area in the case where the value of the NDVI is less than zero, area of low vegetation in the case where the value of the NDVI has been in the range from (0 to 0.20), and area of high vegetation in the case where the value of the NDVI is higher than 0.20. The results of the classification have shown that the class of the low-density vegetation cover has reduced by a rate of 20 % and the class of the high-density vegetation cover has been decreased by a rate of 80 %, which is why, the land percentage that has been categorized under non-vegetation class has been increased as shown in Figure 6. The NDVI digital analysis for between 1992 and 2016 has shown four classes (which are: non-vegetation cover, water, high-density vegetation cover and low-density vegetation cover) where vegetation cover class's area has been reduced by 17.70 % from 592.3 km^2 in 1992 to 487.46 km^2 in 2016. The rate of the area that has been transferred to the non-vegetation cover (i.e., the area of the drought) category which has increased from 251.10 km^2 in 1992 to 355.94 km^2 in 2016. This percentage of decrease is expected to accelerate as a result of the effects of the phenomenon of the desertification in the area, besides other reasons, which include the approaches of the uncontrolled overgrazing, which is utilized by the breeders of the livestock as well as many inhabitants that have turned away from the agricultural life and migrated to cities.

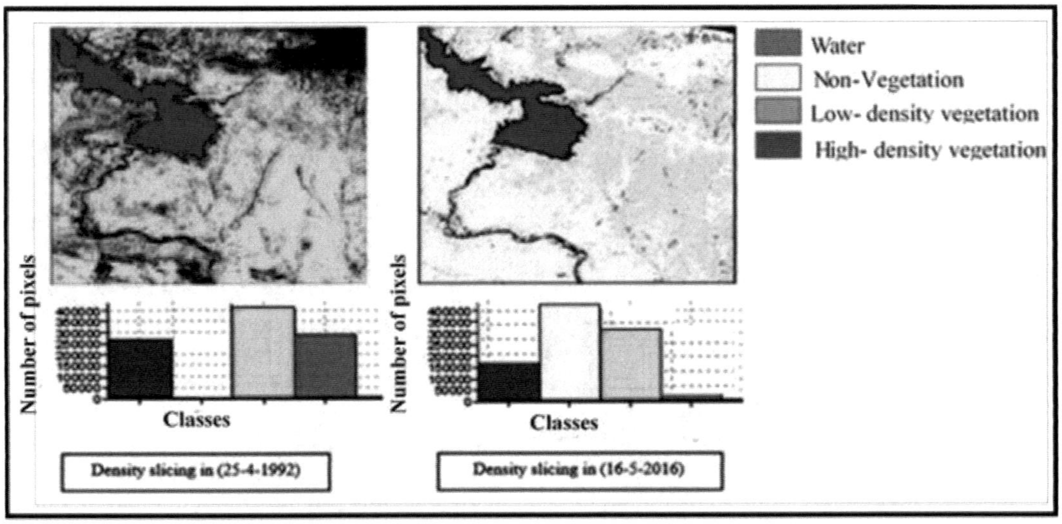

Figure 6. The maps of the NDVI density slicing.

3.2. Develop the ANFIS Model

The initial step in the development of a fuzzy model has been identified by the definition of linguistic variables with a function of the membership (i.e., the phase of fuzzification). The benefits of linguistic variables are quantitatively and qualitatively describing the fuzzy sets for the purpose of making them beneficial and giving the understandable meaning to the experts and for the computer processors to obtain correct values for outputs. The data-set of the inputs based on the fuzzy sets for the speed of the wind, evaporation and temperature variables have been categorized as high, moderate and low, whereas the rainfalls have been classified into very high, high, moderate, semi-moderate, low, and very low. The Membership Functions (MF) have been characterized for every one of the input and output variables. Those functions have to be designed for the purpose of verifying the main condition in which the designed Membership Functions have to be overlapping with the nearest Membership Function as can be seen from Figure 7.

In order to test the capability of the generalization of ANFIS at every one of the epochs, the command genfis-1 function has been utilized. A procedure of validation has been performed for the overfitting of data that has been utilized for the training of the ANFIS model, for every training epoch. Following 52 batch learning epochs, the developed model of the ANFIS has shown sufficient validity and stability. In order to test ANFIS's capability of generalization at every one of the epochs, the function of validation data has been utilized. A process of validation has been performed for the purpose of the overfitting of data that has been utilized for the training of the ANFIS model, for every training epoch. Following 52 batch learning epochs, the developed model of the ANFIS has shown sufficient validity and stability. For the purpose of ascertaining ANFIS's efficiency performance, the Determination Coefficient R^2=0.905, Nash-Sutcliffe coefficient CE=0.830, Root Mean Square Error RMSE=0.935, and Mean Absolute Percentage MAPE=17.390 % where the accuracy of the performance has been 82.61 %, and that made this system sufficient and reliable for the study area's SPI prediction. MAPE value has been compared statistically against the actual and the predicted values of the SPI-12.

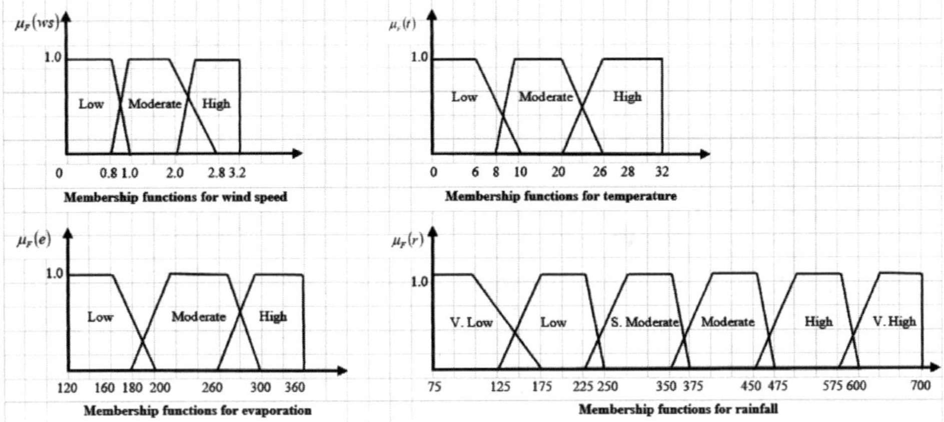

Figure 7. Designed MFs for the inputs dataset.

The result has shown a sufficient indication of reliability of advanced ANFIS. Such results may be of a higher effectiveness in the case where the MAPE has been decreased through the engagement of other meteorological factors, having a strong association with the data of the actual SPI or through the addition of more training data to a model (prediction of the values of the SPI-12 for a while prior to 1970). Following the testing and the running of the model, a considerable convergence has been exhibited between the values of the SPI-12 that have been predicted by the model and the actual values of the SPI. The average error of the prediction has been found 27.78 % as can be seen from Figure 8.

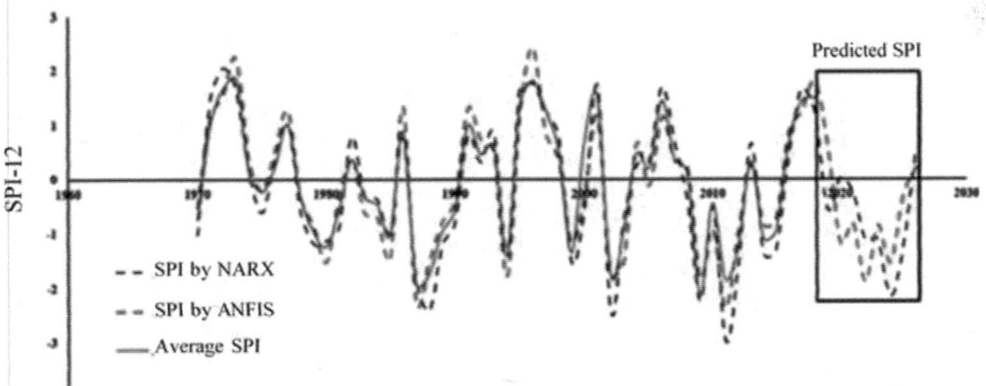

Figure 8. Comparison of SPI-12 values by NARX and ANFIS.

3.3. Drought Analysis and assessment

The drought was measured in 3 periods between 1970 and 1992, 1992 to 2017 and for a predicted period between 2017 and 2026. The average yearly rainfalls for the area of the study have been 512.60 mm. The results of the drought evaluation between 1970 and 1992, and between 1992 and 2017 has shown that the area of the study has been influenced by different drought where the rainfall deficit amount has been in the ranged between 27 % and 34 %. The calculated drought duration (DD), drought magnitude (DM), average drought intensity (ADI), and years of the drought values in two periods have been computed as can be seen from table1, where the years of the drought have been increased by 1.8 % which has caused an increase in the DM by 106.5 mm throughout the two studied time periods. Such analyses had confirmed the analysis results of remote sensing about the vegetation declines and the increase in the land of the desertification in the area of the study. After the records of the predicted rainfalls between 1970 and 2026, which have been obtained from the model of the ANFIS have been involved in this comparison, and based on table 2 the maximum drought magnitude value (i.e., 4655 mm) had occurred in the station of Hadher with a148.62 mm/year ADI and a 1.90 per year DD.

Table1. Assessment of the drought in two studied periods.

Drought characteristics	Analysis period (1970-1992)	Analysis period (1992-2017)
Drought years %	46.5	48.3
DM (mm)	3433	3539.5
DD (year)	2.3	2.4
ADI (mm/year)	95.74	99.67
Increased drought magnitude (mm)		106.5

Table 2. Actual and predicted rainfall records comparison for drought analysis.

	Actual rainfall records				Predicted rainfall records				
Stations	Drought years %	DM mm	DD year	ADI mm/year	Drought years %	DM mm	DD year	ADI mm/year	Increased drought magnitude mm
Mosul	46.5	3433	2.3	95.74	48.3	3539.5	2.4	99.67	106.5
Sinjar	52.7	3134	3.8	128.36	53.9	3368.3	3.9	137.94	234.3
Tel-Afar	54.0	3083	3.2	131.21	56.3	3271.7	3.6	139.24	188.7
Rabiaa	51.4	2709	2.9	101.63	52.8	2911.8	2.4	109.24	202.8
Bashiqah	48.7	2007	2.7	85.23	49.4	2136.7	2.9	90.47	129.7
Hadher	58.8	4655	1.9	148.62	61.5	5051.5	2.0	161.28	396.5
Total DM 19021 mm					Total DM 20279.5 mm				1258.5

4. Discussion

The drought years' percentage has been increased by 10.10 % with the increase in the magnitude of the drought by 1258.5 mm throughout the predicted nine years up to 2026 in the entire area of the study. In the case where that percentage keeps increasing each nine years, the area of the study will be suffering from an exceptional drought that will definitely result in affecting the environment in general and the vegetation cover and agricultural structure in particular. In addition to that, the predicted results' analysis has shown that the levels of the drought in the western and the southern regions of the area of the study (i.e. Tel-Afar, Hadher and Sinjar) have been increased throughout the predicted period (between 2017 and 2026) by 6.20 % as the year of the drought with an increase in the magnitude of the drought by 819.5 mm, in comparison with the eastern and the northern parts (i.e. Rabiaa, Baashiqah and Mosul) where the year of the drought is 3.90 % with expected increase in the magnitude of the drought that has been 439 mm. All of the conditions

IOP Conf. Series: Materials Science and Engineering 1152 (2021) 012024 doi:10.1088/1757-899X/1152/1/012024

of the drought have been observed that the whole area of the study has been influenced by a variety of the levels of drought, where average wet years and average drought years have been estimated as 45.60 % and 54.40 % respectively. In addition to that, the analysis of the drought has shown that the conditions of the severe drought ranged between -1.521 and -1.862, moderate drought range between -1.053 and -1.495, and the extreme drought ranged between -2.107 and -2.865. For the long-term assessment of the drought, the SPI-12 values have been directly associated with the state of stream flows, ground-water levels and reservoirs levels. Those values may be considered as an explanation of environmental changes that result from the drought and the desertification which may take place in a region throughout a long time. After the analysis of SPI12 for the area of the study, utilizing the predicted and the actual SPIs, it has been discovered that, in general, the area will experience conditions of extreme drought except for the region of Sinjar, where it is predicted that the condition of the drought will be severely dry as has been listed in table 3.

Table3. Classification of the predicted SPI-12 for the period between 2017 and 2026.

Station	SPI-12 classification					
	Moderate drought	Year	Severe drought	Year	Extreme drought	Year
Mosul	-1.158	2018-2019	-1.862	2015-2016	-2.267	2020-2026
Sinjar	-1.48 -1.073	2015-2016 2018-2019	-1.685	2020-2021		
Tel-Afar	-1.319	2018-2019	-1.540	2014-2015	-2.82	2020-2026
Hadher	-1.192 -1.158	2015-2016 2017-2018			-2.267	2020-2026
Rabiaa					-2.107 -2.865	2017-2018 2020-2026
Bashiqah	-1.373	2017-2018	-1.521	2014-2015	-2.380	2020-2026

GIS was used to prepare a long-term spatial distribution map based upon the actual and the predicted SPI-12 as shown in Figure 9. This distribution map was addressed based on the levels of the drought (severe, moderate and extreme which have possibly taken place in the area of the study in predicted period between 2017 and 2026). As a result of drought years' increase and the decrease in mean annual rain-fall (512.6 mm to 396.7 mm), a condition of the moderate drought has been noticed, which had an impact on the whole area of the study, which ranged between (-1.053 and -1.495) as can be seen from Figure 9a. The severe drought will have an impact on the area of the study as well, where it ranges between (-1.521 and -1.862) as can be seen from Figure 9b. The eastern part will be adversely influenced by the condition of the extreme drought that ranges between (-2.107 and -2.865). The area of the study is part of Tigris Riverfeeding area in Iraq over several tributaries, which indicates lower levels of the river water in future; which is why, the system of the ground-water recharging id going to be damaged and impacted adversely on ground-water wells' supply that have been scattered in the area of the study as can be seen from Figure 9c.

5. Conclusions

There are numerous advantages to the assessment of a region's drought conditions with the use of SPI, due to the fact that it only requires one input type in a form of the rain-fall records and simple mathematical formulas. The index of the SPI has been identified with its capability in the classification of the droughts into extreme, severe, or moderate for the purpose of giving a more explicit indication of conditions of any of the areas in the terms of drought effects. In addition to that, there is a possibility in calculating the characteristics of the drought through the calculation of its intensity, duration, average and magnitude. The drought conditions were studied in the study area during three periods. In the first and second period, 1970 to 1992, 1992 through 2017, the rainfall records of six meteorological stations in the study area were used for assessment the drought conditions while the third period included the extension of the rainfall records up to the year 2026 by a designed artificial intelligence model by ANFIS for drought assessment.

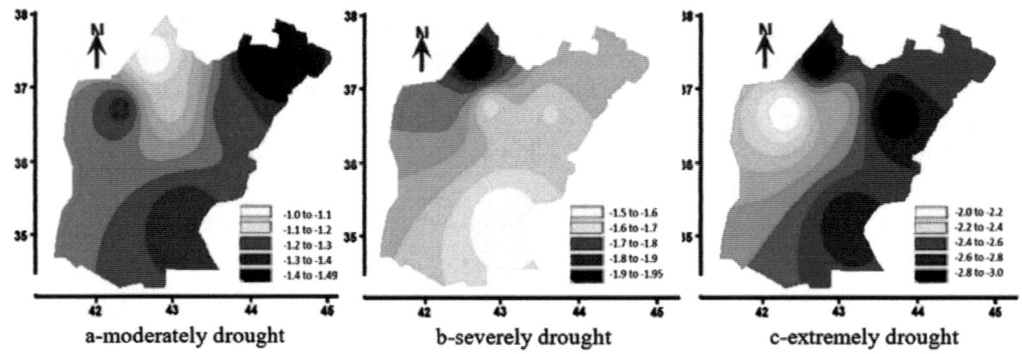

Figure 9. Long term SPI-12 spatial distribution in the area of the study.

The use of this model in predicting future values of short-term SPI-3 and long-term SPI-12 can be useful for the monitoring and analysis of all future cases of drought scenarios and can assist in the management and evaluation of agricultural and environmental projects in the study area. The effects of such drought are starting to appear through the decline of natural vegetation cover and encroaching dunes, particularly in the western and the southern regions of the area of study. To prevent the dangers of drought in the study area, effective solutions and rapid plans must be implemented to reduce the growth of this phenomenon. Such plans may include exploitation of precipitation in the rainy years and harvesting it with one of the rainwater harvesting techniques according to the geological and geomorphological nature of the region. Irrigations systems should be developed in the region and decrease the irrigation losses to the lowest possible extent. The remaining natural vegetation should be preserved, and natural drought-tolerant plants should be cultivated to stabilize the soil and prevent its erosion and thus turning into sand storms. All of these actions together can contribute to reducing the drought phenomenon and its impact on the study area.

6. References

[1] Alam A, Rahman M and Saadat A 2013 Monitoring meteorological and agricultural drought dynamics in Barind region Bangladesh using standard precipitation index and Markov chain model. International Journal of Geomatics and *Geosciences*, **3**(3), pp.511-524.
[2] Al-Bakri J, Al-Khreisat A, Shawash S, Qaryouti E and Saba M 2018 Assessment of Remote Sensing Indices for Drought Monitoring in Jordan. Asian Journal of Geoinformatics, **17**(3).
[3] Morid S, Smakhtin V and Bagherzadeh K 2007 Drought forecasting using artificial neural networks and time series of drought indices. *International Journal of Climatology*, **27**(15), pp.2103-2111.
[4] UNEP, How Environmental Damage Causes Food Insecurity in IRAQ. World Environmental Day, Technical Assessment Report, Accessed June 2013. http://www.uniraq.org/index.php?option=com_k2&view=item&task.
[5] Al-Faraj F, Scholz M and Tigkas D 2014 Sensitivity of surface runoff to drought and climate change: Application for shared river basins. *Water*, **6**(10), pp.3033-3048.

[6] Walczak, S. (2019). Artificial neural networks. In Advanced Methodologies and Technologies in Artificial Intelligence, Computer Simulation, and Human-Computer Interaction (pp. 40-53). IGI Global.

[7] HungQ, Babel S, Weesakul S and Tripathi K2009. An artificial neural network model for rainfall forecasting in Bangkok, Thailand. *Hydrology and Earth System Sciences*, **13**(8), pp.1413-1425.

[8] Traore S, Luo Yand Fipps G2016 Deployment of artificial neural network for short-term forecasting of evapotranspiration using public weather forecast restricted messages. *Agricultural Water Management*, **163**, 363-379.

[9] Hsu K, Gupta V, and Sorooshian S 1995 Artificial neural network modeling of the rainfall-runoff process. *Water resources research*, **31**(10), 2517-2530.

[10] Ghimire S, Deo R, Downs Nand Raj N2019 Global solar radiation prediction by ANN integrated with European Centre for medium range weather forecast fields in solar rich cities of Queensland Australia. *Journal of cleaner production*, **216**, 288-310.

[11] Zadeh, L. A. (1965). Fuzzy sets. Information and Control 8:3. doi:10.1016/S0019-9958(65) 90241-X.

[12] Jang S, 1993 ANFIS: adaptive-network-based fuzzy inference system. *IEEE transactions on systems, man, and cybernetics*, **23**(3), pp.665-685.

[13] Nayak C, Sudheer P, Rangan M and Ramasastri K 2004 A neuro-fuzzy computing technique for modeling hydrological time series. *Journal of Hydrology*, **291**(1-2), pp.52-66.

[14] Yousif, J. H., Al-Balushi, H. A., Kazem, H. A., &Chaichan, M. T. (2019). Analysis and forecasting of weather conditions in Oman for renewable energy applications. Case Studies in Thermal Engineering, 13, 100355.

[15] Yahya B and Seker D 2019 Designing weather forecasting model using computational intelligence tools. *Applied Artificial Intelligence*, **33**(2), 137-151.

[16] Jain A, Mao J and Mohiuddin K 1996 Artificial neural networks: A tutorial. *Computer*, **29**(3), pp.31-44.

[17] Basheer I and Hajmeer M 2000 Artificial neural networks: fundamentals, computing, design, and application. *Journal of microbiological methods*, **43**(1), pp.3-31.

[18] Lara L, Rasche L and Schneider U 2017Modeling land suitability for Coffeaarabica L. in Central America Environ Model. *Software*,**95**, 196 – 209.

[19] Jamshidi R, Karimi A and Haghshenas M 2018 Risk assessment of particulate matters in a dentistry school using fuzzy inference systems. *Measurement***116**, 257 – 263.

[20] Lambin F and Ehrlich D 1997 Land-cover changes in sub-Saharan Africa (1982–1991): Application of a change index based on remotely sensed surface temperature and vegetation indices at a continental scale. *Remote sensing of environment*, **61**(2), pp.181-200.

[21] Ahmad A, Upadhyay R, Lal B and Singh D 2018 Change Detection of Sodic Land in Raebareli District Using Remote Sensing and GIS Techniques*In Environmental Pollution* Singapore pp. 487-498.

[22] Rouse Jr, Haas H, Schell Aand Deering W 1973 Monitoring the vernal advancement and retrogradation (green wave effect) of natural vegetation. Prog. Rep. RSC 1978-1, Remote Sensing Center, Texas A&M Univ., College Station, E73-106393, 93 (NTIS No. E73- 106393).

[23] Salahedin M, Temeliye Kand Montaseri M 2014 Analysis of hydrological drought classes' transition using SPI (a case study: Urmia Lake watershed). *International Journal of Biosciences* (IJB), **4**(1), pp.452-462.

[24] McKee B, DoeskenJ and Kleist J 1993 The relationship of drought frequency and duration to time scales. In Proceedings of the 8th Conference on Applied Climatology,*American Meteorological Society*,Vol. **17**, No. 22, pp. 179-183.

[25] Yu, J., Lim, J. and Lee, K.S., 2018. Investigation of drought-vulnerable regions in North Korea using remote sensing and cloud computing climate data. *Environmental monitoring and assessment*, **190** (3), p.126.
[26] Hellwig N, Graefe U, Tatti D, Sartori G, Anschlag K, Beylich A, Gobat J and Broll G 2017Upscaling the spatial distribution of enchytraeids and humus forms in a high mountain environment on the basis of GIS and fuzzy logic. *European Journal of Soil Biology*, **79**, pp.1-13.
[27] Belal A, El-Ramady H, Mohamed E and Saleh A 2014. Drought risk assessment using remote sensing and GIS techniques. *Arabian Journal of Geosciences*, **7**(1), pp.35-53.
[28] Iraqi Meteorological Network Data. 2017, Documentation. https://www. http://www.agromet.gov.iq.
[29] Zhang, G., Patuwo, B.E. and Hu, M.Y., 1998. Forecasting with artificial neural networks: The state of the art. *International journal of forecasting*, **14** (1), pp.35-62.
[30] Aghelpour, P., Bahrami-Pichaghchi, H., & Kisi, O. (2020). Comparison of three different bio-inspired algorithms to improve ability of neuro fuzzy approach in prediction of agricultural drought, based on three different indexes. *Computers and Electronics in Agriculture*, **170**, 105279.
[31] Ross J 2005 Fuzzy logic with engineering applications. John Wiley & Sons.
[32] Sivanandam N, Sumathi Sand Deepa N 2007. Introduction to fuzzy logic using MATLAB (Vol. 1). Berlin: Springer.
[33] Türkeş MandTatlı H 2009 Use of the standardized precipitation index (SPI) and a modified SPI for shaping the drought probabilities over Turkey. *International Journal of Climatology*: A Journal of the Royal Meteorological Society, **29** (15), 2270-2282.

Acknowledgments
The authors are grateful to the Iraqi Ministry of Higher Education and Scientific Research and Mosul University for providing financial support. The authors are also grateful to the Iraqi Ministry of Transport for providing the rainfall data.

Hybrid DCT-SVD Based Digital Watermarking Scheme with Chaotic Encryption for Medical Images

Ahmed A Mohammed[1*], **Bilal A Jebur**[1] **and Karam M Younus**[1]

[1]College of Electronics Engineering, Ninevah University, Nineveh, Iraq

[*]E-mail: ahmed.mohammed@uoninevah.edu.iq

Abstract. The quality of medical image is crucial due to the sensitivity of medical diagnoses. Nowadays medical images are commonly shared between various entities across the medical sector to improve the quality of diagnoses. Thus, watermarking process is usually applied on these images to embed patient's private information. One of the key challenges is preserving image quality, keeping it clean of distortions after the watermarking process, and sustaining robustness. In this work, a hybrid Discrete Cosine Transform (DCT) and Singular value decomposition (SVD) based image watermarking scheme with chaotic encryption is proposed. The obtained results have demonstrated that the proposed scheme is secure and robust compared to the present schemes. In particular, the results show decent robustness and excellent imperceptibility of watermarked host medical images. Moreover, Peak signal to noise ratio (PSNR), and Normalized Correlation (NC) are used to evaluate the performance of the scheme. The scheme has shown a minimum PSNR of 59 dB and an NC value of 1 with no attacks applied, and not less than 0.5 with the presence of various attacks.

1. Introduction

Recently, Medical images are being circulated between two or more doctors for accurate diagnoses and exchange of expertise. On the other hand, access to multimedia has witnessed a noticeable increase due to the Internet growth. Consequently, the copyright protection of digital images, video, etc., becomes more vulnerable to be unauthorized access. One of the most recent solutions to the aforementioned problem is digital watermarking. Two types of watermarking could be used, visible and invisible, however, invisible watermarking is preferred for privacy and security[1]. Subsequently, invisible watermarking will be used in this research. The use of watermarking has been adopted since 1990. The early algorithms were based on two main categories, spatial domain and transform domain. However, the spatial domain watermarking is less robust than the transform domain watermarking, the main reason behind that is the image under attack is usually prone to pixel value alteration which leads to loss of embedded payload information, and that makes it irretrievable[2], on the contrary, using the transform domain as a watermarking embedding environment will provide more robustness since the original payload is inserted within the modified coefficients of the transform and not directly to the image pixels[3]. There are several main watermarking requirements in which the designer takes into account when designing a watermarking algorithm. The most common is robustness and transparency[4]. The watermarked image has to be robust against attacks like signal processing attacks, geometric attacks amongst other attacks.

The SVD has been used by researchers for watermarking, although the SVD has the problems such as false-positive, still very reliable in watermarking when combined with other transforms (i.e. hybrid domain) like the discrete wavelet transform (DWT), DCT and discrete Fourier transform (DFT).[5]. Transparency on the other hand is crucial and must be taken into consideration in parallel with robustness, depending on the type of data and application, a fair trade-off between robustness and transparency must be realized[6]. The main issue with medical images is that they are case-sensitive when it comes to medical decisions. Most present medical image watermarking schemes focus on

Content from this work may be used under the terms of the Creative Commons Attribution 3.0 licence. Any further distribution of this work must maintain attribution to the author(s) and the title of the work, journal citation and DOI.
Published under licence by IOP Publishing Ltd

robustness alongside with transparency but still not paying enough attention to keeping the medical image fully undistorted. In this study we propose a hybrid domain watermarking scheme in DCT-SVD domain and our aim is to achieve highest transparency possible with robust algorithm against various attacks.

2. Related Work

The authors in [7] employed a Hybrid Transform scheme, which comprises DWT, DCT, and SVD, to ensure the security of the electronic patient record (EPR) transfer between the health sector institutions. Although their scheme robustness is high, low transparency is noticed regarding the embedded binary payload. The DWT-SVD was proposed in [8] as a dual watermarking technique to secure the transmitting of the medical images. Moreover, Hamming error correction code was used to reduce the effect of channel noise, which in return improved the security level as Hamming code usually adds extra bits for error detection and correction to data to be transmitted, in this case watermarked image, results were acceptable transparency and robustness-wise compared to the scheme proposed in [9]. A wavelet-based watermarking technique was proposed in [10] to conceal the ECG signals in a QR code, which has shown performance that surpasses in terms of retrieval quality and embedding capacity compared with works proposed in [11] and [12]. An optimized and robust watermarking technique has been proposed in [13] combined with a firefly algorithm of Lifting Wavelet Transform (LWT). Additional Security has been added by scrambling the watermark before embedding the fusion of DWT, sanitization Algorithm and Oppositional Grasshopper Optimization Algorithm (OGAO) embedded the secret bits in an optimized locations and shown that this technique offers enhanced imperceptibility when put into comparison with proposed works in [14] and [15]. Ali et al. presented a reliable for the transmission-secure watermarking scheme of images [16], The Redistributed Invariant DWT, Artificial Bee Colony (ABC) optimization with SVD has been used to obtain transmission secure watermarking technique and the human visual system has been used to embed the Watermarks in blocks. Authors in [17] presented an SVD robust watermarking algorithm using colored images and also used the human visual system along with Particle Swarm Optimization (PSO) for automatic selection of embedding position. In [18] a mix between LWT and SVD with Arnold Cat Map (ACM) for upgrading the security algorithms have been used in for color images with secure watermarking for copyright protection. For improving the trade-off amid perceptual quality and robustness, Grey Wolf Optimization (GWO) technique has been used also. in [19] the presented scheme used SVD-PSO with Improved Discrete Wavelet Transform (IDWT) for a reliable embedding of the watermark , the entropy adapted to locate the suitable embedding environment. the compensation between robustness and perceptual quality is managed using PSO as an optimization algorithm.

From the forementioned works, we notice that the main concern with medical images is perceptual quality and being case-sensitive which emphasizes that distortions are not allowed i.e., preserving the perceptual quality of the image is crucial and that was the major drawback in all previously listed works since they've always focused on robustness as an equal match for imperceptibility and tried to sustain that state of equilibrium. In this developed work, we combined the SVD with DCT for a hybrid embedding domain to achieve better transparency.

3. Preliminaries and Performance Evaluation

3.1. Discrete Cosine Transforms (DCT):

DCT is a transformation technique that is usually used in a large number of image processing schemes. In essence, DCT and IDCT transforms are used to transform images from the spatial domain to the DCT domain. Moreover, the two-dimensional DCT and IDCT are given as

$$F(m,n) = \frac{2}{\sqrt{MN}} C(m)\, C(n) \sum_{x=0}^{M-1} \sum_{y=0}^{N-1} f(x,y) \times COS \frac{(2x+1)m\pi}{2M} COS \frac{(2y+1)n\pi}{2N}$$
$$F(x,y) = \frac{2}{\sqrt{MN}} \sum_{m=0}^{M-1} \sum_{n=0}^{N-1} C(m)\, C(n) f(m,n) \times COS \frac{(2x+1)m\pi}{2M} COS \frac{(2y+1)n\pi}{2N} \quad (1)$$

3.2. Singular Value Decomposition (SVD)
Singular value decomposition (SVD), is a matrix decomposition technique that has been employed in various signal processing schemes such as (a general tool for numerical linear algebra, data reduction, Digital Watermarking, etc.). The SVD converts any $N \times M$ rectangular matrix into three matrices, i.e., U, S, and V such that [19]

$$A = USV^H \quad (2)$$

where U is an $N \times M$ matrix that satisfies:

$$U^H U = I_{N \times N} \quad (3)$$

and V is an $N \times M$ matrix that satisfies:

$$V^H V = I_{M \times M} \quad (4)$$

This has made the SVD very useful in size reduction algorithms along with several other algorithms.

3.3. Peak Signal to Noise Ratio (PSNR)
PSNR is one of the most common performance metrics that is usually used to evaluate the effectiveness of image processing algorithms. In particular, it is an empirical measurement that is obtained by dividing the peak signal power by the noise power, which is given as

$$PSNR = 10log\left(\frac{255^2}{\mathrm{MSE}}\right) \quad (5)$$

where, MSE denotes the mean-square-error, which is expressed as

$$f(x) = \frac{1}{MN} \sum_{y=1}^{M} \sum_{x=1}^{N} |C(x,y) - WC(x,y)|^2 \quad (6)$$

3.4. Normalized Correlation (NC)
The Normalized-Correlation (NC) is a quantitative measurement that is mainly used to compute the similarity between the original and the extracted watermarks. Consequently, it is used as an indication of the effectiveness and robustness of the watermarking algorithms. The NC can be computed as in equation (7).

$$NC\left(W, W' = \frac{\sum_{x=0}^{M} \sum_{y=0}^{N} [W(x,y) W'(x,y]}{\sqrt{\sum_{x=0}^{M} \sum_{y=0}^{N} W(x,y)} \sqrt{\sum_{x=0}^{M} \sum_{y=0}^{N} W'(x,y)}} \right) \quad (7)$$

4. Proposed Scheme
In this work, we present a non-blind scheme for medical image watermarking, the medical image is transformed into the DCT domain to get different frequency coefficients, the mid-frequency coefficients are selected using the zig-zag method, next the selected coefficients vector is reshaped into an

$N \times N$ matrix then SVD is applied on the selected coefficients. The SVD diagonal matrix is selected as the embedding domain but before embedding the payload, we encrypt it with a special key along with chaotic scrambling to add another security level to the proposed scheme. The embedding and extracting procedures are explained in the next sub-sections.

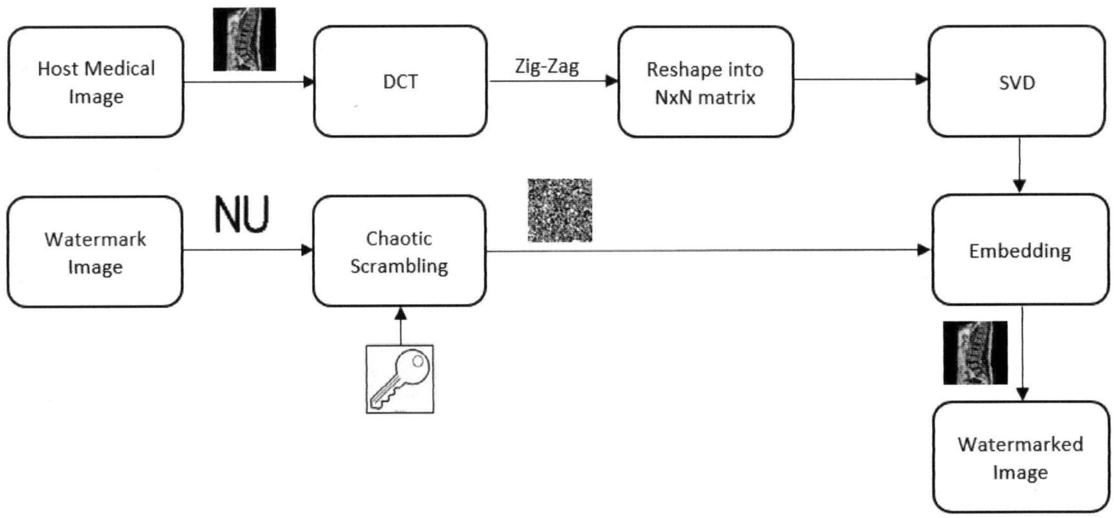

Figure 1. Proposed Scheme

4.1. Watermark Encoding Process

- Read input host image I of size 1024X1024
- Read watermark image W, size 64X64
- Encrypt W using chaotic scrambling + key
- Apply DCT on host image: $DCT_{Image} = dct2(I)$
- Perform SVD on selected mid-frequency coefficients:
$$[U\ S\ V] = SVD(selected\ mid - frequencies\ of\ DCT_{Image})$$
- Embed encrypted W within the diagonal matrix S with a scaling factor of 0.1:
$$S_{modified\ Diagonal\ matrix} = S_{digonal\ matrix} + scaling\ factor\ k * W_{encrypted}$$
- Perform the inverse SVD: $SVD_{modified} = U\ S_{modified\ Diagonal\ matrix}\ V^{Transposed}$
- Finally, apply the Inverse DCT to get the watermarked image I^* :
$$I^* = idct2(DCT_{modified})$$

4.2. Watermark Decoding Process

- Read I^* (could be attacked)
- Apply DCT on host image: $DCT = dct2(\ I^*)$
- Apply SVD on mid-frequencies:
$$[U_2\ S_2\ V_2\ ] = SVD(selected\ mid - frequencies\ of\ DCT\ of\ )$$
- Extract watermark $W_{extracted}$:
$$W_{extracted} = \frac{(S_2 - S)}{scaling\ factor\ k}$$

5. Results and Simulation

This section presents the simulation results and analysis. First, two medical images were selected to be the main test host. These images are a Magnetic Resonance Imagining (MRI) of the spine and a Computed Tomography (CT) of the brain. The watermark image is a binary logo that represents an abbreviation of Ninevah University. figure 2 shows the watermark image before and after scrambling. On the other hand, figure 3 shows the original test images and their watermarked versions for various PSNR values after the embedding process. In general, A 45 dB of PSNR is the performance threshold above which the performance is considered to be good. In the proposed scheme we attained a minimum of 59 dB in both test images. This implies high transparency and therefore undistorted medical images.

(a) (b)

Figure 2. (a) Watermark, (b) Chaotic Scrambled Watermark

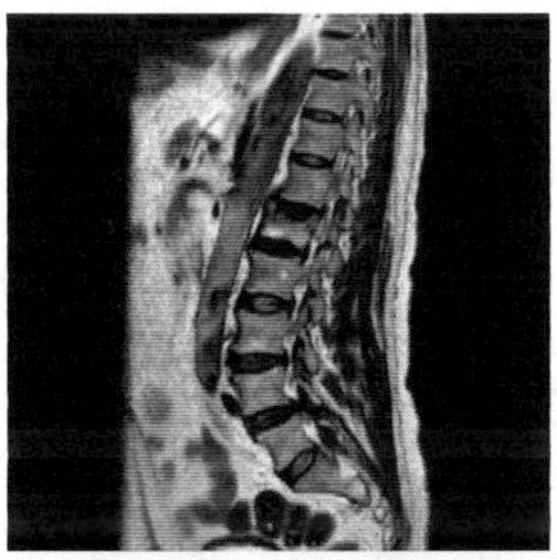

(a) Host Image 1

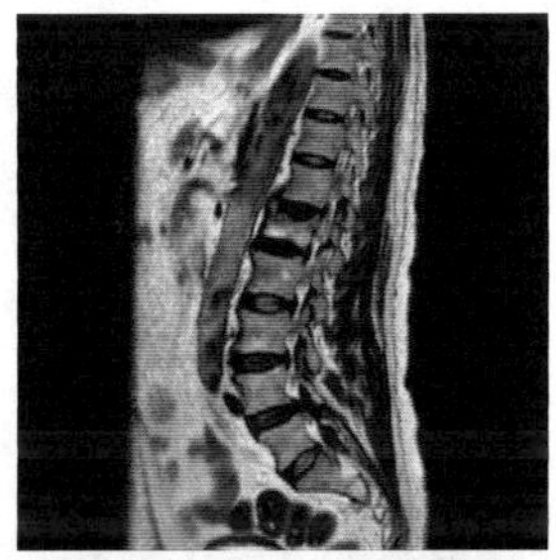

(c) Watermarked Image1, PNSR= 59.98 dB

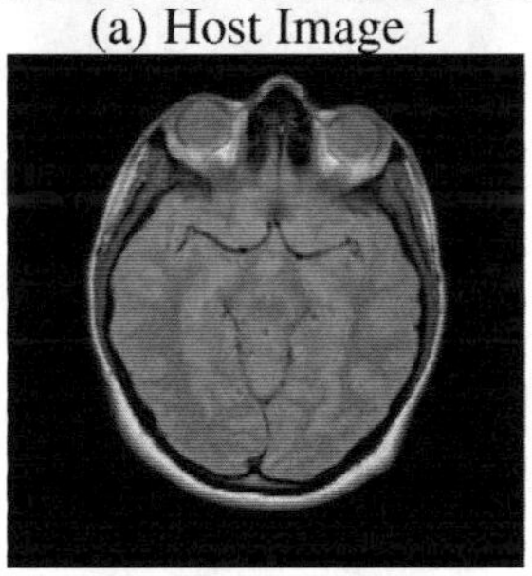

(b) Host Image 2

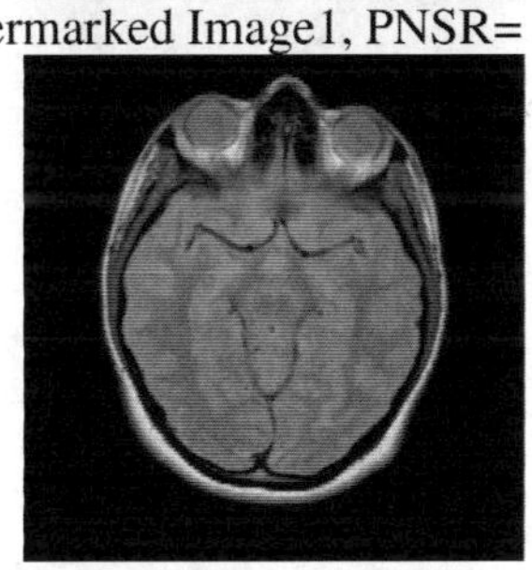

(d) Watermarked Image 2, PNSR= 59.01 dB

Figure 3. Original Host Images and Watermarked images

Moreover, figure 4 and 5 show that the attacked images and the extracted watermarks after each attack. They also show the extracted watermark with both objective and evaluations. In essence, different attacks were chosen to test the robustness of the proposed scheme, some are geometric attacks, others are common signal processing attacks.

Gaussian noise attack is applied twice, the variance of the first one was 0.01 and the extracted watermark has an NC value of 0.9774, while the second gaussian attack had a variance of 0.1 and yet the extracted watermark was visible with an NC value of 0.6367. Next, JPEG lossy compression, which is commonly used by most of today's apps such as; Facebook messenger, Viber, etc., to compress the images before sending them to other parties, can be considered as one of the signal processing attacks. Hence, we investigated the impact of the JPEG lossy compression on the performance of the proposed

IOP Conf. Series: Materials Science and Engineering 1152 (2021) 012025 doi:10.1088/1757-899X/1152/1/012025

scheme, and the extracted watermark after the JPEG attack with a quality factor of 25% has shown an NC value of 0.9951, which implies the robustness of our scheme.

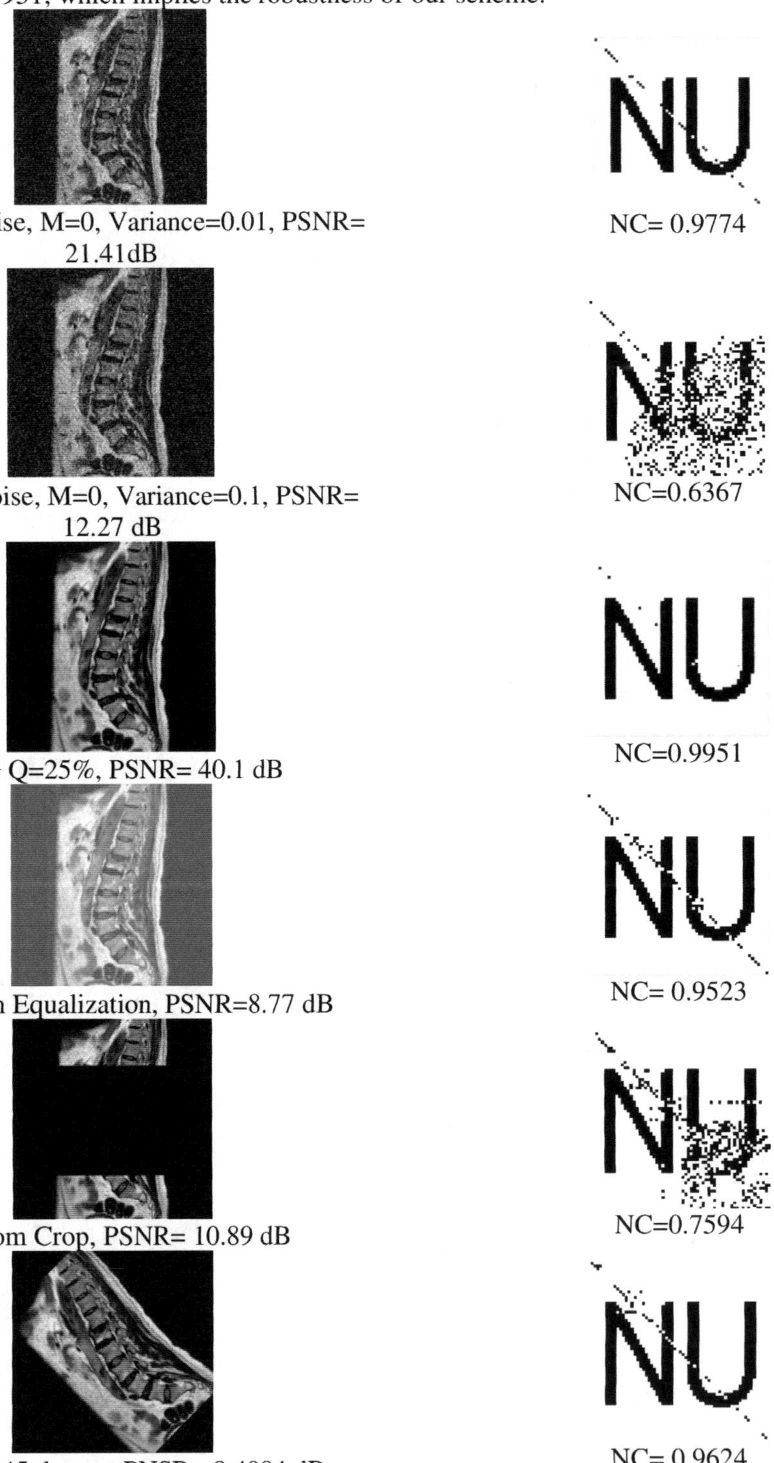

Figure 4. Attacks vs. Extractions

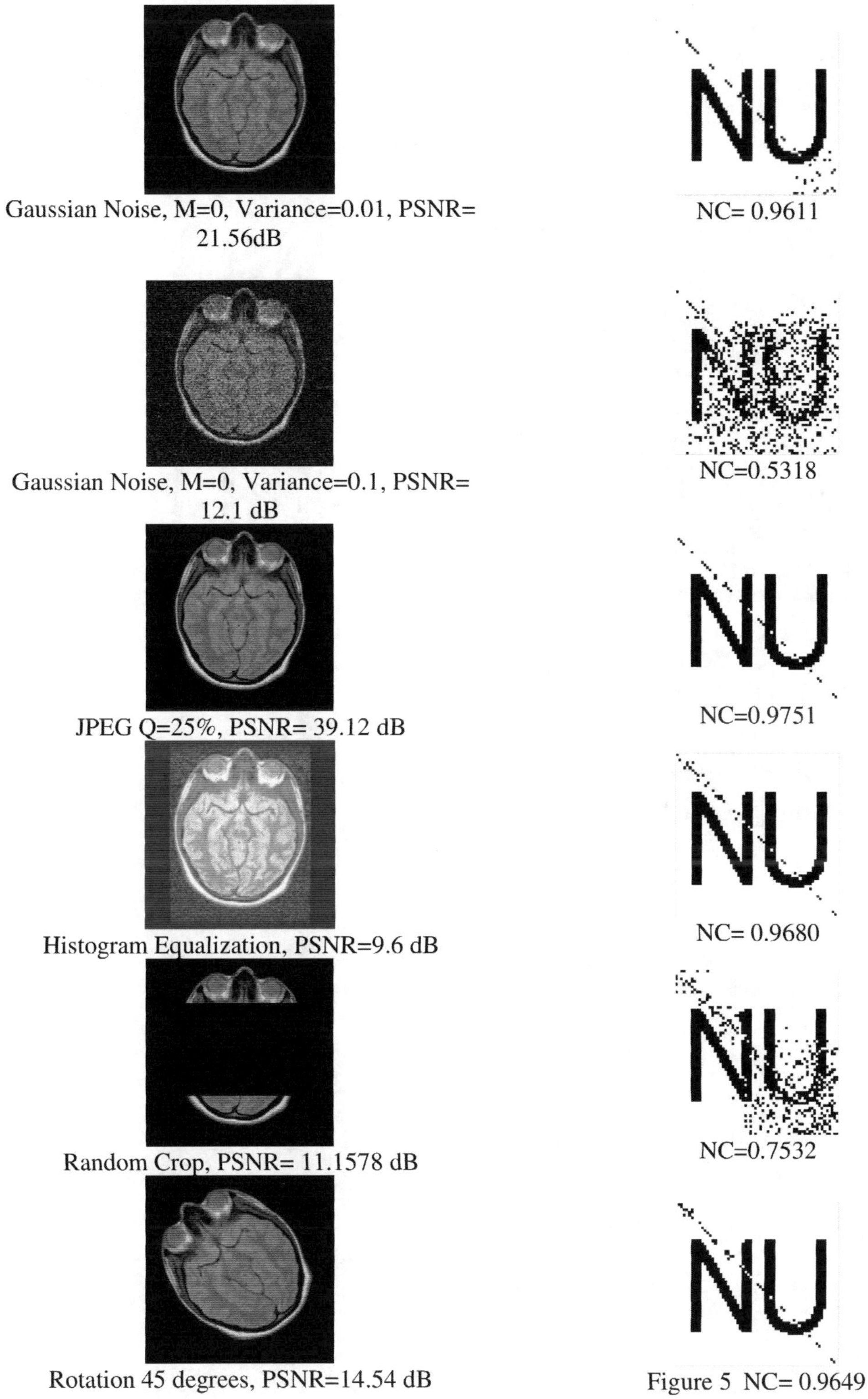

Figure 5. Attacks vs. Extractions

Finally, the random crop attack was also considered to thoroughly examine the performance of the proposed scheme as the attacker could remove any part/s of the image. The attained results have proved the robustness of the proposed scheme against such a random attack with a minimum NC of o.7532. More random crop results are illustrated in figure 6.

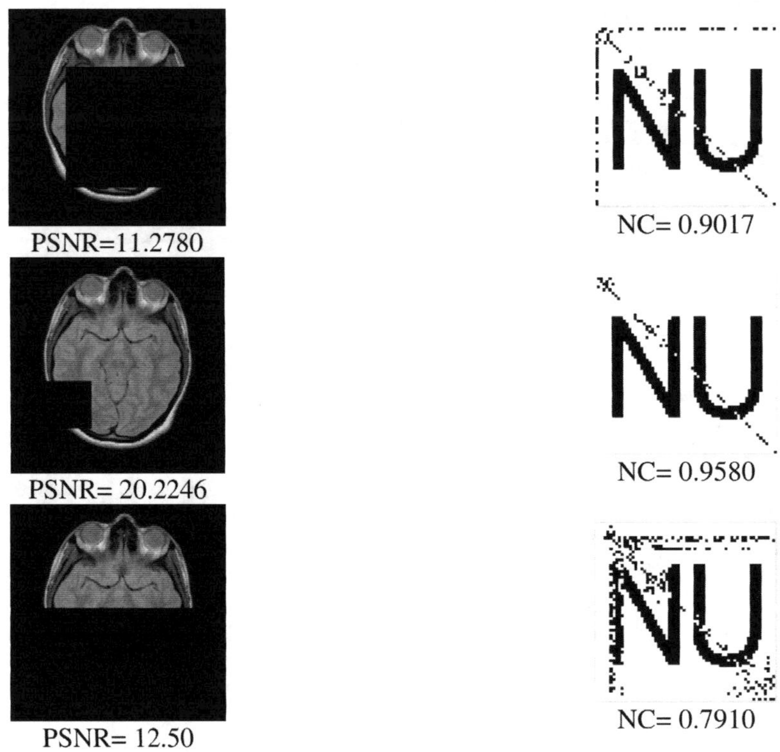

Figure 6. Attacks vs. Extractions

Tables 1 and 2 show the obtained results for various attack scenarios. A closer inspection of these results shows that the proposed scheme has proved to be resistive against these attacks.

Table 1. Proposed Scheme against various Attacks, Test Image 1

Attack	PSNR in dB	NC
Salt & Pepper Noise, Density=0.1	13.8907	0.7573
Mean Filter 3X3	41.46	0.9726
Gemma correction	21.72	0.9774
Motion Blur	31.25	0.9702

Table 2. Proposed Scheme against various Attacks, Test Image 2

Attack	PSNR in dB	NC
Salt & Pepper Noise, Density=0.1	14.05	0.6067

Mean Filter 3X3	50.59	0.9726
Gemma correction	20.7194	0.968
Motion Blur	36.762	0.9661

Next, Table 3 shows the comparison of the performance of the proposed scheme with those proposed in [7], [10], and [19] in terms of PSNR and NC values.

Table 3. Performance comparison

Scheme	PSNR in dB
Our Scheme	59.98
[10]	58.85
[19]	48.25
[7]	50.88

A closer look at table 3, shows that our scheme has compromising results with the PSNR value of more than 59 dB in both test images. Subsequently the proposed scheme has managed to keep higher transparency with adequate robustness.

6. Conclusion

Digital Watermarking plays an important role in image copyright protection, for the medical application it can be used to embed patient information within the medical data while keeping high privacy of that information. In this paper, we proposed a robust, reliable, and transparency preserver technique for medical image watermarking. In terms of transparency, the obtained results have shown a minimum PSNR of 59 dB, which is excellent to keep the medical data as original as possible and guarantees no medical decision interference. Moreover, almost all extracted watermarks have shown NC values of 0.9 and more after these attacks, which demonstrate the robustness of the proposed scheme.

7. References

[1] Narayan R 2018 Encyclopedia of Biomedical Engineering. Elsevier

[2] Preedanan W, Kondo T, Bunnun P and Kumazawa I 2018 Image quality assessment for medical images based on gradient information *in 5th Int. Conf. on Business and Industrial Research (ICBIR)* pp 189-94.

[3] Nai Y, Schaefferkoetter J D, Fakhry-Darian D, Conti M, Shi X, Townsend D W, Sinha A K, Tham I, Alexander D C and Reilhac A 2018 Improving Lung Lesion Detection in Low Dose Positron Emission Tomography Images Using Machine Learning *in IEEE Nuclear Science Symp. and Medical Imaging Conf. Proc. (NSS/MIC)* pp 1-3.

[4] Mohammed A, Maraş H H and Elbasi E 2014 A new robust binary image embedding algorithm in discrete wavelet domain *in IEEE 8th Int. Conf. on Application of Information and Communication Technologies (AICT)* pp 1-7.

[5] Segars W P, Tsui B M, Cai J, Yin F-F, Fung G S and Samei E 2017 Application of the 4-D XCAT phantoms in biomedical imaging and beyond *IEEE Transactions on Medical Imaging* **37** 680-92.

[6] Sullivan B J, Ansari R, Giger M L and MacMahon H 1994 Relative effects of resolution and quantization on the quality of compressed medical images *in Proc. of the 1st Int. Conf. on Image Processing* pp 987-91.

[7] Kumar G P, Saranya M, Tamilselvan K, JL M I and Kavitha S 2020 Investigation on Watermarking Algorithm for Secure Transaction of Electronic Patient Record by Hybrid Transform *in Fourth Int. Conf. on I-SMAC (IoT in Social, Mobile, Analytics and Cloud)(I-SMAC)* pp 379-83.

[8] Anand A and Singh A K 2020 An improved DWT-SVD domain watermarking for medical information security *Computer Communications* **152** 72-80.

[9] Singh A K, Kumar B, Dave M and Mohan A 2015 Robust and imperceptible dual watermarking for telemedicine applications *Wireless Personal Communications* **80** 1415-33.

[10] Sanivarapu P V, Rajesh K N, Reddy N R, Reddy N C 2020 Patient data hiding into ECG signal using watermarking in transform domain *Physical and Engineering Sciences in Medicine* **43** 213-26.

[11] Jero S E and Ramu P 2016 Curvelets-based ECG steganography for data security *Electronics Letters* **52** 283-5.

[12] Mathivanan P, Jero S E, Ramu P, Ganesh A B 2018 QR code based patient data protection in ECG steganography *Australasian Physical & Engineering Sciences in Medicine* **41** 1057-68.

[13] Chaturvedi A K, Shukla P K 2020 Effective watermarking technique using optimal discrete wavelet transform and sanitization technique *Multimedia Tools and Applications* **79** 13161-77.

[14] Ingale S P, Dhote C 2016 Digital watermarking algorithm using DWT technique *International Journal of Computer Science and Mobile Computing* **5** 1-9.

[15] TN S, Ramesha K and Raj C P 2014 A New Technique to Digital Image Watermarking Using DWT for Real Time Applications *International Journal of Engineering Research and Applications* **4** 102-107 .

[16] Ali M, Ahn C W, Pant M and Siarry P 2015 An image watermarking scheme in wavelet domain with optimized compensation of singular value decomposition via artificial bee colony *Information Sciences* **301** 44-60.

[17] Ahmadi S B B, Zhang G and Jelodar H 2019 A robust hybrid SVD-based image watermarking scheme for color images *in IEEE 10th Annual Information Technology, Electronics and Mobile Communication Conf. (IEMCON)* pp 0682-8.

[18] Pandey M K, Parmar G, Gupta R, Sikander A 2019 Lossless robust color image watermarking using lifting scheme and GWO *International Journal of System Assurance Engineering and Management* **11** 320-31

[19] Gangadhar Y, Akula V G, Reddy P C 2018 An evolutionary programming approach for securing medical images using watermarking scheme in invariant discrete wavelet transformation *Biomedical Signal Processing and Control* **43** 31-40.

Synthesis and Implementation of IIR Filter using VHDL Language

N. T. Gadawe[1*], T. A. Fathi[1], S. L. Qaddoori[1] and R. W. Hamad[1]

[1]Electronic Engineering Department, Electronics Engineering College, Ninevah University, Mosul, Iraq

[*]E-mail: noor.gadawe@uoninevah.edu.iq

Abstract. VHDL is a hardware description language of different hardware architectures at divers' levels of abstraction. The language is used for design, modelling, and investigation of hardware. In this paper, it is presented hardware implementation architectures of IIR filter using synthesizable VHDL. The MATLAB Filter Design and Analysis Tool (FDA Tool) is usedto calculate filter coefficients. Then, the filter is simulated and implemented on the Xilinx xc3s500efg320-4. A fixed-point number representation is used in the implementation of the filter. The proposed design of the filter includes two architectures: parallel architecture and programmable multiplexing architecture, each type has its advantages and disadvantages. In the results, some figures depending on the size of design (filter length) are introduced in the synthesis report.

1. Introduction

Field programmable gate array (FPGA) possessadequate performance and logic capacity to applysomeDigital Signal Processing (DSP) algorithms efficiently [1].

FPGA has two advantages that make it suitable for DSP, the first one is that its architecture is well appropriated for vastly parallel implementation of DSP purposes with high performance, the second is that designers can make a tradeoff between performance vs. device area by choosing the suitable level of parallelism to employ their tasks. Designers can achieve higher performance and optimization design with low cost [2]

Very High-Speed Integrated Circuit Hardware Description Language (VHDL) is one prevalent programming language that supports Register Transfer Level RTL hardware descriptions [3]. Theessential motivation to utilize VHDL is thatit is a regular, vendor/technology independent language, and is hencereusable and portable [4].

Digital filter is a very significant part of DSP, thereforethe need to implement it has been appeared. The two basic types of digital filtersareinfinite impulse response (IIR)and finite impulse response (FIR). Where, An IIR needs an infinite sum to be achieved. An FIR filter involves a finite number of sample values, where the filter equation convolutionis used to find a finite sum per output sample [5]. Digital filter design is an uncomplicated task because it isimplemented using influential software tools [6].

In [7], Savadi A., and others designed an IIR filter using the convolution method, which is called convolution filter, they showed the effectiveness of the Urdhava-Tiryagbhyamtechnique for multiplication that carries a difference in the real process of multiplication itself. This multiplicationwasachieved in Xilinx 13.4 ISE and applied on vertex-5 FPGA.

In [8] Toledo-Pérez et al, implemented a developed IIR filter on FPGA, and used MATLAB functions to obtain and prove the filter coefficients then transferred these coefficients to an FPGA filter design. The filter was applied within a Basys 3 Artix-7 FPGA Trainer kit of the Xilinx family. In [9] P. Jubair Ahamed, M. Abdul Haseeb, presented hybrid programming methods which use MATLAB and VB in digital filter design. The proposed algorithm decreases the difficulty of computing as well as partlypreserves the multiplier combinationarchitecture, which leads to effective construction in FPGA.

Content from this work may be used under the terms of the Creative Commons Attribution 3.0 licence. Any further distribution of this work must maintain attribution to the author(s) and the title of the work, journal citation and DOI.
Published under licence by IOP Publishing Ltd

Debarshi D. and Himadri S. D. introduced an enhanced design of reconfigurable IIR filter that isbroadlyutilized in real-time applications in [10]. The designed filter is designed using hardware description language thenverified on Virtex-5 board from Xilinx family.
In this paper, IIR digital filter design using the FDA tool is presented. parallel and programmable multiplexing IIR filter architectures are simulated and implemented on an FPGA kit using ISE14.7i and VHDL.The other sections of the paper is structured as follows: Section 2 describes the digital filter design methodology. IIR filter is synthesized using VHDL in section 3. The simulation and implementation results using a Xilinx FPGA device are given in sections 4 and 5. Finally, section 6 concludes this paper.

2. Digital Filter Design Methodology

Digital filters are type of electronic filters, these filters perform digital mathematical operations on a middle form of a signal (could be represented using floating-point or fixed point) [11]. The digital filter difference equation general form is:

$$y(n) = \sum_{i=0}^{N} a_i x(n-i) - \sum_{i=1}^{N} b_i y(n-i) \tag{1}$$

where the x(n-i)'s is recent or previous filter inputs, y(n) is the recent filter output, the y(n-i)'s are previous filter outputs, N is the filter's order, the a_i 's are the filter's feed-forward coefficients corresponding to the zeros of the filter, the b_i 's are the filter's feedback coefficients corresponding to the poles of the filter, and [12].
IIR digital filter is widely applied in recent signal processing systems such as high-speed telecommunication/RF applications, audio equalization, and biomedical sensor signal processingIIR filters have at least one nonzero feedback coefficients. That is to say, due to the feedback term, the filter will be excited with an impulse and there is always an output if the filter has one or more poles. The output is computed from two sets, a set of input samples and a set of previous outputs, these inputs and outputs are multiplied by a set of coefficients and then accumulated together to yield the output as shown in Figure 1 [12].
The large amounts of memory and arithmetic processing required are one of the major disadvantages of FIR filters. This makes them less used in several applications. Alternatively, IIR filters require much less memory and fewer arithmetic operations, their disadvantages are the design difficulty and they suffer from stability problems. In spite of the design complexity and Although the design is much more, IIR filter usage may led to a higher performance and lower system cost [13]. The filter's behavior is determined by the filter coefficients. From equation (1), a general IIR filter is determined by:

$$y(n) = a_0 x(n) + a_1 x(n-1) + \ldots + a_N x(n-N) + b_1 y(n-1) + b_2 x(n-2) + \ldots + b_N y(n-N) \tag{2}$$

As it can be noticed from the above equation, the digital filter design relies on the coefficients of the filter. Constant coefficients calculation of a digital filter includes an extensive amount of computation, these computations are generally accomplished using a software tool. MATLAB has a packed tool for filter design, The Filter Design and Analysis (FDA) tool [14]. The (FDA) is a graphical user interface (GUI) tool available in the signal processing toolbox of MATLAB, it could be used for designing and analyzing filters. Filter specifications are fed as inputs to the FDA. Figure 2 shows the filter design window of the FDA tool, when the filter design is completed from the FDA window, one can select filter type (FIR or IIR) and all specifications of the filter design. For hardware implementation, the lowpass filter has been selected and designed with specification design as shown in Table 1.

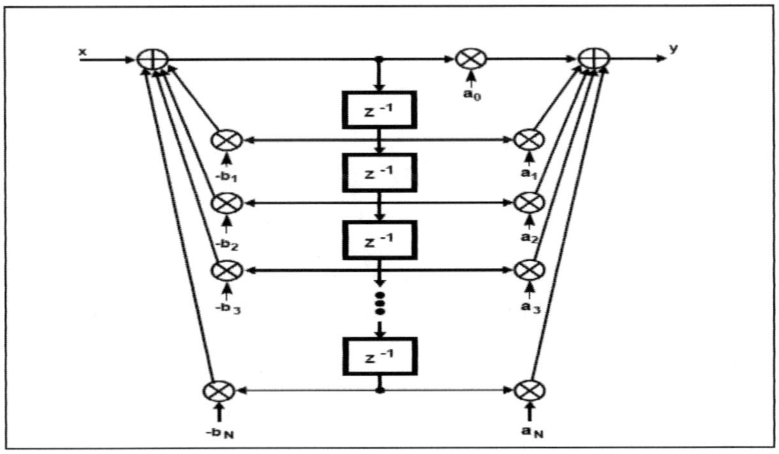

Figure 1. IIR filter (Direct form II)

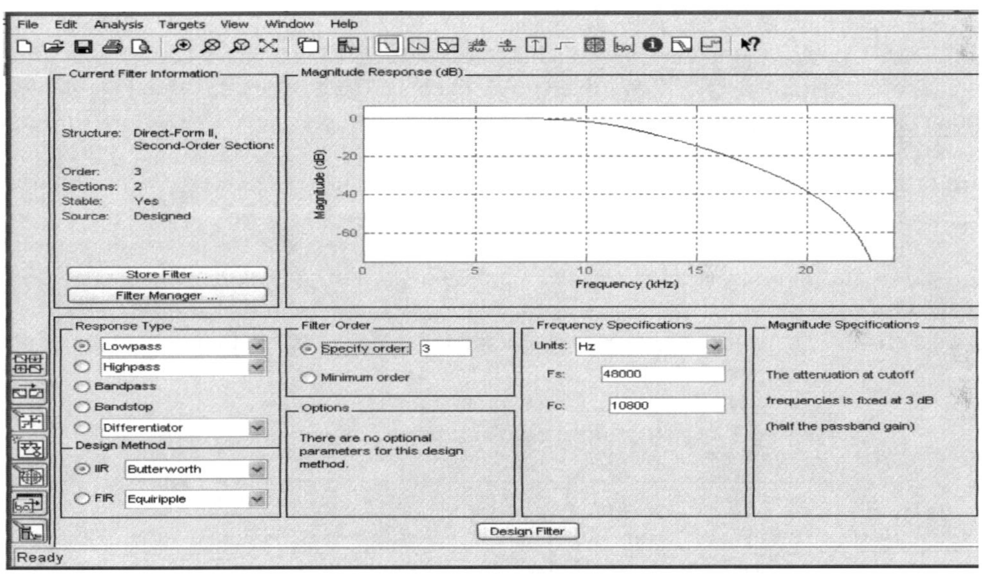

Figure 2. Filter design using MATLAB FDA tool.

Table 1. The Designed Filter Specifications.

Filter Specifications	Options
Design Method	IIR Butterworth
Response Type	Lowpass Filter
Frequency Specifications	
Sampling Frequency (Fs)	48000Hz
Cutoff Frequency (Fc)	10800Hz
Specify filter order	3
Magnitude Specifications	The attenuation at cutoff frequencies is fixed at 3 dB

3. Synthesis of IIR Filter using VHDL

In general, DSP algorithms are realized in two main number representations: floatingand fixed-point. In practicalDSP applications, the extra precision and range provided by floating-point are not necessary. In other words, floating-point representations tend to more costly and power-consumingbecause of additional hardware difficulty, Consequently, the requirement to lower power consumption anddesign cost led the inventors to utilize the fixed-nonrepresentational place of floating-point [10]. It can be noted that the FDA tool utilizes double-precision floating-point representation for the design calculations. The double-length floating-point representation for the filter coefficients poses huge challenges in terms of resourcesandcost.

To overcome this, the filter coefficientsneed to be quantized to fixed point representations, this can be done by selecting the quantization parameter seton the sidebarbutton of the FDA tool and selecting the suitable word length of the coefficients and input outputs. The filter coefficients are fractional values as shown in Table 2. In the implementation, it is not possible to implement those values exactly due to the use of finite word length representation, by using a large number of bits, the results will be more accurate and near the actual values, of course, this requires additional hardware area and costs. 8-bits representation for the coefficients and input data is selected. The data type used for representing the data and coefficients are "signed" representation, because the results may be negative values.

3.1 The architecture of IIR Filters Parallel Form

In general, an IIR has fewer coefficients, but the compulsory output feedback can perform the employment of the circuit more difficult. If the coefficients are not selected correctly, a stable IIR filter can become unstable. The most straightforward implementation is a parallel structure with fixed-point coefficients. There are several possibilities to implement the structure. One of them is the direct form II as shown in Figure 3. From Figure 3, the hardware implementation requiresa D-flip flop register for delay elements, latches to store coefficients, and different arithmetic operations (composed of multipliers and adders). The filter is compiled in Xilinx ISE14.7 version. The timing simulation is shown in Figure 4. For the entity declaration clk, rst, x are inputs of the filter, y is an output of the filter, x1, x2, x3, and x4 are tap delay lines which refer to x(n), x(n-1), x(n-2), and x(n-3). a0=a3, a1=a2, b1, b2, and b3 are the coefficients of the filter in hexadecimal. y1, y2, y3 are output delay y(n-1), y(n-2), and y(n-3).

Table 2. Coefficients of IIR (low-pass) filter from FDA tool.

Coefficients	Decimal Values	8bits Representations
a0	0.13006301	0.12890625
a1	0.39018903	0.390625
a2	0.39018903	0.390625
a3	0.13006301	0.12890625
b1	0.28814049	0.2890625
b2	-0.35476049	-0.35546875
b3	0.02666632	0.2734375

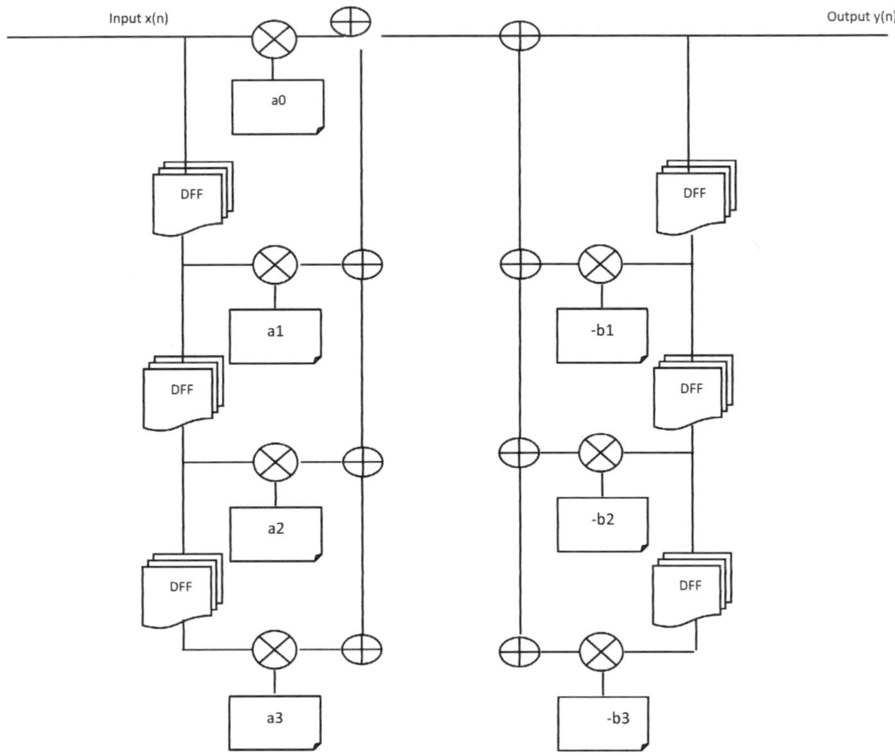

Figure 3. RTL Schematic of parallel form third orderIIR filter

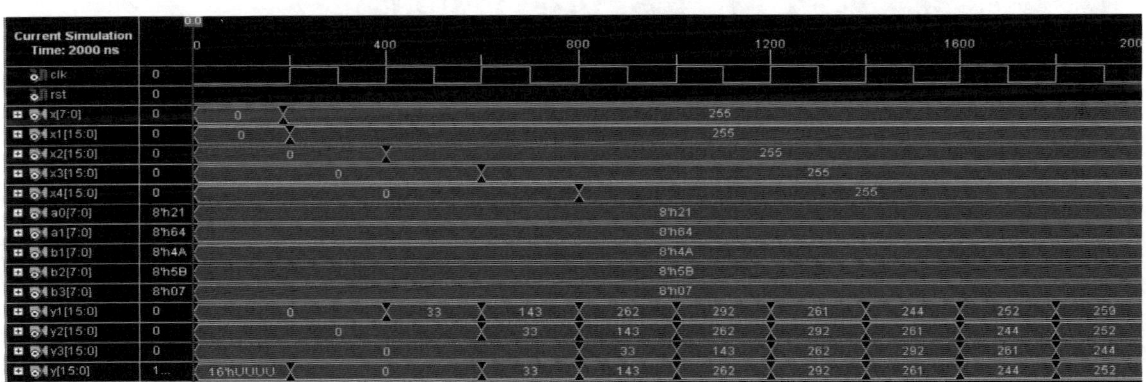

Figure 4. Timing Simulation of parallel formIIR third order

3.2 The architecture of IIR Filters Programmable Multiplexing Form

The implementation of programmable multiplexed IIR filter is built a dataflow structure using one multiplier and one adder which is constituting a multiply and accumulate (MAC) unit for arithmetic operations. The elements are kept in suitable synchronous register banks. The dataflow representation of the implementation architecture for the third-order IIR filter is shown in Figure 5. The timing simulation is shown in Figure 6.

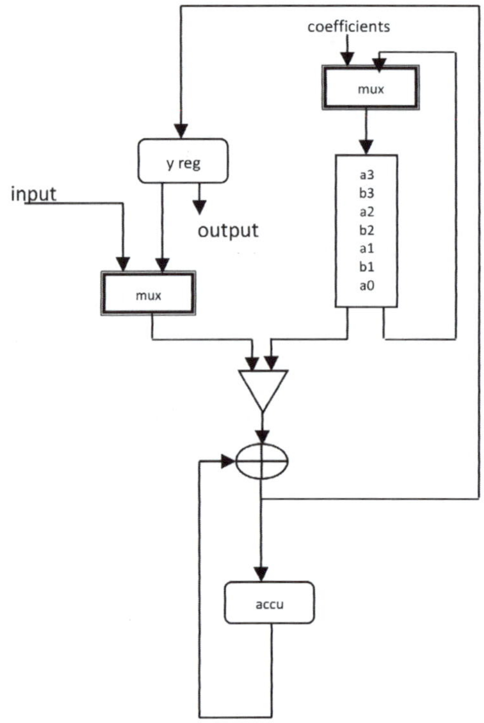

Figure 5. RTL schematic of multiplexing formIIR Filter

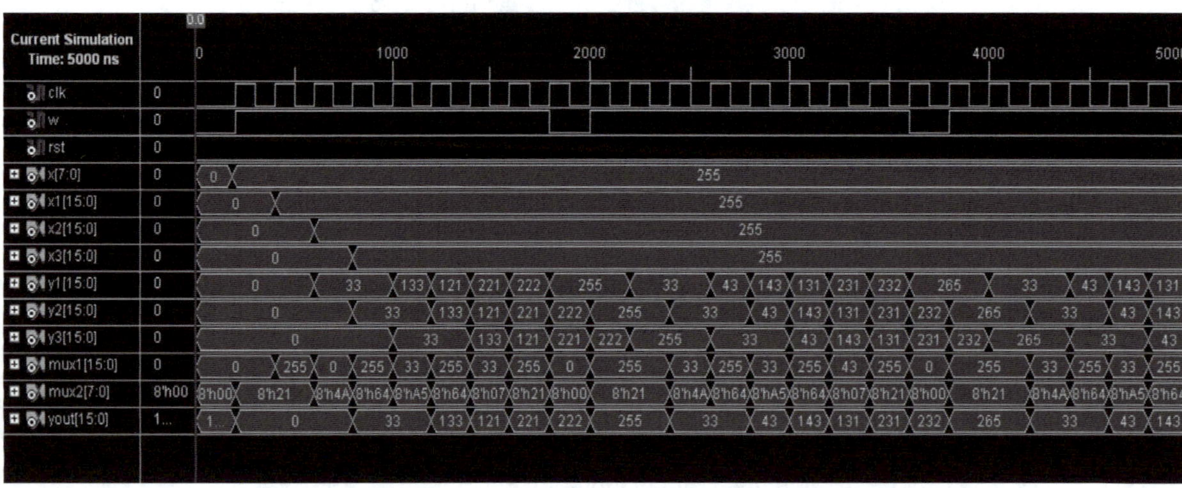

Figure 6. Timing Simulation of programmable multiplexing form

4. FPGA Implementation of IIR Filters

The filters are synthesizedin VHDL using ISE 14.7 tool. The FPGA implementations are realized on SPARTAN3E XC3S500EFG320-4[15]. All implemented filters require 8-bit input samples and their coefficients need 8-bits precision. The major challenge in parallel form is the large number of multipliers utilized, these increased as soon as filter length is increased, compared with programmable multiplexing form which required one multiplier for different filter lengths as shown in Figure (7), but the programmable multiplexing formarchitecture consumed more slices as shown in Figure (8).

Figure 9 demonstrates the relationship between the required number of gates and the filter length, where the number of gatesis increased asthe filter length increased, but this increasing is greater for the parallel form compared with the multiplexing form. The same thing in 4-input (Look Up Tables) LUT utilization as shown in Figure (10). From simulation results, the throughput in parallel form equal maximum frequency/1clk cycles, for multiplexing form the throughput equal maximum frequency/8 clk cycles. therefore, the parallel approach high performance with no constraint in hardware consumed, in contrast, the programmable multiplexed approach is more efficient in terms of resources needed and utilized, but the performance is lower.

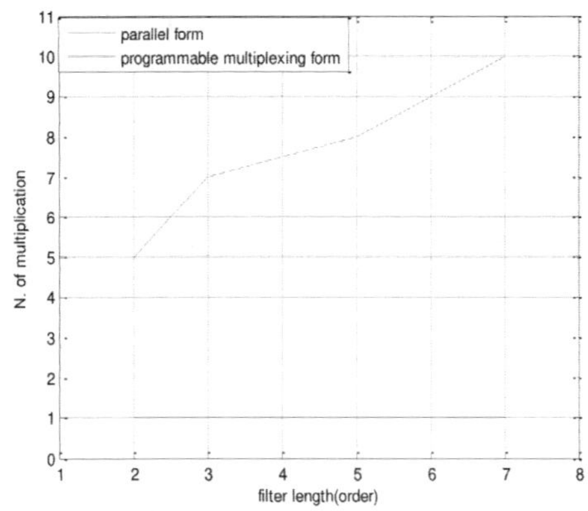

Figure 7. N. of multiplier as function of taps

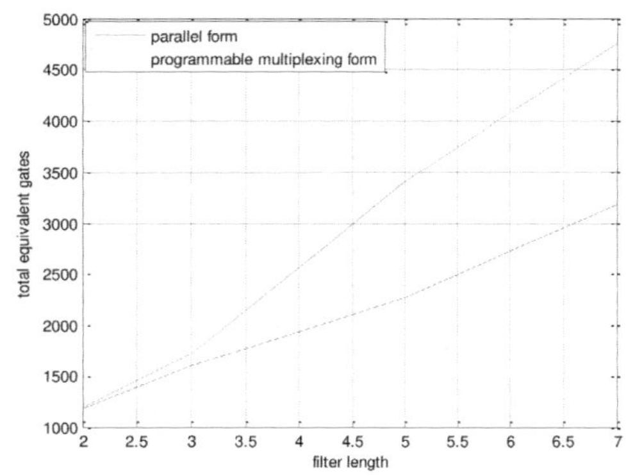

Figure 8. Slices as function of taps

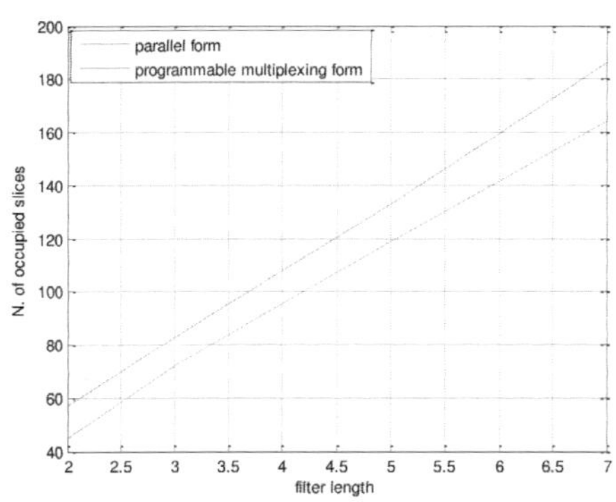

Figure 9. N. of Gates as function of taps

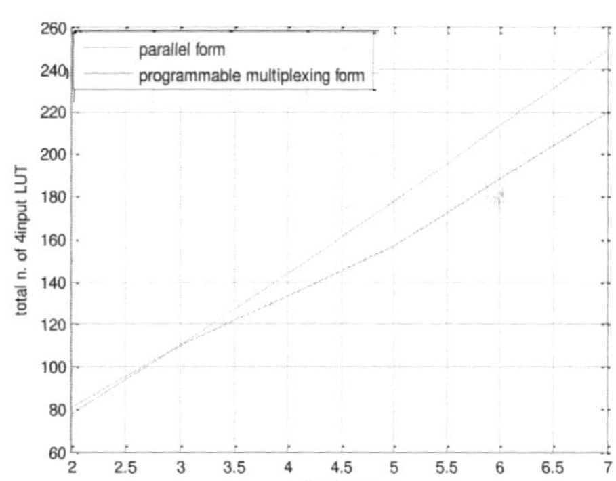

Figure 10. N. of 4-input LUT as a function of filter order

5. Conclusions

Today digital filter design is greatly facilitated by the availability of numerous computer software programs. FDA package is such tool, which used for designing and analyzing filter in this paper. Synthesis, simulated IIR lowpass Butterworth filter using ISE14.7 and VHDL language, then implemented on the FPGA to achieve hardware verification. The target device is spartan3e xc3s500efg320. The research is focused on the architecture part of the design. The proposed design of the filter includes two architectures parallel architecture, and programmable multiplexing architecture, with different filter lengths. The direct approach high performance with no constraint in hardware, in contrast, the programmable multiplexing approach minimizes hardware requirement while yielding less performance. In future works, practical application of IIR filter can be introduced and other architectures of IIR filters can be implemented which can be compared with FIR filter architecture.

6. References

[1] Turney R D, Dick C, David BP and HwangJModeling and implementations of DSP FPGA Solution *XilinxInc* san Jose, CA95124, USA 1-16

[2] WoodR andMcullisterJ2008FPGA based Implementation of Signal Processing Systems*john Wileyand sons* ISBN 978-0-470-03009-7.

[3] HauckS andDehonA2008Reconfigurable Computing, the theory and practice of FPGA based Computation*ElsevierInc* ISBN 978-0-12-370522-8.

[4] PedroniV2004Circuit Design with VHDL*Massachusetts institute of technology (MIT) press* ISBN 0-262-16224-5.

[5] Kunchevo A andYanchev G2008 Synthesis and Implementation of DSP Algorithms in advanced Programmable Architectures *Proceeding InInternational Scientific Conference Computer ScienceISCCS.*

[6] Poučki VM, Žemva A, Lutovac MD and Karčnik T 2008 Chebyshev IIR filter sharpening implemented on FPGA *transfer* 100 1-11

[7] Savadi A,Yanamshetti R and Biradar S 2016Design and implementation of 64 bit IIR filters using Vedic multipliers*Procedia Computer Science* 85 790-797

[8] Toledo-Pérez DC, Martínez-Prado MA, Rodríguez-Reséndiz J, Arriaga ST andMárquez-Gutiérrez MÁ2017 IIR digital filter design implemented on FPGA for myoelectric signals*In 2017 XIII International Engineering Congress (CONIIN)*IEEE 1-7

[9] JubairAhamed P and Abdul Haseeb M2020 Implementation of Digital IIR Filter Design Based on Field Programmable Gate Array*Materials Today: Proceedings*,ISSN 2214-7853.

[10] Datta D and Dutta HS 2021 High performance IIR filter implementation on FPGA *Journal of Electrical Systems and Information Technology* 8(1)1-9

[11] Vinger KA and Torresen J 2003Implementing evolution of FIR-filters efficiently in an FPGA *In NASA/DoD Conference on Evolvable Hardware* IEEE26-29

[12] DemirsoyS,DempsterA and KaleI2004Efficient implementation of digital filters using novel reconfigurable multiplier blocks (REMB)*In Conference Record of the Thirty-Eighth Asilomar Conference on Signals, Systems and Computers* IEEE. (1) 461-464

[13] MokhtariN andRahmanianS2006Hardware Implementation Analysis for Digital Filters*In Proceedings of the 14th Iranian Conference on Electrical Engineering (ICEE'06).*

[14] WanhammarL1999DSP Integrated Circuits*academic press series in engineering* ISBN 0-12-734530-2, UK.

Filter Design for A Fractional N-PLL Frequency Synthesizer At 2.4 GHz

Nour Zaid Naktel[1*], K. K. Mohammed[1] and A. Z. Yonis[1]

[1]Electronics Engineering Department, College of Electronics Engineering, Ninevah University, Mosul, Iraq

[*]E-mail: Kalid.kaleel@yahoo.com

Abstract. Radio frequency transmitter and receivers usually needs an efficient frequency synthesizer, and this synthesizer must use a digital electronic circuit to perform different modern digital application which contain a frequency synthesizer circuit as a main part of the circuits. The fractional-N phase locked loop (PLL) has been used widely in the last twenty year ago because it gives a large bandwidth with fine frequency resolution for various electronic applications. This research presents the design of second order low pass filter at a frequency bandwidth of (50,75,125,250,500) kHz and for each value of the phase margin (30,35,40,47,57,65,70,80) degree, the component values of the filters which the capacitors and resistance were found at each values of the frequency bandwidth and for each phase margin, also the filter impedance, open loop and closed loop gain was calculated, it is found that by proper design of the second order low pass filter component, the system can have more stability operation and the spurs noise in the output can be reduced efficiently.

1. Introduction

There are a two type of phase locked loop (PLL) which are the integer PLL and the fractional one, integer PLL has a disadvantage of the relation between the channel spacing and the value of loop bandwidth high frequency resolution needs the reference frequency to be small as possible, and to obtain high stability in the system, the loop filter must have a value smaller than reference frequency, this means that the reference frequency must be reduced to a small value which leads to very slow switching time and increasing the values of phase noise which make the system has undesirable characteristics operation. So, in order to overcome these problems, fractional-N PLL usually used and dismiss the disadvantages found in the integer PLL. The low pass filter is the one of the essential necessities in many applications to diminish commotion at high frequency [1]. It is imperative that the charge pumps signal can't be interfaced straightforwardly to the Voltage Control Oscillator (VCO) because the output of the Charge Pump contain harmonics that affect the output spectrum of the Voltage Control Oscillator (VCO) and the current output of the Charge Pump cannot be used to control the Voltage Control Oscillator (VCO).Therefore, the Loop filter is used to convert the output current from the charge pump to a voltage that control the Voltage Control Oscillator (VCO) as in Figure.(2), As well as removing the harmonics in the output of the Charge Pump[2]. There are two types of Loop filter, the passive filter which has no active element in its circuit like diode, transistor and integrated circuits as in Figure (١-b), while the second types which are the active filter can contain active elements like diode, transistor and integrated circuits (operational amplifiers) as in Figure (١-a), as well as resistor, capacitor and inductor. Inductors are no longer used in the construction of filters because they make a lot of noise as opposed to resistors

Content from this work may be used under the terms of the Creative Commons Attribution 3.0 licence. Any further distribution of this work must maintain attribution to the author(s) and the title of the work, journal citation and DOI.
Published under licence by IOP Publishing Ltd

IOP Conf. Series: Materials Science and Engineering 1152 (2021) 012027 doi:10.1088/1757-899X/1152/1/012027

and capacitors, and they take up a lot of space when used in practice [3]. the readers can refer to [4-10] for more information about phase locked loop (PLL).

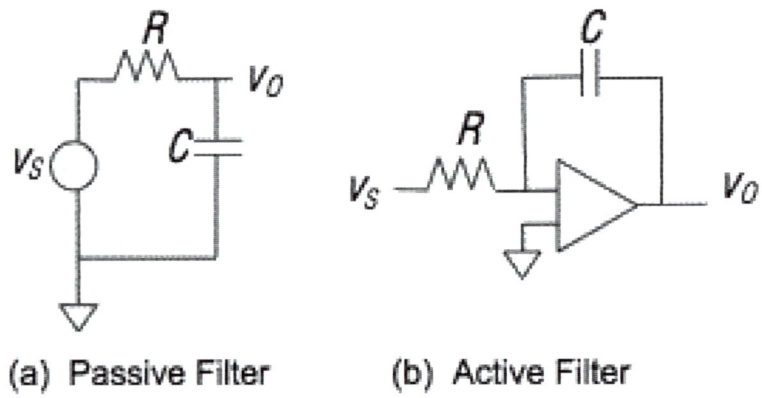

Figure ۱. Passive and active loop filter [3]

2. The Method

Passive Filters were used to increase the complexity of the filter, increase the cost, and reduce the system's stability [11]. Due to the ripple in the loop filter's output voltage, burrs occur in the output range of frequency synthesizers that use first order loop filters. To get a clean output range without burrs, a second order or higher loop filter is used, although this increases the filter's complexity [1]. Usually, the bandwidth of the loop filter determined the bandwidth of the frequency synthesizer [12-15].

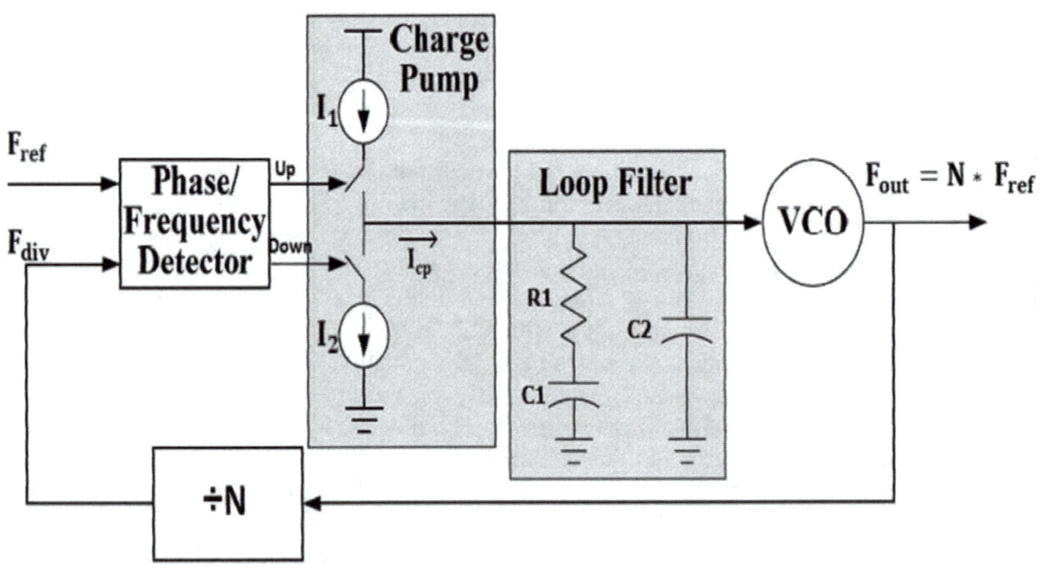

Figure 2. Phase Locked Loop Frequency Synthesizer [16]

The second order filter of the fractional N-PLL synthesizer were designed using the equations shown in the Table 1, also the labels define all the parameters used in the design.

Table 1. The equations of second order loop filter design [17]

Abbreviations	Definition	Equation	Coefficients
Z(s)	The transfer function or impedance of the 2nd order filter	$\frac{1 + s.T2}{s.A0.(1 + s.T1)}$	T1: the zero in the loop filter transfer function. T2: the first pole in the loop filter transfer function. A0: the loop filter coefficient.
N	Division ratio	F_{out}/F_{ref}	F_{ref}: Input signal frequency. F_{out}: Output Frequency
T1	The zero in the loop filter transfer function	$\frac{\sqrt{(1+\gamma)^2.\tan^2\emptyset + 4\gamma} - (1+\gamma).\tan\emptyset}{2.\omega c}$	Ø: the phase margin γ: the improvement factor ωc: the Carrier frequency
T2	The first pole in the loop filter transfer function	$\frac{\gamma}{wc^2.T1}$	
A0	The loop filter coefficient	$\frac{K_\emptyset.K_{vco}}{N.wc^2} * \sqrt{\frac{(1 + wc^2.T2^2)}{(1 + wc^2.T1^2)}}$	$k_\emptyset$: the charges emitting current k_{vco} : the gain of the voltage-controlled oscillator N : the Division ratio
C1	Capacitor 1 of the loop filter	$A0.\frac{T1}{T2}$	
C2	Capacitor 2 of the loop filter	A0-C1	C1:the Capacitor value
R1	Resistor 1 of the loop filter	$\frac{T2}{C2}$	C2: the Capacitor value
A(s)	Open loop gain function	$\frac{Z(s).K_\emptyset.K_{vco}}{N.s}$	Z(s): impedance of the 2nd order filter
G(s)	Close loop gain function	$\frac{A(s)}{1 + A(s)}$	A(s): the open loop gain function

3. Simulations Results

Simulations results of frequency response for open loop and closed loop gain for fractional N-PLL circuit using MATLAB program.

1. A second-order filter is designed for the fractional N-PLL and the filter is designed with more than one value of the bandwidth (50,75,125,250,500 - kHz) and Phase margin=30° and $K_{\phi} = 50\ \mu A$, $K_{vco} = 210\ MHz$, input frequency signal $f_{ref} = 19.2MHz$. Based on the equations of a second-order filter, the values of the filter components were found, and the results of the frequency response were as follows:

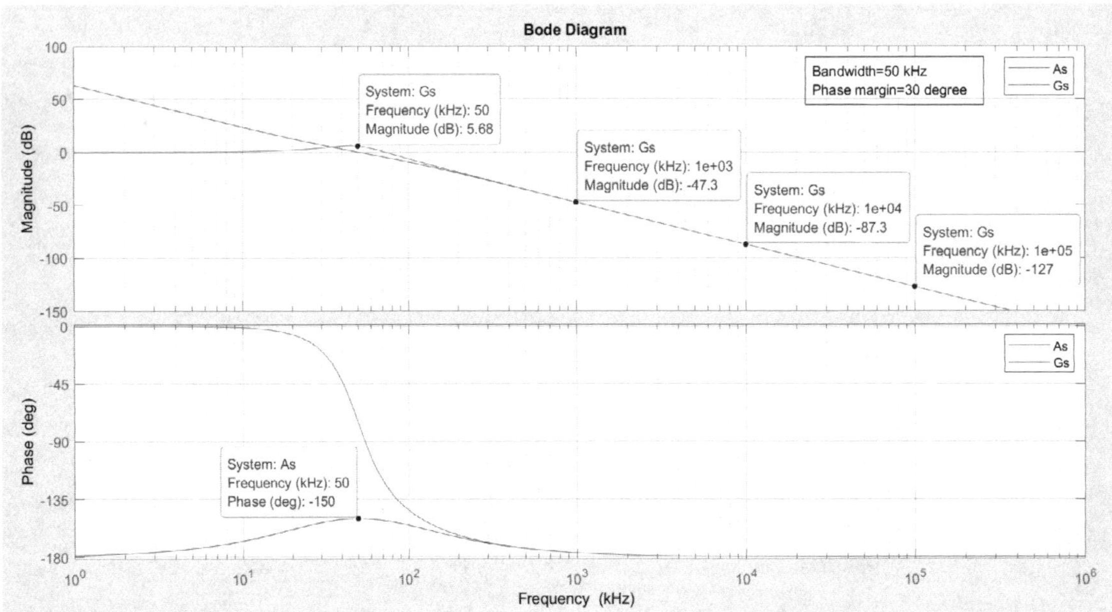

Figure .3 The frequency response of the 2nd order filter at a 30 ° phase margin and a bandwidth of 50 kHz

We note from the Table 2 that the response gain decreases with increasing frequency (the frequency is higher than the value of the bandwidth frequency) and this is one of the characteristics of the low-pass filters, it should be noted that whenever the slope of the gain frequency response faster to go down the, loop filter removes the spurs noise at higher frequencies better.

Table .2 Comparison of the frequency response gain of the second-order filter at 30 ° phase margin

Frequency (MHz)	Closed loop T.F magnitude (dB)				
	B.W=50kHz	B.W=75kHz	B.W=125kHz	B.W=250kHz	B.W=500kHz
1	-47.3	-40.2	-31.3	-19.1	-6.31
10	-87.3	-80.2	-71.4	-59.3	-47.3
100	-127	-120	-111	-99.3	-87.3

2. A second-order filter is designed for the fractional N-PLL and the filter is designed with more than one value of the bandwidth (50,75,125,250,500 - kHz) and Phase margin=47° and $K_{\phi} = 50\ \mu A$, $K_{vco} = 210\ MHz$, input frequency signal $f_{ref} = 19.2MHz$ .Based on the equations of second-order filter, the values of the

filter components were found, and the results of the frequency response were as follows:

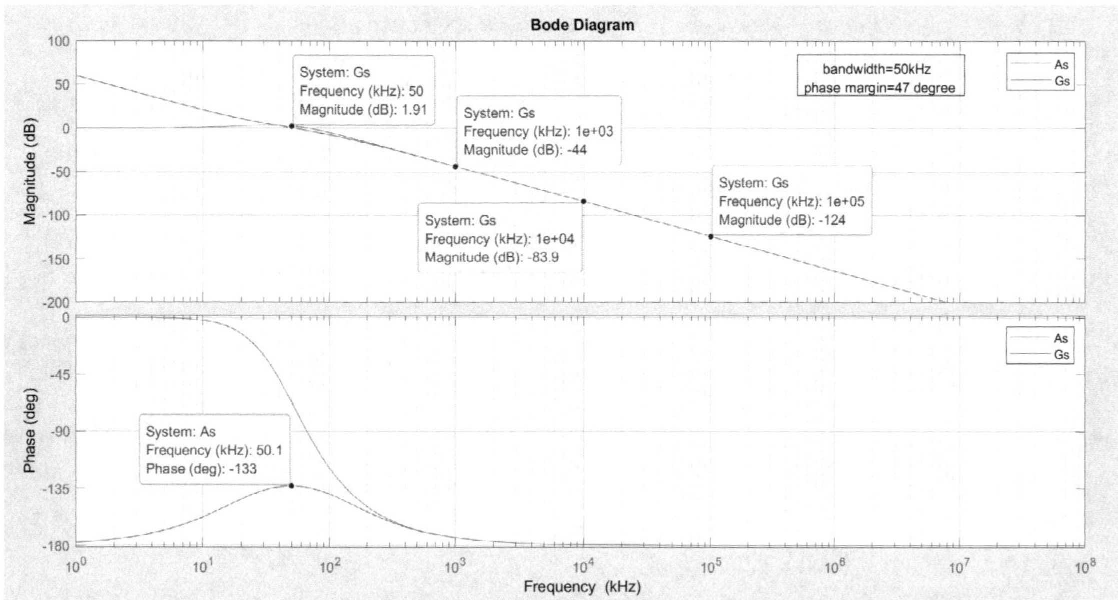

Figure .4 The frequency response of the second-order filter at a 47 ° phase margin and a bandwidth of 50 kHz

It is noted from the Table 3 that the response gain also decreases with increasing frequency (the frequency is higher than the value of the bandwidth frequency). The spurs noise is removed more when the frequency is increased and the frequency response is decreased fast due to the 2nd order filter.

Table .3 Comparison of the frequency response gain of the second-order filter at 47 ° phase margin

Frequency (MHz)	Closed loop T.F magnitude (dB)				
	B.W=50kHz	B.W=75kHz	B.W=125kHz	B.W=250kHz	B.W=500kHz
1	-44	-36.9	-28.1	-16.3	-5.38
10	-83.9	-76.9	-68	-56	-44
١٠٠	-124	-117	-108	-96	-83.9

4. Conclusions

It is concluded from the results obtained that increasing the phase margin will not improve the gain of the frequency response at high frequencies at any value for the bandwidth. For example, when comparing the Gain of the frequency response at a bandwidth of (50 kHz), the value of the response gain was equal to (- 37.2 dB) at a frequency (1 MHz) and a phase margin (70 °) while the value of the frequency response Gain is equal to (-44 dB) for the same frequency value (1 MHz) and a phase margin (47 °) while the value of the frequency response Gain is equal to (-47.3 dB) for the same frequency value (1 MHz) and phase margin (30 °). Consequently, there is an increase in the level of spurs noise affecting the output signal spectrum of the fractional N-PLL.

5. References

[1] Brian D 2008 *Analysis and Design of High Order Digital Phase Locked Loops* (Doctoral Thesis, National University of Ireland, Maynooth, Department of Electronic Engineering).

[2] Zhiming D 2007 *Building Blocks for A Wide-Band Phase-Locked Loop (*Master thesis, University of California, Department of Electrical Engineering and Computer Sciences).

[3] Michael H. P 2005 *PLL Design Using the PLL Design Assistant Program* http://cppsim.org, pp.1-33.

[4] Song, j. Huang, R. Tsao, R. and Wu, Y. 1992 *A new phase-locked loop used in a frequency synthesizer (*IEEE Transactions on Instrumentation and Measurement) vol. 41, no. 3, pp. 432-437.

[5] X. Zhang, "A Fractional-N Frequency Synthesizer with No Fractional Spurs," in IEEE Transactions on Circuits and Systems II: Express Briefs, vol. 53, no. 9, pp. 986-990, Sept. 2006.

[6] K. -C. Peng, W. -L. Wu and J. -H. Lin, "Reduction of Phase Noise in Fractional-N Frequency Synthesizer Using Self-Injection Locking Loop," in IEEE Transactions on Microwave Theory and Techniques, vol. 68, no. 9, pp. 3724-3731, Sept. 2020.

[7] H. Arora, N. Klemmer, J. C. Morizio and P. D. Wolf, "Enhanced phase noise modeling of fractional-N frequency synthesizers," in IEEE Transactions on Circuits and Systems I: Regular Papers, vol. 52, no. 2, pp. 379-395, Feb. 2005.

[8] Y. W. Kim and J. D. Yu, "Phase Noise Model of Single Loop Frequency Synthesizer," in IEEE Transactions on Broadcasting, vol. 54, no. 1, pp. 112-119, March 2008.

[9] S. K. Romanov, A. V. Grechishkin, M. P. Savchenko, I. A. Markov, M. V. Maturazov and T. V. Maturazova, "Reducing the level of spurious components in the phase-locked loop frequency synthesizer," 2017 Systems of Signal Synchronization, Generating and Processing in Telecommunications (SINKHROINFO), Kazan, 2017, pp. 1-4.

[10] R. Gatti et al., "Design of frequency synthesizer for wireless communication," 2017 2nd IEEE International Conference on Recent Trends in Electronics, Information & Communication Technology (RTEICT), Bangalore, India, 2017, pp. 1794-1798.

[11] Curtis B 1999 *Fractional/Integer-N PLL Basics* (Texas Instruments Incorporated) ,pp.2-54.

[12] Alaa H 2002 *Design and Analysis of Fractional-N Frequency Synthesizers for Wireless Communications* (Doctoral thesis, University of Waterloo, Canada).

[13] V. Valenta, M. Villegas and G. Baudoin, 2008 *Analysis of a PLL Based Frequency Synthesizer using Switched Loop Bandwidth for Mobile WiMAX* (18th International Conference Radioelektronika), Prague, Czech Republic, pp. 1-4.

[14] X. Xu, Huihua Liu and S. Ao, 2011 *Design of low phase noise and fast locking PLL frequency synthesizer*, International Conference on Electric Information and Control Engineering, Wuhan, China, pp. 4113-4116.

[15] Y. Feng and G. Chen 2011 *A fractional-N synthesizer for multi-mode positioning system with constant loop bandwidth* IEEE International Conference of Electron Devices and Solid-State Circuits, Tianjin, China, pp. 1-2.

[16] Zhao, B., Mao, X., Yang, H. et al.2009, *A 1.41–1.72 GHz sigma-delta fractional-N frequency synthesizer with a PVT insensitive VCO and a new prescaler.* Analog Integr Circ Sig Process 59, 265–273.

[17] Dean B 2006 *PLL Performance, Simulation, and Design* (Dog Ear Publishing, LLC) 4th Edition, ISBN-10: 1598581341, ISBN-13: 978-1598581348.

Design of Fractional-order Sobel Filters for Edge Detections

Ibtisam Edress[1], Emad A. Al-Sabawi[1] and Majid Dherar Younus[1]

[1]Computer and Information Engineering Department, College of Electronics Engineering, Ninevah University

[*]Email: ibtisamedrees3@gmail.com

Abstract. Most image processing applications use edge detection to extract information as a preliminary step to object segmentation and feature extraction. Edge accuracy is one of the edge detector challenges; The current work presents the design of the fractional-order Sobel filters based on Yi_Fei-1 to Yi_Fei-5.A comparison among the proposed filters has been implemented for edge detection based on a supervised assessment, using mean square error, misclassification error, and symmetric distance.Many images with their ground truths have been used in the evaluation. Results showed that the fractional-order of Sobel-based Yi_Fei-2 has the best edge map among the proposed filters.

1. Introduction

Image edge detection plays an important role as one of the precursory standards in pattern recognition applications and computer vision [1]. Edge detection considers one of the vital research works and a very key step towards identification, segmentation, and object recognition which can occur when edges of an object are detected [2].

An edge is well-Known as a discontinuity; that may be represented as texture, color, or gray level of the image. Detection of edges tends to extract vital features about the objects of the image, like illumination, reflectivity, and geometry [3]. For example, the pixel's gray-level which value is identical to the other around pixel's gray-level; there is maybe not an edge at that position. Actually, if neighbors of a pixel have widely varying gray levels, the pixel possibly an edge pixel [4].

Many operators that have been used to extract edges from the images, the classical approaches are grouped as gradient-based techniques and Gaussian-based techniques, which adopted the first and second derivatives to detect the edges in an image. But, thicker edges usually result from the first-order derivative techniques which cause a lack of image details. The second-order derivative techniques have a strong response to fine specifics, but high noise sensitivity [5].

To reduce this disadvantage, fractional calculus has been involved in edge detection techniques. Due to the frequency characteristic of the fractional-order derivative, it can nonlinearly preserve more low-frequency outline characteristics in flat areas; save the aloft frequency marginal function wherever the grayscale varies frequently, and also improve mid-frequency texture specifics [6]. Caputo,Riemann-Liouville (R-L), and Grümwald-Letnikov (G-L)are the most common definitions of fractional-order derivative [7][8]. Considering these benefits, Yi. Pu et al. [9] proposed six fractional differential masks and presented the structures and parameters of all masks individually in the eight directions.

Experiments showed that, for a digital image which rich in detail, the capability of fractional differential-based approaches seemed obvious better than classical methodologies. Dan, J. Wu and Y. Yang [8] proposed a novel fractional-order gradient detector to extract features from the structure of a medical image. They were generalizing Sobel operators depending on the GL definition of non-integer order derivative. Experiments appeared that the modified Sobel operator yields well observable effects.

In this work, we generalize the classical Sobel [10], using the five fractional-order operators from the six in [9]. The proposed operators are evaluated based on objective metrics to get the minimum difference between the ground truth images and the candidate.

Content from this work may be used under the terms of the Creative Commons Attribution 3.0 licence. Any further distribution of this work must maintain attribution to the author(s) and the title of the work, journal citation and DOI.
Published under licence by IOP Publishing Ltd

2.Edge Detection using Fractional Derivative of Sobel

Let us consider *I (x, y)* as a digital image, whereas *(x, y)* denote spatial coordinates. The magnitude of the image gradient is well-defined as:

$$G = \sqrt{(Gx^2 + Gy^2\ )} \tag{1}$$

Where *Gy* and *Gx* are components of image gradient in *y* and *x* directions respectively.

$$G_y = \frac{\partial I(x,y)}{\partial y} \tag{2}$$

$$G_x = \frac{\partial I(x,y)}{\partial x} \tag{3}$$

The image *I(x, y)* is filtering using Sobel kernels *Kx*and *Ky*, *Gx* and *Gy* are obtained as shown in equation (4).

$$G_x = I \cdot K_x \quad , \qquad G_y = I \cdot K_y \tag{4}$$

$$K_x = \begin{bmatrix} -1 & 0 & 1 \\ -2 & 0 & 2 \\ -1 & 0 & 1 \end{bmatrix} K_y = \begin{bmatrix} -1 & -2 & -1 \\ 0 & 0 & 0 \\ 1 & 2 & 1 \end{bmatrix}$$

$$g_x(x,y) = I(x+1,y+1) + 2I(x+1,y) + (x+1,y-1) - I(x-1,y+1) - 2I(x-1,y) - I(x-1,y-1) \tag{5}$$

$$g_y(x,y) = I(x+1,y+1) + 2I(x,y+1) + I(x-1,y+1) - I(x+1,y-1) - I(x,y-1) - I(x-1,y-1) \tag{6}$$

To get the fractional derivative of the Sobel operator, the five Yi_Fei operators illustrated in table 1[9] are used to generalize the classical one. The first operator was used in paper [11]. The fractional Sobel mask using operator 2 is illustrated in equations (8) and (9).

$$D^p G_x = G_x^{1+p} \tag{7}$$

$$\begin{aligned} g_x^{1+p}(x,y) = \{ & a_{+1}[I(x+2,y+1) + 2I(x+2,y) + I(x+2,y-1) - I(x,y+1) - 2I(x,y) \\ & - I(x,y-1)] \\ & + a_0[I(x+1,y+1) + 2I(x+1,y) + I(x+1,y-1) - I(x-1,y+1) \\ & - 2I(x-1,y) \ - I(x-1,y-1)] \\ & + a_{-1}[I(x,y+1) + 2I(x,y) + I(x,y-1) - I(x-2,y+1) - 2I(x-2,y) \\ & - I(x-2,y-1)] \\ & + a_{-2}[I(x-1,y+1) + 2I(x-1,y) + I(x-1,y-1) - I(x-3,y+1) \\ & - 2I(x-3,y) \ - I(x-3,y-1)] \\ & + \ a_{-3}[I(x-2,y+1) + 2I(x-2,y) + I(x-2,y-1) - I(x-4,y+1) \\ & - 2I(x-4,y) \ - I(x-4,y-1)]\} \end{aligned} \tag{8}$$

$$G_x^{1+p} = I \cdot M2_x, \qquad G_y^{1+p} = I \cdot M2_y \tag{9}$$

The mask using Yi_Fei-2 is given below:

$$M2_x = \begin{bmatrix} \left(\frac{p^2}{8}-\frac{3p^3}{16}+\frac{p^4}{16}\right) & \left(\frac{p^2}{4}-\frac{3p^3}{8}+\frac{p^4}{8}\right) & \left(\frac{p^2}{8}-\frac{3p^3}{16}+\frac{p^4}{16}\right) \\ \left(\frac{5p}{4}+\frac{p^2}{8}+\frac{p^3}{4}\right) & \left(\frac{5p}{2}+\frac{p^2}{4}+\frac{p^3}{2}\right) & \left(\frac{5p}{4}+\frac{p^2}{8}+\frac{p^3}{4}\right) \\ \left(\frac{5p}{4}+\frac{p^2}{8}-\frac{p^3}{2}\right) & \left(\frac{5p}{2}+\frac{p^2}{4}-v^3\right) & \left(\frac{5p}{4}+\frac{p^2}{8}-\frac{p^3}{2}\right) \\ \left(-1-\frac{5p}{4}+\frac{5p^2}{8}+\frac{3p^3}{8}\right) & \left(-2-\frac{5p}{2}+\frac{5p^2}{4}+\frac{3p^3}{4}\right) & \left(-1-\frac{5p}{4}+\frac{5p^2}{8}+\frac{3p^3}{8}\right) \\ \left(\frac{6p}{4}+\frac{p^2}{8}-\frac{5p^3}{16}-\frac{p^4}{16}\right) & \left(\frac{6p}{2}+\frac{p^2}{4}-\frac{5p^3}{8}-\frac{p^4}{8}\right) & \left(\frac{6p}{4}+\frac{p^2}{8}-\frac{5p^3}{16}-\frac{p^4}{16}\right) \\ \left(1-\frac{p^2}{2}-\frac{p^3}{8}\right) & \left(2-p^2-\frac{p^3}{4}\right) & \left(1-\frac{p^2}{2}-\frac{p^3}{8}\right) \\ \left(\frac{p}{4}+\frac{p^2}{8}\right) & \left(\frac{p}{2}+\frac{p^2}{4}\right) & \left(\frac{p}{4}+\frac{p^2}{8}\right) \end{bmatrix}$$

$M2_y = [M2_x]^T$

Where$\alpha = \Gamma(-p) * (-2 * p)$

Table 1. The five operators of Yi_Fei [9].

	Yi_FeiPU-1	*Yi_FeiPU-2*	*Yi_FeiPU-3 , p < 0*	*Yi_FeiPU-4 , p < 0*	*Yi_FeiPU-5,* $0 \leq p < 1$
C_{-1}	0	$\frac{p}{4}+\frac{p^2}{8}$	0	0	0
C_0	1	$1-\frac{p^2}{2}-\frac{p^3}{8}$	$\frac{1}{\Gamma(-p)(-2p)}$	$\frac{1}{\Gamma(-p)(p^2-p)}$	$\frac{1}{\Gamma(2-p)}$
C_1	$-p$	$-\frac{5p}{4}+\frac{p^3}{16}+\frac{p^4}{16}$	$\frac{2^{-p}}{\Gamma(-p)(-2p)}$	$\frac{2^{1-p}-2}{\Gamma(-p)(p^2-p)}$	$\frac{2^{1-p}-2}{\Gamma(2-p)}$
$\vdots$	$\vdots$	$\vdots$	$\vdots$	$\vdots$	$\vdots$
C_i	$\frac{\Gamma(i-p)}{(i)!\,\Gamma(-p)}$	$\frac{\Gamma(i-p-1)}{(i-1)!\,\Gamma(-p)}\cdot\left(-\frac{p}{4}+\frac{p^2}{8}\right)$	$\frac{i^{-p}-(i-1)^{-p}}{\Gamma(-p)(-2p)}$	$\frac{(1-p)i^{-p}-i^{1-p}+(i-1)^{1-p}}{\Gamma(-p)(p^2-p)}$	$\frac{(1-v)i^{-p}-i^{1-p}(i-1)^{1-p}}{\Gamma(2-p)}$

The other masks using Yi_Fei-3 to five operators are as follow:

$$M3_x = \begin{bmatrix} \left(\frac{-2^{-2p}+3^{-p}}{\alpha}\right) & \left(\frac{-2^{1-2p}+2*3^{-p}}{\alpha}\right) & \left(\frac{-2^{-2p}+3^{-p}}{\alpha}\right) \\ \left(\frac{1-2*3^{-p}+2^{-2p}}{\alpha}\right) & \left(\frac{2-2^2*3^{-p}+2^{1-2p}}{\alpha}\right) & \left(\frac{(1-2*3^{-p}+2^{-2p})}{\alpha}\right) \\ \left(\frac{3^{-p}-2^{-p}-1}{\alpha}\right) & \left(\frac{2-2^2*3^{-p}+2^{1-2p}}{\alpha}\right) & \left(\frac{3^{-p}-2^{-p}-1}{\alpha}\right) \\ \left(\frac{-1+3^{-p}-2^{-p}}{\alpha}\right) & \left(\frac{-2+2*3^{-p}-2^{1-p}}{\alpha}\right) & \left(\frac{3^{-p}-2^{-p}-1}{\alpha}\right) \\ \left(\frac{2^{-p}}{\alpha}\right) & \left(\frac{2^{1-p}}{\alpha}\right) & \left(\frac{2^{-p}}{\alpha}\right) \\ \left(\frac{1}{\alpha}\right) & \left(\frac{2}{\alpha}\right) & \left(\frac{1}{\alpha}\right) \end{bmatrix}$$

Where $\alpha = \Gamma(-p) * (p^2 - p)$, $M3_y = [M3_x]^T$

$$M4_x = \begin{bmatrix} \left(\frac{(1-p)*2^{-2p} - 2^{2-2p} + 3^{1-p}}{\alpha}\right) & \left(\frac{(1-p)*2^{1-2p} - 2^{3-2p} + 2*3^{1-p}}{\alpha}\right) & \left(\frac{(1-p)*2^{-2p} - 2^{2-2p} + 3^{1-p}}{\alpha}\right) \\ \left(\frac{2^{2-p} - 2^{2-2p} + (1-p)*2^{-2p} - 1}{\alpha}\right) & \left(\frac{2^{2-p} - 2^{2-2p} + (1-p)*2^{-2p} - 1}{\alpha}\right) & \left(\frac{2^{2-p} - 2^{2-2p} + (1-p)*2^{-2p} - 1}{\alpha}\right) \\ \left(\frac{3 - 3*2^{1-p} + 3^{-p}}{\alpha}\right) & \left(\frac{6 - 3*2^{2-p} + 2*3^{-p}}{\alpha}\right) & \left(\frac{3 - 3*2^{1-p} + 3^{-p}}{\alpha}\right) \\ \left(\frac{3^{1-p} - 2^{2-p}}{\alpha}\right) & \left(\frac{2*3^{1-p} - 2^{3-p}}{\alpha}\right) & \left(\frac{3^{1-p} - 2^{2-p}}{\alpha}\right) \\ \left(\frac{2^{1-p} - 2}{\alpha}\right) & \left(\frac{2^{2-p} - 2^2}{\alpha}\right) & \left(\frac{2^{1-p} - 2}{\alpha}\right) \\ \left(\frac{1}{\alpha}\right) & \left(\frac{2}{\alpha}\right) & \left(\frac{1}{\alpha}\right) \end{bmatrix}$$

Where $\alpha = 1/\Gamma(2-p)$, $M4_y = [M4_x]^T$

$$M5_x = \begin{bmatrix} \left(\frac{2^{2-2p} - (1-p)*2^{-2p} - 3^{1-p}}{\alpha}\right) & \left(\frac{2^{3-2p} - (1-p)*2^{1-2p} - 2*3^{1-p}}{\alpha}\right) & \left(\frac{2^{2-2p} - (1-p)*2^{-2p} - 3^{1-p}}{\alpha}\right) \\ \left(\frac{-2^{2-p} + 3^{1-p} + 1}{\alpha}\right) & \left(\frac{-2^{3-p} + 2*3^{1-p} + 2}{\alpha}\right) & \left(\frac{-2^{2-p} + 3^{1-p} + 1}{\alpha}\right) \\ \left(\frac{2^{2-p} - 2^{2-2p} + (1-p)*2^{-2p} - 1}{\alpha}\right) & \left(\frac{2^{3-p} - 2^{3-2p} + (2-2p)*2^{-2p} - 2}{\alpha}\right) & \left(\frac{2^{2-p} - 2^{2-2p} + (1-p)*2^{-2p} - 1}{\alpha}\right) \\ \left(\frac{-2^{1-p} + 3^{1-p} + 3}{\alpha}\right) & \left(\frac{-2^{2-p} + 2*3^{1-p} + 6}{\alpha}\right) & \left(\frac{-2^{1-p} + 3^{1-p} + 3}{\alpha}\right) \\ \left(\frac{3^{1-p} - 2^{2-p}}{\alpha}\right) & \left(\frac{2*3^{1-p} - 2^{3-p}}{\alpha}\right) & \left(\frac{3^{1-p} - 2^{2-p}}{\alpha}\right) \\ \left(\frac{2^{1-p} - 2}{\alpha}\right) & \left(\frac{2^{2-p} - 2^2}{\alpha}\right) & \left(\frac{2^{1-p} - 2}{\alpha}\right) \\ \left(\frac{1}{\alpha}\right) & \left(\frac{2}{\alpha}\right) & \left(\frac{1}{\alpha}\right) \end{bmatrix}$$

Where $\alpha = \Gamma(3-p)$, $M5_y = [M5_x]^T$

The threshold is based on the mean of gradient magnitude in equation (10).

$$Thresold = Scale \cdot \bar{G} \quad (10)$$

Where $Scale \geq 1$, and$\bar{G}$ is the average of $|G|$. The pixel is considered as an edge

$$\text{if}(Dr = |G| > Threshold). \quad (11)$$

3. Numerical Result

Test images and theirground truth have been used in the evaluation process.Ground truth image is an edge map acquired from human judgment or synthetic data.Test images have been filtered using the proposed filters with orders from 0 to 1in steps of 0.01. The mean square error, misclassification error, and symmetric distanceare illustrated in equations (11),(12), and (13) respectively [12][13][14].

$$MSE = \frac{1}{MN}\sum_{i=1}^{M}\sum_{j=1}^{N}\left(Gt_{i,j} - Dr_{i,j}\right)^2 \quad (12)$$

$$ME = 1 - \frac{TN+TP}{TP+TN+FN+FP} (13)$$

Where
False Negative (*FN*), missing edge points of *Dr*: $FN = |Gt \cap \neg Dr|$,
False Positive (*FP*), fake detected edges of *Dr*: $FP = |\neg Gt \cap Dr|$,
True Negative (*TN*), mutual non-edge points: $TN = |\neg Gt \cap \neg Dr|$,
True Positive (*TP*), mutual points of *Gt* and *Dr*: $TP = |Gt \cap Dr|$.

$$SDM\,(Gt, Dr) = \sqrt[l]{\frac{\sum_{px \in Dr} d^{l}_{Gt}(p) + \sum_{px \in Gt} d^{l}_{Dr}(px)}{|Gt \cup Dr|}} \tag{14}$$

Where $l \in R^{+}$, for a pixel *px* belonging to the detector result *Dr*, $D_{Gt}\,(px)$ is the minimal Euclidean distance between *px* and *Gt*. If *px* belongs to the ground truth *Gt*, $D_{Dr}\,(px)$ is the minimal distance between *px* and *Dr*.

Figure 1, Figure 2, and Figure 3 show the mean square error, misclassification error, and symmetric distance respectively for the classical Sobel (*p=0*) and the five fractional-order Sobel filters. As shown in the figures, the fractional-order Sobel filter based Yi_Fei-2 has the minimum scores and the behavour of the three figures approximately the same.

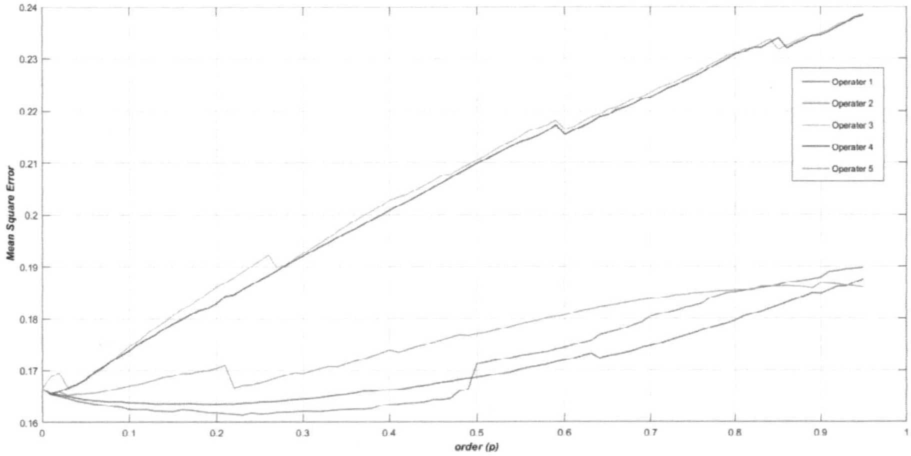

Figure 1. Mean square error for the five fractional-orderSobel filters.

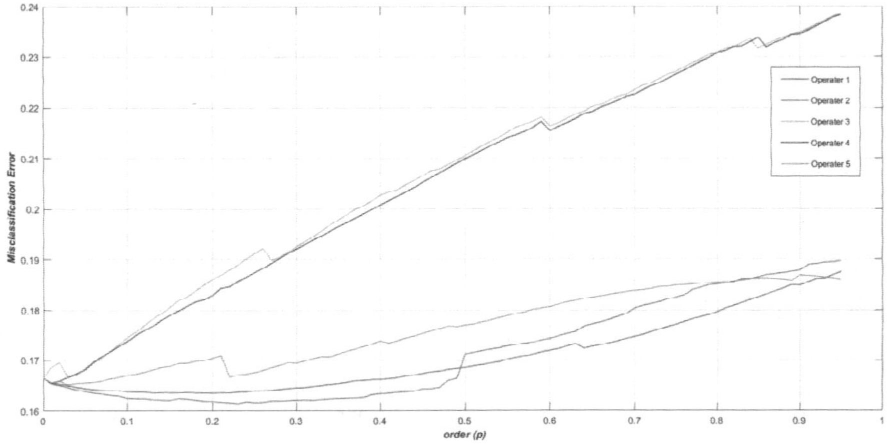

Figure 2. Misclassification errorfor the five fractional-orderSobel filters.

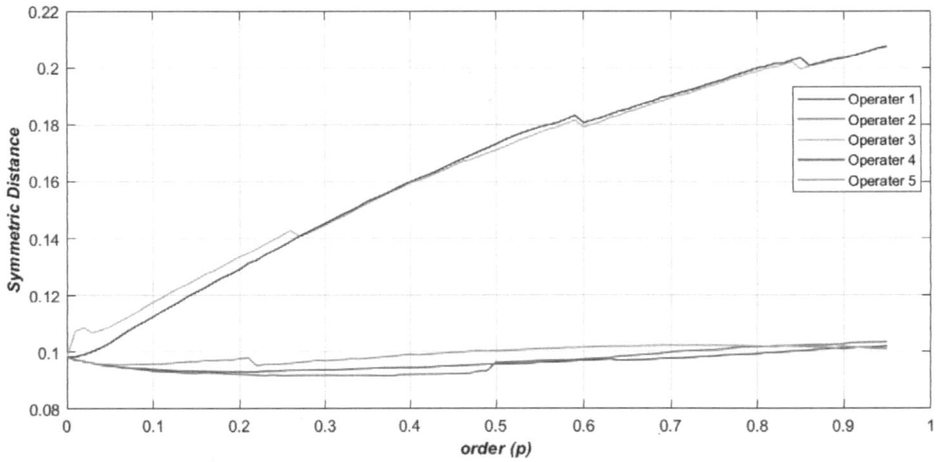

Figure 3. Symmetric distance for the five fractional-orderSobel filters.

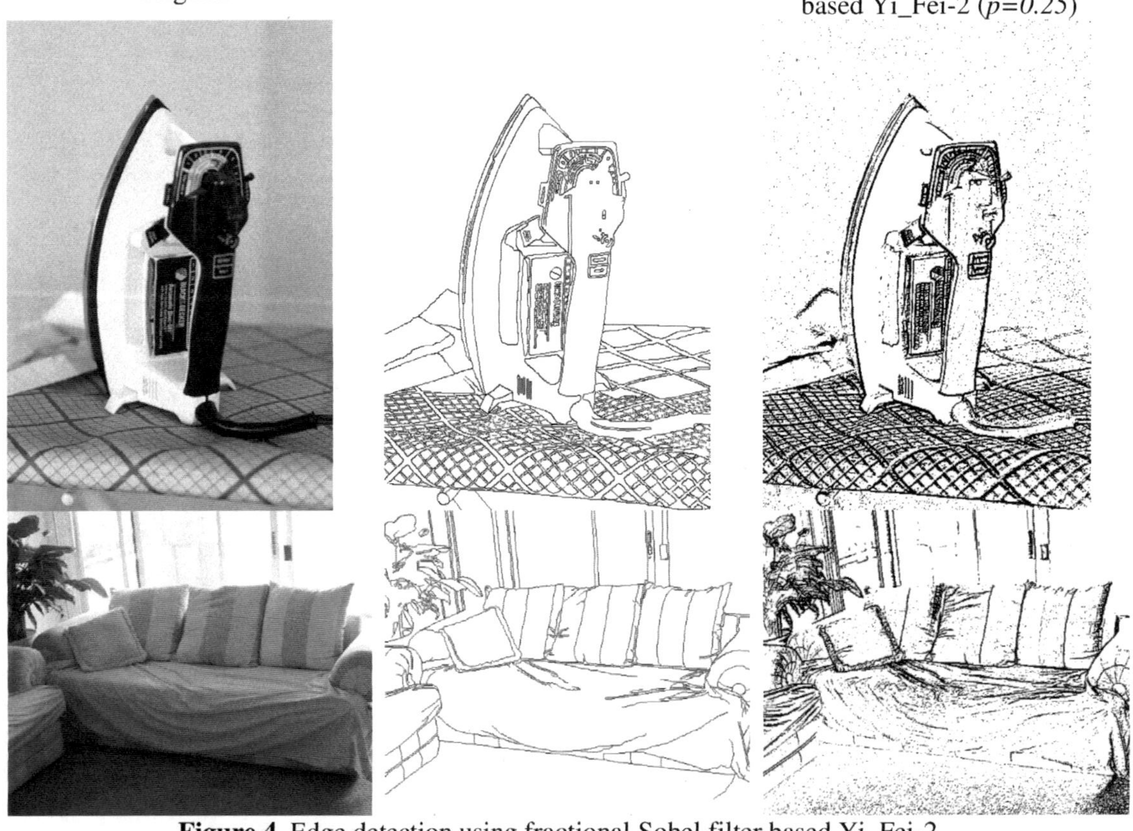

Figure 4. Edge detection using fractional Sobel filter based Yi_Fei-2.

4. Conclusion

The fractional-order Sobel filters based on Yi_Fei-2, Yi_Fei-3, Yi_Fei-4, and Yi_Fei-5 operators have been designed. According to assessment which based on mean square error, misclassification error, and symmetric distance, the traditional Sobel filter (p=0) has better performance than the fractional-order Sobel filter based on Yi_Fei-4, and Yi_Fei-5. Whereas fractional order Sobel filter based Yi_Fei-2 has better performance than classical Sobel, fractional-order Sobel filter based Yi_Fei-1, and Yi_Fei-3 when the order of differentian is greater than 0 and less than 0.47.

5. References

[1] BurgerW and Burge M. J. 2013*Principles of Digital Image Processing: Advanced Methods*Spri. Sci. &Busi. Med. p 369.

[2] Kumar, Indrajeet, Rawat J and Bhadauria H. S. 2014 A conventional study of edge detection technique in digital image processing *Intern. J. of Comput. Sci. and Mob.* **3.4** 328-334.

[3] Marr, David and Hildreth E 1980 Theory of edge detection *Royal Society* **207.1167** 187-217.

[4] Othman, Zolqernine, et al. 2009 Comparison of Canny and Sobel edge detection in MRI images *Comput. Sci., B. & T., and Inf. Sys.* 133-136.

[5] Jain, Ramesh, KasturiR, and SchunckB. G 1995*Machine vision***Vol. 5.**McGraw-hill.

[6] Tian, Dan, L Dapeng, and ZhangY 2015Medical image segmentation based on fractional-order derivative *Asia-Pacific Energy Equi. E. Conf.* Atlantis Press.

[7] Hacini, Meriem, Hachouf F, and CharefA2020A bi-directional fractional-order derivative mask for image processing applications *IET Image Proc.* **14.11** 2512-2524.

[8] Yazgaç, GörkemB, and KırcıM 2019Fractional order calculus based fruit detection *2019 8thInter. Conference on Agro-Geoin. IEEE*1-4.

[9] Pu, Yi-Fei, ZhouJ, and YuanX2009Fractional differential mask: a fractional differential-based approach for multiscale texture enhancement *IEEE trans.***19.2** 491-511.

[10] Mehra, Rajesh, and VermaR 2012Area efficient fpga implementation of sobel edge detector for image processing applications *Inter. J. of Comp. App.* 56.16.

[11] Yaacoub, Charles, and AbiZeidDaouR2019 Fractional order sobel edge detector *2019 Ninth Inter. Conference IPTA IEEE*1-5.

[12] Lavín-Delgado, J. E., et al 2020A new fractional-order mask for image edge detection based on Caputo–Fabrizio fractional-order derivative without singular kernel *Circuits, Sys., and Signal Pro.***39.3** 1419-1448.

[13] BMagnier, AbdulrahmanH, and MontesinosP2018A review of supervised edge detection evaluation methods and an objective comparison of filtering gradient computations using hysteresis thresholds *J.of Imaging* **4.6** 74.

[14] Abdulrahman H, MagnierB, and MontesinosP2017From contours to ground truth: how to evaluate edge detectors by filtering *J. of WSCG***25** 133-142 .

IOP Conf. Series: Materials Science and Engineering 1152 (2021) 012029 doi:10.1088/1757-899X/1152/1/012029

Improved resolution for separation between acoustical transmitters with their locations using Eigenvector algorithm

Azzah T Qaba[1*] and Mujahid F AL-Azzo[1]

[1]Electronic Engineering Department, Electronic Engineering collage, Ninevah university, Mosul-IRAQ.

[*]E-mail: azza.salim20182018@stu.uoninevah.edu.iq

Abstract. Through the mathematical model that is derived to localize and separate acoustic sources. We used two methods and compared the results of them: classical method (Fourier transform), with the newer method (Eigenvector) or what represents (super resolution). The results of the comparison revealed the superiority of (Eigenvector) over (Fourier transform) in several aspects, the most important of which are contiguous sources in addition to the location of the sources. The study relied on a set of different criteria directly related to the problem.

1. Introduction

There is a problem facing many applications such as wireless communication, estimating the direction of arrival, etc., which is the estimation of frequencies, angles, locations, with their magnitudes. In order to obtain the data, a synthetic process of aperture scanning of the acoustic receiver was used, which collects a set of data. The non-parametric Fourier transform technique (FT), also known as (traditional spectral estimation) technique, is considered one of the effective techniques in the calculation. Nevertheless, it has some flaws, the most important of which are: the side lobes that may overcome a weak, desired, transmitted signal. Moreover, the side lobes seem to be an additional, undesired, transmitting signals, where it may cause a confusion in discriminating the acoustical sources, as well as the accuracy as it is low resolution achievement when compared with other modern methods [1]. To solve this problem, it requires more data, and for this we need a "long slot" to overcome such spectral leakage. Resolution is the ability of a spectral method to separate the points that make up the object, that is the scene of transmitters. The main goal of the paper is to compare the traditional (Fourier transform) low-resolution method with the modern (super resolution) method (Eigenvector method). Eigenvector based method has an important role in solving the side-lobes problem as well as it has a high resolution [1] The process of localizing the sources along with their separation process has two steps:

1- The recording of the data in the form of amplitude and phase samples of acoustic waves, and then,
2- processing the data and applying one of the spectral estimation methods, either the Fourier transform or the Eigenvector one.

Eigenvector method has different applications [2, 3], as high-resolution technique. The contribution of this paper is the application of EV method for discriminating the small separation between the transmitting sources and their locations with high resolution. Other relating work can be found in [4,5].

2. Field analysis at the Fresnel region.

The problem to be solved is represented by two sound sources, or we can call them (object), which have a field distribution $q(p)$. Therefore, $S(X)$represents the spread of this field towards the registration axis by[6 -8]

Content from this work may be used under the terms of the Creative Commons Attribution 3.0 licence. Any further distribution of this work must maintain attribution to the author(s) and the title of the work, journal citation and DOI.
Published under licence by IOP Publishing Ltd

$$S(X) = \frac{B}{Z \Lambda} \int_p d(p) \exp(\, jK r(p, X)) dp \tag{1}$$

The previous equation is very similar to the (Fresnel principle). Where K represents the wave number, and B can be considered as constant. Z Indicates the distance between the object plane and the recording plane. There is a distance connecting a certain point on the recording axis X and another point on the object p, which is r that can be calculated through

$$r(p, X) = \sqrt{Z^2 + (X - p)^2} \tag{2}$$

When we used approximation, we get

$$((X - p)^2 / Z^2) \ll 1 \tag{3}$$

The $r\ (p, X)$ is given as

$$r(p, X) = Z + \frac{(X - p)^2}{2\, Z} - \frac{(\, X - p)^4}{8Z^3} + - - - - - \tag{4}$$

and since we are in the Fresnel region, we will only use two terms from the equation (4) for r , i.e.

$$r(p, X) = Z + \frac{X^2}{2Z} + \frac{p^2}{2Z} - \frac{pX}{Z} \tag{5}$$

Substituting equation (5) in equation (1) produce

$$S(X) = B_1 \exp\, (\frac{jKX^2}{2Z}) \int_p d(p) \exp(\frac{jKp^2}{2Z})\, \exp(\frac{-jKX\ \ p}{2Z}) dp \tag{6}$$

B1: a composite constant producing from substituting equation (5) in equation (1).

3. Eigenvector Method principle

In fact, Eigen-vector (EV) method, can be considered as a method derived from the multiple signal classification (MUSIC) method and among several methods, the method Eigen-vector can estimate complex exponential frequencies in a noise. The EV method estimates the exponential frequencies from the peaks of the Eigen spectrum. Whereas, the basis of its action depends on the processes of the autocorrelation matrix Qs of the observed data [9, 10].

Assuming that the signal contains of g exponential of a complex form and b denotes the number of recorded samples, provided that b is subject to (b> g + 1) then the (EV) will estimate the exponential frequencies from the peaks of the eigen spectrum[11].

$$\widehat{P}_{eig}(e^{j\omega_1}) = \frac{1}{(\sum_{l=g+1}^{b} \left(\frac{1}{\Lambda_l}\right) |e^{H_1} V_l|^2} \tag{7}$$

where the eigenvectors (V_l), l =1, 2... gassociated with the eigenvalue (Λ_l) equations (8)

$$\mathbf{e}\ \ = \left[1, e^{i\omega}, e^{i2\omega}, \ldots, e^{i(q-1)\omega}\right] \tag{8}$$

Where H_1 denotes the conjugate transpose [12].

4. Sources localization

Multiplication of the recorded data S(X) by the quadratic phase factor $(\frac{KX^2}{2Z})$ and then applying one of the spectral resolution methods: either a Fourier transform method, or a super resolution (Eigenvector) method.

5. Results and conclusions

To solve the research problem, computer programs were used. Two sources acoustic transmission were used as (two points). This was done by placing the first source at a lateral distance $\rho 1$=2cm from the assumed main axis, from which the future acoustic transducer starts. After this step, we placed the second transmitter source at a sideways distance $\rho 2$ from the main axis, after which we took a set of samples from the received wave from the two sources at the receiving (recording) axis, according to the synthetic aperture scanning and across samples (N_o). Where Z represents the axial distance between the receiving axis and the transmitter axis. As for the wavelength it is λ, there is a sampling interval, which is ΔX . Represents the difference between the two sources location $\Delta\rho$,, which means that it is equal to the difference as shown ($\rho 2$- $\rho 1$).
A set of different values were used for each of Z , ΔX , λ and $\rho 2$ in light of this, we obtained a set of results by applying two methods to estimate the spectrum: Fourier transform (FT) and method Eigenvector (EV).
Where the minimum separation between sources (accuracy) is given by the following equations[13]

$$\sigma_1 = \frac{\lambda Z}{a_0} \tag{9}$$

a_0: it is half of the aperture length.

$$a_0 = \frac{(N_o - 1)}{2} \Delta X \tag{10}$$

(P.E.) represents the percentage of error of $\Delta\rho$, since it is equal to ($\rho 2$- $\rho 1$) and we assumed that this ratio is acceptable at (P. E< 20%).

5.1. Results without noise data

Figure 1 shows the results (implementation) of the two methods when applied. According to equations (9) and (10), σ_1> 4.4 cm and thus (FT) was not able to separate the two sources (two objects), rather it appeared as a single source (object), especially for the values. Less than σ_1 = 5, and this is in contrast to what happened with the (EV) method where it was able to separate the twosources at $\rho 2$ = 3 cm, i.e., $\Delta\rho$ = 1 cm, not only that, but the percentage of error (PE) compared to (FT) method. Much less, as shown in Figure 1.

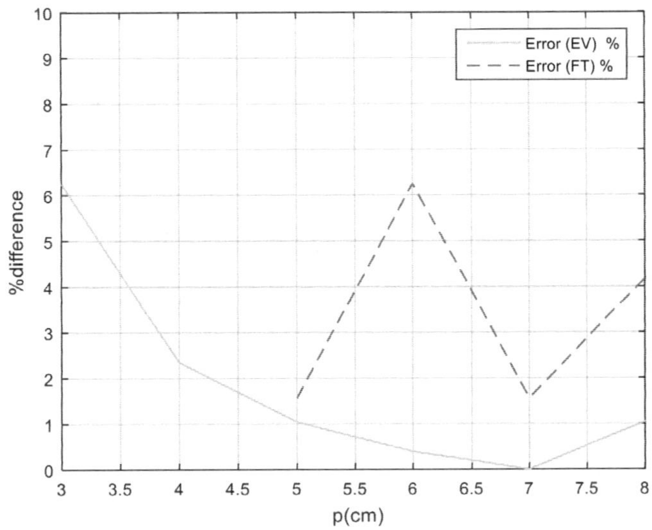

Figure 1. Two points without noise (Ƶ =25, N_o=10, ΔX =0.5, ƙ=0.4, Ꝓ1=2cm).

The results of the figure 2, show the implementation of the two methods when applied. According to equations (9) and (10), $б_1$> 2.8 cm and thus (FT) was not able to separate the two sources (two objects), rather it appeared as a single source (object), especially for the values. Less than $б_1$ = 4, and this is in contrast to what happened with the (EV) method where it was able to separate the two sources at Ꝓ2 = 3 cm, i.e., Δ Ꝓ = 1 cm, not only that, but the percentage of error (PE) compared to (FT) method) Much less, as shown in figure 2.

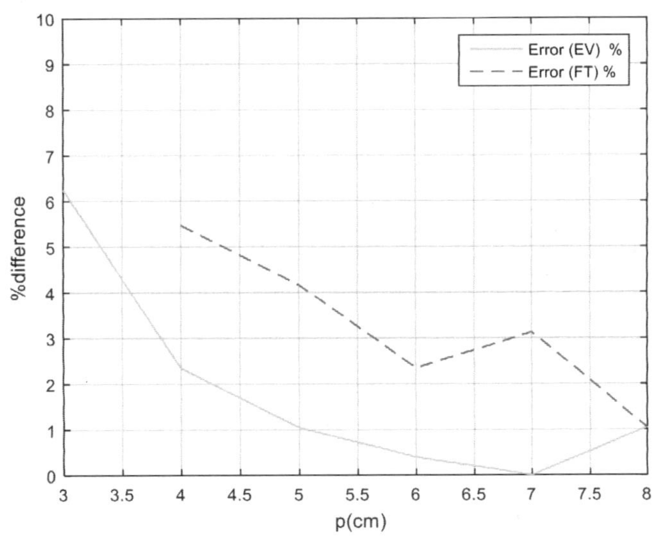

Figure 2. Two points without noise (Ƶ=25, N_o=15, ΔX=0.5, ƙ =0.4, Ꝓ1=2cm).

Figure 3 shows the results (implementation) of the two methods when applied. According to equations (9) and (10), $б_1$> 3.4 cm and thus (FT) was not able to separate the two sources (two objects), rather it appeared as a single source (object), especially for the values. Less than 4cm, and this is in contrast to what happened with the (EV) method where it was able to separate the two sources at Ꝓ2 = 3 cm, i.e., Δ Ꝓ= 1 cm, not only that, but the percentage of error (PE) compared to (FT) method. Much less.

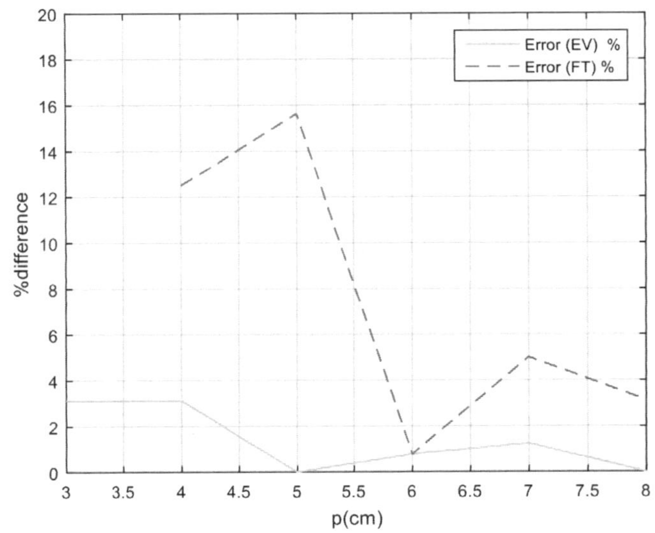

Figure 3. Two points without noise (Ƶ=30, N_o=15, ΔX =1, ʎ =0.8, ƿ1=2cm).

5.2. Results with noise data

In Figure 4 we see the performance of the two methods with adding noise, where the results show the method of (EV) the appearance of the two peaks of the two acoustics sources and clearly with Δƿ=2, while the results of applying the method of (FT) do not show the two sources, but they appear as one source at 6_1> 5.71.

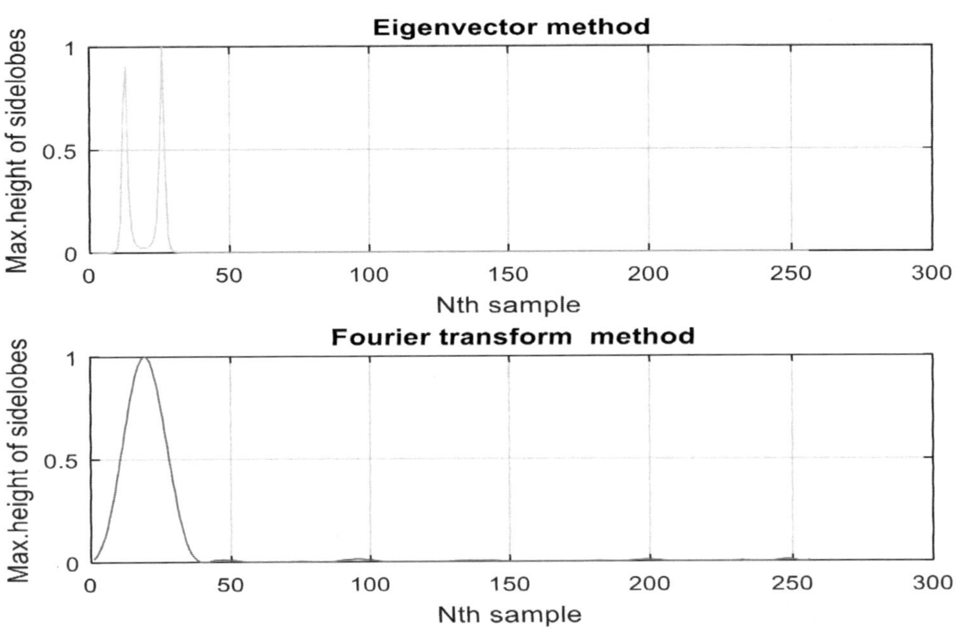

Figure 4. Two points with noise (Ƶ =25; ƿ1=2cm; ƿ2=4; N_o=15; ΔX =0.5; ʎ =0.8).

In Figure 5 we see the performance of the two methods with adding noise, where the results show the method of (EV) the appearance of the two peaks of the two acoustics sources and clearly with Δƿ=1, while the results of applying the method of (FT) do not show the two sources, but they appear as one source at $6_1 > 2.8$.

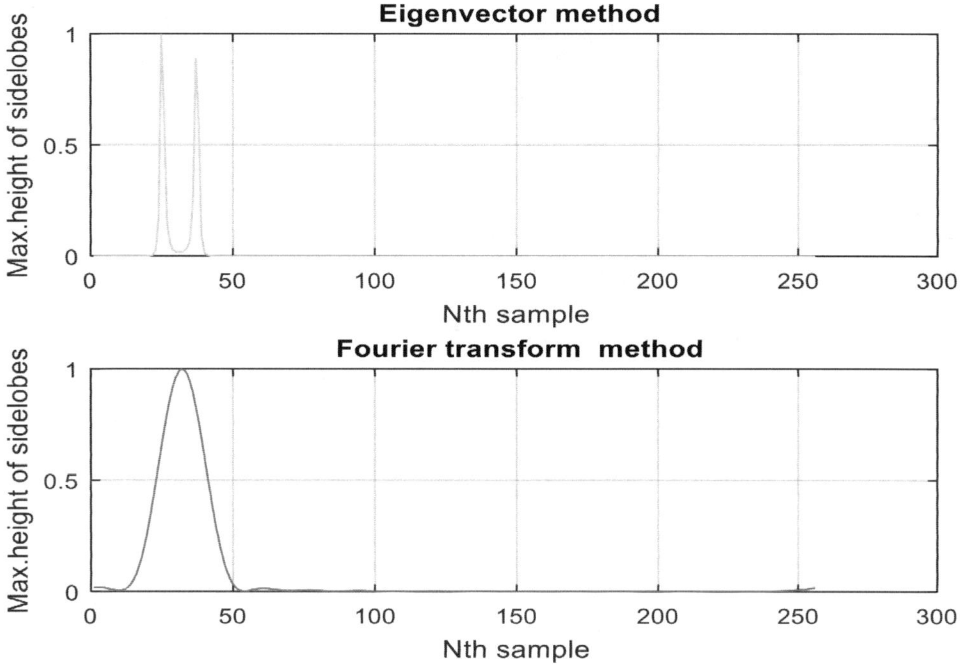

Figure 5. Two points with noise (Ƶ =25; Ꝑ1=2; Ꝑ2=3;N_0=15; ΔX =0.5; ʎ =0.4).

6. Conclusions

Through our apparent results, it is possible to prove the effectiveness of the method (EV) in detecting sources close to each other when the values of ΔꝐ are small, with high accuracy and clarity, and not only that, but the error rate. P.E. Is small and this is the opposite of what was observed in the method (FT) where it failed to discover sources close to each other, and in addition, the error rate P.E was higher than the method (EV). In addition, one of the most important conclusions that must be pointed out is accuracy, as it was noted in figure 1, that the difference between Ꝑ1 and Ꝑ2 in (EV) is 1 while in (FT) it is equal to 4, according to the laws of precision that clarify the preference for (EV). When observing the error rate shown in the graphs (percentage difference), we will notice a difference between (EV) and (FT), as the (FT) score most of the time is much greater than what is found in (EV).

7. References

[1] Proakis J G and Dimitris. G, M 2006*Digital Signal Process*Prentice Hall Inc
[2] Übeyli E D 2009 Analysis of EEG signals by implementing eigenvector methods/recurrent neural networks *Digit. Signal Process. A Rev. J.***19** 134–43
[3] Übeyli E D and Güler I 2003 Comparison of eigenvector methods with classical and model-based methods in analysis of internal carotid arterial Doppler signals *Comput. Biol. Med.***33** 473–93
[4] Tan W and others 2017 Covariance matrix reconstruction for source localization over impaired uniform linear array *2017 IEEE International Conference on Signal Processing, Communications and Computing (ICSPCC)* pp 1–5
[5] Morency M W, Vorobyov S A and Leus G 2018 Joint detection and localization of an unknown number of sources using the algebraic structure of the noise subspace *IEEE Trans. Signal Process.***66** 4685–700

[6] Anderson A P 1977 Microwave holography *Proceedings of the Institution of Electrical Engineers* vol 124 pp 946–62
[7] Wang L 2016 *Basic principles and potential applications of holographic microwave imaging* A S M E
[8] Goodman J W 20017 *Introduction to Fourier optics* (Roberts and Company Publishers)
[9] Althaf C I M and Prema S C 2018 Covariance and eigenvalue based spectrum sensing using USRP in real environment *2018 10th International Conference on Communication Systems & Networks (COMSNETS)* pp 414–7
[10] Wang S, Bao J, Shen B, Huang Q and Chen Q 2014 Eigenvector based cooperative wideband spectrum sensing for cognitive radios *2014 Sixth International Conference on Ubiquitous and Future Networks (ICUFN)* pp 346–51
[11] Hayes M H 2009 *Statistical digital signal processing and modeling* (John Wiley & Sons)
[12] Marple Jr S L and Carey W M 1989 Digital spectral analysis with applications
[13] Marie K H S, Anderson A P and Bennett J C 1982 Digital in-line holographic techniques for long wavelength imaging *IEE Proceedings H (Microwaves, Optics and Antennas)* vol 129 pp 211–20

Flux Observation of Induction Machine Based on the Enhanced Sensorless Voltage Model

Harith Al-Badrani[1*]
[1]Electronic Department, College of Electronics Engineering, Ninevah University, Mosul, Iraq.

[*]E-mail: harith.mohammed@uoninevah.edu.iq

Abstract. This paper presents the identification of the rotor flux of the induction machine. In the field-oriented control topology, since the terminal stator voltages are a pulsating voltage, so the only way is to calculate the reference values of the voltages from the inverter states phase. At very low speed, the reconstructed voltages are nearly not real, so that, the range of speed rotation of an encoder-less field-oriented control is not quite enough to cover all the speed range. However, the design of a 3-phase inverter which is prepared by silicon carbide (SiC)-MOSFETs with inserting an output filter between the inverter and the machine may supply sinusoidal voltages and currents at the output. The high switching frequency which can be applied with the SiC MOSFETs permits the proposal and implementation of a small low pass LC filter that can be equipped with the inverter output side. With this structure, it is possible to reduce the harmonic of the output voltages in addition to reduce the content of the current harmonics causing in a practically sinusoidal voltage as confirmed in the experimental results of this work. The main goal of this outline is two points, on the first side it can be reduce the winding machine stress, therefore, it is minimizing the harmonic in the voltages at the machine terminals. And on the other hand, is to use the measured voltages as an alternative of the recalculated ones to enhance the voltage model and the performance of the control scheme. It is predictable that the range of machine speed of an encoder-less field-oriented control can be extended.

1. Introduction

In the field of control, the e most important technique, which is used to control the induction machine, is the sensor-less field-oriented control. Due to the increased reliability and the lower cost of a system by minimizing the sensor cost, a research focuses on the removal the sensors of the speed or position in the variable-frequency drives was done [1] and [2]. For the deriving of a three-phase machine, the high switching frequency is dispensable in traditional tasks.

To reduce the undesirable effects of the non-sinusoidal voltage; a high-frequency inverter is used, whereas the output voltage can be filtered and turned into a nearly sinusoidal voltage which may enhance the act of the sensorless control. Moreover, the two-level, three-phase, Voltage Source Inverter (VSI) prepared with

Content from this work may be used under the terms of the Creative Commons Attribution 3.0 licence. Any further distribution of this work must maintain attribution to the author(s) and the title of the work, journal citation and DOI.
Published under licence by IOP Publishing Ltd

Silicone Carbide SiC-MOSFETs in addition to inserting an output LC filter have represented several advantages such as [3] and [4]:

- reducing the machine voltages harmonic content
- lowering the produced noise and the stress on the machine winding, and
- ranging the operation values of a sensor-less Field Oriented Control (FOC) scheme.

2. Methods controlling the IM based on the fundamental model

In a real application, the non-sinusoidal power reduces the efficiency of the motor and it is lifetime. The mathematical model in a complex form for the induction machine has been widely proposed for the controlling of the machine [5]. Vector control that also called field oriented control (FOC), was considered in the seventies of the last century [5]. Field orientation and vector control of an asynchronous motor, among other things, are explained in the publications [6]-[9]. In order to avoid the using of a speed sensor, a sensor less voltage model control the induction machine is applied [8], [9] and [10]. The most considerable disadvantage of this method is the attendance of the two integrators, which are used for the determination of the rotor fluxes. To overcome the problem of the integrator, many different modified voltage models were presented [11]. Some of the sources for the error reading include errors of the measurement in the current sensor such as offsets or noise, in addition to the change of temperature that effect on the value of the stator resistance R1 [12]. All these parameters lead to an inaccurate estimation of the stator flux, according to the integration, in particular at very low stator frequencies and unloaded machine. Various approaches pursue a way to make a pure integration less sensitive to these effects, e.g., a first-order delay element sets the integrator.

The classical field-oriented control block diagram shown in Figure 1. All the control process in addition to the transformation and inverse transformation are mentioned.

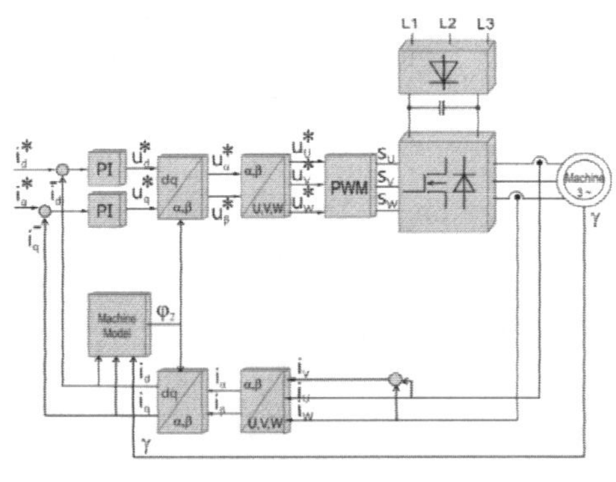

Figure 1. Block diagram of the classical FOC-scheme

3. Field Oriented Control

The control techniques of the induction machine use the rotating reference frame aligned with the rotor flux space phasor. So, it is important to identify the angular position of the stator flux space phasor for field-oriented control [13] and [14]. The rotor flux space phasor angular must be calculated from the corresponding machine models. In sensor-less mode, the models may only be determined by on the electrical measurable quantities, such as the currents and voltages, for calculation of the magnetic flux. The mechanical speed can be involved in the calculation at the operation with a speed sensor, as well. Normally, to calculate the position of the rotor flux space phasor there are two possible models, which are well known and widespread in use: the current and the voltage model.

4. Current Model

This model is thought out as a simple approach to assume the rotor flux space phasor by using the position of the rotor angular along with the measured stator currents. In addition, the angular position of the rotor may acquire by employing an encoder that is involved to the induction machine shaft [15]. Figure 2 shows the block diagram of the current model. It is clear to notice that at constant flux, the quadrature current component is relative to the angular frequency that represents the rotor flux rotation. Due to the proportionally of the quadrature current component to the develop torque, is normally named the torque producing factor.

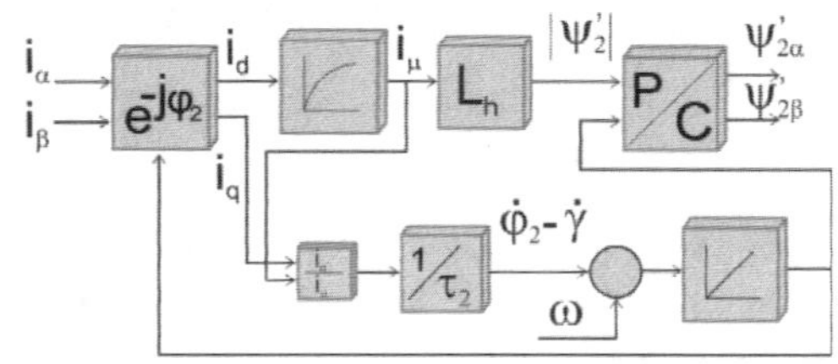

Figure 2. block diagram of the current model

The current model depends on the rotor voltage equation:

$$0 = R_2 \cdot \underline{i}_2^{'} + j \cdot (\dot{\varphi}_{2} - \dot{\gamma}) \cdot \underline{\psi}_2^{'} + \frac{d\underline{\psi}_2^{'}}{dt} \tag{1}$$

Regarding the imaginary part equation, tracks to the rotor flux angle:

$$\varphi_2 = \int\left(\dot{\gamma} + \frac{i_{1q}}{i_{\mu 2}^{'} \cdot \tau_2}\right)dt = \int \dot{\gamma} dt + \int\left(\frac{i_{1q}}{i_{\mu 2}^{'} \cdot \tau_2}\right)dt = \gamma + \int\left(\frac{i_{1q}}{i_{\mu 2}^{'} \cdot \tau_2}\right)dt \tag{2}$$

The magnetizing current $i_{\mu 2}$ can be calculated from i_{1d} and the rotor time constant τ_2 as follows:

$$i_{\mu 2}^{'} = i_{1d} - \tau_2 \cdot \frac{di_{\mu 2}^{'}}{dt} \tag{3}$$

By transformation from the uvw system to the $\alpha\beta$ coordinate system, i_α and i_β may be obtained, and then directly to the second transformation between the $\alpha\beta$ and the dq coordinate systems by using the rotor flux angle φ_2:

$$i_{1d} = i_{1\alpha} \cdot \cos(\varphi_2) + i_{1\beta} \cdot \sin(\varphi_2) \tag{4}$$

$$i_{1q} = i_{1\beta} \cdot \cos(\varphi_2) - i_{1\alpha} \cdot \sin(\varphi_2) \tag{5}$$

These currents could be employed with the rotor angular position in order to calculate the rotor flux as follows:

$$\underline{\psi}_1 = \underline{i} \cdot \left(L_1 - \frac{L_{1h}^2}{L_2'} \right) + \underline{\psi}_2' \cdot \left(\frac{L_{1h}}{L_2'} \right) \tag{6}$$

While this model has a wide range along with various velocity values, it shows some serious disadvantages such as the sensitivity to machine parameters which means more sensitivity to the rotor time constant that can be changeable due to the temperature and saturation. Furthermore, the presence of an additional speed sensor had been expected to reduce system reliability and increase its cost.

5. Voltage Model

The rotor flux and angular position can be obtained by using a simple method and neglecting the additional speed sensor. It can be obtained by using the voltage model technique that depends on voltages of the stator terminals and currents of the induction machine. As a consequence, this flux estimator model is considered to be a convenient method due to its simplicity as it is only a stator resistance parameter dependent. To understand the operation of this model, it's useful to illustrate the stator windings voltage equation:

$$\underline{u}_1 = R_1 \cdot \underline{i} + \dot{\underline{\psi}}_1 \tag{7}$$

The stator flux space phasor could be written as follows:

$$\underline{\psi}_1 = \int (\underline{u}_1 - R_1 \cdot \underline{i})\, dt \tag{8}$$

The stator voltage is composed of both the induced voltage and the voltage drop across the winding resistance:

$$u_{1\alpha}(t) = R_1 \cdot i_{1\alpha}(t) + \frac{d\psi_{1\alpha}(t)}{dt} \tag{9}$$

$$u_{1\beta}(t) = R_1 \cdot i_{1\beta}(t) + \frac{d\psi_{1\beta}(t)}{dt} \tag{10}$$

The stator flux $\alpha\beta$ coordinate system can be stated as follows:

$$\psi_{1\alpha =} \int (\underline{u}_{1\alpha} - R_1 \cdot \underline{i}_\alpha)\, dt \tag{11}$$

$$\psi_{1\beta =} \int (\underline{u}_{1\beta} - R_1 \cdot \underline{i}_\beta)\, dt \tag{12}$$

Where $u_{1\alpha}, u_{1\beta}, i_{1\alpha}, i_{1\beta}, \psi_{1\alpha}, \psi_{1\alpha}$ are the α, β components of the stator voltage, stator current, and stator flux phasors respectively, R1 is the stator winding resistance. By using the stator voltages and current in addition to the stator resistance, the stator flux can be simply calculated and then the rotor flux could be calculated from the stator flux as illustrated in Figure 3 and the machine equations:

$$\psi'_{2\alpha} = \frac{\sigma}{\sigma - 1} \cdot L_{1h} \cdot i_{1\alpha} + (1 + \sigma_2) \cdot \psi_{1\alpha} \tag{13}$$

$$\psi'_{2\beta} = \frac{\sigma}{\sigma - 1} \cdot L_{1h} \cdot i_{1\beta} + (1 + \sigma_2) \cdot \psi_{1\beta} \tag{14}$$

The main advantages of the voltage model are represented by obtaining the rotor angular position φ_2 by using only the voltages and currents of the stator machine. However, it is easy to compensate the stator resistance for changes in the temperature.
In the traditional (VSI), nevertheless, the output voltage of this kind of inverters is quasi-square waveform and not sinusoidal, so, the output of the controller which represents the reconstructed voltages are supposed to be the stator voltages of the machine and addressed to the modulator.

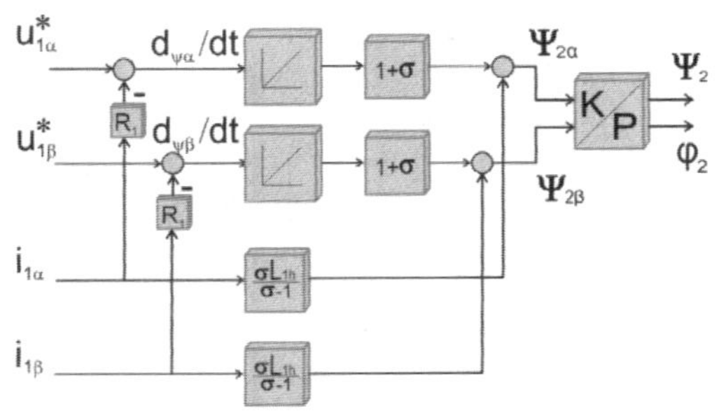

Figure 3. The block diagram of the standard (commercial) voltage model (SUM)

6. Enhanced voltage model

It has been analyzed that is using a sensorless control topology based on the commercial voltage model of the 3-phase induction machine is a proper way to avoid using an angular sensor. As this model is used to detect the flux space phasor of the rotor and is considered to function until minimum stator frequency; different improvements have been applied as shown in Figure 4.
The main goal of this work is to evaluate the effect of replacing the reconstructed voltage with measured ones. As depicted in Figure 3, the voltage reference values (with a star), which are known as the Standard Voltage Model (SUM) is fed by the voltage model, while, the measured voltages of the induction machine (without a star) are forwarded the enhanced one and known as the enhanced voltage model (EUM) that is engaged to the machine after the low pass filter. As a result, considering the smooth measured sinusoidal voltage, it is predictable that the sensorless control will have the ability to function on a widely speed range.

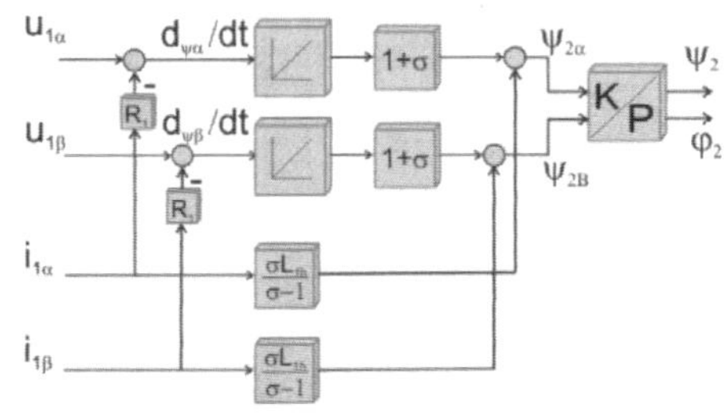

Figure 4. The block diagram of the enhanced (modified) voltage model (EUM)

Due to the machine speed signal is not obligatory, the voltage model is an acceptable flux estimator for the induction machine drives. The stator fluxes are calculated by integrate the back electromagnetic force (back EMF), which means, integrate the value, which found after the calculation of the voltage drop over the resistance of the stator is subtracted from the stator voltages. These two integrators are representing the main problem in the voltage model in charge of the rotor flux calculation. Due to these two integrators, a slight DC offset in measured voltages and currents causes drift problem. Therefore, this problem considered. This is the first problem, whereas the second one that is at zero speed and because of there is no value of the inductance characteristics, the machine appears as an ohmic character. So, for overcoming the first problem, there are some modified models are offered [16], these voltage models are as a weakening method, that in its place of using a pure integrator, the modified one could be used in some many forms such as:

1. Low gain feedback
2. PI element feedback

It is important to note that the correction of the angle error, which resulted by the attenuation should be done with confident limits by appropriate methods [17].

The frequency converter presents an output voltage as sequencing rectangular pulses or quasi-square waveform that differ in the width (PWM). A low pass filter has been added in order to eliminate the switching frequency noise when frequency converter feds a machine. This noise is caused by the pulse width modulated (PWM) voltage that is applied to the machine. In addition to the switching noise, the switching frequency harmonics, (especially the double of the switching frequency) has also resulted. However, adding a sine-wave filter will clean the pulse fashioned voltage from the frequency converter and yield a sinusoidal line to line terminal voltage of the induction machine. Otherwise, when no filters are used, the main source of the high-frequency noise is the pure voltage overshoot that appear at the motor terminals. This can be seen in Figures 5 and 6 that shows the inductor filter voltage and the terminal stator voltage respectively.

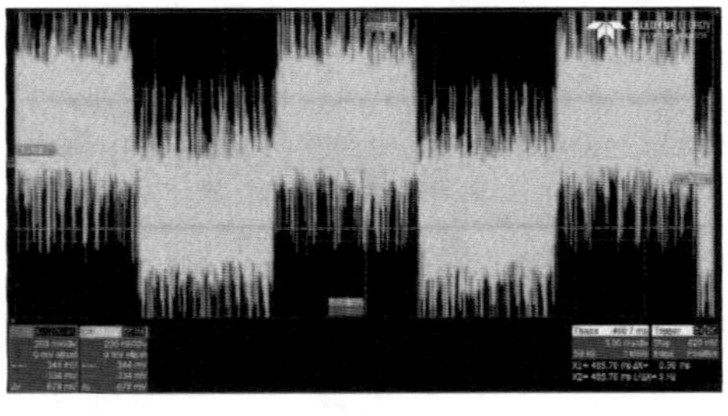

Figure 5. Inductor filter voltage

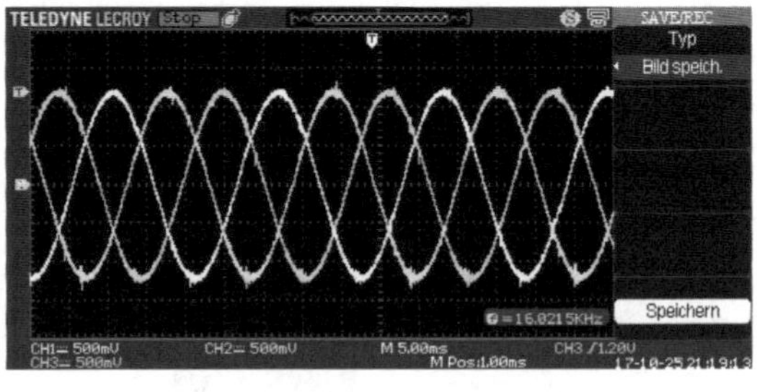

Figure 6. Terminal stator voltage

7. Implementation of the setup

The behavior of a system that consists of a fast-switching silicon carbide SiC-Inverter, which is equipped with an output low pass filter that feeds an asynchronous machine (IM) was experimentally confirmed. The three-phase induction machine is controlled by a digital signal processor (DSP) controller (TMS320F28335 eZdsp). The pulse width modulation was carried out with a modulation frequency of 50 kHz. The conventional two-level inverter topology that is equipped with silicon carbide switches was implemented in the laboratory is shown in

Figure .

Figure 7. Conventional two-level inverter

The most important matters in the designing of this inverter are the choice of the power SiC MOSFETs that is used in the inverter, thus, eventually CCS020M12CM2 SiC MOSFET 1,2kV/20A module by CREE company was carefully chosen due to their fitness in commerce with the essential design characteristics that are listed in Table 1.

Table 1. Electrical specifications of the inverter

Parameter	Value
DC link voltage, $\boldsymbol{U_{dc}}$	$560V$
Output power	$7.5kW$
Switching frequency $\boldsymbol{f_{sw}}$	$50kHz$
Fundamental frequency $\boldsymbol{f_o}$	$50Hz$

In order to exam the enhancement of a field oriented sensorless control scheme, the laboratory induction machine has an installed encoder on the shaft, which allows easy appointing and run of the machine and sensorless drive concept. Furthermore, the encoder delivers a reference measurement signal of the frequency response as a comparison value between identification and encoder. The SiC inverter supplies a 7.5 kW of power to the induction motor (400 V, 17 A). A braking 21 kW field-oriented controlled Siemens motor drive was used as a loaded machine in the current work.

For the inverter that working with the high switching frequency, the values of the LC output filter inductor and capacitor could be smaller, thus, the total size and cost of the inverter is decreased. Also, it has a lower loss since the number of turns is decreased. In this work, the filter inductors were implemented by using a copper litzen-wire, while the capacitors that are used in a filter of six polypropylene capacitors for high switching frequency applications. Since the switching frequency which used in the inverter is $f_s = 50kHz$, the capacitor and the inductor values were fixed to $C = 1.32\ \mu F$ and $L = 0.1\ mH$ respectively. The filter was considered to handle a $15\ A$ as a rated current while the saturation current was set to $35\ A$

8. Experimental results

The investigations in the following results were made based on the input parameters are given in

Table 2. The commercial two-level inverter usually delivers pulsating voltages as the output. For the high switching inverter that is integrated with a small filter at the output, the output voltages can be nearly sinusoidal as shown in Figures 6 and 7. Because of the 50 kHz switching frequency, the output voltage quality was improved and the total harmonic distortion would be reduced as shown in Figures 8 and 9.

Table 2. Experimental input quantities

Parameter or variable	Value
Udc	300 V
Reference speed	500 min^{-1}
Switching frequency f_{SW}	$50kHz$
Fundamental frequency	$f_o = 50Hz$
Sampling period	60

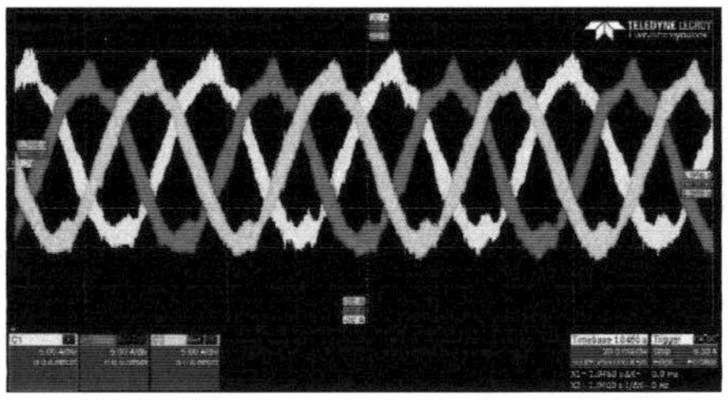

Figure 8. Currents of the inductor filter

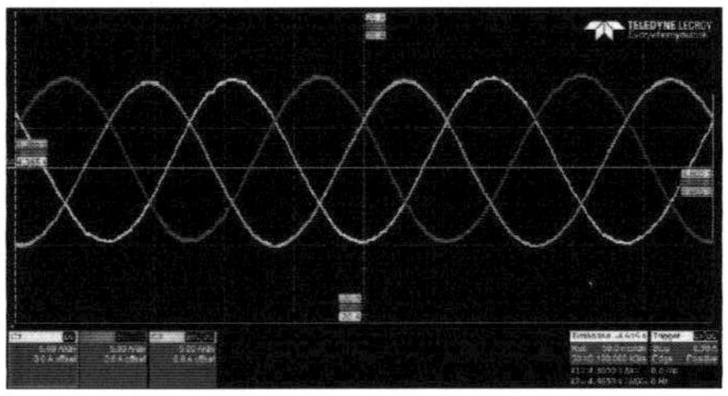

Figure 9. Currents of the stator

9. Rotor flux space phasor

The field-oriented control technique using the voltage model in high-speed range is achieve the target. Based on this consideration, the achievement option was validated experimentally in this section. For this purpose, the induction machine at no load was controlled with the FOC. Figure 10 to Figure 13 are shows the rotor flux in the coordinate system of the induction machine at constant speed of n*= 500 min^{-1} and at low speed of n*= 10 min^{-1} as well. The results of the current model were used as a reference model in order to compare all other models' behavior.

In addition to the fluxes that calculated by the voltage model, the figures contain the rotor fluxes in a coordinate system which are obtained from the current model as well. At n*= 500 min^{-1}, which is high speed, the voltage model runs similar to a current model, i.e., the likeness in the performance becomes clearer from the unillustrated experimental results shown in Figure 10. The alpha, beta rotor fluxes were obtained by using the standard voltage model are in the same phase with the fluxes which are already obtained from the current model.

On the other hand, the SUM shows a very week behavior with an ununiformed rotor fluxes at low reference speed, n*= 10 min^{-1} as showed in Figure 11. The physical explanation for that is based on the fact that there is no induced voltage at low speed while the voltages which were used in the machine model were the reconstructed voltages obtained from the current controller output and not the measured voltages.

A measured sinusoidal voltage was gotten by inserting a small low pass filter between the inverter output and the machine which were used as an alternative to the recalculated ones in order to improve the voltage model. An Enhanced voltage model that presented in Figure 4 can be used for obtaining the rotor fluxes, at high speed, n*= 500 min^{-1}, the EUM acts an excellent result which are look similar to the current model, i.e., the experimental consequences shown in Figure 12 show that the rotor fluxes in alpha, beta coordinate system were calculated from the enhanced voltage model are a very good agreement calculation.

At low reference speed and with the low pass filter used at the output of the fast-switching inverter, the terminal voltages could be measured and, therefore, can be used in a machine model in the calculation of the angle of the rotor flux space phasor. Different the results which are calculated from the commercial model, which is presented in Figure 13, a well performance at low speed, n*= 10 min^{-1} can be obtained by using the modified voltage model, the rotor flux that calculated from the Enhanced UM at 10 min^{-1} were presented in Figure .

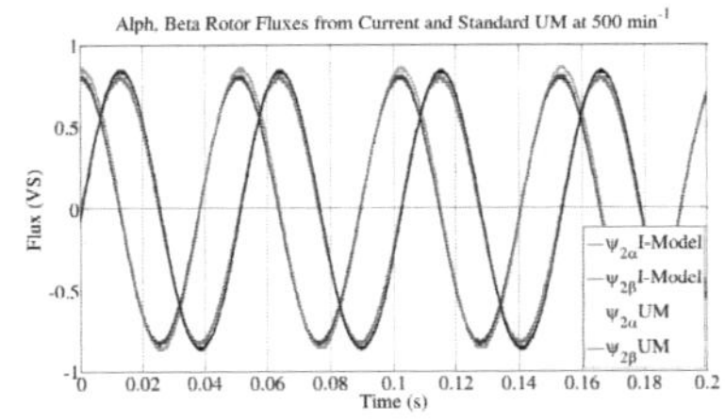

Figure 10. Rotor flux in alpha beta frame for the current and the UM at high speed

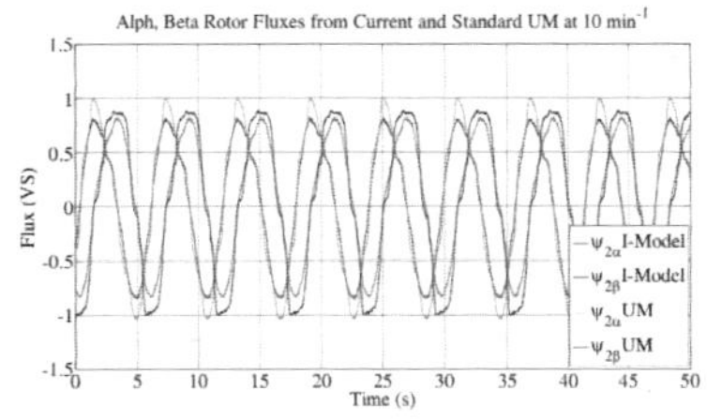

Figure 11. Rotor flux in alpha beta frame for the current and the UM at low speed

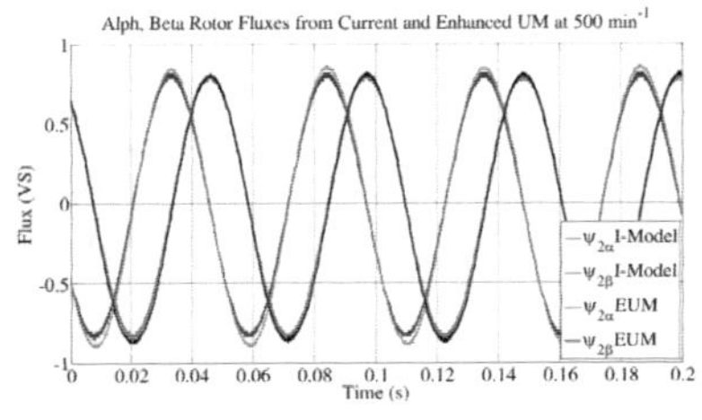

Figure 12. Rotor flux in alpha beta frame for the current and EUM at high speed

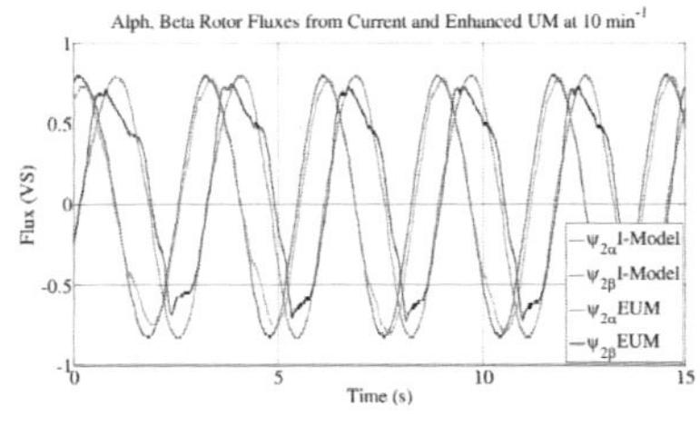

Figure 13. Rotor flux in alpha beta frame for the current and the EUM at low speed

10. Conclusion

The electric drive consists of the electric machine or the motor, and according to the modern understanding; it consists of converters, sensors, control, and other instrumentation. The pulsed voltage outputs on the converters which are fed the machines produce a broad harmonic spectrum and lead to undesirable losses, which can be reduced by increasing the switching frequency.

A SiC MOSFET inverter with a small filter for driving a 3-phase induction machine is presented in this work. The goal is to get a sinusoidal on the machine terminals. The quality improvement of the voltages used for the improvement of the voltage model, which is employed for the rotor flux space phasor estimation. The experimental results were approved on a laboratory setup, which confirms that this control scheme can also be enhanced and showed a good steady state and dynamic performance in the torque control mode.
In order to control the induction machine, the signal of the rotor position or the position of a machine flux is needed. Since the mechanical sensors are highly sensitive to the pollution that can be resulted from aggressive environment, it is desirable to replace the speed sensor by the already existing resources in the drive. Using the sensorless control method is an alternative to the conventional method with a mechanical sensor (encoder system). Thus, using a speed estimation will increase the reliability of a drive system and eliminate the cost of a speed sensor usage.

11. References

[1] B. K. Bose, Power Electronics and Motor Drives : Advances and Trends. Amsterdam: Academic. p. 22., 2006.
[2] B. Drury, The Control Techniques Drives and Controls Handbook (2nd ed.). Stevenage, Herts, UK: Institution of Engineering and Technology. p. xxx., 2009.
[3] Schaffner, "Sine wave filter solutions for motor drive applications (Application Note)". Retrieved Schaffner Website "https://www.schaff-ner.com/fileadmin/media/downloads/applica-tion_note/Schaffner_AN_Filter_solutions_for_mo-tor_drives_applications.pdf"
[4] J. Holtz, M. Höltgen, and J. O. Krah, May 2014, A Space Vector Modulator for the High-Switching Fre-quency Control of Three-Level SiC Inverters, *IEEE Trans. Power Electron.*, **vol. 29, no. 5**, pp. 2618–2626.
[5] S. Wall, R. h Ruan, C. g Wang, and J. r Xie, 2016, valuation of three-phase solar inverters using SiC devices, in 2016 *18th European Conference on Power Electronics and Applications (EPE'16 ECCE Europe),* pp. 1–10.
[6] D. W. Novotny and T. A. Lipo, 1999, Vector Control and Dynamics of AC Drives. Oxford University Press, UK.
[7] P. Vas, 1998, Sensorless Vector and Direct Torque Control. *Oxford University Press*,.
[8] H. Al-Badrani, S. Feuersaenger, and M. Pacas, 2018, VSI with Sinusoidal Voltages for an Enhanced Sensorless Control of the Induction Machine, *in PCIM Europe*, pp. 1–7.
[9] H. Al-Badrani, S. Feuersaenger, and M. Pacas, 2018, SiC-VSI with Sinusoidal Voltages for an Enhanced Sensorless Control of the Induction Machine, *IEEE 4th Southern Power Electronics Conference (SPEC).*
[10] J. Holtz, Feb. 2006, Sensorless Control of Induction Machines #8212;With or Without Signal Injection?, *IEEE Trans. Ind. Electron.*, **vol. 53, no. 1**, pp. 7–30.
[11] Y. B. Zbede, S. M. Gadoue, and D. J. Atkinson, Jun. 2016, Model Predictive MRAS Estimator for Sensorless Induction Motor Drives, *IEEE Trans. Ind. Electron.*, **vol. 63, no. 6,** pp. 3511–3521.
[12] K. Wang, W. Yao, K. Lee, and Z. Lu, Jul. 2014, Regenerating Mode Stability Improvements for Combined Voltage and Current Mode Flux Observer in Speed Sensorless Induction Machine Control, *IEEE Trans. Ind. Appl.*, **vol. 50, no. 4**, pp. 2564–2573.
[13] P. Brandstetter and M. Kuchar, 2015, Rotor flux estimation using voltage model of induction motor,*16th International Scientific Conference on Electric Power Engineering (EPE)*,pp. 246–250
[14] J. Holtz and J. Quan, Sensorless, 2001, vector control of induction motors at very low speed using a nonlinear inverter model and parameter identification, *in Conference Record of the 2001 IEEE Industry Applications Conference. 36th IAS Annual Meeting (Cat. No.01CH37248),* **vol. 4**, pp. 2614–2621.

[15] K. D. Hurst, T. G. Habetler, G. Griva, and F. Profumo, Jul. 1998, Zero-speed tacholess IM torque control: simply a matter of stator voltage integration, *IEEE Trans. Ind. Appl.*, **vol. 34, no. 4**, pp. 790–795,.

[16] J. Holtz, 1995, State of the art of controlled AC drives without speed sensor, *in Proceedings of 1995 International Conference on Power Electronics and Drive Systems*. **vol.1,** PEDS 95, pp. 1–6.

[17] M. Depenbrock and C. Evers, Feb. 2005, Model-based speed identification for induction Machines in the whole operating range, *IEEE Trans. Ind. Electron.*, **vol. 53, no. 1,** pp. 31–40.

IOP Conf. Series: Materials Science and Engineering 1152 (2021) 012031 doi:10.1088/1757-899X/1152/1/012031

Analysis and Simulation of a Single Phase Nine Level Inverter with Reduction in Power Switching Devices

Saif S. Mahmood[1] and Shawkat M. Younus[1*]

[1]Department of Electronic, Collage of electronics, University of Ninevah, Mousl-iraq

*E-mail: shawkat.younus@uoninevah.edu.iq

Abstract. This paper present analysis and modelling of single phase nine level inverter, a Wide range of Modulation index (M) varied between 0.05-1.05 is achieved by two switching strategies (Optimized harmonic stepped waveform (OHSW), (Optimal minimization of total harmonic distortion (OTHD)) the first strategy focusses on harmonic elimination and the other on reducing output voltage total harmonic distortion (THD). Calculation of switching angles for both equal and unequal dc sources is done by Genetic Algorithm (GA). simulation in Matlab, PSpice and PSIM is presented in this work.

1. Introduction

Multilevel inverters (MLI) are emerging as a new breed of power electronics converters for high power applications such as Motor drive, Static VAR compensators, Uninterruptible power supply and renewable energy systems, etc. [1]. Comparing to the traditional 3-level inverter the staircase output voltage of MLI is the major advantage which results in higher power quality, better electromagnetic compatibility (EMC), high efficiency with low frequency control method, lower switching losses higher voltage capability without need of transformer which reduce the costs. [2]. MLI can be divided into three categories: Diode clamped, Flying capacitor and Cascaded H-bridge. Among these categories the cascaded H-bridge with separate dc source appears to be superior to the other MLI structure due to its structure that is not only simple, modular and require the least number of switching devices without need extra voltage balancing capacitor or clamping diode [3]. Switching strategies can be divided in five types: Optimized Harmonic Stepped Waveform technique (OHSW), Optimal Minimization of the Total Harmonic Distortion (OTHD), Space Vector Modulation (SVM), Selective Harmonic Elimination Pulse Width Modulation (SHEPWM) and Sinusoidal Pulse Width Modulation (SPWM). Comparing with high frequency switching strategies (OHSW, OTHD) offer a controllable harmonic, lower switching losses higher efficiency since its switching frequency equal to fundamental frequency. [4,5]. Given a constant step of DC voltages the control variable is only switching angles while if the DC sources can be regulated more harmonic can be eliminated and also lower total harmonic distortion (THD) can be achieved. The topology of MLI used in this paper proposed in [2]. Which has the advantages of reducing the number of switches and gate drive circuits and so the overall system to be less expensive. [7]

2. Structure of the Nine Level Cascaded Inverter

Figure (1) shows the configuration of the conventional nine level inverter which uses 16 switches and fig. (2) shows the used topology which has the advantages of less number of switches i.e. it has 12 switches. the principle of operation of this circuit is explained in Table (1) which indicates the values of

Content from this work may be used under the terms of the Creative Commons Attribution 3.0 licence. Any further distribution of this work must maintain attribution to the author(s) and the title of the work, journal citation and DOI.
Published under licence by IOP Publishing Ltd

output voltages. This circuit consist of two parts: Level creator and full bridge inverter, the first part is responsible for creation of stepped waveform and full bridge inverter used to make the stepped waveform into bipolar state (+ve and -ve). It is clear that the two switches in each cell cannot be ON simultaneously because short circuit across the voltage would produce. The DC sources of this topology may be available through renewable sources such as photovoltaic panels or fuel cells or with energy storage devices such as capacitor or batteries. [6].

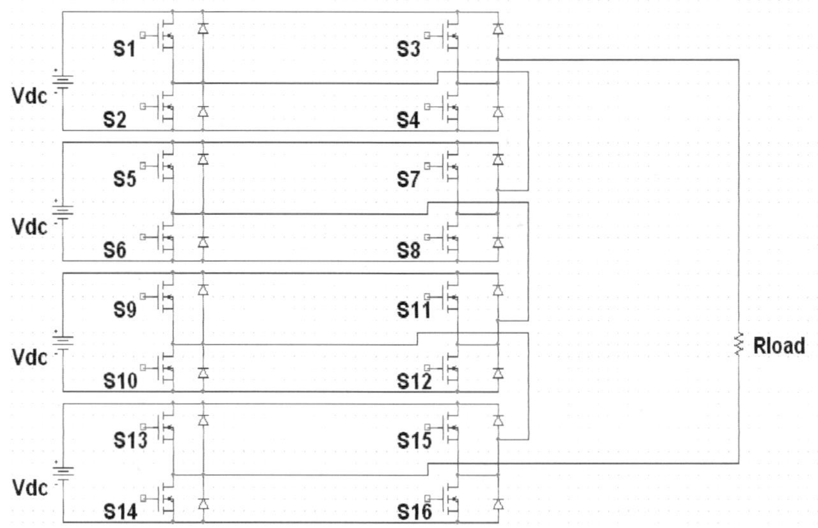

Figure 1. Conventional Nine Level Cascaded H-Bridge

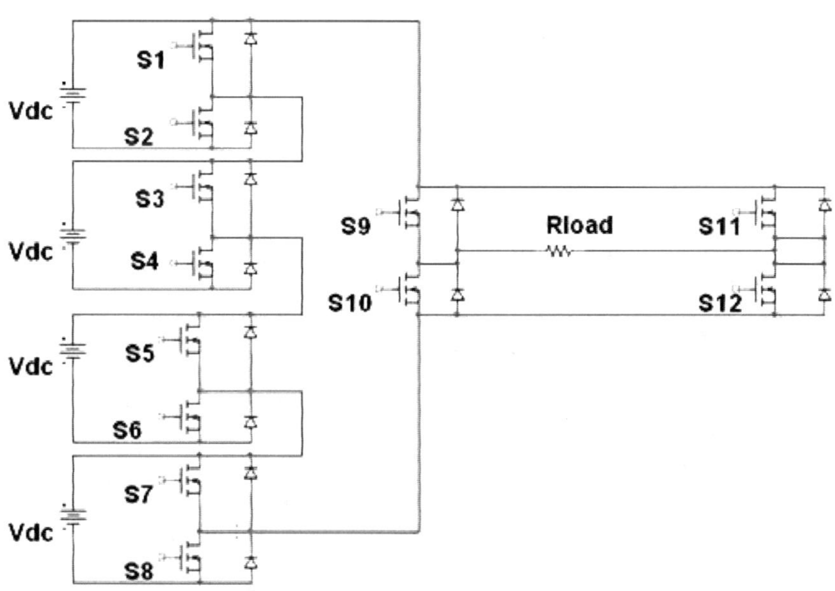

Figure 2. Nine Level Cascaded H-Bridge with Less No. of Switches

Table 1. Switching States of Nine Level Cascaded H-Bridge with Less No. of Switches

Output Voltages	S1	S2	S3	S4	S5	S6	S7	S8	S9	S10	S11	S12
4Vdc	0	1	0	1	0	1	0	1	1	0	0	1
3Vdc	0	1	0	1	0	1	1	0	1	0	0	1
2Vdc	0	1	0	1	1	0	1	0	1	0	0	1
Vdc	0	1	1	0	1	0	1	0	1	0	1	0
0	1	0	1	0	1	0	1	0	1	0	1	0
-Vdc	0	1	0	1	0	1	0	1	0	1	1	0
-2Vdc	0	1	0	1	0	1	1	0	0	1	1	0
-3Vdc	0	1	0	1	1	0	1	0	0	1	1	0
-4Vdc	0	1	1	0	1	0	1	0	0	1	1	0

3. Theoretical Analysis

Figure (3) shows positive part of output voltage of single phase nine level inverter where (Vdc_1=wVdc), (Vdc_2=xVdc), (Vdc_3=yVdc), (Vdc_4=zVdc) and (α_1, α_2, α_3, α_4) are the Switching Angles by using Fourier series expansion the expression of the amplitude of the fundamental and all harmonic content are given as:

$$H_n = \frac{4V_{dc}}{n\pi}(w\cos n\alpha_1 + x\cos n\alpha_2 + y\cos n\alpha_3 + z\cos n\alpha_4) \quad (1)$$

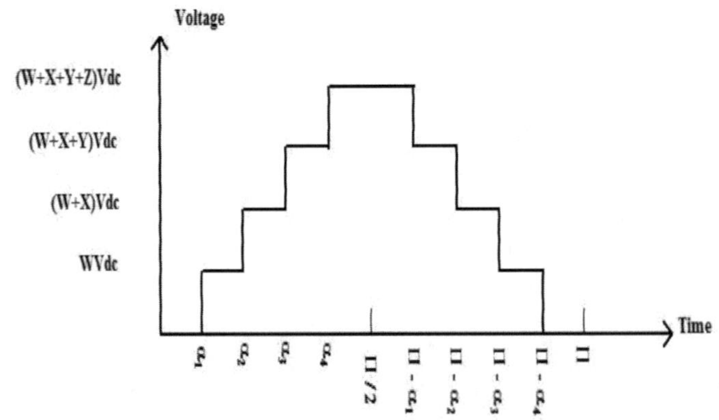

Figure 3. Positive Part of nine level output voltage

3.1. OHSW Analysis

In Nine level inverter with adjustable DC sources eight degree of freedom are used to eliminate (H_3, H_5, H_7, H_9, H_{11}, H_{13}, H_{15}) while keeping the relative fundamental component at a desired value equation (2) describe this the appropriate method to do this is by converting it to an optimization problem so Genetic Algorithm as a Powerful technique is used.

$$\begin{aligned}
&w\cos(\alpha 1) + x\cos(\alpha 2) + y\cos(\alpha 3) + z\cos(\alpha 4) = M * \pi \\
&w\cos(3*\alpha 1) + x\cos(3*\alpha 2) + y\cos(3*\alpha 3) + z\cos(3*\alpha 4) = 0 \\
&w\cos(5*\alpha 1) + x\cos(5*\alpha 2) + y\cos(5*\alpha 3) + z\cos(5*\alpha 4) = 0 \\
&w\cos(7*\alpha 1) + x\cos(7*\alpha 2) + y\cos(7*\alpha 3) + z\cos(7*\alpha 4) = 0 \\
&w\cos(9*\alpha 1) + x\cos(9*\alpha 2) + y\cos(9*\alpha 3) + z\cos(9*\alpha 4) = 0 \\
&w\cos(11*\alpha 1) + x\cos(11*\alpha 2) + y\cos(11*\alpha 3) + z\cos(11*\alpha 4) = 0 \\
&w\cos(13*\alpha 1) + x\cos(13*\alpha 2) + y\cos(13*\alpha 3) + z\cos(13*\alpha 4) = 0 \\
&w\cos(15*\alpha 1) + x\cos(15*\alpha 2) + y\cos(15*\alpha 3) + z\cos(15*\alpha 4) = 0
\end{aligned} \tag{2}$$

3.2. OTHD Analysis

The total harmonic distortion for stepped waveform can be written as follows:

$$THD = \frac{\left(\sum_{n=3}^{\infty} {H_n}^2\right)^{\frac{1}{2}}}{H_1} \tag{3}$$

Where n: harmonic index
H_1: fundamental component
H_n: the harmonics

The THD for periodic voltage waveform also can be written as:

$$THD = \left(\frac{Vrms^2}{V1^2} - 1\right)^{\frac{1}{2}} \tag{4}$$

Where V_1 is the RMS value of the fundamental component and Vrms is written as:

$$V_{rms} = \left(\frac{1}{2\pi}\int_0^{2\pi} f(\omega t)^2 \, d\omega t\right)^{\frac{1}{2}} \tag{5}$$

The RMS value of the fundamental component and the RMS value of the output voltage of a nine-level inverter with adjustable dc sources can be written as follows:

$$V_{rms} = V_{dc}\left(\frac{2}{\pi}\left[\begin{matrix}(\alpha_2 - \alpha_1)w^2 + (\alpha_3 - \alpha_2)(w+x)^2 + (\alpha_4 - \alpha_3)(w+x+y)^2 \\ +\left(\frac{\pi}{2} - \alpha_4\right)(w+x+y+z)^2\end{matrix}\right]\right)^{\frac{1}{2}} \tag{6}$$

IOP Conf. Series: Materials Science and Engineering 1152 (2021) 012031 doi:10.1088/1757-899X/1152/1/012031

$$V_{rms1} = \frac{4V_{dc}}{\pi\sqrt{2}}[\mathrm{w}\cos\alpha_1 + x\cos\alpha_2 + \mathrm{y}\cos\alpha_3 + \mathrm{z}\cos\alpha_4] \tag{7}$$

By substituting Eq. (5) and Eq. (6) into Eq. (3) the THD of output voltage with adjustable dc sources is given by:

$$THD = \left(\left[\frac{\pi}{4} * \frac{(\alpha_2-\alpha_1)w^2 + (w+x)^2(\alpha_3-\alpha_2) + (w+x+y)^2(\alpha_4-\alpha_3) + (w+x+y+z)^2\left(\frac{\pi}{2}-\alpha_4\right)}{[w\cos\alpha_1 + x\cos\alpha_2 + \mathrm{ycs}\,\alpha_3 + \mathrm{zcos}\,\alpha_4]^2}\right] - 1\right)^{\frac{1}{2}} \tag{8}$$

GA is used to minimize Eq. (8). i.e. reducing total harmonic distortion.
Note: 1- when dc sources is equal (w=x=y=z=1). So equation is eliminated
2-The correct solution must satisfy

$$0 < \alpha_1 < \alpha_2 < \alpha_3 < \alpha_4 < \frac{\pi}{2} \tag{9}$$

$$0 \le w, x, y, z \le 1 \tag{10}$$

4. Discussion and Result

G.A is applied in OHSW technique to compute:1- the four variables (α_1, α_2, α_3, α_4) versus desired M. which is result in cancelling (3rd,5th,7th) harmonic. 2-the seven variables (x, y, z, α_1, α_2, α_3, α_4) versus desired M. Which is result in cancelling (3rd, 5th, 7th, 9th, 11th, 13th) harmonic.3-the eight variable (w, x,y,z,α_1,α_2,α_3,α_4) versus M. which is result in cancelling (3rd,5th,7th ,9th,11th,13th,15th) harmonic. While GA is applied to minimize the total harmonic distortion in OTHD technique. figure (4) and figure (5) show the solution of switching angles with M.in case of OHSW(4-variables) and OTHD (4- variables) whereas figure (6) shows a comparison between the two techniques in term of THD with M. figure (7), figure (8) figure (9), figure (10) show the solution of switching angles and the amplitude of dc sources with M.in case of OHSW(7-variables) and OTHD (7-variables). Here for simplicity the first source is assumed constant(w=1), whereas figure (11) shows a comparison between the two techniques in term of THD with M. figure (12), figure (13) , figure (14) and figure(15)show the solution of switching angles and the amplitude of dc sources with M.in case of OHSW(8-variables) and OTHD(8-variables) whereas figure (16) shows a comparison between the two techniques in term of THD with M from the results obtained as M increase THD decrease and the minimum THD is achieved by (OTHD(8-variables)) when the dc sources and switching angles share the same goal which is the minimum THD also (THD_{OHSW}>THD_{OTHD}) since the aim of the second technique is minimize THD , ($THD_{equal\ dc\ source}$>$THD_{adjustable\ dc\ sources}$).Table 2 shows optimal results in OTHD for nine level inverter also for calculating THD it is impossible to take all harmonics into account .usually some specific low order harmonic beginning from 3rd up to a limited order like (49th,63th ,199th) and comparison listed in table 3 and table 4 with the integrating method .Simulation is done by three programs (Matlab PSB),(Pspice) and(PSIM)with(input voltage equal to 40v , M=0.85). The output voltage and its FFT analysis is shown in figure (17-18-19).

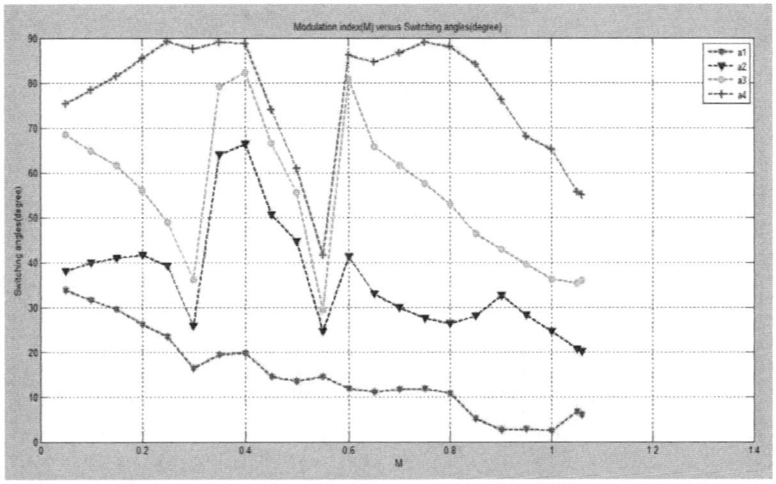

Figure 4. Switching Angles Versus M in (OHSW (4- variables))

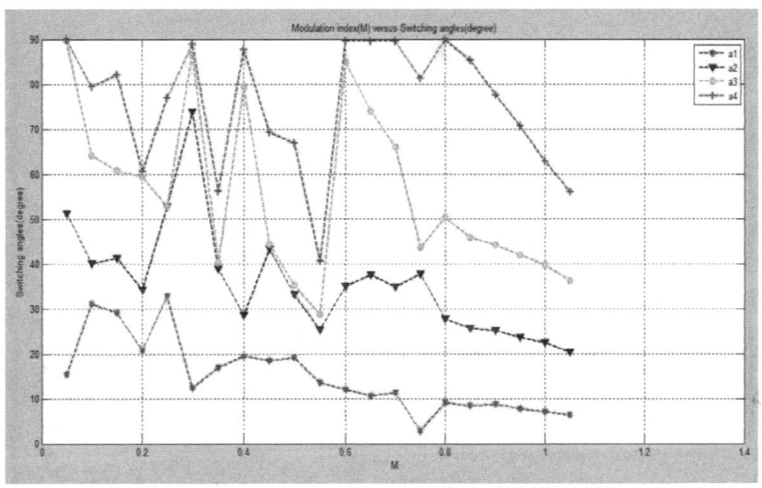

Figure 5. Switching Angles Versus M in (OTHD (4- variables))

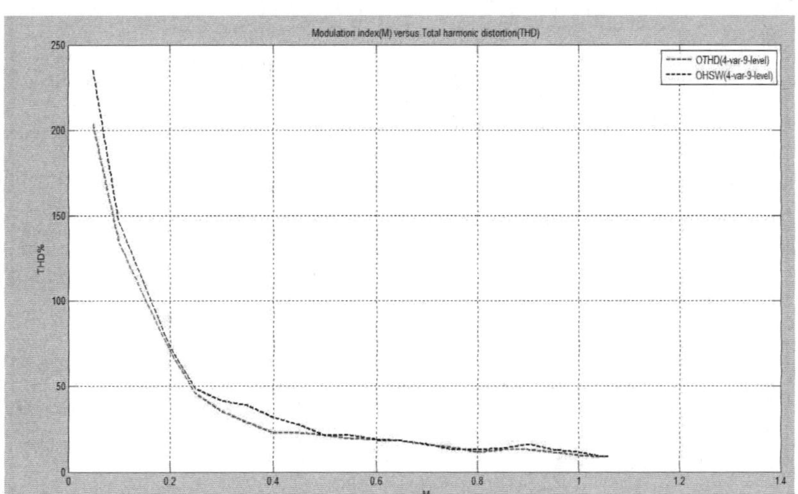

Figure 6. Comparison between OHSW (4- variables) and OTHD (4- variables)

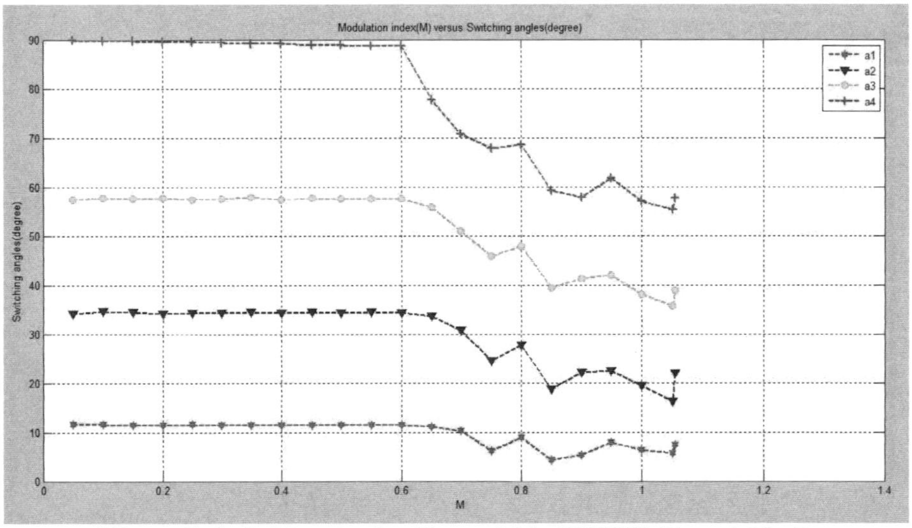

Figure 7. Switching Angles Versus M in (OHSW (7- variables))

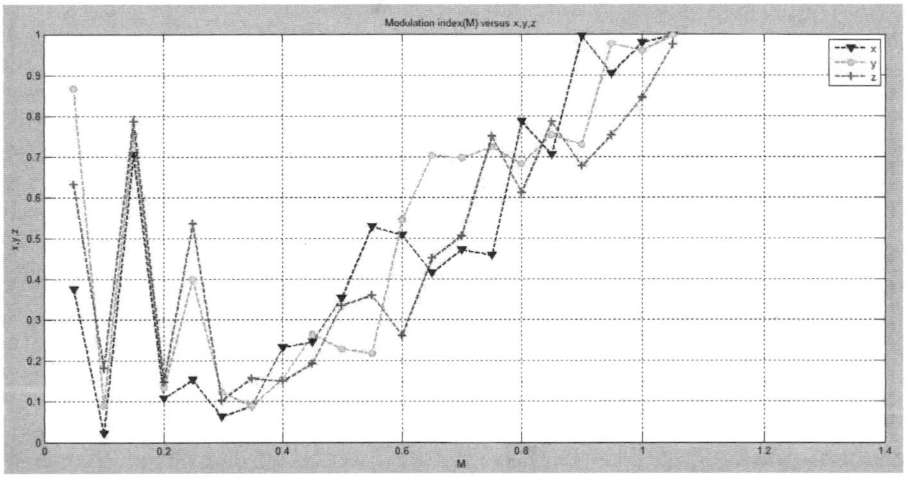

Figure 8. Amplitude of DC sources Versus M in (OHSW (7- variables))

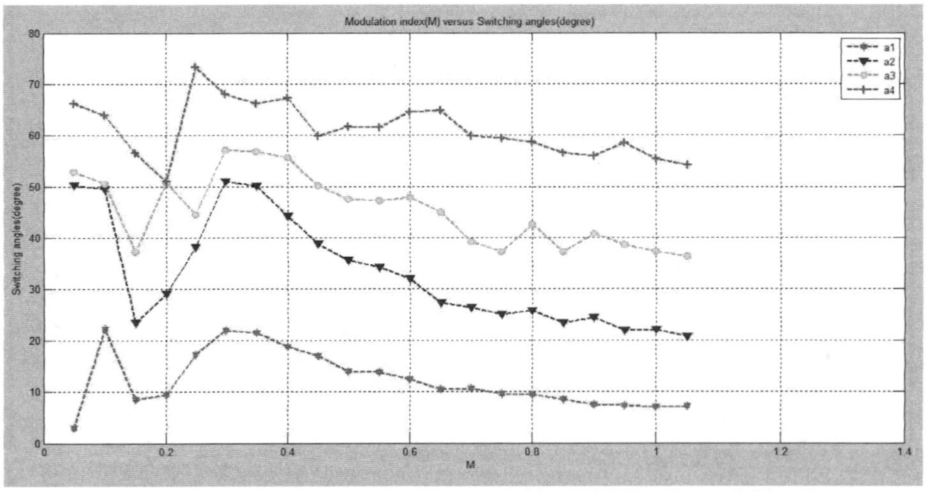

Figure 9. Switching Angles Versus M in (OTHD (7- variables))

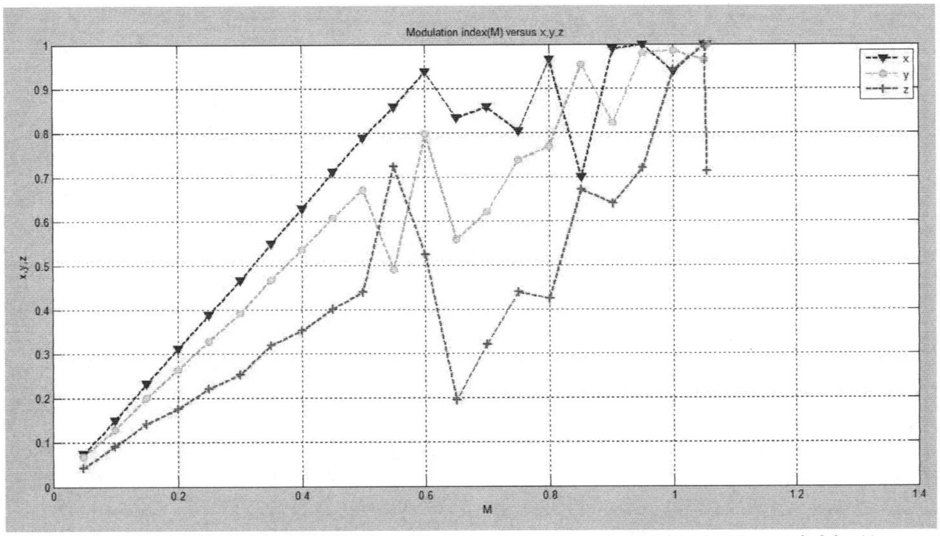

Figure 10. Amplitude of DC sources Versus M in (OTHD (7- variables))

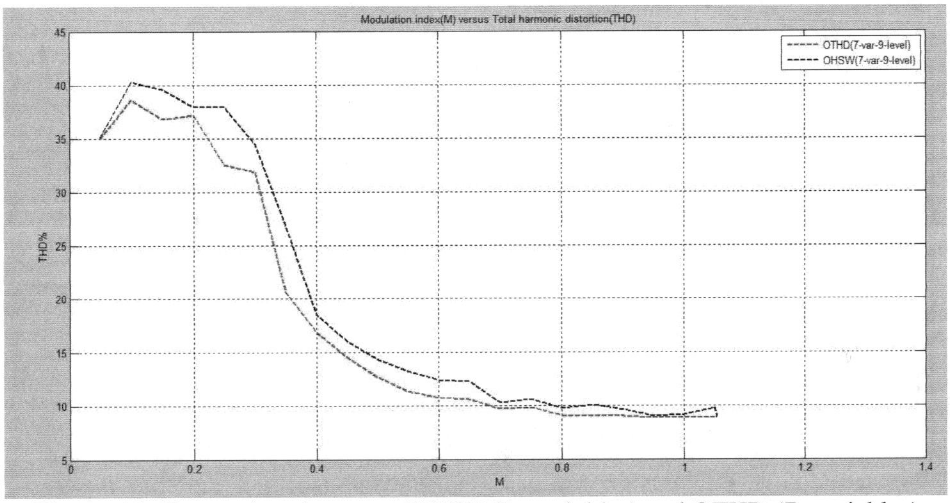

Figure 11. Comparison between OHSW (7- variables) and OTHD (7- variables)

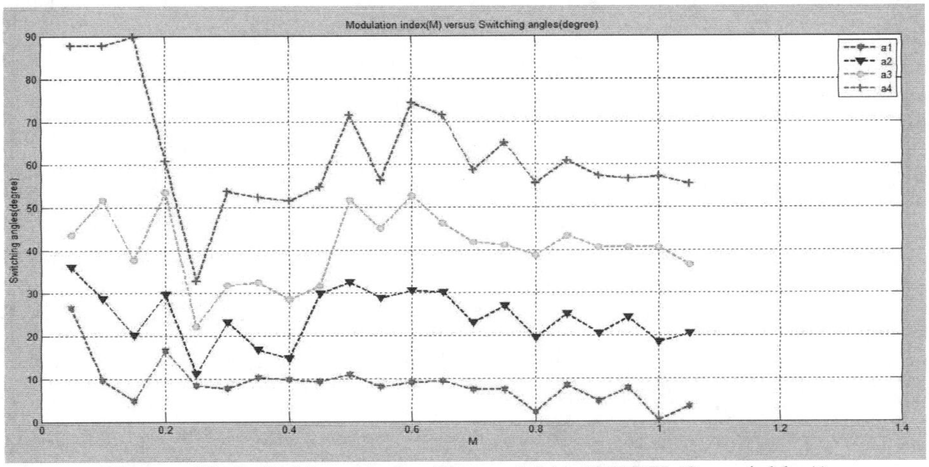

Figure 12. Switching Angles Versus M in (OHSW (8- variables))

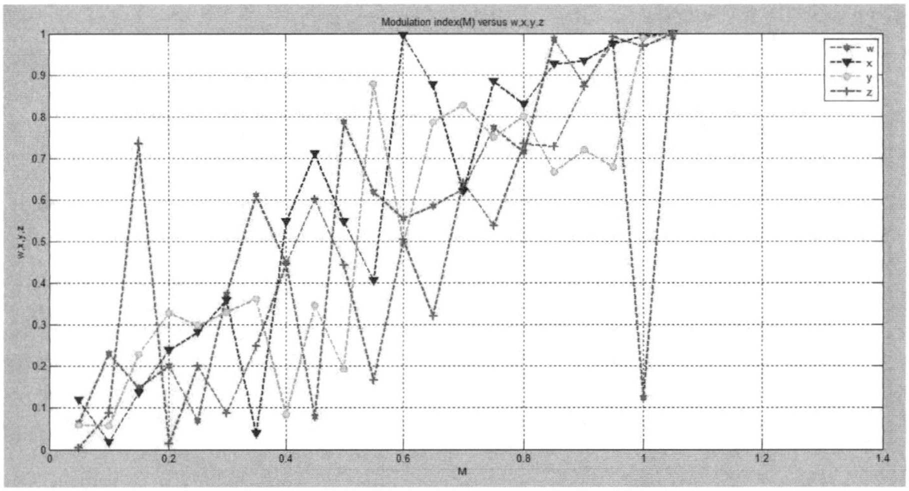

Figure 13. Amplitude of DC sources Versus M in (OHSW (8- variables))

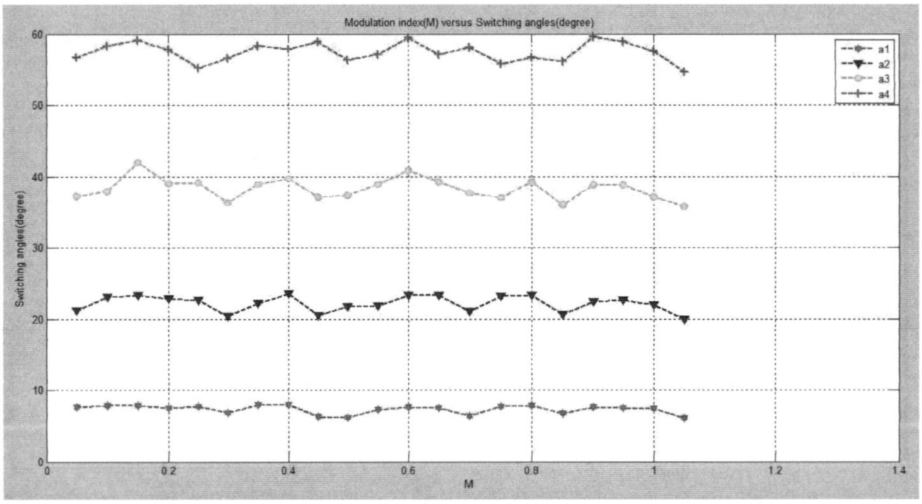

Figure 14. Switching Angles Versus M in (OTHD (8- variables))

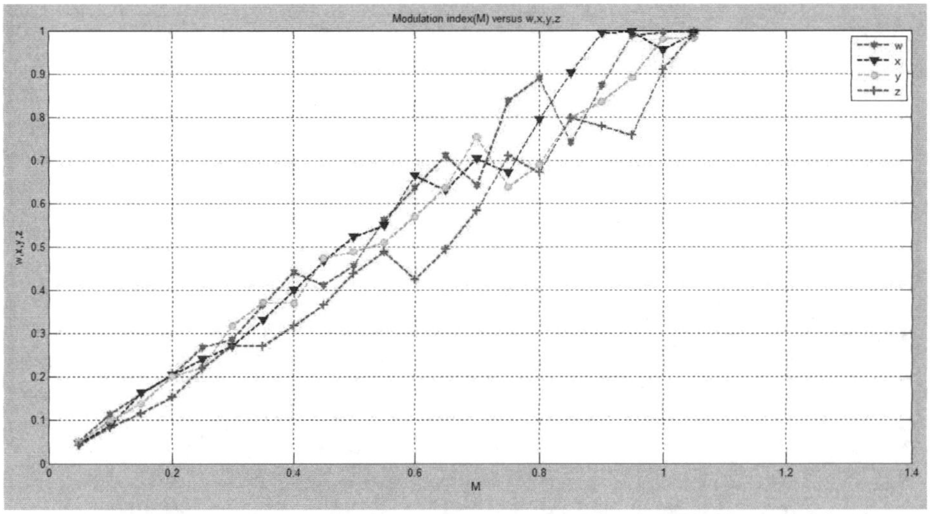

Figure 15. Amplitude of DC sources Versus M in (OTHD (8- variables))

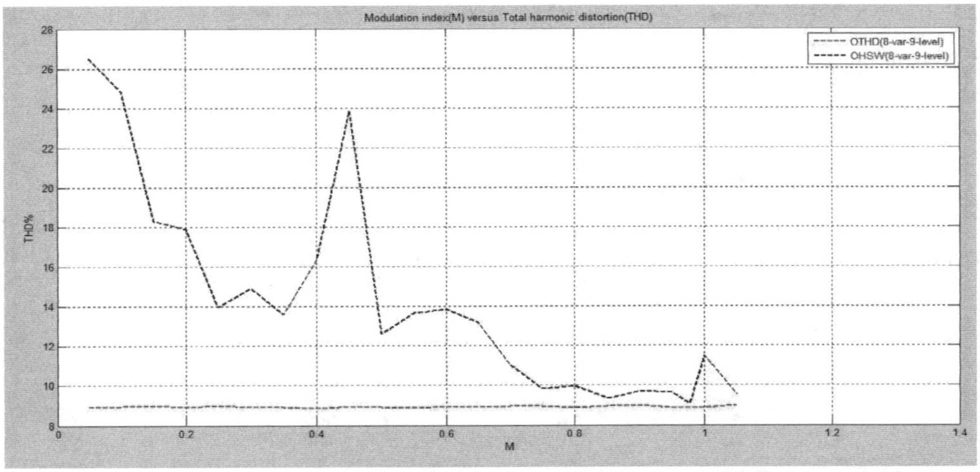

Figure 16. Comparison between OHSW (8- variables) and OTHD (8- variables)

Table 2. Optimal Results obtained by OTHD technique in 9-level inverter

No. of Variables	THD%	1α	2α	3α	4α	w	x	y	z
8	8.83	7.5630	22.6318	38.7892	58.9001	0.95	0.987	0.998	0.89
7	8.88	7.339	21.9443	38.6747	58.5563	1	0.903	0.977	0.752
4	8.9	7.101	22.459	39.82	62.7960	1	1	1	1

Table 3. THD Calculation obtained by OHSW technique in 9-level inverter

No.of Harmonics	OHSW (8-var-9-level)	OHSW (7-var-9-level)	OHSW (4-var-9-level)
49th	8.1073	8.983	12.6018
63th	8.3574	9.2798	12.8277
199th	9.0471	10.0103	13.4794
Integrating method	9.1231	10.1357	8.9149

Table 4. THD Calculation obtained by OTHD technique in 9-level inverter

No.of Harmonics	OTHD(8-var-9-level)	OTHD(7-var-9-level)	OTHD(4-var-9-level)
49th	7.752	7.9636	11.7117
63th	8.1916	8.1958	11.9683
199th	8.8847	8.9174	12.7026
Integrating method	8.9716	9.0268	12.8295

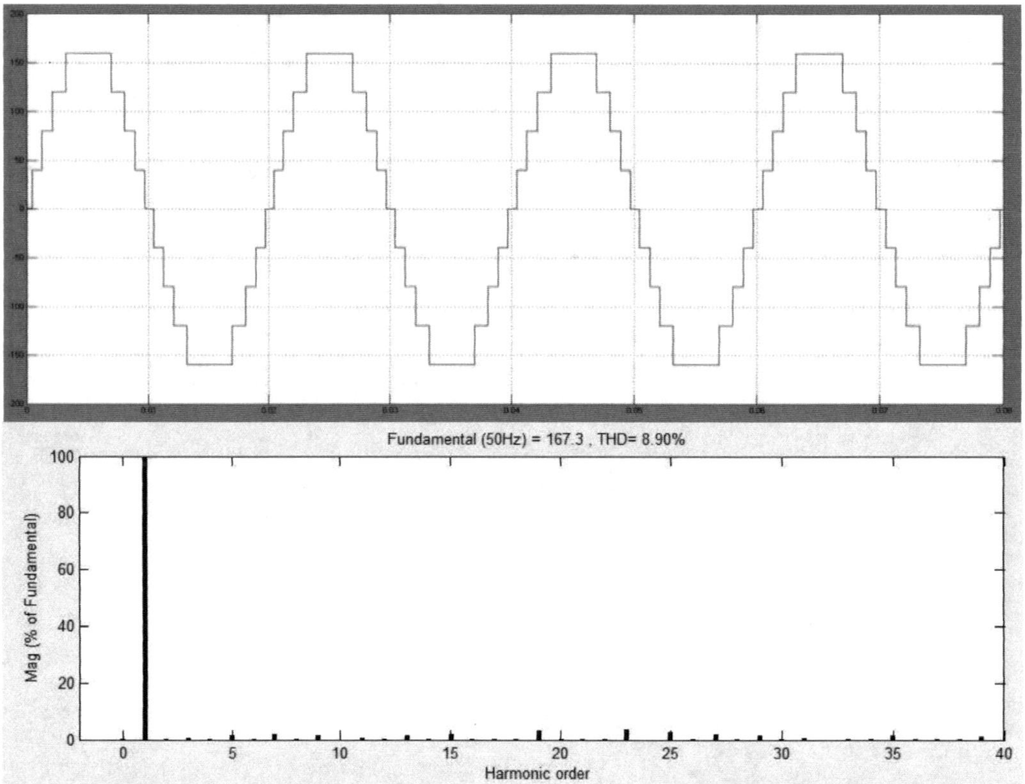

Figure 17. Matlab Simulation Result output voltage and its FFT at M=0.85

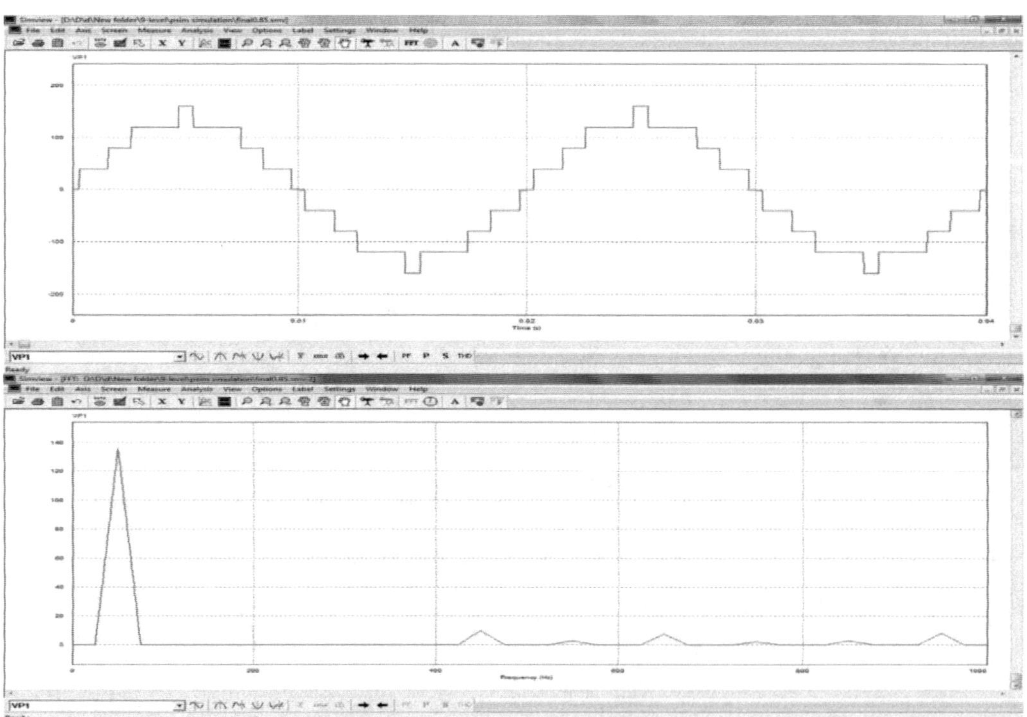

Figure 18. Psim Simulation Result output voltage and its FFT at M=0.85

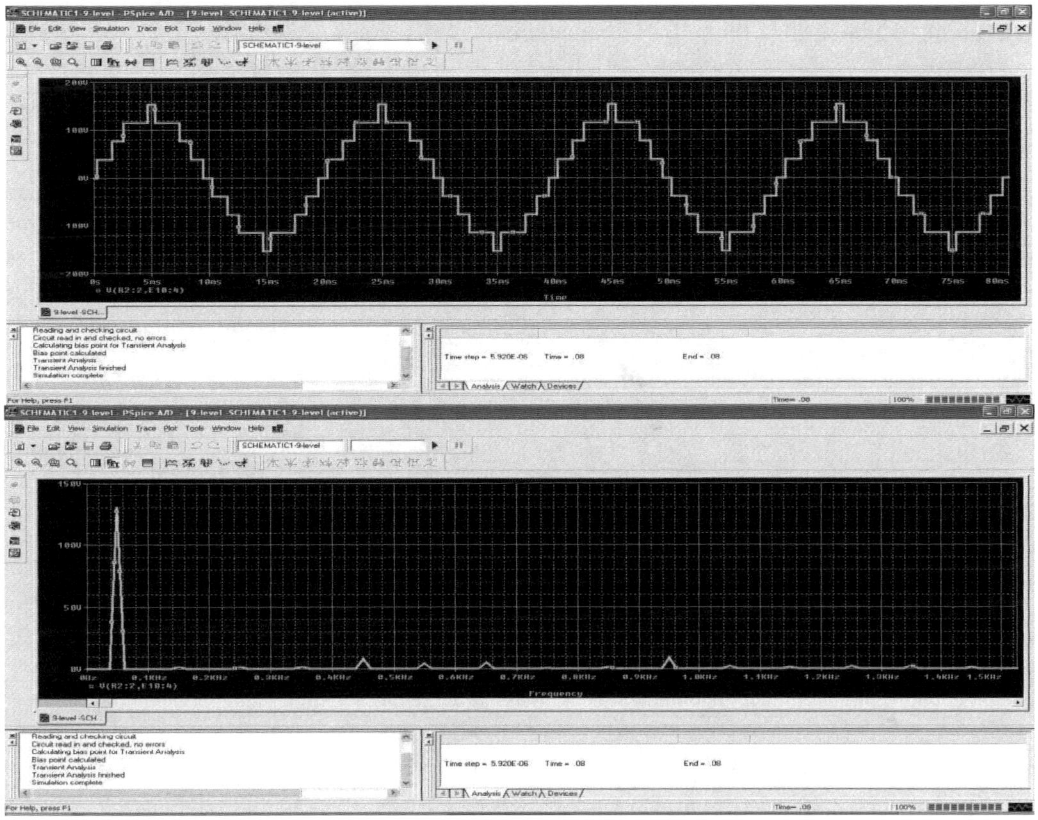

Figure 19. Pspice Simulation Result output voltage and its FFT at M=0.85

5. Conclusions

In this paper two control strategies OHSW and OTHD techniques are applied to single phase nine level inverter with equal and adjustable dc sources. a wide range of modulation index is achieved by solving non-linear equations and the results is shown in the figures. Comparison shows that employing adjustable dc sources which can be controlled results in reducing Total harmonic distortion in OTHD and eliminates more harmonics in OHSW but increasing number of variables results in increasing solution complexity. Simulation is done to validate the calculation of switching angles. Three model were built in three different programs (Matlab, Pspice and PSIM) which deals with mathematical model, actual model and thermal model.

6. References

[1] lhami Colak, Ersan Kabalci and Ramazan Buyindir, 2010, Review of Multilevel Voltage Source Inverter Topologies and Control Scheme, *Elsevier, Energy Conversion and Management* 2010.

[2] Ebrahim Babaei, Mohammad Farhadi Kangarlu and Farshid NajatyMazgar,2012, Symmetric and asymmetric multilevel inverter topologies with reduced switching devices, *Elsevier, Electric Power Systems Research* 86-122-130, 2012.

[3] N. Yousefpoor, S.H. Fathi, N. Faroknia and J. Moghani, 2009, Developed Single Phase OMTHD Technique for Cascaded Multilevel Inverter by Considering Adjustable DC Sources, *IEEE International Conference on Electric Power and Energy Conversion Systems (EPECS)* pp 1-6.

[4] Ebrahim Babaei and Seyed Hossein Hosseini, 2009, New cascade multilevel inverter topology with minimum number of switches, *Elsevier, Electric Power Systems Research* 50-2761-2767.

[5] Y. Sahali and M.K. Fellah, 2006, Application of The Optimal Minimization of the THD Technique to the Multilevel Symmetrical Inverters and Study its Performance in Comparision with the Selective

Harmonic Elimination Technique, *International Symposium on Power Electronics, Electrical Drives, Automation and Motion* 1-4244-0194.

[6] Sirisukprasert, Siriroj, 1999, Optimized Harmonic Stepped Waveform for Multilevel Inverter, MSC Thesis. Department of electrical and computer engineering, *Virginia Polytechnic Institute and State University (Virginia tech).*

[7] Ebrahim Babaei, Mohammad Farhadi Kangarlu, Mehran Sabahi & Mohammad Reza Alizadeh Pahlavani, 2013, Cascade Multilevel inverter using sub-Multilevel cell, *Elsevier, Electric Power Systems Research* 96-101-110.

Study of defects in $CH_3NH_3PbI_3$-based perovskite solar cells

Hajar Kh Ibrahim[1*], Ahmed M A Sabaawi[1] and Qais Th Algwari[1]

[1]College of Electronics Engineering, Nineveh University, Mosul, Iraq

[*]E-mail: hajerkhib90@gmail.com

Abstract. The present work will focus on the investigation of the influence of defects density and carriers capture cross-section are in absorption layer on the performance of perovskite solar cells. Furthermore, the impact of defect density in the electron transport material *ETM*/absorber interface layer will also be studied. The hybrid halide perovskite solar cells are chosen in this work and will be simulated by using one-dimension solar cells capacitance simulation SCAPS-1D. The proposed structure of the perovskite solar cell is a planar *FTO*/ *ETM*/ perovskite/ *HTM* based onTiO_2 and Spiro-OMeTAD as electron transport layer *ETM* and hole transport layer *HTM*, respectively, and on $CH_3NH_3PbI_3$ as absorber layer (perovskite).The results showed that when the defect density in the absorber layer was varied from 10^{10}cm^{-3}to 10^{18} cm^{-3}, the cell parameters, open circuit voltage (V_{oc}), the short circuit current (J_{sc}), the fill factor (*FF*),and the efficiency (*PCE*)were significantly decreased. At high defect density (10^{18} cm^{-3}), the defect affects negatively on the cell performance due to the high recombination rate of Shockley Read Hall. Similarly, the capture cross section exhibits similar behaviour in the active layer. By increasing the defect density in the *ETM*/absorber interface layer, the main cell parameters *PCE*,*FF*, and J_{sc} are slightly reduced and there has been no noticeable impact on V_{oc}.

1. Introduction

Photovoltaic solar is becoming one of the challenging solutions to the problems of securing a clean and sustainable energy source. This includes Si-based solar power cells and photovoltaic cells that are sensitized as well as the organic solar cells [1].Recently, inorganic-organic solar cells of the metal halide Perovskite have gained considerable global attention as a candidate for low-cost and high-performance photovoltaic over the last few years. Organ lead halide perovskite materials properties, such as its broad duration of electron-hole diffusion, good electron-hole mobility, and large coefficient of absorption have several advantages for photovoltaic device application [2]. There has been a discovery of the existence of perovskite substances in solar panels with an efficiency (*PCE*) lower than 4% [3]. In the short time in the history of solar cells, where the newly reported power conversion efficiency is 22.1%, rapid chronological progress is seen [4]. Perovskite, such as methyl ammonium lead iodide ($MAPbI_3$), shows different intrinsic defects caused by vacancies, interstitial, and antisites. Despite its past problem with perovskite is good with defects tolerance, this recent research has demonstrated that deep level defects are present in the Shockley Quiesser (*SQ*) restricted approach that affects the performance of perovskite solar cell (*PSC*). Different studies demonstrated the impact of defect density on hybrid Organic-inorganic perovskite (*HOJPS*) solar cells efficiency [5], however, there is no comprehensive study available yet. To make Perovskite solar cells a winner against traditional technology of solar cells, a deep understanding is thus needed. It is possible to classify the defect based on its location in the band gap energy, where they can be detected at either deep or shallow levels [5]. Si Fengjuanet al studied the effects of defects in *PSC* in particular the defect states in the

Content from this work may be used under the terms of the Creative Commons Attribution 3.0 licence. Any further distribution of this work must maintain attribution to the author(s) and the title of the work, journal citation and DOI.
Published under licence by IOP Publishing Ltd

absorption layer and adjacent interfaces [6]. A.S.Chouhan et al studies to simulate self-consistent system model with SCAPS using data of the experimental from 17.5% efficiency *PSC*. Experiential results, such as doping consideration of compactness TiO_2, densities of the defect and doping in perovskite. The only reasonable variables in the model are the interface layer defect density at the *HTM*/perovskite and perovskite/*ETM*. For simulation, it was shown that the *HTM* interface layer defect density strongly affects the V_{oc}, whereas *ETM* interface layer has a significant influence on J_{sc} [7]. M.S. Jamal et al [8] presented SCAPS-1D's numerical analysis on inverted planer NiO-based *PSC* as a hole transportation material (*HTM*) and investigated the impact of the absorption layer and *HTM* /absorber interface layer's defect density and energy levels on output. The result is obtained when various energy levels of up to 0.9 eV have been studied for different defect density values from 10^{13} cm^{-3}to10^{19}cm^{-3}. Deeper defects, less than 0.2eV, and higher defect density were found to produce a low efficiency. The perovskite layer has been examined in regards to its optical and structural characteristics be deficient. Then numerical investigations of the effect of perovskite absorption layer thickness on related solar cell parameters have been done using SCAPS-1Das well as the defect density of the absorption layer. Since the active layer thickness changes from 200-800 nm and shallow level defect changed from 10^{13} cm^{-3} to 10^{18} cm^{-3}, the efficiency reduced sharply from 26.7% to 0.9%. However, the *PCE* decrease from approximately 19.3% to approximately 6.15% as the thickness and the deep density increase from 200-800 nm and 10^{13} cm^{-3} to 10^{16} cm^{-3}, respectively [9]. Until nowadays, there are not enough studies investigated the effect of defect mechanism in solar perovskite cells. In this paper, the aim is to investigate various aspects of defects in planer *PSC* structure: Glass/*FTO*/TiO_2/$CH_3NH_3PbI_3$/Spiro-OMeTAD/ back contact, which is simulated using the 1-D capacitance simulation for solar cells (SCAPS-1D) [10]. The work studies the impact of defects on the *PSC* performance, where the focus will be on the carriers capture cross-section area, defect density in the absorption layer, and defect density in *ETM* / perovskite interface layer.

2. Device simulation parameters

SCAPS 1D software is employed in this work as it is considered one of the most effective and user-friendly numerical simulations of solar cell systems. The proposed structure of perovskite solar cell is a planer structure of glass /*FTO* (Fluorine doped Tin Oxide) /TiO_2 (Electron Transportation layer)/$CH_3NH_3PbI_3$(active layer)/Spiro-OMeTAD (Hole Transportation layer)/metal back contact. Figure 1 illustrates the basic structure for the studied solar cell, whereas the values of the simulation parameters are summarized based on values found in the literature as shown in table 1. Table 2 summarizes defects parameters of the absorber layer and interfaces *ETM*/absorber and *HTM*/absorber of the perovskite solar cell.

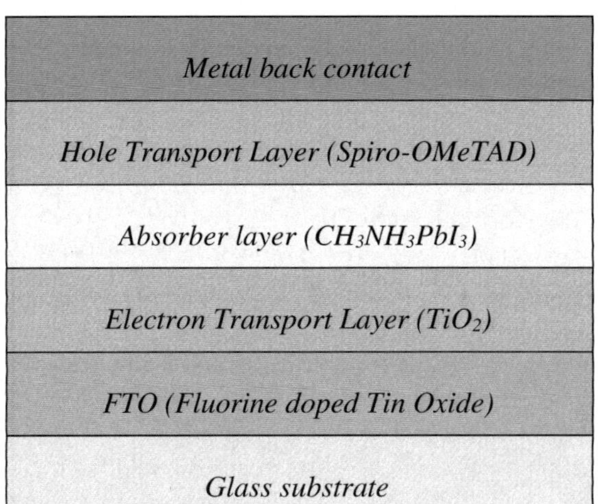

Figure 1. The basic structure of the perovskite solar cell.

Table 1. Simulation parameters of perovskite solar cells devices.

Parameter	*FTO*[13]	TiO_2[12]	$CH_3NH_3PbI_3$[7]	*Spiro-OMeTAD*[11]
Thickness t (μm)	0.4	0.001	0.4	0.4
Band gap Eg (eV)	3.5	3.26	1.55	3.06
Electron affinity χ (eV)	4	3.7	3.9	2.05
Dielectric permittivity ε_r (relative)	9	10	6.5	3
CB effective density of state N_c (cm^{-3})	2.2×10^{18}	2.2×10^{18}	2.2×10^{18}	2.2×10^{18}
VB effective density of state N_v (cm^{-3})	2.2×10^{18}	2.2×10^{18}	2.2×10^{18}	2.2×10^{18}
Electron Mobility μ_n (cm^2/Vs)	20	20	2	2×10^{-4}
Hole Mobility μ_p(cm^2/Vs)	10	10	2	2×10^{-4}
Donor density N_D (cm^{-3})	1×10^{19}	1×10^{17}	0	0
Acceptor density N_A (cm^{-3})	0	0	1×10^{15}[2]	3×10^{18}
Electron thermal velocity v_{thn} (cm/s)	1×10^7	1×10^7	1×10^7	1×10^7
Hole thermal velocity v_{thp} (cm/s)	1×10^7	1×10^7	1×10^7	1×10^7

Table 2. Defect parameters of interface and absorber [14].

Parameter	Defect of absorber	Defect of *HTM*/Perovskite	Defect of *ETM*/Perovskite
Defect type	Neutral	Neutral	Neutral
Capture cross section electrons σ_n(cm^2)	2×10^{-15}	2×10^{-15}	2×10^{-16}
Capture cross-section holes σ_p(cm^2)	2×10^{-15}	2×10^{-15}	2×10^{-16}
Energetic distribution	Gaussian	Single	Single
Characteristic energy (eV)	0.1	0.1	0.1
Energy with respect to reference (eV)	0.5	0.65	0.65
Total density (cm^{-2})	(10^{10}-10^{18})	1×10^{18}	1×10^{18}

Other parameters such as defect density Nt in absorption layer is 2.5×10^{13} cm^{-3}using Gaussian distribution and in the other layers are set to 10^{15} cm^{-3}. The energy level of the defect is located at the middle of band gap energy with the single distribution in the layers at characteristics energy level 0.1 eV, the type of the defect is neutral, the hole and electron capture cross-section area are 1×10^{-15} cm^2, and the absorption coefficient for each layer is set to $10^5 cm^{-1}$. In the simulation, the normal atmospheric conditions are used (*AM*1.5 G, 300 K) spectrum. The default value is used for the operating point and numerical setting values. The voltage is varied from 0 to 1.3 V. The current-voltage *J-V* characteristics of reference cell is shown in Figure 2 and the initial performance parameters are: PCE = 12.93%, FF =74.99%, J_{sc} = 21.03 mA·cm^{-2}, and V_{oc} = 0.8202 V.

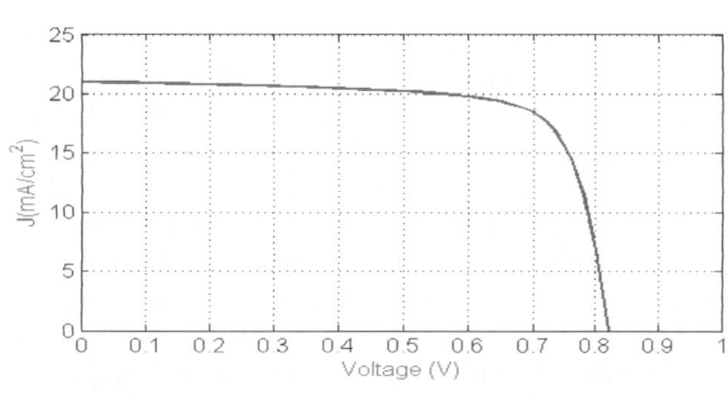

Figure 2. J-V characteristics of the reference cell.

3. Results simulation and Discussions

3.1. Impact of defects in the absorption layer

The performance of perovskite solar cells involved two major procedures which are: one is the generation of a charge carrier, while another is a recombination of charge carrier. In the event of sunlight, the perovskite layer absorbs solar radiation, where the photons are absorbed to produce charged carriers (electrons-holes pair). The electrode sequentially extracts these photo-generated carriers and passes them into the outer in a form of electrical current. Due to poor perovskite layer consistency, a significant number of carriers are lost. In poor perovskite films, a higher value of defect density results in a higher recombination rate, which decreases the propagation length of the carriers and also the lifetime of charge carriers, which can comprehend the fact that there are defects in this layer as a result, it can be decided to investigate recombination as a result of perovskite defects.

3.1.1. Defect density

The impact of defects density in the absorption layer on the performance of the solar cell has been studied by varying the density of the defect from 10^{10} cm^{-3} to 10^{18}cm^{-3} as shown in Figure 3.

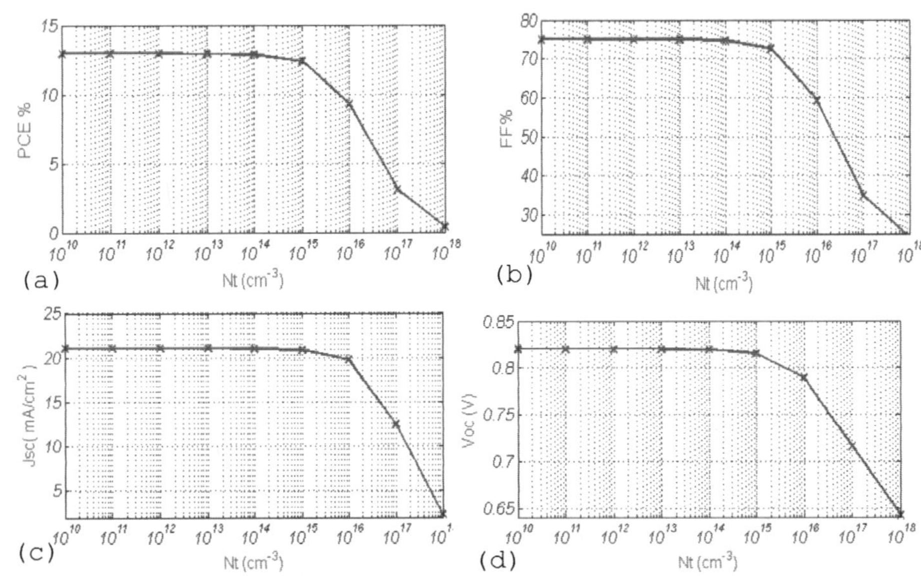

Figure 3. Performance parameters versus defect density: (a) PCE, (b) FF, (c) J_{sc} , (d) V_{oc}.

As seen 'in Figure 3' the power conversion efficiency remains unchanged over the defect density range from 10^{10} cm^{-3} to 10^{15}cm^{-3}. Above the value of 10^{15} cm^{-3}, the *PCE* decreased sharply from 12.35% to 0.35%. This can be ascribed to the higher defect density that causes a higher state levels that in turn leads to increasing the trapped region assisted recombination carrier rate. Once a defect densityexceeds10^{16} cm^{-3}, the J_{sc} reduced rapidly from 21.03 mA·cm^{-2}to 2.29 mA·cm^{-2}as revealed in figure 2d, which gives an indication to the losses of charge carrier where the trap and recombination processes become significant, while there has been no change in J_{sc} as the density of the defect was below 10^{16} cm^{-3}. Similar behavior for V_{oc} was observed, it is decreased from 0.82 V to 0.64 V over the defect's density variation from 10^{15}cm^{-3} to10^{18} cm^{-3}with no noticeable change at defect densities below 10^{15} cm^{-3}. On the other hand, the fill factor (*FF*) was degraded from 75.02% to 23.99%across the tested defect density range, where the *FF* remains 75.02%at defect density below 10^{13} cm^{-3}. It can be deduced from figure3 that the density of the defect above 10^{15} cm^{-3}has a significant impact on the performance of the *PSC* device. This interpretation of this phenomenon is due to the fact that because the defect density affects only the active layer which is at the same time the absorber region. It can be observed that as a defect density increases, the lifetime of charge carrier decreases and recombination rate increases, while low defect density results in lesser recombination carrier in perovskite cell.

3.1.2. Defects of capture cross-section

The impact of carriers capture cross-section area at absorption layer on the performance of the solar cell has also been investigated, where the capture cross-section was changed from 2×10^{-18}cm^2 to 2×10^{-10} cm^2 as shown in figure 4.

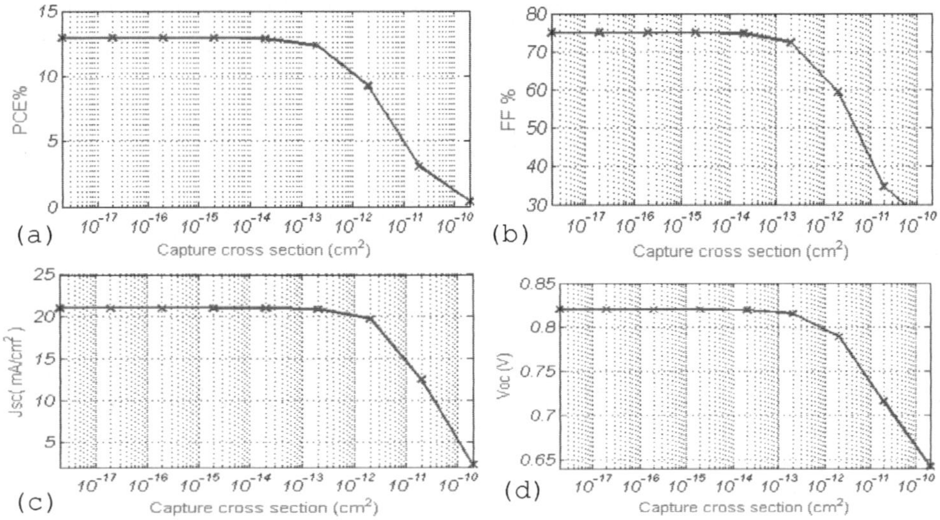

Figure 4. The effect of capture cross-section on the cell performance: (a) PCE, (b) FF, (c) J_{sc}, (d).V_{oc}

As seen in figure 4, that all cell performance parameters had almost the same behavior as that of defect density variation. At capture-cross section values below 2×10^{-14} cm^2, the power conversion efficiency (*PCE*) is almost constant, whereas, at higher values, the *PCE* drops rapidly from 12.94% to 0.35%. The fill factor (*FF*) is strongly influenced by increasing the capture cross-section area, where *FF* decreased slightly below 2×10^{-13}cm^2, however, it was reduced sharply from 72.48% to 23.99% for higher capture cross-section area. The J_{sc} degradation from 21.03 mA·cm^{-2} to 2.29 mA·cm^{-2}is noticed when the capture cross-section was varied from 2×10^{-14}cm^{2}to2×10^{-10} cm^2. Also, it was observed that when the capture cross-section area was increased the V_{oc} reduced slightly from 0.82 V to 0.64 V. The *PCE* reduces from the maximum value from efficiency about 12.59 % and *FF* decrease from 75.02% below 51.1% and J_{sc} around 18.74% V_{oc} about 0.18%. The results in the figure4 showed that the impact of the capture cross-section exhibited similar behavior to the influence of defects density in the absorption layer. This behavior can be interpreted as the lifetime of carriers is determined by the trap density and carriers capture cross-section, thus when the capture cross-section increase, the carrier lifetime decrease due to dimension of defect trap increase leads to significantly reduced solar cells performance.

3.2. Impact of defects in the interface layer

Two interfaces (*ETM* /absorber) and (absorber/*HTM*) were presented in the proposed structure of the cell and the impact of defects in *ETM*/absorber interface layer on the performance of the cell was investigated. The defect parameters of the two interface layers are shown in table 2. The influence of the defect density at *ETM* /absorber interface layer on the solar cell parameters has been changed from 10^{11}cm^{-2}to 10^{20} cm^{-2}while the other variables were kept unchanged. The results are shown in figure 5.

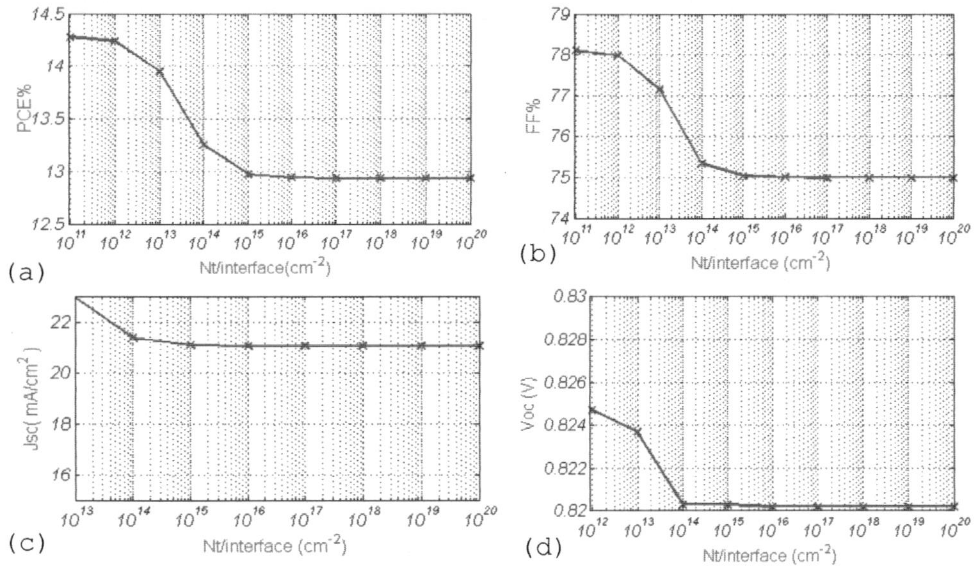

Figure 5. Change in cell parameters (a) PCE, (b) FF, (c) J_{sc}, (d) V_{oc} in relation to the defect density on the ETM/ absorber interface layer.

It is seen in figure 5 that the increase in the defect density results in a slight decrease in the *PEC, FF,* and J_{sc}, while the V_{oc} has shown no significant change. It can also be noticed that the cell parameters exhibited a small change for defect densities greater than10^{14}cm^{-2}below the defect density of 10^{14} cm^{-2}, the *PCE* reduced from maximum efficiency by about 1.35% and *FF* decreased from 78.10 by around 3%, and J_{sc} decrement was less than1.14% while V_{oc} remained unchanged. The solar cell performance was trivially influenced by the density of the defect at the *ETM* interface. In *HTM*/absorber interface layer, it is observed no prominent impact of the defect density on the performance parameters in *PSC*. Due to the interfaces are represented as defects when the defect density was varied over the tested range, the cell performance suffered a slight reduction.

4. Conclusion

In the present work, the effect of defect density and carriers capture cross section area at the absorber layer and defect density at the *ETM*/ Perovskite interface layer in perovskite solar cells with planer structure $FTO/TiO_2/CH_3NH_3PbI_3$/Spiro-OMeTAD were studied by one dimensional Solar Cells Capacitance Simulator (SCAPS-1D). The results showed that when the defect density increased in the absorption layer, the *PCE, FF,* and J_{sc} are dropped sharply, however, V_{oc} was slightly decreased. Furthermore, it was noticed that a higher capture cross-section area causes significant degradation in the cell performance and has shown similar behavior of the defect density effect. The cell parameters such as *PCE, FF,* and J_{sc} decreased slightly when the defect density at *ETM*/absorber interface layer was changed from 10^{11}cm^{-2}to10^{20} cm^{-2}and showed that the V_{oc} has no significant impact. It was also shown that the influence of defect density in the interface layer was trivial when compared with the absorption layer. Further investigation is needed to explore the effect of high defect densities in other materials rather than Spiro-OMeTADsuchCu_2O and *CuI*.

5. References

[1] Rai S, Pandey B K and Dwivedi D K 2020 Modeling of highly efficient and low cost CH3NH3Pb (I1-xClx) 3 based perovskite solar cell by numerical simulation *Opt. Mater. (Amst).* **100** 109631

[2] Lin L, Jiang L, Qiu Y and Yu Y 2017 Modeling and analysis of HTM-free perovskite solar cells based on ZnO electron transport layer *Superlattices Microstruct.* **104** 167–77

[3] Kojima A, Teshima K, Shirai Y and Miyasaka T 2009 Organometal halide perovskites as visible-light sensitizers for photovoltaic cells *J. Am. Chem. Soc.* **131** 6050–1

[4] Yang W S, Park B-W, Jung E H, Jeon N J, Kim Y C, Lee D U, Shin S S, Seo J, Kim E K and Noh J H 2017 Iodide management in formamidinium-lead-halide–based perovskite layers for efficient solar cells *Science (80-.).* **356** 1376–9

[5] Pathak C and Pandey S K 2020 Design, Performance, and Defect Density Analysis of Efficient Eco-Friendly Perovskite Solar Cell *IEEE Trans. Electron Devices* **67** 2837–43

[6] Fengjuan S, Fuling T, Hongtao X and Rongfei Q 2016 Effects of defect states on the performance of perovskite solar cells *J. Semicond.* **37** 72003

[7] Chouhan A S, Jasti N P and Avasthi S 2018 Effect of interface defect density on performance of perovskite solar cell: Correlation of simulation and experiment *Mater. Lett.* **221** 150–3

[8] Jamal M S, Shahahmadi S A, Wadi M A A, Chelvanathan P, Asim N, Misran H, Hossain M I, Amin N, Sopian K and Akhtaruzzaman M 2019 Effect of defect density and energy level mismatch on the performance of perovskite solar cells by numerical simulation *Optik (Stuttg).* **182** 1204–10

[9] Mahjabin S, Haque M M, Sobayel K, Jamal M S, Islam M A, Selvanathan V, Assaifan A K, Alharbi H F, Sopian K, Amin N and Akhtaruzzaman M 2020 Perceiving of Defect Tolerance in Perovskite Absorber Layer for Efficient Perovskite Solar Cell *IEEE Access* **8** 106346–53

[10] Tan K, Lin P, Wang G, Liu Y, Xu Z and Lin Y 2016 Controllable design of solid-state perovskite solar cells by SCAPS device simulation *Solid. State. Electron.* **126** 75–80

[11] Casas G A, Cappelletti M A, Cedola A P, Soucase B M and y Blancá E L P 2017 Analysis of the power conversion efficiency of perovskite solar cells with different materials as Hole-Transport Layer by numerical simulations *Superlattices Microstruct.* **107** 136–43

[12] Azri F, Meftah A, Sengouga N and Meftah A 2019 Electron and hole transport layers optimization by numerical simulation of a perovskite solar cell *Sol. energy* **181** 372–8

[13] De Los Santos I M, Cortina-Marrero H J, Ruíz-Sánchez M A, Hechavarría-Difur L, Sánchez-Rodríguez F J, Courel M, Hu H, Montoya De Los Santos I, Cortina-Marrero H J, Ruíz-Sánchez M A, Hechavarría-Difur L, Sánchez-Rodríguez F J, Courel M and Hu H 2020 Optimization of CH3NH3PbI3 perovskite solar cells: A theoretical and experimental study *Sol. Energy* **199** 198–205

[14] Haider S Z, Anwar H and Wang M 2018 A comprehensive device modelling of perovskite solar cell with inorganic copper iodide as hole transport material *Semicond. Sci. Technol.* **33** 35001

IOP Conf. Series: Materials Science and Engineering 1152 (2021) 012033 doi:10.1088/1757-899X/1152/1/012033

The influence of the interface layer between the electron transport layer and absorber on the performance of perovskite solar cells

Dena N Qasim Agha[1*] **and Qais Th Algwari**[1]

[1]College of Electronics Engineering, Department of Electronic Engineering, Ninevah University, Mosul-Iraq

[*]E-mail: dina.abduljabbar20192019@stu.uoninevah.edu.iq

Abstract. High efficiency, lightweight, and cost-effectiveness put the perovskite solar cells to the top of the focus researches of solar cells. The architecture of the cell especially the energy band alignment at the interface is a critical issue in the cell performance. In the current paper, the solar cell structure under investigation consists of TiO_2, $CH_3NH_3PbI_3$, and Spiro OMe TAD as electron transport layer, absorber and hole transport layer respectively. A 3C-SiC material with energy gap of 2.420 eV was used as interface layer. The role of the interface layer between the perovskite and electron transport layer was considered. Before inserting the interface layer, a parametric study including the thickness and doping of each layer, was achieved. The results showed that the best performance of the cell at a thickness of 400 nm, 300 nm, 200 nm for absorber layer, ETL, and HTL respectively, with a doping concentration of 10^{14} cm^{-3}, 10^{19} cm^{-3}, 10^{19} cm^{-3} for the same layers. These parameters give a V_{oc}, J_{sc}, FF, and PCE of 1.11 V, 28.9 mA·cm^{-2}, 83.19%, and 26.88% respectively. Inserting an interface layer improved the performance of the cell where the PCE increased over 29% at a thickness of 90 nm. The results showed that the parameters of the interface layer play a significant role in cell performance.

1. Introduction

The endless expansion in world population and improved lifestyle continuously brings up a larger demand for electricity to power produces. With global warming being a major climate issue worldwide, the use of renewable energy could represent a valuable alternative to help to reduce the effect of pollution, which is the main drawback of conventional power generating devices. Natural sources like water and wind are still recruited in electrical power generation. However, the sun is still one of our greatest sources of energy, which could be utilized through the introduction of solar cells. The latter's mode of action is usually based on the conversion of light or photons into electricity; for that reason, it is often referred to as a photovoltaic device [1].

The perovskite solar cells (PSCs) are the most recent photovoltaic devices which use the light-absorbing layer made of perovskite material. A long carrier diffusion length, high light absorption, and tunable bandgap are the main features that made perovskite solar preferable over the conventional solar cells [2]. Since their introduction, they have dragged considerable attention due to their high efficiency (>15%) and remarkable ease of processing over large areas [3]. PSC is composed of organic metal halide as an absorber. The lead halide perovskite

Content from this work may be used under the terms of the Creative Commons Attribution 3.0 licence. Any further distribution of this work must maintain attribution to the author(s) and the title of the work, journal citation and DOI.
Published under licence by IOP Publishing Ltd

material is promising to use in the production of perovskite solar cells due to its superior optoelectronic properties [4].
In addition to the cathode and anode, the PSC comprises three main layers which are the hole transporting material layer (HTL), the absorber perovskite layer, an electron transporting material layer (ETL). From their names, the function of the HTL is to extract and transfer holes and block electrons while the function of ETL is to block the holes and transfer the electrons [5]. Upon absorption of light, photo-generation of free charge carriers occurs in the absorber, and these carriers, under the built-in potential will drift to HTL-perovskite and ETL-perovskite interface region. The weak charge transport and/or the energy barrier of the HTL or ETL will produce the charges accumulation at the border of the interface of these layers and this will cause an increase in charges recombination and hence reduce the solar cell efficiency. Therefore, to avoid the aforementioned problem, the interfacial materials should be engineered carefully. In the present work, a layer of n-3C-SiC was added as an interface layer between ETL and the perovskite layer. The added layer would hopefully improve the charge transfer between the layers due to band match and decreases the energy losses of the electrons during transportation which in turn lead to enhance the photovoltaic performance.

2. Simulation details

In the present work, the perovskite solar cell simulation was achieved using SACPS (Solar Cell Capacitance Simulator). SCAPS is a 1D simulation program with a multi-layers (up to seven) input. This software was developed by a group at the Department of Electronics and Information System, University of Gent, Belgium [6]. To simulate the proposed device, the individual parameters for each layer should be well defined. In terms of the energy bandgap (E_g), electron affinity (χ), dielectric permittivity (ε), conduction and valence band density of states (N_C and N_V), electron and hole mobility (μ_n and μ_p), and donor and acceptor density ($N_D$and N_A), the property of each layer were used as an input to the simulator. These parameters were taken from the literature [4,7-9] and are summarized in Table 1.

Table 1. Materials property for the perovskite cell layers.

Parameters	TiO_2	n-3C-SiC	$CH_3NH_3PbI_3$	Spiro OMe TAD
Bandgap E_g (eV)	3.2	2.420	1.55	3.0
Electron affinity (eV)	4.0	3.830	3.9	2.45
Dielectric permittivity	9.0	9.720	10.0	3.0
CB effective density of states ($1/cm^3$)	2×10^{18}	1.553×10^{19}	2×10^{18}	2.2×10^{18}
VB effective density of states ($1/cm^3$)	2×10^{19}	1.163×10^{19}	2×10^{19}	1.9×10^{19}
Thermal velocity of electron (cm/s)	10^7	1×10^7	10^7	10^7
Thermal velocity of hole (cm/s)	10^7	1×10^7	10^7	10^7
Electron mobility (cm^2/Vs)	0.2	650	1.0	2.0×10^{-4}
Hole mobility (cm^2/Vs)	0.1	40	1.0	2.0×10^{-4}
Shallow uniform donor density N_D ($1/cm^3$)	10^{19}	10^{18}	10^{14}	_
Shallow uniform acceptor density N_A ($1/cm^3$)	_	_	_	10^{19}

As shown 'in Figure 1', the current device having a structure of TiO_2, $CH_3NH_3PbI_3$, and Spiro OMe TAD as ETL, absorber, and HTL respectively. The initial values of thickness and doping for each layer are given in Table 2 [4]. The thickness and doping concentration were changed to obtain the solar cell structure with the best performance. A thin layer of 3C-SiC was inserted in the optimum structure between ETL and absorber as an interface layer. The standard conditions, illuminated under AM1.5G solar spectrum with 100 $mW.cm^{-2}$ incident power density and temperature of 300 °K, are considered in the current study. The absorption coefficient of the absorber, HTL, and ETL are defined in the program for simulation.

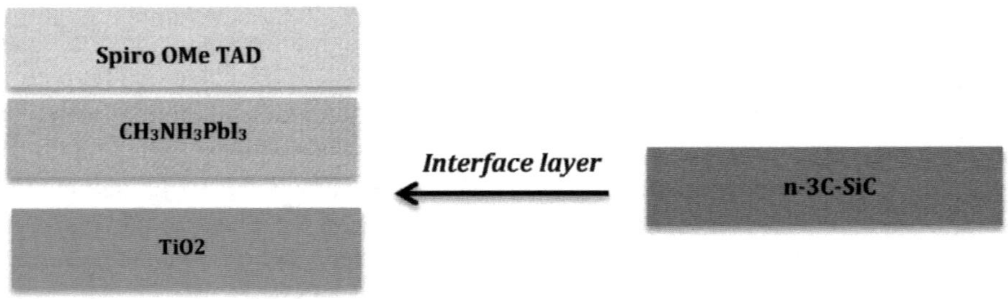

Figure 1. Perovskite solar cell structure.

Table 2. Initial values of thickness and doping of the proposed cell.

Layer	Thickness (nm)	Doping (cm^{-3})
Absorber	200	10^{13}
HTL	700	10^{18}
ETL	100	3×10^{19}

3.Results and discussion

To obtain the optimum structure of the solar cell that gives the best performance the current study involved two main parts, the first is the optimum thickness and doping density of each layer of the cell while the second part focused on the role of the interface layer between the perovskite and ETL on the cell parameters.

3.1 Effect of layers thickness

To obtain best absorption of the solar light in the perovskite layer and hence increase the generation process of electron-hole pairs the thickness of the absorber layer has been varied from 100 nm to 900 nm. Figure 2 illustrated the variation of the solar cell parameters with the absorber layer thickness. As shown 'in Figure 2' the J_{sc} increases from 19.88 $mA\cdot cm^{-2}$to 43.61$mA\cdot cm^{-2}$ over the examed thickness range, while V_{oc} slightly decreases from 1.18 V to 1.08 V. The absorber thickness effect noticeably on the efficiency although the FF drop from 84.5% to 75.98%. As represented in figure 2 the cell efficiency increased from 20% to 36%. This increase can be signed to the photo absorption enhancement in this layer. This means that the number of photo-generated carriers is associated with the total amount of absorbed photons, resulting in a J_{sc} increase with layer thickness increases. It can be noticed that above 300 nm the increase in J_{sc} trend to be slow compare to the J_{sc} variation at low absorber thickness. The efficiency has the same behavior in this range. This could be ascribed to the saturation in the light absorption occurs. As the absorber becomes thicker, the photo-generated carriers at the center of the absorber layer will recombine when the thickness of this layer exceeds the diffusion length. This is reflected on the V_{oc} due to the decrease in the effective band gap, also, to increase in the recombination processes [9-12]. Although the maximum efficiency was found to be 36% at absorber thickness of 900 nm the results show that the better thickness of 400 nm where the efficiency significantly improved to 31% with J_{sc}, V_{oc}, FF of 33 mA·cm^{-2}, 1.12 V, and 83%, respectively.

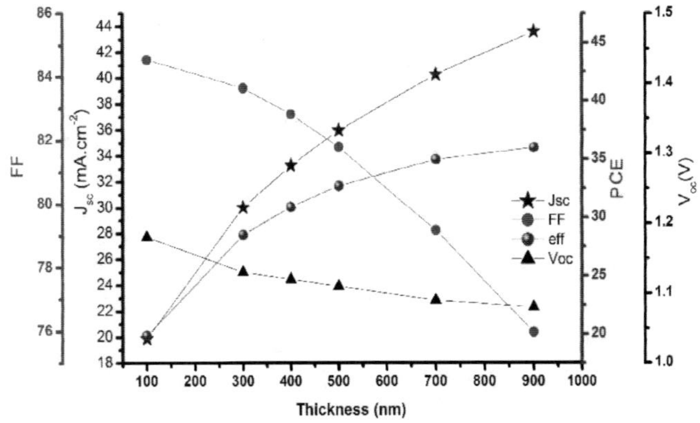

Figure 2. The variation of solar cell parameters with absorber

'Figure 3.' illustrates the thickness influence of the electron transport layer on J_{sc}, V_{oc}, FF, and PCE. This figure reveals a dramatic decrease in the J_{sc} and PCE with increasing the electron transport layer thickness while no significant influence of the thickness increasing on the V_{oc} and FF. As shown in figure 3 the J_{sc} decreases from 33.2mA·cm^{-2} to 10.29mA·cm^{-2}and PCE decreases from 31 % to 9.1 % over the ETL thickness from 100 nm to 1000 nm. The impact of the electron transport layer on the solar cell parameters can be attributed to the fact that the TiO_2 has a critical drawback of intrinsic low mobility and increasing the thickness of this layer will results in a charge accumulation and hence increases the recombination. The best performance was found at a thickness of 300 nm [4].

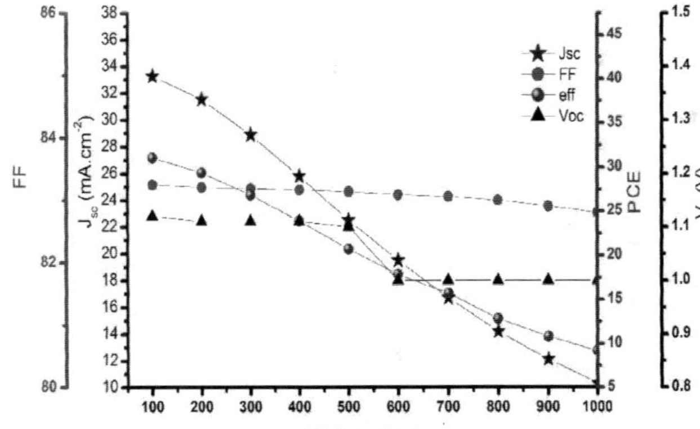

Figure 3. The variation of solar cell parameters with ETL thickness.

Keeping the thickness of the absorber and ETL at the optimum thicknesses which gave the best cell performance, the effect of the HTL thickness on the cell performance was also studied by varying the thickness of this layer over the range from 100 nm to 1000 nm with step of 200 nm. 'Figure 4' shows that the thickness of HTL does not have a critical impact on the solar cell performance and the values of V_{oc}, J_{sc}, FF, and PCE are around 1.12 V, 33.27 mA·cm^{-2}, 83.25%, and 31% respectively. This result agrees with what was obtained by Hima et. al., and it can be attributed to the constant number of charge carriers produced in the absorber layer [4]. We used a 200 nm as a reference thickness for this layer.

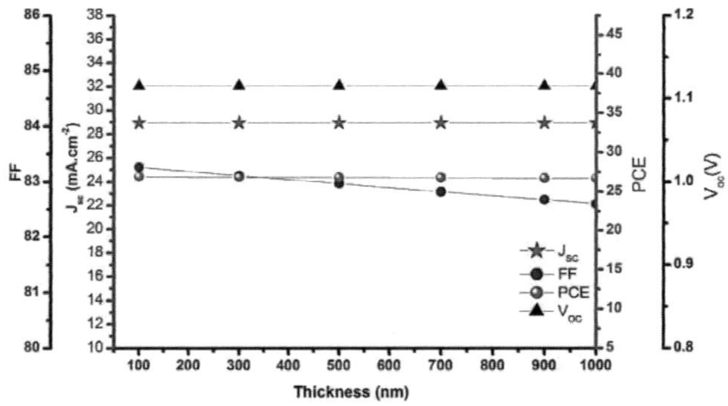

Figure 4. The variation of solar cell parameters with HTL thickness.

3.2 Effect of layers doping

Layer doping plays a critical issue in decides the electrical behavior and hence the performance of the solar cell. To improve the cell performance proper doping of ETL, HTL and absorber are necessary. The impact of the doping concentration of the absorber layer was investigated by varying the doping level from 10^{12}cm^{-3} to 10^{22} cm^{-3}. 'Figure 5' shows the change in cell parameters with doping density, it can be seen that the cell parameters are almost unchanged with increasing the absorber density until the doping concentration of 10^{17} cm^{-3}. Above the doping level of 10^{17} cm^{-3}, a dramatic decrease in the J_{sc}, FF, and PCE was observed while an increase in the V_{oc} from 1.18 V to 2.6 V can be noticed. The reason which stands behind this behavior is that high doping leads to a reduction in the width of the depletion region and hence increases the recombination process at the bulk of the absorber [13]. The current results show that doping density for the absorber layer (>10^{16}cm^{-3}) is not suitable for high efficiency.

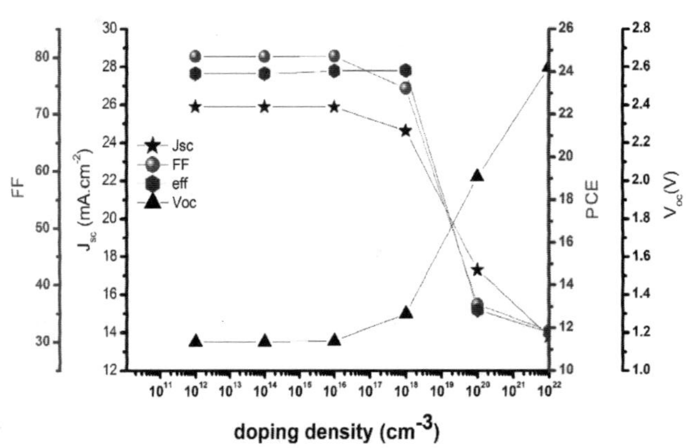

Figure 5. The variation of solar cell parameters with absorber doping.

The HTL doping impact on solar cell performance was studied by varying the acceptor concentration from 10^{17}cm^{-3} to 10^{22} cm^{-3}. As shown 'in Figure 6' the difference overdoping concentration are FF and PCE while both J_{sc} and V_{oc} are almost constant and independent of acceptor concentration which indicates that the recombination rate is unrestrained of doping density. The recombination rate affects minority carrier's density and at high doping of acceptor in HTL the density of minority deceases which in turn blocks the electrons from absorber [14]. A dramatic increase in the FF and PCE at doping density above 10^{18} cm^{-3}could be attributed to the sheet resistance and conductivity enhancement in HTL at a high doping rate [15]. Doping over $10^{18}cm^{-3}$ gives a best cell performance with efficiency of 25.3 %.

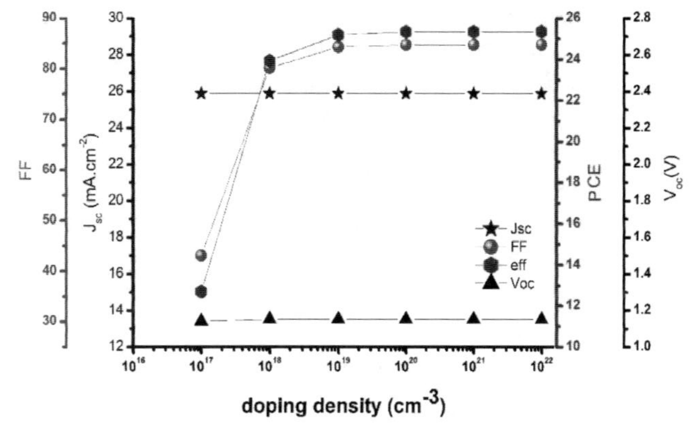

Figure 6. The variation of solar cell parameters with HTL doping.

By keeping the doping density of perovskite and HTL at 10^{14} cm^{-3} and 10^{19} cm^{-3} respectively the doping concentration of ELT was varied from 10^{17} cm^{-3} to 10^{22} cm^{-3} to investigate the impact of the ETL doping on the solar cell performance. 'Figure 7' shows that the cell parameters remain unaffected by donor doping variation and this can be attributed to the Moss-Burstein effect [16]. The results of the doping rate in the ETL suggest that ETL doping density at 10^{19} cm^{-3} is the better value for best cell performance.

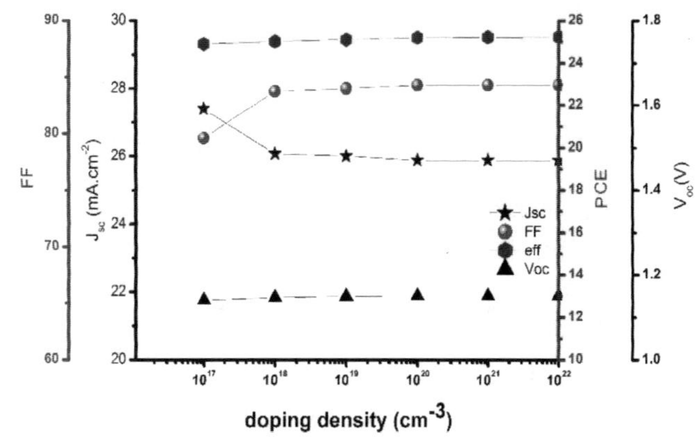

Figure 7. The variation of solar cell parameters with ETL doping.

From the aforementioned parametric study of the thickness and doping of each layer, one can conclude the optimum structure parameters are 400 nm, 200 nm, 300 nm as a thickness of absorber layer, HTL, and ETL respectively, while the doping of 10^{14} cm^{-3}, 10^{19} cm^{-3}, 10^{19} cm^{-3} for the same layers. These parameters give a V_{oc}, J_{sc}, FF, and PCE of 1.11 V,28.9mA·cm^{-2},83.19%, and 26.88% respectively.

3.3 Interface layer:

To boost the PCE, the charge losses at the interface between the perovskite/ETL and /or perovskite / HTL should be reduced. One of the interface engineering strategies is to use a thin layer between the active layer and HTLs (ETLs) to improve the energy alignment and hence the charge transport across the interfaces. In the current study, the interface layer of 3C-SiC was used in the optimum structure between ETL and absorber. To study the impact of this layer on the cell parameters the thickness and doping density were taken into consideration. The thickness was varied from 30 nm to 100 nm, with doping density of 10^{16} cm^{-3}, and the results are presented 'in Figure 8'. It can be seen that the J_{sc} increases slightly from 29.6mA·cm^{-2} to 31.5mA·cm^{-2}and this can be ascribed to the reduction in the charge recombination. The interface layer makes a drop of energy barrier at the perovskite/interfacial layers which in turn facilitates the charge transfer. This explanation enhances the stability in V_{oc} over the examed interface thickness. An enhancement in the PCE from 27.6 % to 29% was also noticed.

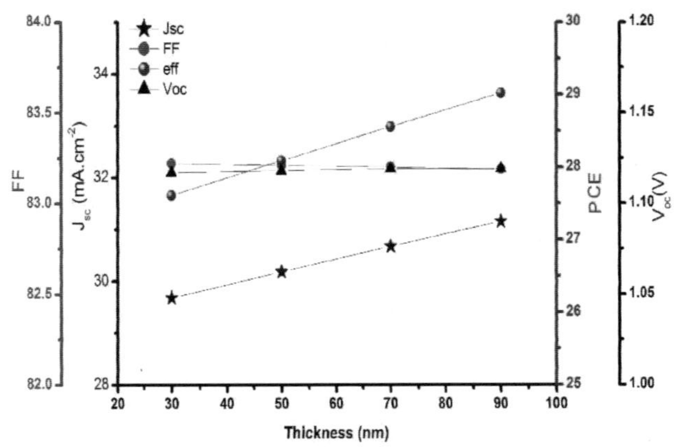

Figure 8. The variation of solar cell parameters with interface thickness.

To study the impact the doping of interface layer on the cell performance the donor doping density of this layer was varied from 10^{2} cm^{-3} to 10^{18} cm^{-3}. The results show that the doping density does not affect significantly on the cell parameter until doping density of $>10^{15}$ cm^{-3} and beyond this value of doping the J_{sc} decreases slightly while an increase in the PCE from 28.8% to 29.25%due to narrowing the depletion region beneficial for charge carrier collection.

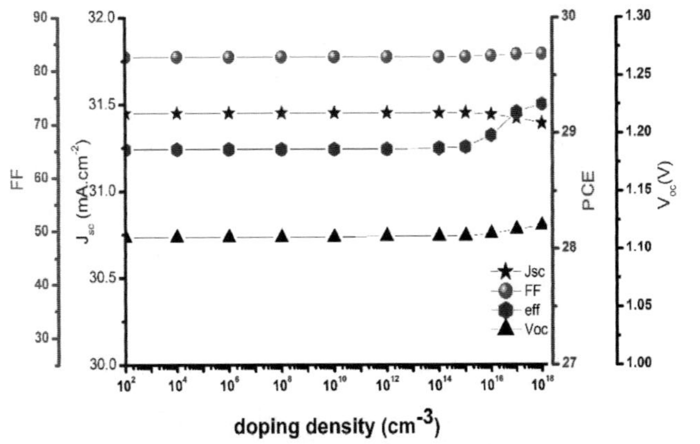

Figure 9. The variation of solar cell parameters with interface doping.

4. Conclusions

In the present work, the impact of layer thicknesses and doping of $CH_3NH_3PbI_3$-based solar cell on different electrical parameters was investigated. It was found that the perovskite layer thickness has a significant influence on electrical parameters in comparison with that of HTL and ETL. It was found that the optimized layer thickness values were 400 nm, 200 nm, and 300 nm for the perovskite layer, HTL, and ETL, respectively which provided the better power conversion efficiency of26.88%.The parametric study of the doping variation on the proposed structure showed that the doping density for absorber layer (> 10^{16}cm^{-3}) was not suitable to obtain a high efficiency while the doping of acceptor in HTL (> 10^{16}cm^{-3}) enhanced the cell performance whilst the cell parameters remained unaffected by ETL donor doping variation. With regards the interface layer, the present work demonstrated that the introduction of n-3C-SiC into the solar cell design evidently had a good effect on enhancing the overall cell performance through achieving an enhancement in efficiency by improving charge transport and reducing recombination; which was in correspondence to the proposed hypothesis. Within the tested thickness range, it was shown that the lower the thickness, the better the device efficiency obtained. In contrast, it was found that the highest doping density within the used range produced the best observed improvement in device efficiency. However, it is worth to consider that these findings need to be further investigated using other interface materials rather than n-3C-SiC to confirm the current findings. This in turn can uncover additional possibilities for future versions of solar cell devices, opening the horizon for an entirely new fields of knowledge and research.

5. References

[1] Lee M, Teuscher J, Miyasaka T, Murakami T and Snaith H. 2012. *Science*. Numerical simulations on perovskite photovoltaic devices. **2(338)**. pp. 643-7.

[2] Zhang F, Castaneda J, Chen S, Wu W, DiNezza M, Lassise, M, Nie W, Mohite A, Liu Y, Liu S and Friedman D. 2020. *Materials Today*. Comparative studies of optoelectrical properties of prominent PV materials: Halide perovskite, CdTe, and GaAs, **36**, pp.18-29.

[3] Di Giacomo F, Fakharuddin A, Jose R and Brown T. 2016. *Energy & Environmental Science*. Progress, challenges and perspectives in flexible perovskite solar cells. **9(10)**, pp.3007-3035.

[4] Hima A, Lakhdar N, Benhaoua B, Saadoune A, Kemerchou I and Rogti, F. 2019. *Superlattices and Microstructures*. An optimized perovskite solar cell designs for high conversion efficiency. **129**, pp.240-246.

[5] Zekry A, Yahyaoui I and Tadeo F. 2019. *10th International Renewable Energy, Congress (IREC)* Generic Analytical Models for Organic and Perovskite Solar Cells.. (pp. 1-6). IEEE.

[6] Niemegeers A, Burgelman M, Decock K, Verschraegen J and Degrave S. 2013. *SCAPS manual*. University of Gent, 13.

[7] Lakhdar N and Hima A, 2020. *Optical Materials.* Electron transport material effect on performance of perovskite solar cells based on CH3NH3GeI3. **99**, p.109517.

[8] Hima A, Le Khouimes, A, Rezzoug, A, Yahkem, M, Khechekhouche A and Kemerchou, I. 2019. *International journal of energetics.* Simulation and optimization of CH3NH3PbI3 based inverted planar heterojunction solar cell using SCAPS software **4(1)**, pp.56-59.

[9] Rahman M, Miah S, Marma M and Sabrina, T. 2019. In 2019 *International Conference on Electrical, Computer and Communication Engineering (ECCE).* Simulation based Investigation of Inverted Planar Perovskite Solar Cell with All Metal Oxide Inorganic Transport Layers. (pp. 1-6).

[10] Momblona C, Malinkiewicz O, Roldán-Carmona C, Soriano A, Gil-Escrig L, Bandiello E, Scheepers M, Edri E and Bolink H. 2014. *Applied Materials.* Efficient methylammonium lead iodide perovskite solar cells with active layers from 300 to 900 nm. **2(8)**, pp.081504.

[11] Jeyakumar R., Bag A, Nekovei R and Radhakrishnan R. 2020. *Journal of Electronic Materials.* Influence of Electron Transport Layer (TiO 2) Thickness and Its Doping Density on the Performance of CH 3 NH 3 PbI 3-Based Planar Perovskite Solar Cells. **(14)** pp.3533–3539.

[12] Mandadapu U, Vedanayakam S and Thyagarajan K. 2017. *Indian Journal of Science and Technology*. Simulation and analysis of lead based perovskite solar cell using SCAPS-1D. **10(11)**, pp.65-72.

[13] An Y, Shang A, Cao G, Wu S, Ma D and Li X. 2018. *Solar RRL.* Perovskite solar cells: optoelectronic simulation and optimization Perovskite. **2(11).** p.1800126.

[14] Jeyakumar R, Bag A, Nekovei R and Radhakrishnan R. 2019. *Solar Energy.* Interface studies by simulation on methylammonium lead iodide based planar perovskite solar cells for high efficiency. **190**, pp.104-111.

[15] Nguyen W, Bailie, C, Unger E and McGehee M.D. 2014. *Journal of the American Chemical Society.* Enhancing the hole-conductivity of spiro-OMeTAD without oxygen or lithium salts by using spiro (TFSI) 2 in perovskite and dye-sensitized solar cells. **136 (31)**, pp.10996-11001.

[16] Trukhanov, V, Bruevich V. and Paraschuk D. 2011. *Physical Review B.* . Effect of doping on performance of organic solar cells. **84(20)**, pp.205318.

Design and simulation of cascaded H-bridge 5-level inverter for grid connection system based on multi-carrier PWM technique

Hadeel S Maaroof[1,2*], Harith Al-Badrani[1] and Ahmad T Younis[1]

[1]Electronic Department, College of Electronics Engineering, Ninevah University, Mosul, Iraq 41002.
[2]State company of the North Distribution Electricity, Ministry of Electricity, Baghdad 10013, Iraq.

*E-mail: hadeel.taher20192019@stu.uoinevah.edu.iq

Abstract. Cascaded H-Bridge (CHB) multi-level inverter has become attractive in medium voltage and grid connection to improve power quality with high efficiency, and low switching losses. Voltage Oriented Control (VOC) regulates the injected power and the connection between the cascaded H-bridge inverter and the utility grid. In the modulation stage for the VOC, there are several techniques such as Space Vector Pulse Width Modulation (SVPWM), Multi-Carrier Pulse Width Modulation (MCPWM), Selective Harmonic Elimination (SHE) used to obtain gating pulses for the IGBTs switches. In this paper, a three-phase 5-level CHB inverter with a grid connection system is present and the technique of MCPWM is used. A comparative study between each method of (MCPWM) using MATLAB/ Simulink environment has been done on the Total Harmonic Distortion (THD) for inverter phase voltage and current with different injected reference current values. It is found that phase current THD is less with the Phase Shift PWM (PS-PWM) technique.

1. Introduction

In the last decades, the need to generate electrical power by using renewable resources such as photo-voltaic and wind turbines has increased since it is considered environmentally friendly and endless sources for electric power in the future[1,2].

Power electronic converters are used to connect these sources to the utility grid, known as inverters that convert DC power to AC power. However, the application of the utility grid is classified as a medium and high power, making the conventional two-level inverter unsuitable for this purpose. Moreover, due to the lack number of output voltage levels, more harmonic waves may be resulted [3,4]. Therefore, a Cascaded H-Bridge Multilevel Inverter (CHBMLI) is the best choice for grid connection and PV systems. Several advantages may be obtained, including high power quality, high voltage capability along with higher efficiency, low switching losses, and reduced magnetic interference [5-7].

The cascaded H-bridge multi-level inverter is easily designed and realized by fewer components compared with other multi-level inverters (diode clamped topology, flying capacitor topology, as figure 1

Content from this work may be used under the terms of the Creative Commons Attribution 3.0 licence. Any further distribution of this work must maintain attribution to the author(s) and the title of the work, journal citation and DOI.
Published under licence by IOP Publishing Ltd

(a) shows 3-level diode clamped, while in (b) three-level flying capacitor. In contrast, (c) shows a single-phase 5-level CHB inverter[5].

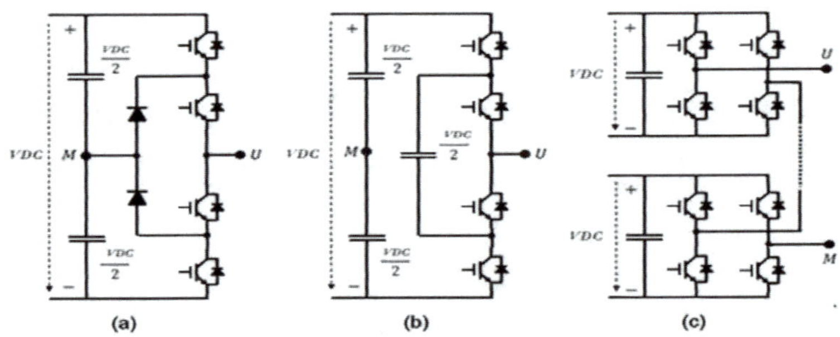

Figure 1. One leg of multi-level inverter topology

This paper presents in section (2) a proposed topology for the multi-level inverter system using cascaded H-bridge. The multi-carrier modulation techniques that control the three-phase 5-level CHB inverter is used in grid connection with control strategy are covered in section (3); section (4) provides simulation results for each (MC-PWM) modulation method. Finally, section (5) summarizes the conclusion of the paper.

2. Proposed topology

The proposed three-phase grid connection system shown in figure 2 consists of six DC isolated sources, a 5-level cascaded H-bridge inverter, RL-filter, and a utility grid.

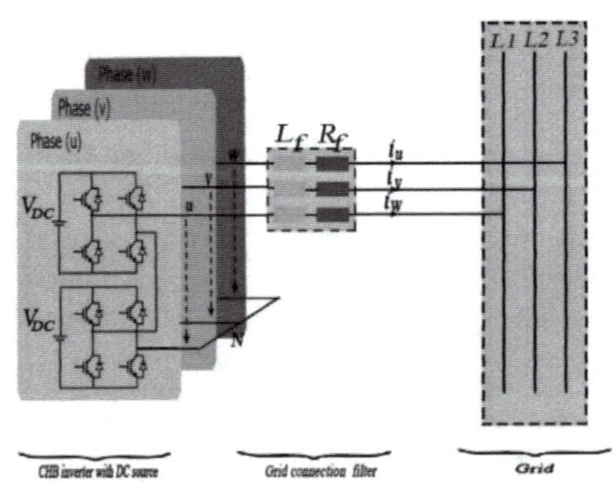

Figure 2. Grid connection system

Three-phase CHBMLI means three single CHBMLI are connected in (Y) form. Each phase consists of two full-bridges (each has a separate DC source) called a cell. In contrast, the bridge consists of four IGBTs switches with an anti-parallel diode; as previously shown in figure 1(c). IGBT is chosen as a power semiconductor switch since it has a more favorable feature, unlike other power semiconductor switches. The number of cells (K) that are used in a single-phase determines the specified output voltage level (L) [8,9].

$$L=2K+1 \tag{1}$$

The total output voltage level of each phase is the sum of the voltages that are generated by the cells, where each cell is capable of producing an output voltage level of -1V, 0, +1V, so the output voltage for a 5-level inverter is -2V, -1V, 0, +1V, +2V DC voltages [10].

3. Modulation and control strategy

3.1 Proposed modulation techniques

To obtain an output voltage with low distortion as smooth as sinusoidal waveform, it is important to design a modulation circuit to generate pulses (gating signals) that control the ON/OFF time (switching frequency) for the semiconductor switch [9]. The modulation techniques used in the multi-level inverter are:

- Space Vector Pulse Width Modulation (SVPWM).
- Multi-Carrier Pulse Width Modulation (MCPWM).
- Selective Harmonic Elimination (SHE)[11].

Multi-carrier modulation PWM comprises two common methods, which are widely used and can be classified as:

3.1.1. Phase Shift Carrier Modulation (PS-PWM)

Phase Shift Carrier Modulation (PS-PWM) is the most common method used in multi-level inverter due to its simple implementation and it has some advantages such as even power distribution and the output current has high quality waveform [12]. Moreover, based on comparing the sinusoidal reference signal with the triangle carrier signal, the amplitude modulation index can be defined as:

$$m_a = \frac{Am}{Ac} \tag{2}$$

Where: *Am* is the peak-to-peak amplitude of the sinusoidal reference signal
And *Ac* is the amplitude of triangle carrier signal [13]

For cascaded H-bridge multi-level inverter, the carrier signals of different cells (which are equal in amplitude and frequency) are phase shifted by $\frac{180°}{n}$, where *n* is the number of cells in each phase [14] and will be 90° for 5-level CHB inverter, as shown in figure 3.

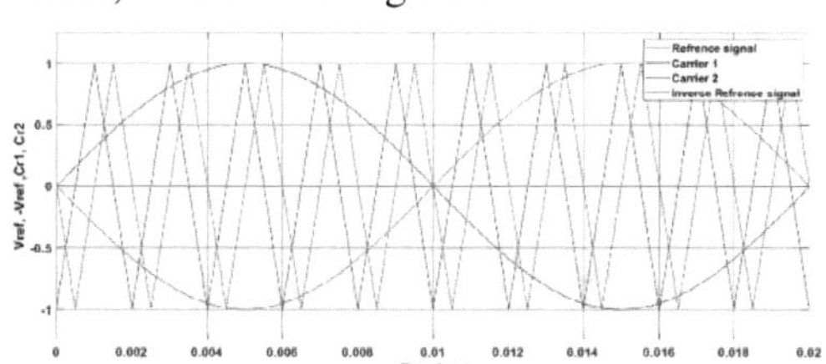

Figure 3. PS-PWM

3.1.2. Level Shift Pulse Width Modulation (LS-PWM)

In the level Shift Pulse Width Modulation (LS-PWM), all carrier signals are equal in amplitude, frequency, and phase but dispositioned vertically relative to the ZERO reference line. For 5-level CHB inverter, it needs (m-1) carrier signals for (m) output voltage levels resulting in four carrier signals, two of them are above the ZERO reference, and the other two signals are below the ZERO reference, the amplitude modulation index is defined as[3,15]:

$$m_a = \frac{Am}{(m-1)Ac} \tag{3}$$

Level shift pulse width modulation can be classified into three main techniques:

- Phase Disposition (PD): in this category of (LP-PWM), all carrier signals are in phase with each other, as shown in figure 4 (a).
- Phase Opposition Disposition (POD): in this technique, the carrier signals above the ZERO reference line are shifted by 180° out of phase with those signals under ZERO-line. Figure 4 (b) shows the carrier signals positioning to the ZERO-line and the sinusoidal reference signal.
- Alternative Phase Opposition Disposition (APOD): in this technique, each carrier signal is dispositioned 180° out of phase with its neighbor carrier, as shown in figure 4 (c)[10].

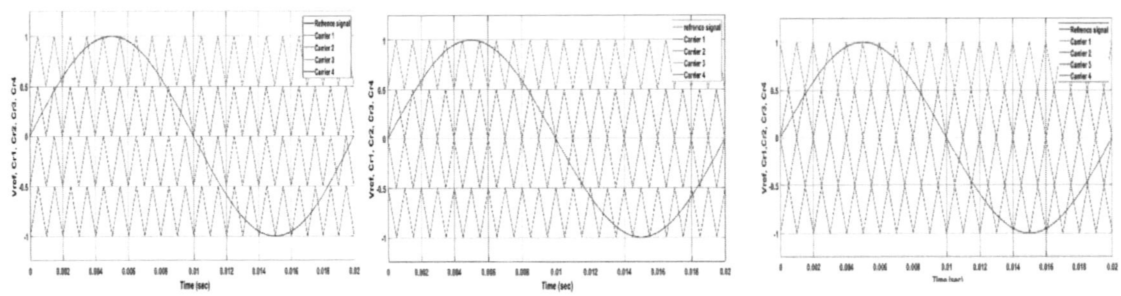

(a) (b) (c)

Figure 4. LS-PWM technique PD, POD, APOD

3.2. Voltage Oriented Control (VOC)

Voltage Oriented Control is the most common control scheme in the grid connection application that consists of the modular stage that regulates the interaction between the CHB and the utility grid [16]. Figure 5 shows the block diagram for the grid connection system with VOC.

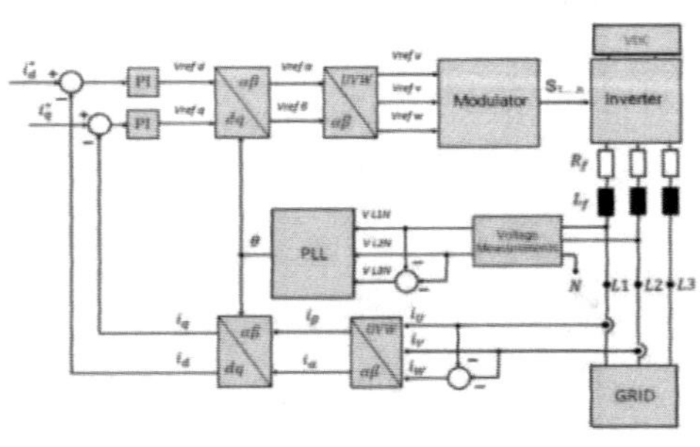

Figure 5. Voltage oriented control diagram

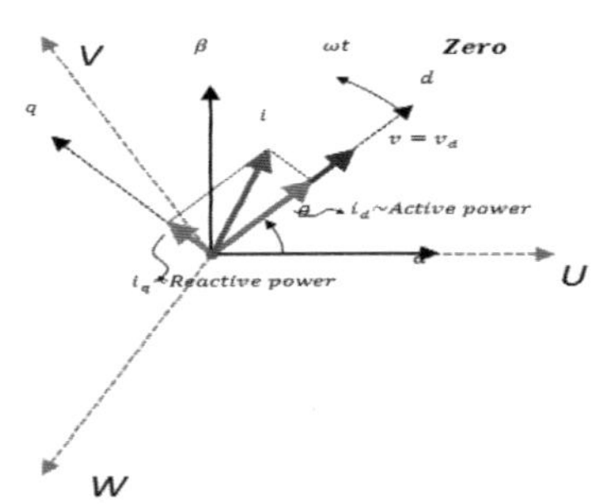

Figure 6. Space phasor diagram of VOC

In VOC, the quantities of the three-phase system are converted to the stationary alpha-beta reference frame, this is similar to field oriented control for an electrical machine [17]:

$$\begin{bmatrix} f_\alpha \\ f_\beta \end{bmatrix} = \frac{2}{3} \begin{bmatrix} 1 & -\frac{1}{2} & -\frac{1}{2} \\ 0 & \frac{\sqrt{3}}{2} & -\frac{\sqrt{3}}{2} \end{bmatrix} \cdot \begin{bmatrix} f_u \\ f_v \\ f_w \end{bmatrix} \tag{4}$$

These new quantities will convert to DC components in the rotating d-q frame because it is easier to deal with DC component as follows:

$$\begin{bmatrix} f_d \\ f_q \end{bmatrix} = \begin{bmatrix} \cos\theta & \sin\theta \\ -\sin\theta & \cos\theta \end{bmatrix} \cdot \begin{bmatrix} f_\alpha \\ f_\beta \end{bmatrix} \tag{5}$$

Where f can be current or voltage.

In stationary α, β frame, the active power P and the reactive power Q can be calculated according to the following equations:

$$P = \frac{3}{2}\left(v_\alpha i_\alpha + v_\beta i_\beta\right) \tag{6}$$

$$Q = \frac{3}{2}(v_\alpha i_\beta - v_\beta i_\alpha) \tag{7}$$

Where v_α and v_β are voltages in the stationary α-β frame
i_α and i_β are currents in the stationary α-β frame

Using space phasor $wt= d\theta/dt$, whereas wt is the angular frequency of the grid voltage, and the active power P is directly related to the i_d. In contrast, the reactive power Q is related to the i_q according to equations (8) and (9), respectively.

$$P = \frac{3}{2}(v_d i_d + v_q i_q) = \frac{3v_d i_d}{2} \tag{8}$$

$$Q = \frac{3}{2}(v_d i_q - v_q i_d) = \frac{3v_d i_q}{2} \tag{9}$$

Where v_d and v_q are voltages in the rotating d-q frame
i_d and i_q are currents in the rotating d-q frame

However, it is noticed that when the rotating reference frame is synchronized with the grid voltage, v_q becomes zero to inject the current with the unity power factor [18]. Figure 6 shows the stationary and the rotating frames of VOC.

The i_d and i_q will be sent to the PI controller to obtain reference voltages ($V_{ref\,d}$, $V_{ref\,q}$), which is converted to α-β coordinate ($V_{ref\,\alpha}$, $V_{ref\,\beta}$) by using equation (10) and then to UVW system ($V_{ref\,U}$, V_{refV}, $V_{ref\,W}$) using equation (11). The resulting voltages will use as reference signals for the modulation stage to get the gating signals for the IGBTs[17].

$$\begin{bmatrix} V_{ref\,\alpha} \\ V_{ref\,\beta} \end{bmatrix} = \begin{bmatrix} \cos\theta & -\sin\theta \\ \sin\theta & \cos\theta \end{bmatrix} \cdot \begin{bmatrix} V_{refd} \\ V_{refq} \end{bmatrix} \tag{10}$$

$$\begin{bmatrix} V_{refu} \\ V_{refv} \\ V_{refw} \end{bmatrix} = \begin{bmatrix} 1 & 0 \\ -\frac{1}{2} & \frac{\sqrt{3}}{2} \\ -\frac{1}{2} & -\frac{\sqrt{3}}{2} \end{bmatrix} \cdot \begin{bmatrix} V_{ref\,\alpha} \\ V_{ref\,\beta} \end{bmatrix} \tag{11}$$

The phase angle (θ) can be obtained by using Phase Lock Loop (PLL) as shown in figure 7, in which the measured voltages from the grid (V_u, V_v, and V_W) are transformed to rotating frame V_d and V_q as in equation

(4) and (5) respectively (V_q) controlled by PI controller which employs as loop filter. Its output will be the grid frequency (ω), which is integrated to obtain the output phase angle (θ) [19].

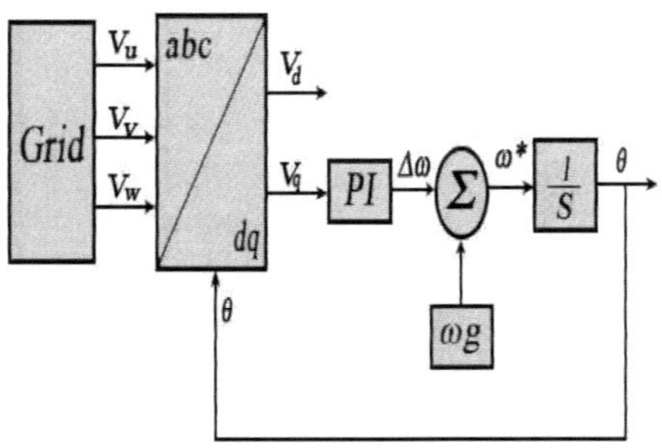

Figure 7. Block diagram for PLL

Table 1. Electrical parameters values of CHB inverter grid connection system

Parameters	Value
CHB DC source	200 v
Switching frequency	5000 Hz
Grid phase to phase voltage	380 v
Grid fundamental frequency	50 Hz
Filter inductance	0.5 mH
Filter resistance	5 mΩ
Reference current	100 A

4. Simulation results

A 5-level CHB inverter grid connection system is simulated using MATLAB/Simulink environment. The simulated results obtained for phase voltage for the inverter and the grid voltage are synchronized in frequency and phase angle, as shown in figure 8(a). Also, the inverter current is in phase with the grid phase voltage, as shown in figure 8(b), which means the power is injected with the unity power factor. Therefore, at the same simulation parameters, as listed in table 1, carrier shift modulation techniques are studied by using Fast Fourier Transform (FFT) analysis.

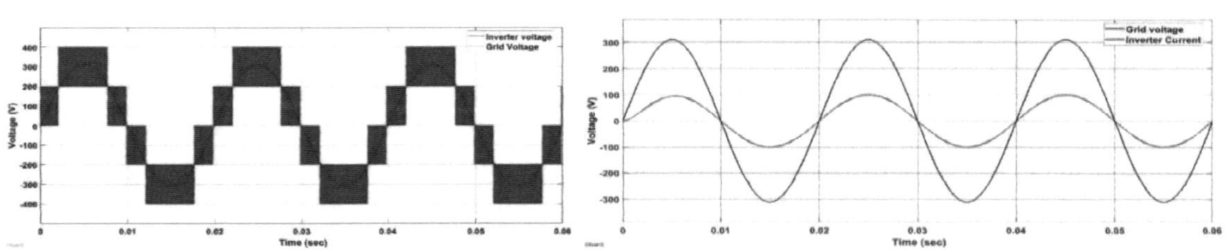

(a) (b)

Figure 8. Grid voltage with inverter voltage and with inverter current in phase (u)

Figure 9(a) illustrates the modulation circuit in the Phase Shift (PS-PWM) technique that provides a train of pulses control the ON/OFF time for each IGBT to get the inverter output voltage by comparing between reference sinusoidal voltage obtained from current control loop for the VOC and tringle carrier generator in the modulation stage. It is observed from figure 9(b) that the modulation index is 0.8 according to the amplitudes of the reference voltage (0.8) and tringle signal that equal to 1 by using equation (2).

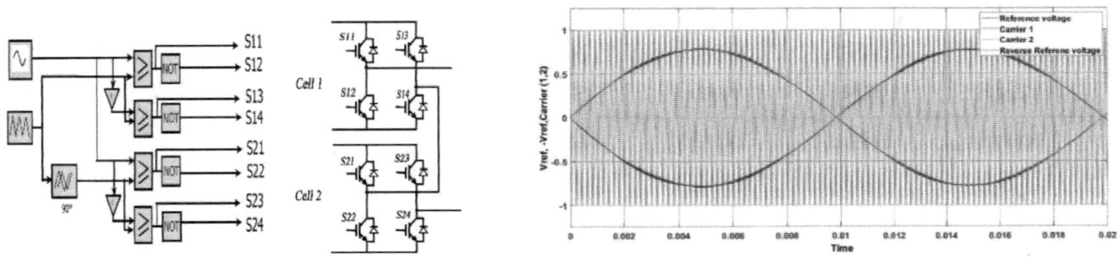

(a) (b)

Figure 9. The circuit block diagram and the waveforms for PS-PWM in phase (u)

The Total Harmonic Distortion (THD) for phase (u) of the inverter current and voltage is presented in figure 10 by using the PS-PWM technique. It can be seen that the THD for the inverter current is 1.21%, while the THD of the inverter phase (u) voltage is 39.36% as shown in figure 10(a) and figure 10(b) respectively.

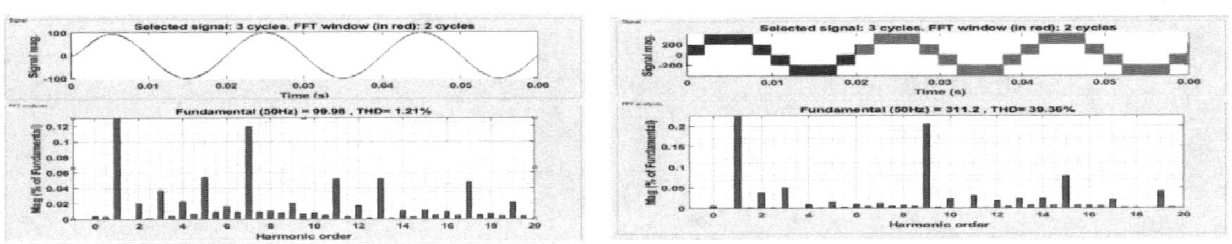

(a) (b)

Figure 10. THD in PS-PWM

The Phase Disposition PWM system illustrated in figure 11(a) is realized by comparing four individual carriers (which all in phase with each another) with the reference voltage that will be changed for the three phases legs depending on its phase angle. However, it can be seen that the modulation index is 0.8 by using equation (3), as shown in figure 11(b).

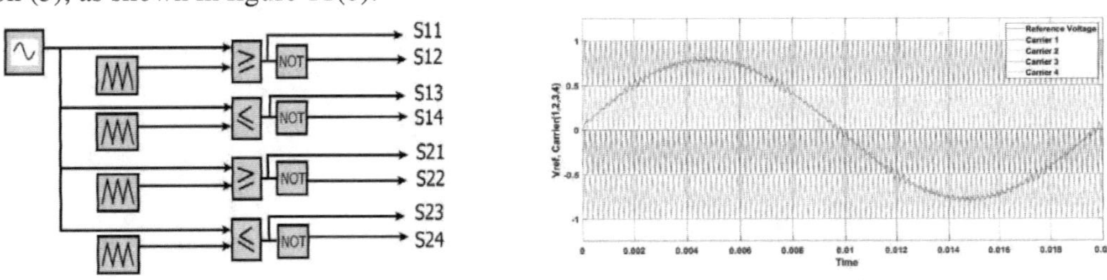

(a) (b)

Figure 11. The circuit block diagram and the waveforms for PD-PWM in phase (u)

Figure 12 illustrates the total harmonic distortion for phase (u) of the inverter current and voltage. The THD for the inverter current is shown to be 2.71%, as illustrated in figure 12(a) and THD for the inverter phase (u) voltage is 39.60% as in figure 12(b).

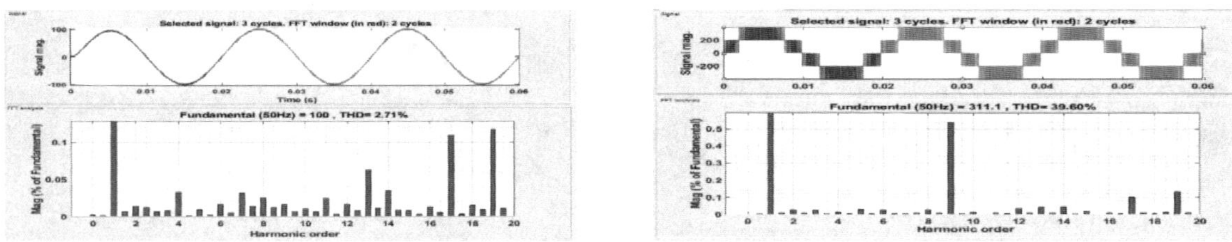

(a) (b)

Figure 12. THD in PD-PWM

Figure 13(a) shows the Phase Opposition Disposition (POD) modulation structure. It illustrates that the lower carriers are 180° degrees out of phase with the upper carrier signal; figure 13(b) shows the modulation index is 0.8 by using equation (3).

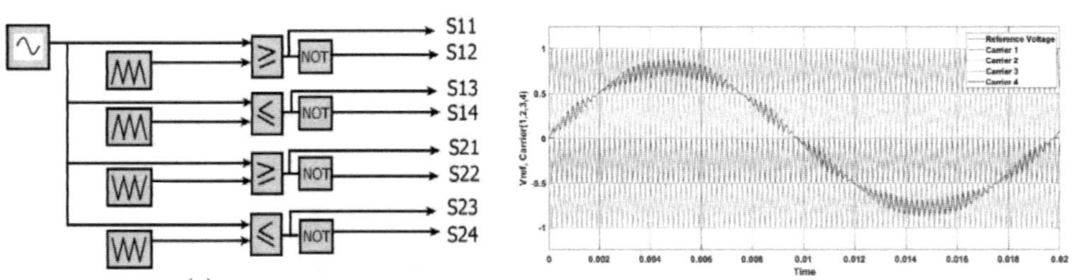

(a)

(b)

Figure 13. The circuit block diagram and the waveforms for POD-PWM in phase (u)

The total harmonic distortion for phase (u) of the inverter current and voltage is illustrated in figure 14, where the THD for the inverter current is 6.44% as in figure 14(a), and THD for the inverter phase (u) voltage is 39.30% as in figure 14(b).

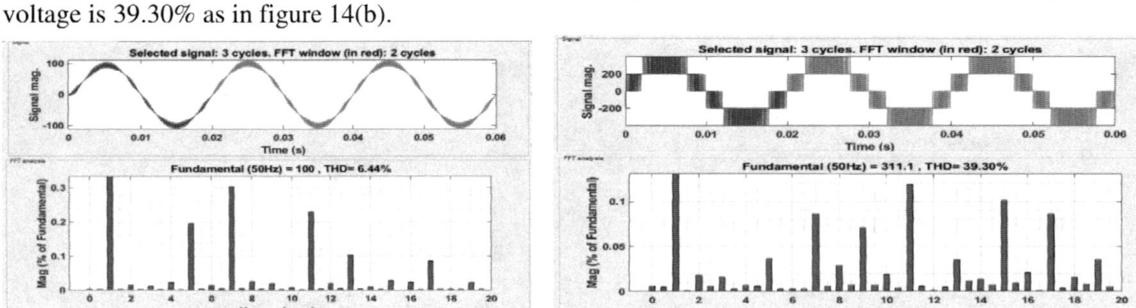

(a) (b)

Figure 14. THD for POD-PWM

An Alternative Phase Opposition Disposition (APOD) modulation topology (each carrier signal is dispositioned 180° out of phase with its neighbor carrier) is presented in figure 15(a). It is clear that the modulation index is 0.8 by using equation (3) as shown in figure 15(b).

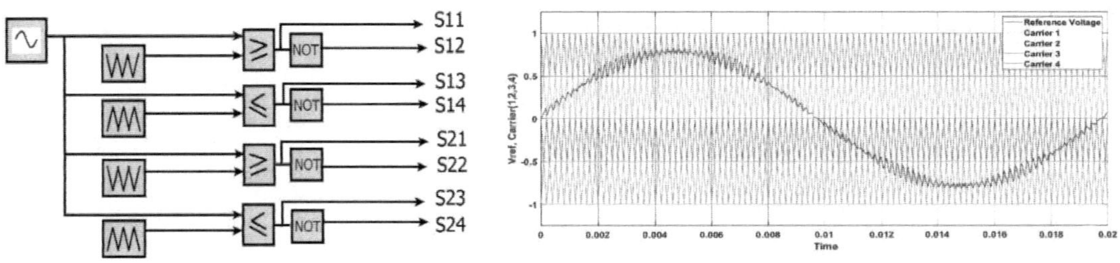

(a) (b)

Figure 15. The circuit block diagram and the waveforms for APOD-PWM in phase (u)

The total harmonic distortion in this technique for phase (u) of the inverter current is 4.80% and THD for the inverter voltage is 39.46% as shown in table 2.

Table 2 listed different values of injected reference current and their THD measured in Fast Fourier Transform (FFT) analysis for inverter line current and voltage for different carrier shift modulation methods.

Table 2. THD of carrier shift methods

Injected current (A)	PS-PWM		PD-PWM		POD-PWM		APOD-PWM	
	Inverter current THD	Inverter voltage THD	Inverter current THD	Inverter voltage THD	Inverter current THD	Inverter voltage THD	Inverter current THD	Inverter voltage THD
100	1.21%	39.60%	2.71%	39.60%	6.44%	39.30%	4.80%	39.46%
125	0.97%	39.30%	2.17%	39.56%	5.15%	39.25%	3.86%	39.41%
150	0.81%	39.29%	1.81%	39.53%	4.29%	39.25%	3.23%	39.37%
175	0.70%	39.26%	1.55%	39.47%	3.68%	39.22%	2.78%	39.34%
200	0.61%	39.18%	1.35%	39.44%	3.22%	39.16%	2.44%	39.28%

5. Conclusion

This paper involves a study and analysis for a three-phase 5-level cascaded H-bridge inverter for grid connection with unity power factor. The system controlled by traditional voltage-oriented control. Carrier shift techniques are used in modulation stag and simulated by MATLAB/Simulink environment. The application of Fast Fourier Transform (FFT) analysis for THD in these techniques shows that the total harmonic distortion for the inverter current by using Phase Shift Pulse Width Modulation (PS-PWM) is small compared with other existing methods. It means that high-quality output current and even high-power distribution are obtained; therefore, this technique is suitable for grid connection.

6. References

[1] Hasan N S, Rosmin N and Musta H 2017 Reviews on multilevel converter and modulation techniques *Renew. Sustain. Energy Rev.* **80** 163–74

[2] Tarek M S I, Siam A, Zia M and Rahman M M 2018 A novel five-level inverter topology with reactive power control for grid-connected PV system *2018 International Conference on Smart Grid and Clean Energy Technologies, ICSGCE 2018* (IEEE) pp 101–5

[3] Sinha A, Chandra Jana K and Kumar Das M 2018 An inclusive review on different multi-level inverter topologies, their modulation and control strategies for a grid connected photo-voltaic system *Sol. Energy* **170** 633–57

[4] Al-Badrani H, Feuersanger S and Pacas M 2018 VSI with Sinusoidal Voltages for an Enhanced

Sensorless Control of the Induction Machine *PCIM Europe 2018, Nuremberg, Germany* pp 1199–205

[5] Shanono I H, Rul N, Abdullah H and Muhammad A 2018 A Survey of Multilevel Voltage Source Inverter Topologies , Controls and Applications *Int. J. Power Electron. Drive Syst.* **9** 1186–201

[6] Coppola M, Guerriero P, Iannuzzi D, Daliento S and Pizzo A Del 2020 Extended operating range of PV module-level CHB inverter *Electr. Power Energy Syst.* **119** 105892

[7] Wang F, Yang L E, Mao W, Shineng Y and Zhang X 2016 Power Balance Control Scheme of Cascaded H- Bridge Multilevel Inverter for Grid-Connection Photovoltaic Systems *2016 IEEE 8th International Power Electronics and Motion Control Conference (IPEMC-ECCE Asia)*

[8] Shuzheng W, Jianfeng Z and Chao S H I 2011 Research on a three-phase cascaded inverter for grid-connected photo- voltaic systems *The International Conference on Advanced Power System Automation and Protection Research* pp 543–8

[9] Bashi S M, Mailah N and Kadir Z 2008 Generation of Triggering Signals for Multilevel Converter *Eur. J. Sci. Res.* **24** 548–55

[10] Ranjan A K, Bhaskar D V and Parida N 2015 Analysis and simulation of cascaded H-bridge multi level inverter using level-shift PWM technique *IEEE International Conference on Circuit, Power and Computing Technologies, ICCPCT 2015*

[11] Nagar S, Khan S and Singh B 2018 Performance of Cascaded Diode Bridge Integrated H-Bridge 13 Level Multilevel Inverter *2017 Recent Dev. Control. Autom. Power Eng. RDCAPE 2017* **3** 399–403

[12] Kim S, Lee K, Kim S and Lee K 2019 A Modified Third Harmonic Pulse-Width Modulation for Reduced Switching Loss in Cascaded H-Bridge Multilevel Inverters *IFAC-PapersOnLine* **52** 472–6

[13] Fduulhu Y R I, Aghay S H, Selvaraj J, Rahim N A, Orz Y H U, Dqg G, Wkh D W and Wlph V 2016 Performance Evaluation of Multi-carrier PWM Methods for Solar Cell Powered Multi-level Inverter *4th IET Clean Energy and Technology Conference (CEAT 2016)*

[14] Sarkar I and Fernandes B G 2014 Modified Hybrid Multi-Carrier PWM Technique for Cascaded H-Bridge Multilevel Inverter *Annu. Conf. Ind. Electron. Soc.* 4318–24

[15] Kale A S, Tamhane A V. and Kalage A A 2018 Comparative study of SPWM and SVPWM cascaded h-bridge multilevel inverter *2017 International Conference on Intelligent Computing and Control, I2C2 2017*

[16] Rivera S, Wu B and Mar F S 2012 Cascaded H-Bridge Multilevel Converter Topology and Three-phase Balance Control for Large Scale Photovoltaic Systems *IEEE Int. Symp. Power Electron. Distrib. Gener. Syst.* 690–7

[17] Al-Badrani H 2018 PHD thesis siegan-Ger. Univ. SiC-VSI with Sinusoidal Voltages for an Enhanced Sensorless Control of the Induction Machine

[18] Zhao T, Zong Q, Zhang T and Xu Y 2016 Study of photovoltaic three-phase grid-connected inverter based on the grid voltage-oriented control *Proc. 2016 IEEE 11th Conf. Ind. Electron. Appl. ICIEA 2016* 2055–60

[19] Sevilmiş F and Karaca H 2018 Comparison of SRF-PLL and E-PLL Techniques for Grid Connected Inverter *6th Int. Conf. Adv. Technol. Sci.*

AUTHOR INDEX